THYRISTOR-BASED FACTS CONTROLLERS FOR ELECTRICAL TRANSMISSION SYSTEMS

IEEE Press
445 Hoes Lane, P.O. Box 1331
Piscataway, NJ 08855-1331

Books of Related Interest

DYNAMICS AND CONTROL OF LARGE ELECTRICAL POWER SYSTEMS
Marija Ilic and John Zaborszky
2000 Hardcover 856 pp ISBN 0-471-29858-1

UNDERSTANDING FACTS: Concepts and Technology of Flexible AC Transmission Systems
Narain Hingorani and Laszlo Gyugyi
2000 Hardcover 452 pp ISBN 0-7803-3455-8

POWER SYSTEM ECONOMICS
Steven Stoft
2002 Hardcover 344 pp ISBN 0-471-15040-1

THYRISTOR-BASED FACTS CONTROLLERS FOR ELECTRICAL TRANSMISSION SYSTEMS

R. Mohan Mathur
Ontario Power Generation
Toronto, ON, Canada

Rajiv K. Varma
Indian Institute of Technology
Kanpur, India

IEEE PRESS SERIES ON POWER ENGINEERING

Mohamed E. El-Hawary, *Series Editor*

IEEE

IEEE Press

WILEY-INTERSCIENCE

A JOHN WILEY & SONS, INC. PUBLICATION

This book is printed on acid-free paper. ♾

For ordering and customer service, call 1-800-CALL-WILEY.

Library of Congress Cataloging-in-Publication Data is available.

ISBN 0-471-20643-1

10 9 8 7 6 5 4 3 2 1

CONTENTS

FOREWORD

Power electronic components began to appear in electric power transmission systems in the 1970s. In high-voltage, direct-current (HVDC) interconnections, thyristors were first used in back-to-back asynchronous ties and soon thereafter in long-distance transmission. Since the late 1970s, thyristors have been the principal conversion devices for HVDC transmission links.

A new generation of power electronic devices, with forced turn-off capabilities, is now appearing in voltage-sourced converter (VSC) configurations for HVDC transmission. These converters have significant potential for weak-system applications, and one can anticipate a growth curve over the next few years.

Parallel developments were occurring in alternating-current (ac) transmission systems in the late 1970s. The thyristor-controlled reactor (TCR) and the thyristor-switched capacitor (TCR) likewise appeared in the 1970s and were applied singly, together, or with mechanically switched capacitors (MSCs) for reactive-power (voltage) control. After many discussions of proper terminology, this device was christened the static var compensator (SVC). The first SVC textbook, *Reactive Power Control in Electric Systems*, appeared in 1982; it was written by T. J. E. Miller and colleagues and published by Wiley-Interscience. Later in the 1980s, the series-connected dual, the thyristor-switched series capacitor (TCSC), was first installed to mitigate inter-area low-frequency oscillations.

The contemporary generation of power electronic–based controllers for ac transmission systems employs gate turn-off (GTO) devices in a bridge configuration. Appropriate use of on- and off-switching creates leading or lagging voltage–current relationships to provide reactive-power/voltage control (shunt configuration) and power-flow control (series configuration) without the need for large reactive energy–storage elements. GTO devices first appeared in Japan in the early 1990s and the United States in the mid-1990s. We are just beginning to tap the potential of these circuit elements.

The term *FACTS controllers* describes power electronic–based circuit configurations applied in ac transmission systems. The acronym, in fact, represents *flexible ac transmission systems*, with the term "flexible" implying the controllability of voltage and/or current. As mentioned previously, shunt configurations can provide voltage/reactive-power control, whereas series configurations can provide current/power-flow control. Combined series/shunt configurations, however, can provide both controls.

This book presents a broad range of detailed information on the characteristics and applications of FACTS controllers. The reader will find circuit diagrams, performance commentary, block diagrams, transfer functions, computer simulations, comparative costs, and numerous references. A chapter discussing the coordination of FACTS controllers is particularly interesting. As they become more integrated into our existing ac transmission networks, FACTS controllers cannot be designed in isolation from one another. In particular, the stable response of one controller to a disturbance may not be an accurate predictor of the response of multiple controllers in an electrical region—of which obvious examples are powerful controllers in a series path and two voltage controllers shunted near each other. However, adverse interactions can also occur between one series and one shunt controller if their design is not coordinated.

Given the heavily stressed state of today's deregulated and restructured electric power industry and the concomitant need for transmission system security and reliability, it seems apparent that sophisticated control elements will play a major role in future system operations. FACTS controllers are representative of these sophisticated control elements; this book makes a valuable contribution to their understanding.

WILLIS F. LONG
PROFESSOR

University of Wisconsin–Madison

This book is
dedicated
to
Gurudev
Sri Paramhansa Ram Mangal Das ji
(1893–1984)
and all those who have inspired us.
Mohan and Rajiv

PREFACE

Rising energy costs and a greater sensitivity to environmental impact of new transmission lines necessitated the search and application of new controllers to minimize losses and maximize the stable power-transmission capacity of existing lines. In the first instance, these objectives were met by reactive-power control on transmission lines. With the availability and application of thyristors, a new breed of thyristor-based reactive-power controllers was realized, which provided a very high speed of response. This rapid-control feature added damping control to voltage and reactive-power control applications. Flexible ac transmission system (FACTS) technology is the application of a variety of new power-electronic controllers for both active and reactive power on selected lines. FACTS controllers are becoming an integral component of modern power-transmission systems.

The electrical power industry is in transition: it is undergoing restructuring to accommodate competing suppliers to reach customers. The aspects of generation, transmission, and distribution are becoming independent businesses. In this new reality, transmission companies and, to a lesser extent, distribution companies must respond to the demands of power wheeling by accepting power injection at any node and load connection at any other node. Also, these companies must not allow power congestion in any transmission corridor to restrain the available transmission capacity for any new business opportunities. Controlling power flow and securing the highest attainable stability margin on selected lines help meet these challenges. FACTS controllers, therefore, have become increasingly important to help transmission companies achieve these objectives.

Most FACTS controllers currently in use are thyristor-based, because the thyristor technology is mature and high-power-rating thyristors are now readily available.

Static var compensators (SVCs) constitute the first generation of FACTS controllers; they have been extensively employed worldwide since the 1970s. The past decade has witnessed the emergence of several new FACTS controllers for transmission applications. Prominent among these FACTS controllers are the thyristor-based thyristor-controlled series compensator (TCSC) and gate turn-off (GTO), as well as the thyristor-based static synchronous compensator (STATCOM), static synchronous series compensator (SSSC), and unified power-flow controller (UPFC). Newer high-power semiconductor devices, such as IGBTs and MTOs, are in the offing to allow for the development of a range of FACTS controllers that will provide greatly improved performance.

In the past three decades, a wealth of FACTS literature has been created by worldwide contributions from academicians, researchers, and utility engineers. However, this literature is widely scattered and, at times, not easily accessible to students, researchers, and engineers. Therefore, in this book the authors have consolidated and presented systematically the basic concepts and application of FACTS controllers in power-transmission systems. Although transmission applications are presented, the material applies equally to distribution systems.

The book is focused primarily on thyristor-based FACTS controllers—namely, the SVC and TCSC—because these controllers are currently in use and likely to be selected for new applications. The book emphasizes operating principles and control techniques for these two FACTS controllers so that different objectives can be achieved, such as enhancement of power-system stability, augmentation of system damping, alleviation of subsynchronous resonance (SSR), prevention of voltage instability, and improvement of high-voltage direct-current (HVDC)–terminal performance.

The book is intended to create an awareness of control issues, which must be addressed in designing or selecting a specific controller. Control-design techniques are illustrated by case studies. The coordination of multiple FACTS controllers of both similar and dissimilar kinds are also presented and illustrated through examples. Several references are cited at appropriate locations for securing additional details or for pursuing further research.

The material in the book is built around phasor and linear analysis methods, and the level of mathematics is kept just sufficient for the understanding of system-design concepts. The book is neither mathematically rigorous nor intensive in providing practical installation details. In this regard, it treads a midline path that may be acceptable to both utility engineers and academicians.

The book evolved from study material of university graduate-level courses and intensive short courses for utility engineers that the authors conducted, individually and over a period of several years, in Canada, Brazil, and India.

The book is organized in 10 chapters, as follows:

Chapter 1 describes the control mechanisms conventionally employed in power-transmission systems. It briefly introduces high-power semiconductor devices and presents essential concepts of FACTS technology and emerging power-transmission systems.

Chapter 2 elucidates the principles of reactive-power control and compares the performance of transmission lines both with and without shunt-series compensation.

Chapter 3 describes the operating principles of a range of reactive-power compensators, including the synchronous condenser, the saturated reactor (SR), the thyristor-controlled reactor (TCR), the thyristor-switched capacitor (TSC), and the TSC–TCR. Also, a comparative evaluation of different compensators is described.

Chapter 4 presents the details of various components of SVC control and protection systems and models of SVCs for different kinds of power-system studies. A highlight of this chapter is the discussion on the demodulation effects of various measurement techniques, which may influence different types of harmonic instability and network resonance.

Chapter 5 discusses the voltage-control function of an SVC and illustrates the basic procedure for designing the voltage controller of an SVC. This chapter provides an in-depth discussion of the influence of network resonance and 2nd and 3rd harmonics on SVC performance. Also, the behavior of an SVC in the presence of different resonant modes in series-compensated systems is described.

Chapter 6 discusses the various power-transmission-system applications of an SVC, including steady-state and transient-stability enhancement, damping improvement, subsynchronous-oscillation suppression, voltage-instability prevention, and dynamic reactive support to HVDC terminals. The design of SVC controllers, especially with auxiliary signals for the foregoing applications, is discussed with case studies.

Chapter 7 introduces the operating principles, capability characteristics, and different models of a TCSC.

Chapter 8 explains the different controls of a TCSC and describes the various applications of TCSCs in power systems. These applications encompass the improvement of system stability, the mitigation of SSR, and the obviation of voltage instability. A special feature of this chapter is a description of the performance of two recent TCSC installations: one in Brazil, the other in Sweden.

Chapter 9 discusses the coordination of different FACTS controllers. The nature of different controller interactions is described, and the interactions between various combinations of controllers—such as SVC–SVC, SVC–TCSC, TCSC–TCSC, and SVC–HVDC—are illustrated. Several linear-control techniques for coordinating FACTS controllers are outlined.

For completeness, Chapter 10 discusses emerging FACTS controllers. These controllers are not thyristor-based but, rather, employ GTO-based voltage-source-converter (VSC) technology. Examples include the STATCOM, the SSSC, and the UPFC. Different FACTS controllers are compared in terms of both cost and performance. Some discussion is given on the possible direction of FACTS technology in future power systems.

ACKNOWLEDGMENTS

We express gratitude to the IEEE, the PES, the CIGRE, the IEE, and the EPRI for use of their invaluable vast literature in our preparation of this book. We also thank the Canadian Electrical Association (now CEA Technologies, Inc.) and Kluwer Academic Publishers (Boston, MA, USA) for granting us permission to reproduce some material from their publications.

We thank Gino Romegialli for allowing us to use some material from his notes on a course on static compensators that he conducted at the University of Manitoba (Winnipeg, Canada), and we thank Krishnat Patil for letting us use some of his PhD-thesis results on STATCOM for our discussion in Chapter 10.

We are indebted to Professor Willis F. Long (University of Wisconsin, Madison, USA) for the foreword to this book. We thank all the reviewers for their constructive suggestions and guidance for this book. In particular, we want to acknowledge Dr. Sasan Jalali of Siemens for reviewing the two chapters on TCSCs. Our thanks are due to C. G. Carlsson and R. S. Moni of ABB (New Delhi, India) for providing the picture of the Brazil TCSC that appears in the cover design of this book.

Our immense thanks are due to the Indian Institute of Technology (Kanpur, India) and the University of Western Ontario (London, Canada) for providing all the support and facilities to us without which this book would not have taken its present shape. In particular, we thank Ajay Srivastava for meticulously typing the manuscript and Sandhya Agnihotri for drawing the figures under a very stringent time frame.

Our sincere thanks are due to the officials at IEEE Press—John Griffin, Marilyn G. Catis, Robert Bedford, and Anthony VenGraitis— for their patience and kind understanding of the inordinate delays on our end.

Our once-independent career paths converged at the University of Western Ontario when Rajiv Varma arrived there on research leave. A casual conversation to work together and consolidate material on static compensators sparked a mutual interest and enthusiasm that finally led to this book. Therefore, we have two separate groups of people to whom we thank.

R. Mohan Mathur During the last 25 years or so, as a logical extension of my interest in HVDC systems, I have pursued aspects of study now known better as FACTS technology. I express my sincere thanks to the Natural Sciences and Engineering Research Council (Canada) for granting me ongoing research funds, of which a part was used for supporting graduate students and research associates who contributed to FACTS literature. I am appreciative of the opportunities for research, development, and teaching provided to me by the University of Manitoba, the University of Western Ontario, and Brown Boveri and Co. (BBC)—now Asea Brown Boveri (ABB) (Baden, Switzerland).

During the course of my studies, I have benefited immensely from my students and associates. In particular, I want to acknowledge my former students Adel Hammad, Reza Iravani, Raghuveer Basati, David Flueckigar, Sandeep Nyati, Sumrit Hungsasutra, Sainath Moorty, Xuegong Wang, Xiaokang Xu, Krishnat Patil, and Christina Terek, and my research associates Gino Romegialli, J. Senthil, S. H. Hosseini, and—indeed—my coauthor, Rajiv Varma. Also, I acknowledge the help of and useful discussions with my former colleagues Robert Menzies, Anir Gole, and Jin Jiang, who worked on various aspects of series and shunt compensators, phase shifters, and VSC applications for realizing the STATCOM, the SSSC, and the UPFC.

I am grateful to Hermann Hoedle, H. Stemmler, and Lucien Terens for providing me with opportunities, on four separate (1-year-each) sabbatical leaves, to work at the BBC/ABB (Baden) and participate in TCR and TSC studies.

I sincerely thank my wife, Aruna, and my children, Tinu and Shikha, for their unwavering encouragement and support throughout my work.

Rajiv K. Varma First, I want to thank immensely my PhD supervisor Dr. K. R. Padiyar, professor, Indian Institute of Science (Bangalore), who, while at the Indian Institute of Technology (IIT) (Kanpur), inspired and introduced me to FACTS technology and the nuances of power-system research. My heartfelt gratitude to Mr. Wayne Litzenberger of the Bonneville Power Administration (Portland, OR, USA), who kindly accepted my assistance for the preparation of the *Annotated Bibliographies on HVDC Transmission and FACTS Devices*, 1994–1995 and 1996-1997 editions. These bibliographies were of tremendous help in preparing this book.

At IIT, I thank Professor S. S. Prabhu for the opportunity to work with him and teach, for several years, the course "Reactive Power Compensation in Power Systems." I am appreciative of the encouragement that I received from Dr. V. Sinha, Dr. Surendra Gupta, Dr. G. K. Dubey, Dr. Ashwini Kumar, Dr. Sachchidanand, Dr. Arindam Ghosh, Dr. K. R. Srivathsan, Dr. S. C. Srivastava (head, EE Department), and Professor K. A. Padmanabhan (former director of IIT).

I express my deepest thanks to Dr. M. A. Pai, professor, University of Illinois (Urbana–Champaign, USA); Dr. Prabha Kundur, president and CEO, Powertech Labs., Inc. (Canada); and Dr. Rambabu Adapa of the EPRI (Palo Alto, CA, USA) for their support and guidance in book writing. I wish to offer my sincere gratitude to Mr. Mata Prasad, consultant, Power Grid Corporation of India, Ltd., and ex-advisor at ABB (New Delhi), for providing me with numerous books and reports that I found highly useful for this book.

I thank the University of Western Ontario (UWO) for providing me, on several occasions, opportunities to work there and research at the Power Engineering Analysis and Research Laboratory (PEARL). At UWO, I wish to express my sincere gratitude to Dr. G. S. P. Castle, Dr. A. R. Webster, Dr. T. Bonnema, and Dr. R. V. Patel (present chair, ECE Department) for providing me with all possible facilities and support for the book-writing activity. Very special thanks are due to Dr. Jin Jiang for his cheerful company, constant support, and encouragement for this book.

The help received from Dr. Xiao Kang Xu, Andrew Cullen, Dr. Qihua Zhao, and everyone at PEARL is deeply appreciated. Jacquie Taylor and Sharon Ling simplified all the tough tasks for me at UWO, for which I am thankful.

I express sincere gratitude to my friends T. Sadagopan, Dr. J. Senthil, and Dr. Narendar Sumukadas, whose help at critical junctures can never be forgotten. I am also grateful to Dr. Krishnat Patil, Nilesh Hiremath, Manoj Kumar, and Sujit Mandal for their unconditional help. I acknowledge with many thanks my doctoral students Dr. N. K. Sharma, Dr. C. P. Gupta, and Ram Prakash Gupta for assisting me in gathering literature.

I am deeply appreciative of the many forms of help given by P. V. K. Reddy, Gopesh Tiwari, Manoj Gupta, Varada Rajan, Sony Farmer, Sharabh Pradhan, and Himanshu Singh as well as the staff of the Technology Development Mission Project at IIT. A special word of thanks goes to all my postgraduate students as well as to all the undergraduate students of the EE 330 Power Systems course that I taught over several years at IIT; their enthusiasm cheered me during the book writing.

Achievements in life do not come alone by one's own efforts but by the blessings of teachers and parents. The guidance and selfless love of Sri Jagat Narain ji and the motivation of Sri Mahesh Dwivedi provided me with the strength to write this book. I express my profound gratitude to my parents, Mr. Satish Chandra Varma and Mrs. Saroj Varma, and to my in-laws, Mr. Avadhesh Chandra Srivastava and Mrs. Anil Srivastava.

I also sincerely thank the families of my wife's and my own brothers and sisters for their encouragement. I greatly appreciate the support received from the family of my cousin Ajay.

I am indeed indebted to Dr./Mrs. Aruna Mathur and the whole family for their affectionate support, which has been a tremendous source of strength for me.

My infinite love for my "little babies" Sarvesh and Ratna, who donated their time with their father to let him write this book. This book owes its completion to my strongest critic, truest well-wisher, and best friend in life—my wife, Malini.

This book is only a stepping stone for entering into the vast realm of FACTS. We will be only too happy if this book can serve as a useful initial reference for academicians and utility engineers embarking on a study of FACTS technology. Any suggestions for improvement of this book are most welcome.

R. Mohan Mathur
Rajiv K. Varma

CHAPTER 1

Introduction

1.1 BACKGROUND

This chapter briefly discusses the growth of complex electrical power networks. It introduces the lack of controllability of the active- and reactive-power flows in energized networks. (These flows tend to diffuse in the network, depending primarily on the impedance of power lines.) This chapter also describes the conventional controlled systems, such as automatic governor control and excitation control employed at generating stations. Transformer tap-changer control is another control feature generally available in transmission networks. Arising from the transformer combinations and the use of on-load tap changers, phase-shifting transformers are realized, which are primarily used to mitigate circulating power on network tie-lines.

This introduction and the recognition of limited controllability provide the basis for introducing the concept of the flexible ac transmission system (FACTS). Since newly developed FACTS devices rely on the advances made in semiconductor components and the resulting power-electronic devices, these, too, are introduced.

This chapter also introduces the basic operating principles of new FACTS devices. (These principles are fully discussed in later chapters of this book.) Finally, the chapter presents a brief commentary on emerging deregulation, competition, and open access in power utilities. In that context, the value of FACTS devices for emerging transmission companies is identified.

1.2 ELECTRICAL TRANSMISSION NETWORKS

The rapid growth in electrical energy use, combined with the demand for low-cost energy, has gradually led to the development of generation sites remotely located from the load centers. In particular, the remote generating stations include hydroelectric stations, which exploit sites with higher heads and significant water flows; fossil fuel stations, located close to coal mines; geothermal stations and tidal-power plants, which are sitebound; and, sometimes, nuclear power plants purposely built distant from urban centers. The generation of bulk power at remote locations necessitates the use of transmission lines to connect generation sites to load centers. Furthermore, to enhance system reliability, multiple lines that connect load centers to several sources, interlink neighboring utilities, and build the needed levels of redundancy have gradually led to the evolution of complex interconnected electrical transmission networks. These networks now exist on all continents.

An electrical power transmission network comprises mostly 3-phase alternating-current (ac) transmission lines operating at different transmission voltages (generally at 230 kV and higher). With increasing requirement of power-transmission capacity and/or longer transmission distances, the transmission voltages continue to increase; indeed, increases in transmission voltages are linked closely to decreasing transmission losses. Transmission voltages have gradually increased to 765 kV in North America, with power transmission reaching 1500 MVA on a line limited largely by the risk that a power utility may be willing to accept because of losing a line.

An ac power transmission network comprises 3-phase overhead lines, which, although cheaper to build and maintain, require expensive right-of-ways. However, in densely populated areas where right-of-ways incur a premium price, underground cable transmission is used. Increasing pressures arising from ecological and aesthetic considerations, as well as improved reliability, favor underground transmission for future expansion.

In a complex interconnected ac transmission network, the source-to-a-load power flow finds multiple transmission paths. For a system comprising multiple sources and numerous loads, a load-flow study must be performed to determine the levels of active- and reactive-power flows on all lines. Its impedance and the voltages at its terminals determine the flow of active and reactive powers on a line. The result is that whereas interconnected ac transmission networks provide reliability of power supply, no control exists on line loading except to modify them by changing line impedances by adding series and/or shunt-circuit elements (capacitors and reactors).

The long-distance separation of a generating station from a load center requiring long transmission lines of high capacity and, in some cases in which a transmission line must cross a body of water, the use of ac/dc and dc/ac converters at the terminals of an HVDC line, became a viable alternative many years ago. Consequently, beginning in 1954, HVDC transmission has grown steadily to the current ±600 kV lines with about 4000 A capacity. Also, direct current (dc) transmission networks, including multiterminal configurations, are already embedded in ac transmission networks. The most significant feature of an HVDC transmission network is its full controllability with respect to power transmission [1]–[5].

Until recently, active- and reactive-power control in ac transmission networks was exercised by carefully adjusting transmission line impedances, as well as regulating terminal voltages by generator excitation control and by transformer tap changers. At times, series and shunt impedances were employed to effectively change line impedances.

1.3 CONVENTIONAL CONTROL MECHANISMS

In the foregoing discussion, a lack of control on active- and reactive-power flow on a given line, embedded in an interconnected ac transmission network, was stated. Also, to maintain steady-state voltages and, in selected cases, to alter the power-transmission capacity of lines, traditional use of shunt and series impedances was hinted.

In a conventional ac power system, however, most of the controllability exists at generating stations. For example, generators called *spinning reserves* maintain an instantaneous balance between power demand and power supply. These generators, in fact, are purposely operated at reduced power. Also, to regulate the system frequency and for maintaining the system at the rated voltage, controls are exercised on selected generators.

1.3.1 Automatic Generation Control (AGC)

The megawatt (MW) output of a generator is regulated by controlling the driving torque, T_m, provided by a prime-mover turbine. In a conventional electromechanical system, it could be a steam or a hydraulic turbine. The needed change in the turbine-output torque is achieved by controlling the steam/water input into the turbine. Therefore, in situations where the output exceeds or falls below the input, a speed-governing system senses the deviation in the generator speed because of the load-generation mismatch, adjusts the mechanical driving torque to restore the power balance, and returns the operating speed to its rated value. The speed-governor output is invariably taken through several stages of mechanical amplification for controlling the inlet (steam/water) valve/gate of the driving turbine. Figure 1.1 shows the basic speed-governing system of a generator supplying an isolated load. The operation of this basic feedback-control system is enhanced by adding further control inputs to help control the frequency of a large interconnection. In that role, the control system becomes an automatic generation control (AGC) with supplementary signals.

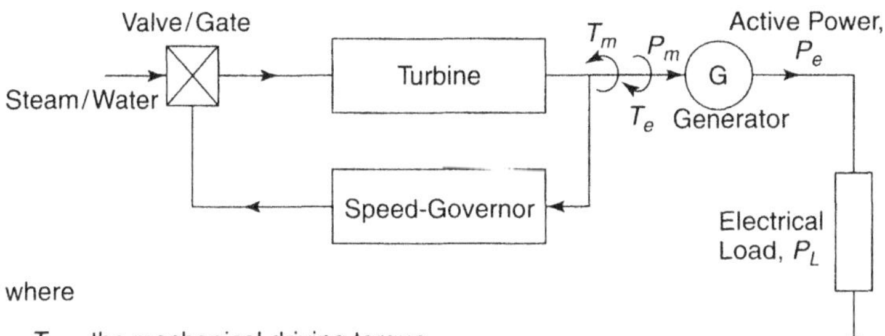

where

T_m = the mechanical driving torque
T_e = the mechanical load torque from the generator electrical output
P_m = the mechanical power input to the generator

Figure 1.1 A speed-governor system.

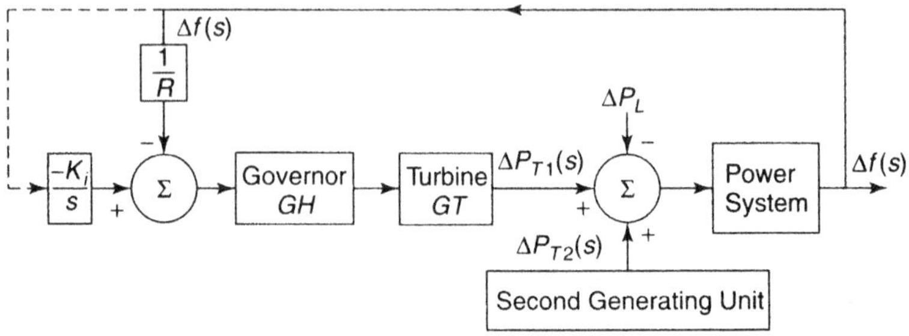

Figure 1.2 An AGC with supplementary control on the principal generating unit.

To avoid competing control actions, in a multigenerator unit station each speed-governor system is provided with droop (R) characteristics through a proportional feedback loop (R, Hz/MW). Figure 1.2 shows an AGC on the principal generating unit with supplementary control. In contrast, the second, third, and remaining generating units in a multiunit station operate with their basic AGCs. In a complex interconnected system, the supplementary control signal may be determined by a load-dispatch center.

1.3.2 Excitation Control

The basic function of an exciter is to provide a dc source for field excitation of a synchronous generator. A control on exciter voltage results in controlling the field current, which, in turn, controls the generated voltage. When a synchronous generator is connected to a large system where the operating frequency and the terminal voltages are largely unaffected by a generator, its excitation control causes its reactive power output to change.

In older power plants, a dc generator, also called an exciter, was mounted on the main generator shaft. A control of the field excitation of the dc generator provided a controlled excitation source for the main generator. In contrast, modern stations employ either a brushless exciter (an inverted 3-phase alternator with a solid-state rectifier connecting the resulting dc source directly through the shaft to the field windings of the main generator) or a static exciter (the use of a station supply with static rectifiers).

An excitation-control system employs a voltage controller to control the excitation voltage. This operation is typically recognized as an automatic voltage regulator (AVR). However, because an excitation control operates quickly, several stabilizing and protective signals are invariably added to the basic voltage regulator. A power-system stabilizer (PSS) is implemented by adding auxiliary damping signals derived from the shaft speed, or the terminal frequency, or the power—an effective and frequently used technique for enhancing small-signal stability of the connected system. Figure 1.3 shows the functionality of an excitation-control system.

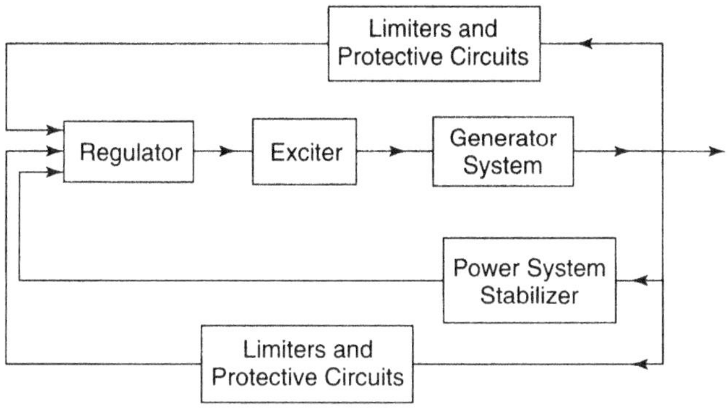

Figure 1.3 A conceptual block diagram of a modern excitation controller.

1.3.3 Transformer Tap-Changer Control

Next to the generating units, transformers constitute the second family of major power-transmission-system apparatuses. In addition to increasing and decreasing nominal voltages, many transformers are equipped with tap-changers to realize a limited range of voltage control. This tap control can be carried out manually or automatically. Two types of tap changers are usually available: off-load tap changers, which perform adjustments when deenergized, and on-load tap changers, which are equipped with current-commutation capacity and are operated under load. Tap changers may be provided on one of the two transformer windings as well as on autotransformers.

Because tap-changing transformers vary voltages and, therefore, the reactive-power flow, these transformers may be used as reactive-power-control devices. On-load tap-changing transformers are usually employed to correct voltage profiles on an hourly or daily basis to accommodate load variations. Their speed of operation is generally slow, and frequent operations result in electrical and mechanical wear and tear.

1.3.4 Phase-Shifting Transformers

A special form of a 3-phase–regulating transformer is realized by combining a transformer that is connected in series with a line to a voltage transformer equipped with a tap changer. The windings of the voltage transformer are so connected that on its secondary side, phase-quadrature voltages are generated and fed into the secondary windings of the series transformer. Thus the addition of small, phase-quadrature voltage components to the phase voltages of the line creates phase-shifted output voltages without any appreciable change in magnitude. A phase-shifting transformer is therefore able to introduce a phase shift in a line.

Figure 1.4 shows such an arrangement together with a phasor diagram. The phasor diagram shows the phase shift realized without an appreciable change in magnitude by the injection of phase-quadrature voltage components in a 3-phase

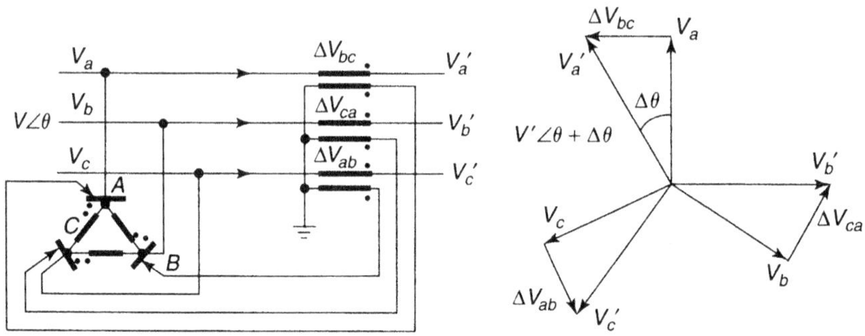

(a) (b)

Figure 1.4 A phase-shifting transformer: (a) a schematic diagram and (b) a phasor diagram.

system. When a phase-shifting transformer employs an on-load tap changer, controllable phase-shifting is achieved. The interesting aspect of such phase shifters is that despite their low MVA capacity, by controlling the phase shift they exercise a significant real-power control. Therefore, they are used to mitigate circulating power flows in interconnected utilities. A promising application of these devices is in creating active-power regulation on selected lines and securing active-power damping through the incorporation of auxiliary signals in their feedback controllers. From this description, it is easy to visualize that an incremental in-phase component can also be added in lines to alter only their voltage magnitudes, not their phase.

The modification of voltage magnitudes and/or their phase by adding a control voltage is an important concept. It forms the basis of some of the new FACTS devices discussed in this book. The injected voltage need not be realized through electromagnetic transformer–winding arrangements; instead, by using high-speed semiconductor switches such as gate turn-off (GTO) thyristors, voltage source inverters (VSIs)—synchronized with the system frequency—are produced. The application of a VSI to compensate the line-voltage drop yields a new, fast, controllable reactive-power compensator: the static synchronous series compensator (SSSC). The application of a VSI to inject a phase-quadrature voltage in lines yields a new, fast, controllable phase shifter for active- power control. Once a synchronized VSI is produced, it is indeed easy to regulate both the magnitude and the phase angle of the injected voltages to yield a new, unified power-flow controller (UPFC).

1.4 FLEXIBLE AC TRANSMISSION SYSTEM (FACTS)

The FACTS is a concept based on power-electronic controllers, which enhance the value of transmission networks by increasing the use of their capacity.

[6]–[15]. As these controllers operate very fast, they enlarge the safe operating limits of a transmission system without risking stability. Needless to say, the era of the FACTS was triggered by the development of new solid-state electrical switching devices. Gradually, the use of the FACTS has given rise to new controllable systems. It is these systems that form the subject matter of this book.

Today, it is expected that within the operating constraints of the current-carrying thermal limits of conductors, the voltage limits of electrical insulating devices, and the structural limits of the supporting infrastructure, an operator should be able to control power flows on lines to secure the highest safety margin as well as transmit electrical power at a minimum of operating cost. Doing so constitutes the increased value of transmission assets.

The search for enhanced controllability of power on ac transmission networks was initiated by newly acquired current and power controllability in HVDC transmission. Replacement of mercury-arc valves by thyristors yielded robust ac/dc converters, minimized conversion losses, and yielded fast control on transmitted power—so much so that line-to-ground fault clearing became possible without the use of circuit breakers. Instead, by rapidly attaining current zero through the use of current controllers and, in addition, by rapidly recovering the electromagnetic energy stored in the energized line, the faulted dc line could be isolated by low interruption–rating isolators.

The very fast power controllability in HVDC systems made them candidates for special applications in back-to-back configurations to control the power exchange between the networks they linked. The rapid control of power led to the added use of HVDC links for enhancing transient stability of connected systems through active-power damping. The enhancement in stability was accomplished by adding auxiliary signals in the current controllers of the converters [16], [17].

1.4.1 Advances in Power-Electronics Switching Devices

As mentioned previously, the full potential of ac/dc converter technology was better realized once mercury-arc valves were replaced by solid-state switching devices called *thyristors*. Thyristors offered controlled turn-on of currents but not their interruption. The rapid growth in thyristor voltage and current ratings accelerated their application, and the inclusion of internal light triggering simplified the converter controls and their configurations even more. Most applications, however, were based on the natural commutation of currents. In special cases where forced commutation was required, elaborate circuitry using discharging capacitors to create temporary current zeroes were employed.

Thyristors are now available in large sizes, eliminating the need for paralleling them for high-current applications. Their voltage ratings have also increased so that relatively few are required to be connected in series to yield switches or converters for power-transmission applications. Actually, the present trend is to produce high-power electronic building blocks (HPEBBs) to configure high-power switches and converters, thus eliminating the custom-design needs at the device level. Availability of HPEBBs should accelerate development of new FACTS devices. The HPEBB thyristors are available in compact packaging and in sufficiently large sizes (e.g., 125-mm thyristors: 5.5 kV, 4 kA or 4.5 kV, 5.8 kA) for most applications. For switching applications, such as that for tap changers or static phase shifters, anti-parallel–connected thyristor modules, complete with snubber circuits, are available. These switches provide sufficiently high transient-current capacity to endure fault currents.

The GTO semiconductor devices facilitate current turn-on as well as turn-off by using control signals. This technology has grown very rapidly; consequently, high-power GTOs are now available (100 mm, 6 kV or 150 mm, 9 kV). Full on–off control offered by GTOs has made pulse width–modulated (PWM) inverters easy to realize [18].

Advances in semiconductor technology are yielding new efficient, simple-to-operate devices. The insulated gate bipolar transistor (IGBT) and the metal-oxide semiconductor (MOS)–controlled thyristor (MCT) control electric power using low levels of energy from their high-impedance MOS gates, as compared to high-current pulses needed for thyristors or GTOs. Unfortunately, the available voltage ratings of these devices are still limited.

The MOS turn-off (MTO) thyristor combines the advantages of both thyristors and MOS devices by using a current-controlled turn-on (thyristor) and a voltage-controlled turn-off having a high-impedance MOS structure [19]. Hybrid MTOs are being proposed that show substantially low device losses relative to GTOs. Because MTOs use nearly half the parts of GTOs, their application promises significant reliability improvement.

The availability of new and significantly improved switching devices in convenient packages (HPEBB) will aid the development of new, more versatile FACTS devices. The symbolic representation and equivalent circuits of a thyristor, GTO, and MCT are shown in Fig. 1.5.

1.4.2 Principles and Applications of Semiconductor Switches

In high-power applications, semiconductor devices are used primarily as switches. To accommodate switching in an ac system, two unidirectional conducting devices are connected in an antiparallel configuration, as shown in Fig. 1.6. Such a switch may be employed per phase to connect or disconnect a shunt-circuit element, such as a capacitor or reactor, or to short-circuit a series-connected–circuit element, such as a capacitor. A reverse-biased thyristor automatically turns off at current zero, for which reason an antiparallel thyristor connection is used to control the current through a reactor by delaying its turn-on instant, as shown in Fig. 1.6(b). It is easy to see that the current through a connected reactor may be controlled from full value to zero by adjusting the delay angle, α, of the gate's firing signal from 90° to 180°.

Thus a thyristor switch offers current control in a reactor, rendering it a controlled reactor. However, because a capacitor current leads the applied voltage by approximately 90°, the capacitor switching always causes transient in-rush

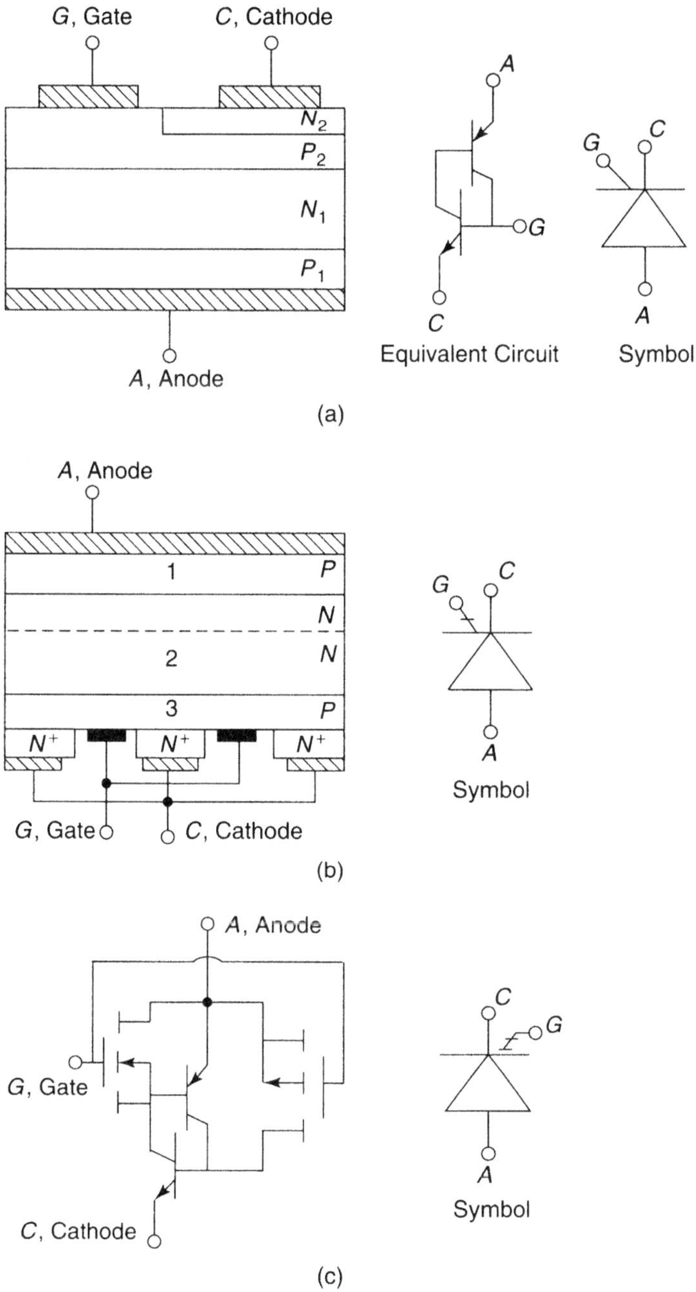

Figure 1.5 Semiconductor switching devices for power-electronics applications: (a) a thyristor (silicon-controlled rectifier); (b) a gate turn-off (GTO) thyristor; and (c) a P-MCT equivalent circuit.

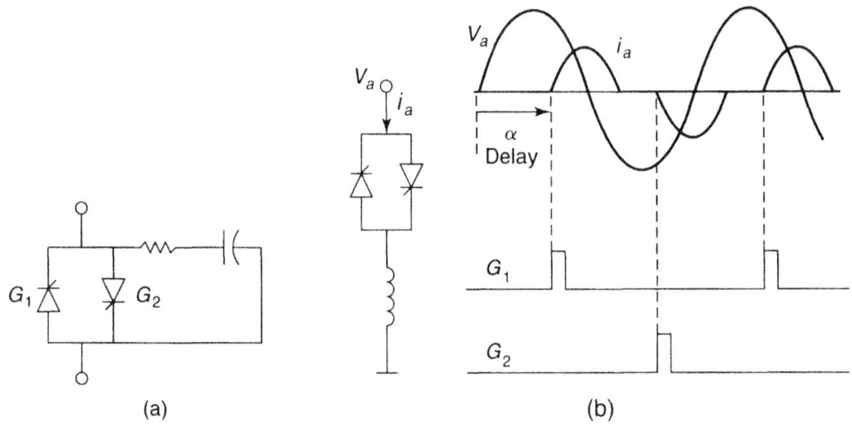

Figure 1.6 A thyristor switch for ac applications: (a) a switch and (b) a controlled reactor current.

currents that must be minimized by switching charged capacitors at instants when the voltage across the switch is near zero. Therefore, a thyristor switch is used only to turn on or turn off a capacitor, thereby implementing a switched capacitor.

Parallel combination of switched capacitors and controlled reactors provides a smooth current-control range from capacitive to inductive values by switching the capacitor and controlling the current in the reactor. Shunt combinations of thyristor-controlled reactors (TCRs) and thyristor-switched capacitors (TSCs) yield static var compensators (SVCs), which are described in detail in Chapters 3–6.

Thyristor switches may be used for shorting capacitors; hence they find application in step changes of series compensation of transmission lines. A blocked thyristor switch connected across a series capacitor introduces the capacitor in line, whereas a fully conducting thyristor switch removes it. In reality, this step control can be smoothed by connecting an appropriately dimensioned reactor in series with the thyristor switch—as shown in Fig. 1.7—to yield vernier con-

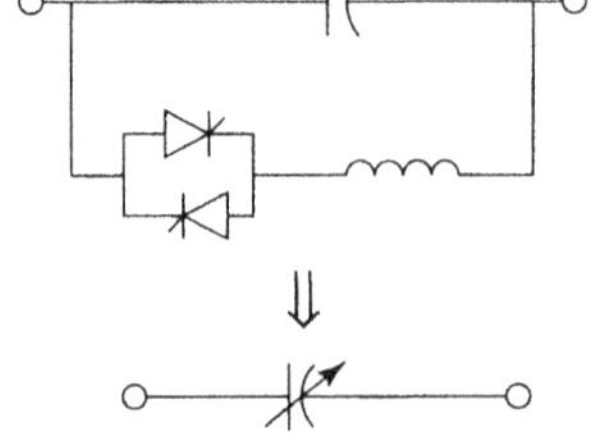

Figure 1.7 A thyristor-controlled series capacitor (TCSC).

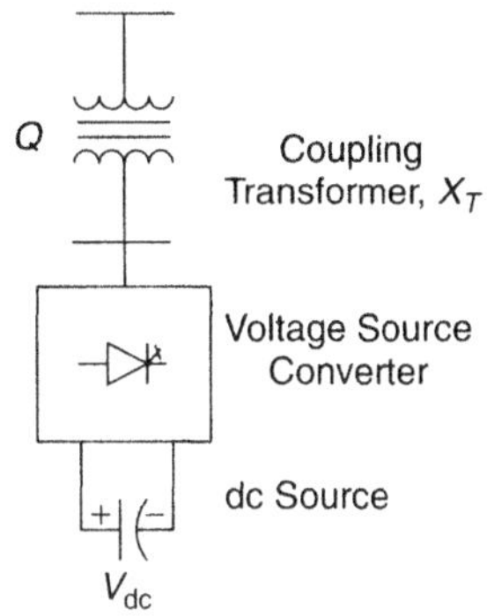

Figure 1.8 A GTO-based static synchronous compensator (STATCOM).

trol. This application of thyristor switches creates the thyristor-controlled series capacitor (TCSC) FACTS controller. A detailed discussion of this controller is presented in Chapters 7–9.

In the foregoing applications, thyristor switches were used to control the current through circuit elements, such as capacitors and reactors. The switches are also used to perform switching actions in on-load tap changers, which may be employed as thyristor-controlled phase-shifting transformers (TCPSTs).

Generally, the use of fully rated circuit elements is expensive, so to perform similar functions, another important class of FACTS controllers is realized by dc/ac converters. The application of GTO devices makes forced commutation possible, and therefore PWM converters offer a more elegant solution. The output voltages of PWM converters contain low-harmonic content. The voltage-source converters (VSCs) form the basic element of this new class of FACTS controllers, and numerous applications of this technology exist.

An alternative to a thyristor-controlled SVC is a GTO-based VSC that uses charged capacitors as the input dc source and produces a 3-phase ac voltage output in synchronism and in phase with the ac system. The converter is connected in shunt to a bus by means of the impedance of a coupling transformer. A control on the output voltage of this converter—lower or higher than the connecting bus voltage—controls the reactive power drawn from or supplied to the connected bus. This FACTS controller is known as a *static compensator* (STATCOM) [20] and is shown symbolically in Fig. 1.8. (This subject is discussed in Chapter 10.)

The use of voltage-source converters to inject a voltage by way of series-connected transformers leads to another interesting group of FACTS controllers: the SSSCs, which inject voltages to compensate for the line-reactance voltage drops [6]–[8]. It is easy to visualize that if the reactive drop of a line is partly compensated by an SSSC, it amounts to reducing the line reactance (X_L), or in other words, it is akin to controlled series compensation. The injected voltage in the line is independent of the line current. Figure 1.9 shows a 1-line diagram of an SSSC, which controls the active-power flow on a line.

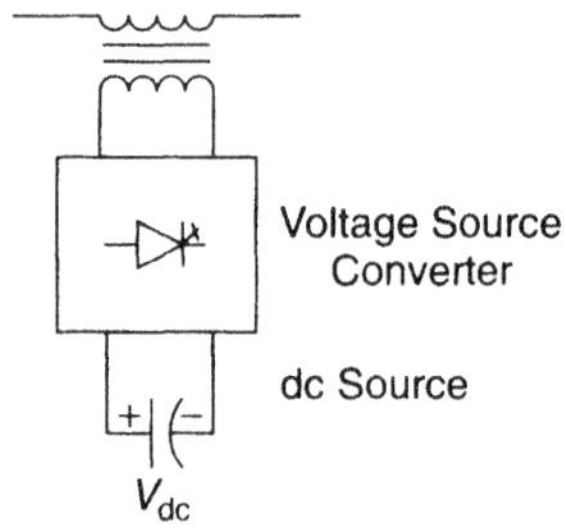

Figure 1.9 A 1-line diagram of a static synchronous series compensator (SSSC).

The functions of an SSSC and a STATCOM, in fact, may be combined to produce a unified power-flow controller (UPFC) [6]–[8], [21] A 1-line diagram of a UPFC is shown in Fig. 1.10. In the UPFC shown, a dc energy source is shared between the STATCOM and SSSC. Normally, no net energy is drawn from this source, but to compensate for the controller losses, the STATCOM can operate so that it draws the compensating active power from the connected ac bus. Thus a UPFC offers a fast, controllable FACTS device for the flow of combined active–reactive power in a line.

Finally, there are FACTS controllers classified as power-conditioning equipment. These controllers are employed as battery-energy–storage systems (BESSs) or superconducting magnetic-energy–storage (SMES) systems [6]–[8], [10]–[13]. These controllers also use GTO-based converters, which operate in dual roles as rectifiers for energy storage and inverters for energy return.

1.5 EMERGING TRANSMISSION NETWORKS

A historic change is overtaking electrical power utility businesses. Customers are demanding their right to choose electrical energy suppliers from competing

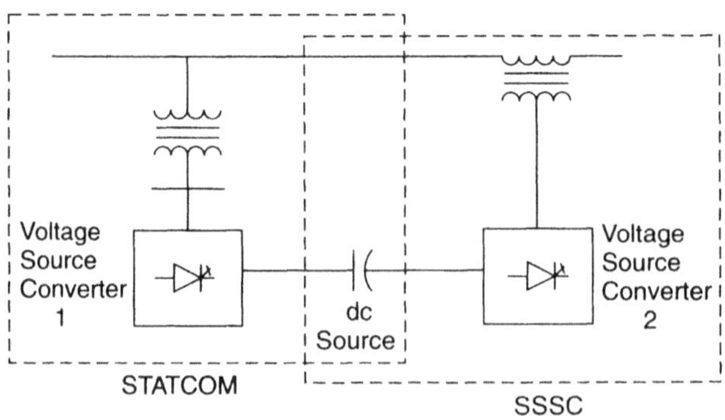

Figure 1.10 A 1-line diagram of a unified power-flow controller (UPFC).

vendors—a movement that has arisen from the benefits of lower costs of such services as long-distance telephone calls, natural-gas purchases, and air travel. The industries embracing these activities have been recently deregulated, and in these sectors, competition has been introduced. The basic belief is that competition leads to enhanced efficiency and thus lower costs and improved services.

For nearly 100 years, electrical power utilities worldwide have been vertically integrated, combining generation, transmission, distribution, and servicing loads. Also, most such utilities have operated as monopolies within their geographic regions. Their method of operation has been "power at cost," and their principal financers have been governments. Therefore, to many people the pressure of electrical power utilities to operate efficiently has been missing.

Operating the electrical energy sector competitively requires the unbundling of generation, transmission, and distribution. Competition is expected to exist among generators as well as retailers. The transmission and distribution (i.e., the controlling wires) must, out of necessity, be regulated. The new order requires new agencies taking the responsibility to link customers (loads) with generators (market operators) and, at the same time, to clearly understand the limitations and capabilities of power-transmission and -distribution networks [22], [23].

On becoming responsible for its own business, a power-transmission company must make the best use of its transmission capacity and ensure that transmission losses are reduced to their lowest values. Also, any loss of transmission capacity means loss of income for the company; therefore, all actions must be taken to ensure that unwanted circulating power is not clogging the available transmission capacity. In addition, energy congestion in critical transmission corridors must be avoided to eliminate the risk of missed business opportunities. Finally, to offer the greatest flexibility to market operators, a transmission company must create the maximum safe operating limits to allow power injection and tapping from its buses without risking stable operation. The success of a transmission company depends on offering the maximum available transmission capacity (ATC) on its lines.

From the foregoing discussion, it is evident that in the emerging electrical energy business, transmission companies have a greater need to make their networks more flexible. Fortunately, advances in power-electronics technology now offer new fast, controllable FACTS controllers to secure the needed flexibility [15], [22], [23].

The subject matter contained in this book is intended to assist engineers seeking FACTS knowledge and help utilities meet the energy challenge.

REFERENCES

[1] C. Adamson and N. G. Hingorani, *High-Voltage Direct Current Power Transmission*, Garraway, 1960.

[2] E. W. Kimbark, *Direct Current Transmission*, Vol. 1, John Wiley and Sons, New York, 1971.

[3] E. Uhlmann, *Power Transmission by Direct Current*, Springer-Verlag, 1975.

[4] J. Arrillaga, *High-Voltage Direct Current Transmission*, Peter Peregrinus, 1983.

[5] K. R. Padiyar, *HVDC Power Transmission Systems*, Wiley Eastern Limited, New Delhi, India, 1990.

[6] IEEE Power Engineering Society/CIGRE, *FACTS Overview*, Publication 95TP108, IEEE Press, New York, 1995.

[7] N. G. Hingorani and L. Gyugyi, *Understanding FACTS*, IEEE Press, New York, 1999.

[8] Y. H. Song and A. T. Johns, Eds., *Flexible AC Transmission Systems (FACTS)*, IEE Press, London, 1999.

[9] IEEE Power Engineering Society, *FACTS Applications*, Publication 96TP116-0, IEEE Press, New York, 1996.

[10] W. H. Litzenberger, Ed., "An Annotated Bibliography of High-Voltage Direct-Current Transmission, 1989–1991," Published by the Bonneville Power Administration (BPA) and the Western Area Power Administration, Portland, OR, 1992.

[11] W. H. Litzenberger, Ed., "An Annotated Bibliography of High-Voltage Direct-Current Transmission and Flexible AC Transmission (FACTS) Devices, 1991–1993," Published by the Bonneville Power Admnistration (BPA) and the Western Area Power Administration, Portland, OR, 1994.

[12] W. H. Litzenberger and R. K. Varma, Eds., "An Annotated Bibliography of High-Voltage Direct-Current Transmission and FACTS Devices, 1994–1995," Published by the Bonneville Power Administration (BPA) and the U.S. Department of Energy, Portland, OR, 1996.

[13] W. H. Litzenberger, R. K. Varma, and J. D. Flanagan, Eds., "An Annotated Bibliography of High-Voltage Direct-Current Transmission and FACTS Devices, 1996–1997," Published by the Electric Power Research Institute (EPRI) and the Bonneville Power Administration (BPA), Portland, OR, 1998.

[14] CIGRE Working Group 14.29, "Coordination of Controls of Multiple FACTS/HVDC Links in the Same System," CIGRE Technical Brochure No. 149, Paris, December 1999.

[15] Electric Power Research Institute (EPRI) Report, "Guide for Economic Evaluation of Flexible AC Transmission Systems (FACTS) in Open Access Environments," EPRI TR 108500, Final report prepared by General Electric Company (GE), Schenectady, NY, August 1997.

[16] C. E. Grund et al., "Dynamic Performance Characteristics of North American HVDC Systems for Transient and Dynamic Stability Evaluations," IEEE Committee Report, *IEEE Transactions on Power Apparatus and Systems*, Vol. PAS–100, No. 7, July 1981, pp. 3356–3364.

[17] IEEE Committee Report, "A Description of Discrete Supplementary Controls for Stability," *IEEE Transactions on Power Apparatus and Systems*, Vol. PAS–97, No. 1, January/February 1978, pp. 146–165.

[18] R. W. De Doncker, O. Demirci, S. Arthur, and V. Temple, "Characteristics of GTOs and High-Voltage MCTs in High Power Soft-Switching Converters," *IEEE Transactions on Industry Applications*, Vol. 30, No. 6, November/December 1994, pp. 1548–1556.

[19] Silicon Power Corporation Product Catalog, "Introduction to MTO," August 1995.

[20] I. A. Erinmez and A. M. Foss, Eds., "Static Synchronous Compensator (STATCOM)," Working Group 14.19, CIGRE Study Committee 14, Document No. 144, August 1999.

[21] CIGRE Task Force 14-27, "Unified Power Flow Controller," CIGRE Technical Brochure, 1998.

[22] CIGRE Task Force 38.02.17, "Advanced Angle Stability Controls," CIGRE Technical Brochure No. 155, Paris, April 2000.

[23] CIGRE Task Force 38.02.16, "Impact of Interactions Among Power System Controls," CIGRE Technical No. Brochure 166, Paris, August 2000.

CHAPTER 2

Reactive-Power Control in Electrical Power Transmission Systems

2.1 REACTIVE POWER

In this chapter, an understanding of reactive power associated with power-transmission networks is developed. To make transmission networks operate within desired voltage limits, methods of making up or taking away reactive-power—hereafter called *reactive-power control*—are discussed. Before proceeding further, however, a thorough understanding of the reactive power in ac systems is necessary.

Upon energization, the ac networks and the devices connected to them create associated time-varying electrical fields related to the applied voltage, as well as magnetic fields dependent on the current flow. As they build up, these fields store energy that is released when they collapse. Apart from the energy dissipation in resistive components, all energy-coupling devices, including transformers and energy-conversion devices (e.g., motors and generators), operate based on their capacity to store and release energy.

For the ac circuit shown in Fig. 2.1(a), instantaneous power from the voltage source to the load $Z\angle\phi$, in terms of the instantaneous voltage v and current i, is given as

$$p = vi \tag{2.1}$$

In the steady state, where $v = V_{\max}\cos(\omega t)$ and $i = I_{\max}\cos(\omega t - \phi)$:

$$\begin{aligned} p &= \frac{V_{\max} I_{\max}}{2}\,[\cos\phi + \cos(2\omega t - \phi)] \\ &= VI\cos\phi(1 + \cos 2\omega t) + VI\sin\phi\sin 2\omega t \end{aligned} \tag{2.2}$$

where V and I are the respective root mean square (rms) values of v and i.

Equations (2.1) and (2.2) are pictorially represented in Fig. 2.1(b). Equation (2.2) comprises two double-frequency (2ω) components. The first term has an average value as well as a peak magnitude of $VI\cos\phi$. This average value is the active power, P, flowing from the source to the load. The second term has

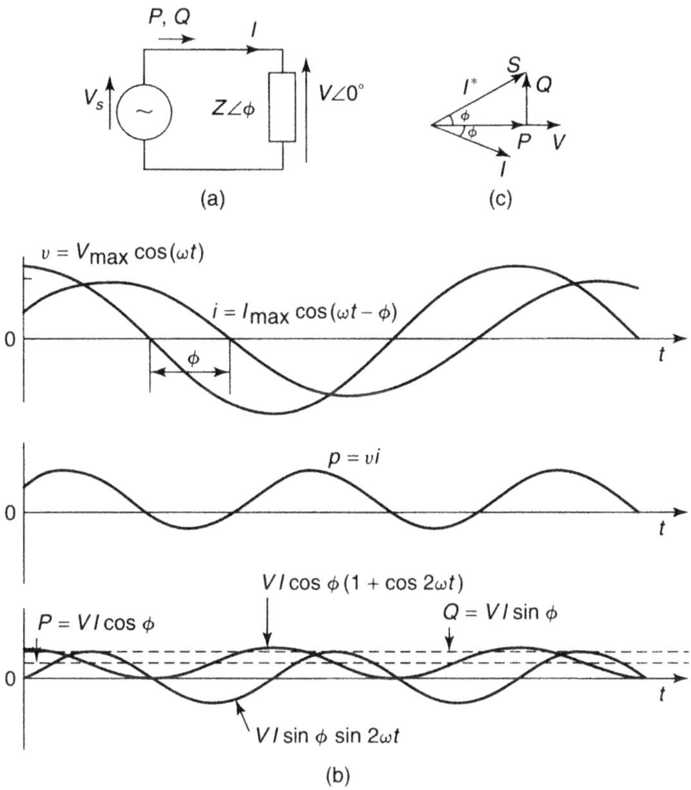

Figure 2.1 The electrical parameters in an ac network.

a zero average value, but its peak value is $VI\sin\phi$. Written in phasor domain, the complex power in the network in Fig. 2.1(a) is given by

$$\begin{aligned} S &= \overline{V}\cdot\overline{I}^{*} \\ &= P + jQ = VI\cos\phi + jVI\sin\phi \end{aligned} \tag{2.3}$$

where P is called the active power, which is measured in watts (W), and Q is called the reactive power, which is measured in volt–ampere reactives (var). Comparing Eqs. (2.3) and (2.2), the peak value of the second component of instantaneous power in Eq. (2.2) is identified as the reactive power.

The reactive power is essential for creating the needed coupling fields for energy devices. It constitutes voltage and current loading of circuits but does not result in an average (active) power consumption and is, in fact, an important component in all ac power networks. In high-power networks, active and reactive powers are measured in megawatts (MW) and MVAR, respectively. Figure 2.1(c) shows a commonly used power triangle.

Electromagnetic devices store energy in their magnetic fields. These devices draw lagging currents, thereby resulting in positive values of Q; therefore, they are frequently referred to as the absorbers of reactive power. Electrostatic devices, on the other hand, store electric energy in fields. These devices draw leading currents and result in a negative value of Q; thus they are seen to be suppliers of reactive power. The convention for assigning signs to reactive power is different for sources and loads, for which reason readers are urged to use a consistent notation of voltage and current, to rely on the resulting sign of Q, and to not be confused by absorbers or suppliers of reactive power.

2.2 UNCOMPENSATED TRANSMISSION LINES

2.2.1 A Simple Case

To develop a good, qualitative understanding of the need for reactive-power control, let us consider a simple case of a lossless short-transmission line connecting a source V_s to a load $Z\angle\phi$. (For simplicity, the line is represented only by its inductive reactance X_l.) Figure 2.2 shows such a network with its parameters, as well as a phasor diagram showing the relationship between voltages and currents. From Fig. 2.2(b), it is clear that between the sending- and the receiving-end voltages, a magnitude variation, as well as a phase difference, is created. The most significant part of the voltage drop in the line reactance ($\Delta V_1 = j\bar{I}_x X_l$) is due to the reactive component of the load current, I_x. To keep the voltages in the network at nearly the rated value, two control actions seem possible:

1. load compensation, and
2. system compensation.

2.2.1.1 Load Compensation

It is possible to compensate for the reactive current I_x of the load by adding a parallel capacitive load so that $I_c = -I_x$. Doing so causes the effective power factor of the combination to become unity. The absence of I_x eliminates the voltage drop ΔV_1, bringing V_r closer in magnitude to V_s; this condition is called *load compensation*. Actually, by charging extra for

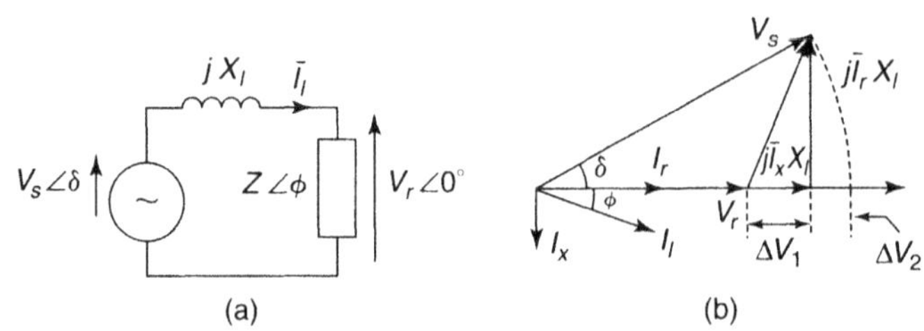

Figure 2.2 A short, lossless transmission line feeding a load.

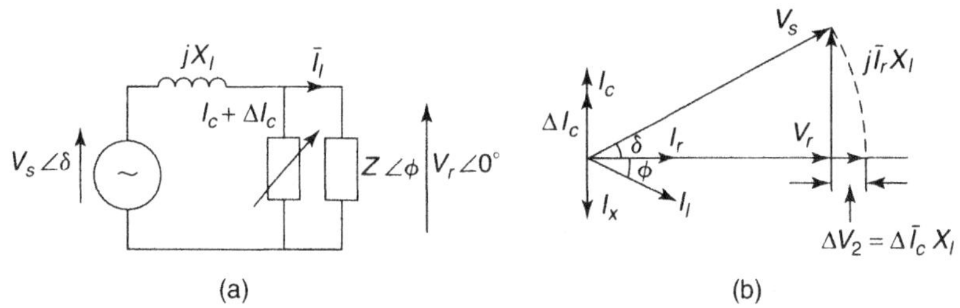

Figure 2.3 The reactive-power control for voltage regulations.

supplying the reactive power, a power utility company makes it advantageous for customers to use load compensation on their premises. Loads compensated to the unity power factor reduce the line drop but do not eliminate it; they still experience a drop of ΔV_2 from $j\bar{I}_r X_l$.

2.2.1.2 System Compensation To regulate the receiving-end voltage at the rated value, a power utility may install a reactive-power compensator as shown in Fig. 2.3. This compensator draws a reactive current to overcome both components of the voltage drop ΔV_1 and ΔV_2 as a consequence of the load current I_l through the line reactance X_l. To compensate for ΔV_2, an additional capacitive current, ΔI_c, over and above I_c that compensates for I_x, is drawn by the compensator. When $\Delta\bar{I}_c X_l = \Delta V_2$, the receiving-end voltage, V_r, equals the sending-end voltage, V_s. Such compensators are employed by power utilities to ensure the quality of supply to their customers [1].

2.2.2 Lossless Distributed Parameter Lines

Most power-transmission lines are characterized by distributed parameters: series resistance, r; series inductance, l; shunt conductance, g; and shunt capacitance, c—all per-unit (pu) length. These parameters all depend on the conductors' size, spacing, clearance above the ground, and frequency and temperature of operation. In addition, these parameters depend on the bundling arrangement of the line conductors and the nearness to other parallel lines.

The characteristic behavior of a transmission line is dominated by its l and c parameters. Parameters r and g account for the transmission losses. The fundamental equations governing the propagation of energy along a line are the following wave equations:

$$\frac{d^2\overline{V}}{dx^2} = zy\overline{V} \tag{2.4a}$$

$$\frac{d^2\overline{I}}{dx^2} = zy\overline{I} \tag{2.4b}$$

where $zy = (r + j\omega l)(g + j\omega c)$.

For a lossless line, the general solutions are given as

$$\overline{V}(x) = \overline{V}_s \cos \beta x - jZ_0\overline{I}_s \sin \beta x \tag{2.5a}$$

$$\overline{I}(x) = \overline{I}_s \cos \beta x - j\,\frac{\overline{V}_s}{Z_0}\,\sin \beta x \tag{2.5b}$$

These equations are used to calculate voltage and current anywhere on line, at a distance x from the sending end, in terms of the sending-end voltage and current and the line parameters. In Eqs. (2.4) and (2.5),

$$Z_0 = \sqrt{\frac{l}{c}}\ \Omega = \text{the surge impedance or characteristic impedance}$$

$$\beta = \omega\sqrt{lc}\ \text{rad/km} = \text{the wave number}$$

$$\beta a = \omega\sqrt{lca}\ \text{rad} = \text{the electrical length of an } a\text{-km line}$$

where l is the line inductance in henries per kilometer (H/km), c is the line-shunt capacitance in farads per kilometer (F/km), and $1/\sqrt{lc}$ is the propagation velocity of electromagnetic effects on the transmission line. (It is less than the velocity of light.)

From Eq. (2.5), we get

$$\overline{I}_s = \frac{\overline{V}_s \cos \beta a - \overline{V}_r}{jZ_0 \sin \beta a}$$

If $\overline{V}_s = V_s\angle 0^\circ$ and $\overline{V}_r = V_r\angle - \delta = V_r(\cos \delta - j \sin \delta)$, then

$$\overline{I}_s = \frac{V_r \sin \delta + j(V_r \cos \delta - V_s \cos \beta a)}{Z_0 \sin \beta a} \tag{2.6}$$

Therefore, the power at the sending end is given as

$$\begin{aligned} S_s &= P_s + jQ_s = \overline{V}_s\overline{I}_s^* \\ &= \frac{V_sV_r \sin \delta}{Z_0 \sin \beta a} + j\,\frac{V_s^2 \cos \beta a - V_sV_r \cos \delta}{Z_0 \sin \beta a} \end{aligned} \tag{2.7}$$

Likewise, power at the receiving end is given as

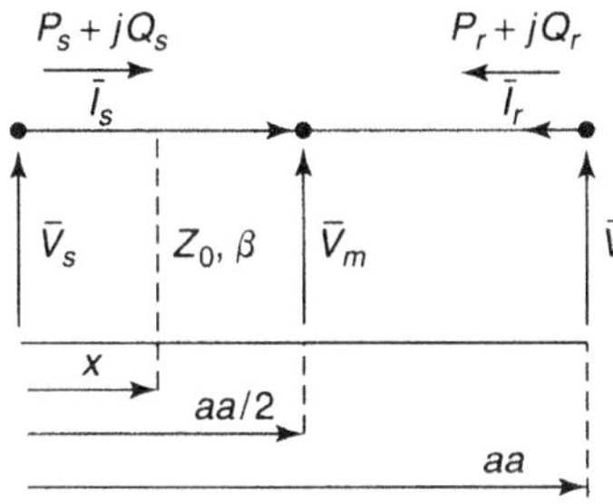

Figure 2.4 The power on a lossless distributed line.

$$S_r = P_r + jQ_r = -\frac{V_s V_r \sin\delta}{Z_0 \sin\beta a} + j\,\frac{V_r^2 \cos\beta a - V_s V_r \cos\delta}{Z_0 \sin\beta a} \tag{2.8}$$

Comparing Eqs. (2.7) and (2.8) and taking the directional notation of Fig. 2.4 into account, it is concluded that for a lossless line, $P_s = -P_r$, as expected. However, $Q_s \neq Q_r$ because of the reactive-power absorption/generation in the line.

From Eqs. (2.7) and (2.8), the power flow from the sending end to the receiving end is expressed as

$$P = \frac{V_s V_r \sin\delta}{Z_0 \sin\beta a}$$

In electrically short power lines, where βa is very small, it is possible to make a simplifying assumption that $\sin\beta a = \beta a$ or $Z_0 \sin\beta a = Z_0\beta a = \omega la$, where $\omega la = X_l$ is the total series reactance of a line. This substitution results in the following well-recognized power equation:

$$P = \frac{V_s V_r}{X_l} \sin\delta \tag{2.9}$$

Accordingly, the maximum power transfer is seen to depend on the line length. When the power-transfer requirement for a given length of a line increases, higher transmission voltages of V_s and V_r must be selected.

This chapter is not intended to provide a comprehensive analysis of transmission lines. Rather, its objective is to examine those aspects that enhance the understanding of the interplay between voltages on the line and the resulting reactive-power flows.

2.2.2.1 Symmetrical Lines When the voltage magnitudes at the two ends of a line are equal, that is, $V_s = V_r = V$, the line is said to be symmetrical. Because power networks operate as voltage sources, attempts are made to hold almost all node voltages at nearly rated values. A symmetrical line, therefore, presents a realistic situation. From Eqs. (2.7) and (2.8) the following relationships are derived:

$$P_s = -P_r = \frac{V^2}{Z_0 \sin \beta a} \sin \delta \tag{2.10}$$

and

$$Q_s = Q_r = \frac{V^2 \cos \beta a - V^2 \cos \delta}{Z_0 \sin \beta a} \tag{2.11}$$

Active and reactive powers of a transmission line are frequently normalized by choosing the surge-impedance load (SIL) as the base. The SIL is defined as $P_0 = V_{\text{nom}}^2/Z_0$, where V_{nom} is the rated voltage. When $V_s = V_r = V_{\text{nom}}$,

$$\frac{P_s}{P_0} = -\frac{P_r}{P_0} = \frac{\sin \delta}{\sin \beta l} \tag{2.12}$$

and

$$\frac{Q_s}{Q_0} = \frac{Q_r}{Q_0} = \frac{\cos \beta a}{\sin \beta a} - \frac{\cos \delta}{\sin \beta a} \tag{2.13}$$

2.2.2.2 Midpoint Conditions of a Symmetrical Line The magnitude of the midpoint voltage depends on the power transfer. This voltage influences the line insulation and therefore needs to be well understood. For a symmetrical line where the end voltages are held at nominal values, the midpoint voltage shows the highest magnitude variation. In terms of the midpoint voltage $\overline{V}_m$, the receiving-end voltage of a symmetrical line, from Eq. (2.4), is given as

$$\overline{V}_r = \overline{V}_m \cos \frac{\beta a}{2} - jZ_0 \overline{I}_m \sin \frac{\beta a}{2} \tag{2.14}$$

For simplification, define $\overline{V}_m = V_m \angle 0^\circ$ as the reference phasor. Because the line is symmetrical and lossless, that is, $P_s = -P_r = P_m = P$ and $Q_m = 0$, the midpoint current I_m is given by $I_m = P/V_m$. Under these conditions, Eq. (2.14) can be rewritten as

$$\overline{V}_r = V_m \cos \frac{\beta a}{2} - jZ_0 \frac{P}{V_m} \sin \frac{\beta a}{2}$$

or

$$V_r^2 = V_m^2 \cos^2 \frac{\beta a}{2} + Z_0^2 \frac{P^2}{V_m^2} \sin^2 \frac{\beta a}{2}$$

Setting $V_r = V_{\text{nom}}$ and $V_{\text{nom}}^2/Z_0 = P_0$, we get

$$\frac{V_r^2}{V_{\text{nom}}^2} = \left(\frac{V_m}{V_{\text{nom}}}\right)^2 \cos^2 \frac{\beta a}{2} + \left(\frac{Z_0}{V_{\text{nom}}^2}\right)^2 P^2 \left(\frac{V_{\text{nom}}}{V_m}\right)^2 \sin^2 \frac{\beta a}{2}$$

If we let $V_m/V_{\text{nom}} = \tilde{V}_m$ (per-unit voltage at the midpoint), then considerng that $(V_r/V_{\text{nom}}) = 1$, we have

$$\tilde{V}_m^4 - \frac{\tilde{V}_m^2}{\cos^2 \dfrac{\beta a}{2}} + \left(\frac{P}{P_0}\right)^2 \tan^2 \frac{\beta a}{2} = 0$$

Therefore

$$\tilde{V} = \left[\frac{1}{2\cos^2 \dfrac{\beta a}{2}} \pm \sqrt{\frac{1}{4\cos^2 \dfrac{\beta a}{2}} - \left(\frac{P}{P_0}\right)^2 \tan^2 \frac{\beta a}{2}}\,\right]^{1/2} \tag{2.15}$$

Equation (2.15) determines the midpoint voltage of a symmetrical line as a function of the power flow P on it.

Practical Considerations In general, the values of line parameters l and c remain reasonably independent of the transmission voltage. For example, typical values of l and c may lie in the following ranges:

l = the line inductance/km = 0.78–0.98 mH/km

c = the line capacitance/km = 12.1–15.3 nF/km

On the basis of these parameters, the surge impedance, $Z_0 = \sqrt{l/c}$, lies in the range of 225 to 285.

2.2.2.3 Case Study To illustrate a number of important considerations, let us choose a 735-kV symmetrical lossless transmission line with $l = 0.932$ mH/km, $c = 12.2$ nF/km, and a line length of 800 km. From the foregoing parameters,

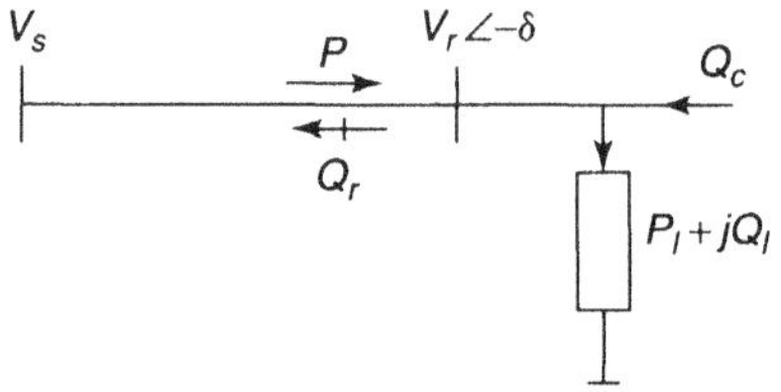

Figure 2.5 The reactive-power balance at the receiving end.

$$Z_0 = \sqrt{\frac{l}{c}} = \sqrt{\frac{0.932}{12.2}}\, 10^3 = 276.4\ \Omega$$

Therefore, the SIL is

$$P_0 = \frac{V_{\text{nom}}^2}{Z_0} = \frac{(735 \times 10^3)^2}{276.4} = 1954.5\ \text{MW}$$

For this line to operate as a symmetrical line, that is, $V_s = V_r = 735$ kV, we have from Eq. (2.10):

$$P_s = \frac{V^2}{Z_0 \sin \beta a} \sin \delta = \frac{735^2}{276.4 \sin[(\omega\sqrt{lc})\, 800]} \sin \delta$$

$$= 2298.5 \sin \delta \text{ MW} = 1.176 P_0 \sin \delta \text{ MW} \tag{2.16}$$

It is important to calculate the required additional reactive power to hold the receiving-end voltage to 1 pu (735 kV). Let us assume that connected at the receiving end is a load of fixed power factor 0.9 lagging. For any load condition, the reactive-power balance at the receiving-end bus shown in Fig. 2.5 is $Q_c = Q_r + Q_l$, where Q_r is the reactive-power flow from the receiving end into the line, Q_l is the reactive-power component of the load, and Q_c is the reactive power needed from the system to hold V_r to the rated value (1 pu).

Figure 2.6 shows Q_r/P_0, Q_l/P_0 and Q_c/P_0 as functions of P/P_0. It should be observed that at no load ($P = 0$), nearly 1090-MVAR or 0.557-pu reactive power must be absorbed to hold the receiving-end voltage to 1 pu. To avoid overinsulating the line so that it might withstand overvoltages under no-load or light conditions, a common practice is to permanently connect shunt reactors at both ends to allow line energization from either end. Unfortunately, this natural protection becomes a liability under increased load conditions, for extra reactive power, Q_c, is needed to hold the terminal bus voltages at the desired level. The midpoint voltage of this line is calculated using Eq. (2.15), and typical voltage distribution on a distributed line is shown in Fig. 2.7.

Alternatively, consider the receiving half of the line. From Eq. (2.7), the power flow on the line is given as

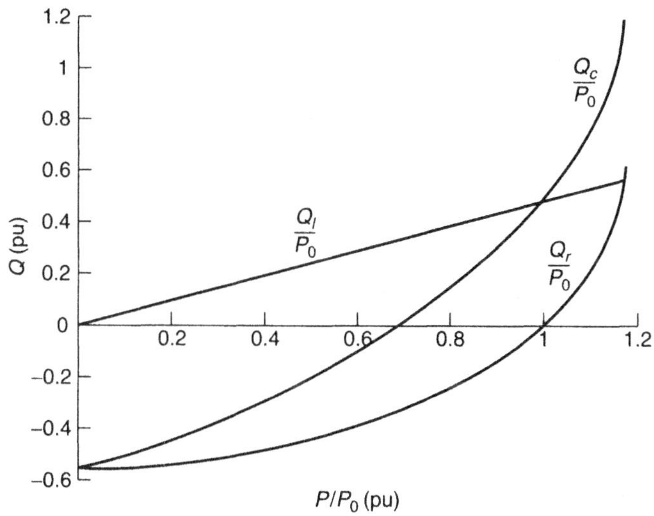

Figure 2.6 The reactive-power flows at the receiving end.

$$P_0 = \frac{V_m V_r}{Z_0 \sin \dfrac{\beta a}{2}} \sin \frac{\delta}{2} = \frac{V_{\text{nom}}^2}{Z_0 \sin \beta a} \sin \delta$$

Because $V_r = V_s = V_{\text{nom}}$,

$$\tilde{V}_m = \frac{V_m}{V_{\text{nom}}} \frac{\sin \dfrac{\beta a}{2}}{\sin \beta a} \frac{\sin \delta}{\sin \dfrac{\delta}{2}}$$

$$= 0.5724 \frac{\sin \delta}{\sin \dfrac{\delta}{2}} \tag{2.17}$$

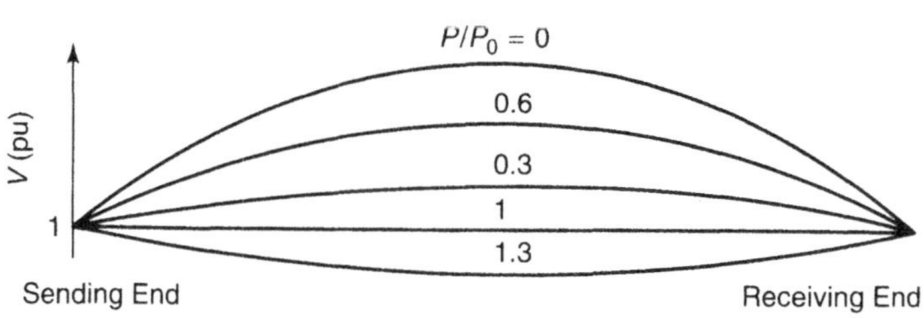

Figure 2.7 The typical voltage distribution on a distributed line.

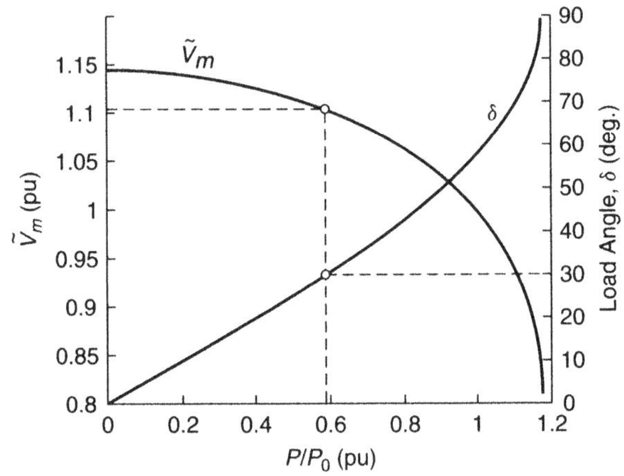

Figure 2.8 The midpoint voltage, $\tilde{V}$, and load angle, δ, as functions of P/P_0.

Figure 2.8 shows the load angle δ and the midpoint-normalized per-unit voltage ($\tilde{V}_m$) as functions of P/P_0 per-unit power transfer on the line. For this line, it is observed that for light loading below the surge-impedance load, the midpoint voltage exceeds V_{nom} and reaches its highest value at no load. Furthermore, for stability considerations should an operating load angle of, say, 30° be chosen to define the full-load rating of the line ($0.588P_0$), the midpoint voltage of the line will be 1.1058 pu. These expected overvoltages in the range 0.1–0.2 pu from full load to no load are not within acceptable limits. Therefore, special techniques must be used to control these overvoltages.

It is possible to control the overall voltage profile of such a line as the one described in this case study by creating a midpoint voltage bus and connecting a controllable reactive-power source, called a *var compensator*, to it so that below the surge-impedance loading P_0, the var compensator absorbs reactive power, and above P_0, it supplies reactive power.

Figure 2.9 shows a midpoint var compensator employed as a voltage controller and the expected voltage profile along the line. Of course, it is not important to hold the midpoint voltage V_{mc} at 1-pu voltage, especially if there is no load connected to it. Also, it is not necessary to have a controllable var source at the midpoint; instead, an adequately sized fixed- or switched-shunt reactor could be used to keep the overvoltage within limits.

To continue this discussion with the aid of the case study, let us hold the midpoint voltage to V_{mc} under all load conditions by employing a continuous var controller of unlimited capacity. Using Eq. (2.7), we have

$$Q_m = \frac{V_{mc}^2 \cos \dfrac{\beta a}{2} - V_s V_{mc} \cos \dfrac{\delta}{2}}{Z_0 \sin \dfrac{\beta a}{2}} \tag{2.18}$$

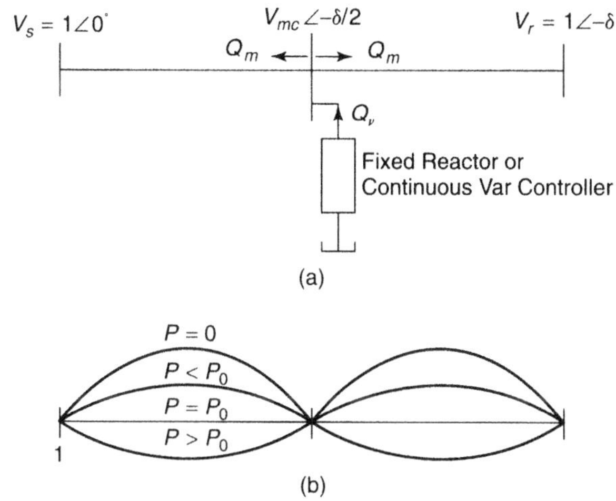

Figure 2.9 The midpoint-overvoltage control: $V_{mc} = 1$ pu.

Therefore, the var requirement from the midpoint var controller is

$$Q_v = 2Q_m \tag{2.19}$$

In terms of the midpoint voltage, we can rewrite the power transfer of the line given by Eq. (2.7) as

$$P_{\text{comp}} = \frac{V_s V_{mc}}{Z_0 \sin \dfrac{\beta a}{2}} \sin \frac{\delta}{2} \tag{2.20}$$

For the 735-kV, 800-km line with $\beta = 1.27 \times 10^{-3}$ rad/km, and assuming $V_s = V_r = 1$ pu and $V_{mc} = 1.05$ pu, we get

$$\beta \frac{a}{2} = \beta \times 400 = 0.508 \text{ rad}$$

$$P_{\text{comp}} = \frac{735^2 \times 1.05}{276.4 \times \sin(0.508)} \sin \frac{\delta}{2}$$

$$= 4215.28 \sin \frac{\delta}{2} = 2.157 P_0 \sin \frac{\delta}{2}$$

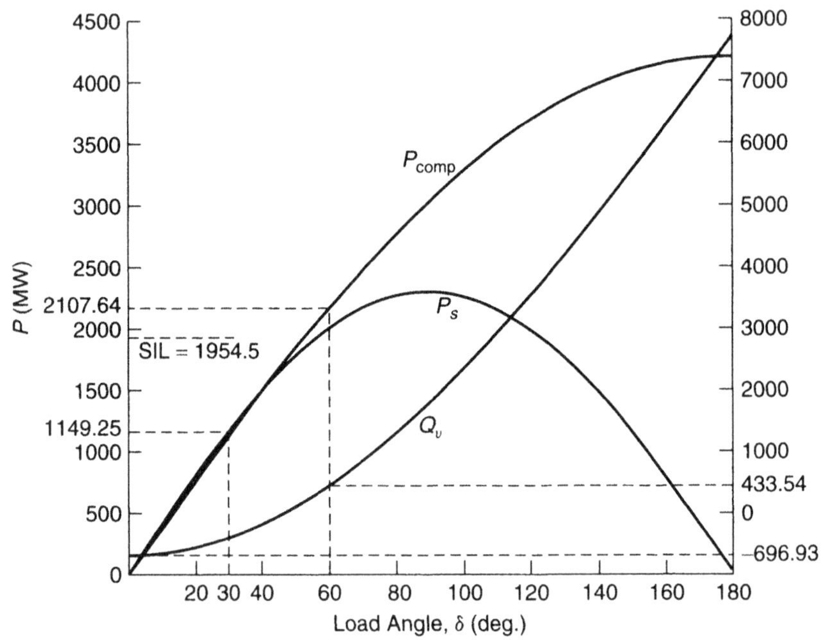

Figure 2.10 The relationship between active power, P, and reactive power, Q, with load angle δ in the 735-kV midpoint-compensated line.

$$Q_v = 2Q_m = 2 \times \frac{(735 \times 1.05)^2 \cos(0.508) - 735^2 \times 1.05 \cos \dfrac{\delta}{2}}{276.4 \sin(0.508)}$$

$$= 7740.98 - 8437.91 \cos \frac{\delta}{2} = \left(3.96 - 4.32 \cos \frac{\delta}{2}\right) P_0$$

Figure 2.10 depicts the following relations:

$P_s = 2298.5 \sin \delta$ MW = the uncompensated line

$P_{comp} = 4215.28 \sin (\delta/2)$ MW = the unlimited midpoint compensation to hold $V_{mc} = 1.05$ pu

$Q_v = 7740.98 - 8437.91 \cos (\delta/2)$ MVAR = the reactive power injected by the compensator SIL of the line

$P_0 = 1954 \cdot 5$ MW

The results of Fig. 2.10 need careful interpretation. Note the following:

1. If the nominal rating of a line might correspond to $\delta = 30°$, the 735-kV, 800-km symmetric line in this case study will be rated at $P_{nom} = 2298.5 \sin(30°) = 1149.25$ MW, which is only 58.8% of the SIL.

2. If the preceding 735-kV symmetric line is provided with unlimited midpoint reactive-power compensation, Q_v, to hold the midpoint voltage, V_{mc}, at 1.05 pu, then the maximum transferable power increases from 2298.5 MW (P_s in the uncompensated case) to 4215.28 MW (P_{comp} in the compensated case), implying a ±83.39% increase. The nominal rating for this line could be 2107.64 MW (or 108% of SIL) for $\delta = 60°$.

3. To maintain the midpoint voltage, V_{mc}, of the 735-kV symmetric line at 1.05 pu, the midpoint compensator's var supply would range from −696.93 MVAR to 7740.98 MVAR. This is a very large operating range for a line with 1954.5 MW SIL.

Therefore, let us search for a workable solution.

1. Assume that the midpoint-compensated line is rated at $\delta = 60°$ for stable operation, that is, $P_{\text{comp}} = 2107.64$ MW, with a nominal rating $(1.08P_0)$.
2. The midpoint-compensator power for $\delta = 60°$ is $Q_v = 433.54$ MVAR.
3. On the basis of entries 1 and 2, select a realistic midpoint var compensator rated to operate from −600 to +400 MVAR.

The performance of the 735-kV, 800-km symmetric line, with a midpoint var compensator designed to operate at 1.05 pu terminal voltage in a controllable range, can be analyzed as follows: Beyond the limit $Q_v > 400$ MVAR, the var compensator behaves like a fixed capacitor of rating

$$X_c = \frac{V_m^2}{Q_v} = \frac{(1.05 \times 735)^2}{400} = 1489\ \Omega$$

In the uncontrollable range, from Eq. (2.18) the corresponding value of Q_m in each half of the line is

$$Q_m = \frac{V_m^2}{2X_c} = \frac{V_m^2 \cos \dfrac{\beta a}{2} - V_s V_m \cos \dfrac{\delta}{2}}{Z_0 \sin \dfrac{\beta a}{2}}$$

or

$$V_m = \frac{V_s \cos \dfrac{\delta}{2}}{\cos \dfrac{\beta a}{2} - \dfrac{Z_0}{2X_c} \sin \dfrac{\beta a}{2}}$$

$$= \frac{735 \cos \dfrac{\delta}{2}}{\cos(0.508) - \dfrac{276.4}{2 \times 1489} \sin(0.508)}$$

$$= 886.778 \cos \frac{\delta}{2} = 1.2065 V_{\text{nom}} \cos \frac{\delta}{2}$$

From this value of V_m (beyond the controllable capacitor range), the power flow on the line is given as

$$P = \frac{V_s V_m \sin \dfrac{\delta}{2}}{Z_0 \sin \dfrac{\beta a}{2}}$$

$$= \frac{735 \times 886.778 \sin \dfrac{\delta}{2} \cos \dfrac{\delta}{2}}{276.4 \sin(0.508)}$$

$$= 2423.9 \sin \delta$$

The start of uncontrollable range on the capacitive side of the var compensator corresponds to $Q_v > 400$ MVAR at a value of δ calculated from

$$Q_v = 400 = 7740.98 - 8437.91 \cos \frac{\delta}{2}$$

or

$$\delta = 59^\circ$$

Likewise, below $Q_v < -600$ MVAR, the inductive limit of the var compensator is reached at a value of δ calculated from

$$Q_v = 400 = 7740.98 - 8437.91 \cos \frac{\delta}{2}$$

or

$$\delta = 17.38^\circ$$

Thus

$$P = 4215.28 \sin(8.69^\circ) = 637.1 \text{ MW}$$

The value of the fixed inductor below $\delta = 17.38^\circ$ is calculated as

$$X_l = \frac{V_m^2}{Q_v} = \frac{(735 \times 1.05)^2}{600} = 992.66\ \Omega$$

since

$$Q_m = \frac{-V_m^2}{2X_l} = \frac{V_m^2 \cos\dfrac{\beta a}{2} - V_s V_m \cos\dfrac{\delta}{2}}{Z_0 \sin\dfrac{\beta a}{2}}$$

or

$$V_m = \frac{V_s \cos\dfrac{\delta}{2}}{\cos\dfrac{\beta a}{2} - \dfrac{Z_0}{2X_c}\sin\dfrac{\beta a}{2}}$$

$$= \frac{735 \cos\dfrac{\delta}{2}}{\cos(0.508) + \dfrac{276.4}{2 \times 992.66}\sin(0.508)}$$

$$= 780.72 \cos\frac{\delta}{2} = 1.0622 V_{\text{nom}} \cos\frac{\delta}{2}$$

Using the foregoing value of V_m, power in the inductive uncontrollable range of the midpoint var compensator is given as

$$P = \frac{V_s V_m \sin\dfrac{\delta}{2}}{Z_0 \sin\dfrac{\beta a}{2}}$$

$$= \frac{735 \times 780.72 \sin\dfrac{\delta}{2} \cos\dfrac{\delta}{2}}{276.4 \sin(0.508)}$$

$$= 2134 \sin\ \delta$$

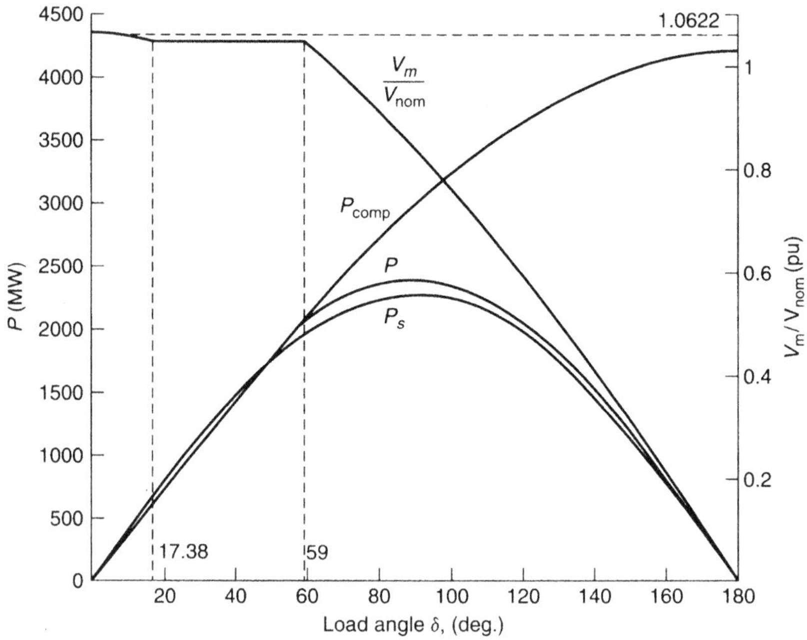

Figure 2.11 The relationship between power P, midpoint voltage V_m, and load angle δ of a 735-kV symmetric line compensated by a −600/+400 MVAR range–controlled var compensator.

Figure 2.11 depicts following relations:

$P_s = 2298.5 \sin\delta$ MW = the uncompensated line

$P_{\text{comp}} = 4215.28 \sin(\delta/2)$ MW = the unlimited midpoint compensation to hold $V_{mc} = 1.05$ pu

$V_m = 1.0622 V_{\text{nom}} \cos(\delta/2)$ pu, for $0 < \delta \leq 17.38^\circ$

$P = 2134 \sin\delta$ MW

$V_m = 1.05 V_{\text{nom}}$ pu, for $17.38^\circ \leq \delta \leq 59^\circ$

$P = 4215.28 \sin(\delta/2)$ MW

$V_m = 1.2065 V_{\text{nom}} \cos(\delta/2)$ pu, for $59^\circ \leq \delta \leq 180^\circ$

$P = 2423.9 \sin\delta$ MW

Figure 2.11 shows the results of a symmetric 735-kV, 800-km lossless line where, at its midpoint, a controllable var compensator is installed that holds the midpoint voltage at 1.05-pu value. The var compensator has a fully controllable range of −600 MVAR to +400 MVAR. Beyond the +400-MVAR limit, the compensator behaves like a fixed capacitor ($X_c = 1489\ \Omega$); below −600 MVAR, it acts like a fixed inductor ($X_l = 992.66\ \Omega$). It should be observed that under these circumstances, the maximum power transfer of the line is modified to 2423.9 MW, and the maximum overvoltage of the midpoint is limited to 1.062 pu.

The conclusions are as follows:

1. As the length of the line increases on account of the line-charging capacitances, the line experiences significant overvoltages at light-load conditions.
2. Overvoltages can be limited by using fixed- or switched-shunt reactors at the line ends as well as at intermediate buses where needed.
3. The application of midpoint or intermediate bus-voltage controllers (var compensators) enhances the power-transmission capacity of a long line.
4. In practical cases, var controllers are sized by carefully selecting their continuous-operating range to hold the connecting-bus voltage within an acceptable range of values in the normal line-loading range.

Active and Passive Var Control When fixed inductors and/or capacitors are employed to absorb or generate reactive power, they constitute passive control. An active var control, on the other hand, is produced when its reactive power is changed irrespective of the terminal voltage to which the var controller is connected.

2.3 PASSIVE COMPENSATION

In the foregoing discussion, a lossless line was analyzed, and the case study presented in Section 2.2 provided many numerical results and highlighted the problems of voltage control and the need to exercise reactive-power control to make a system workable. Reactive-power control for a line is often called *reactive-power compensation.* External devices or subsystems that control reactive power on transmission lines are known as *compensators.* Truly speaking, a compensator mitigates the undesirable effects of the circuit parameters of a given line. The objectives of line compensation are invariably

1. to increase the power-transmission capacity of the line, and/or
2. to keep the voltage profile of the line along its length within acceptable bounds to ensure the quality of supply to the connected customers as well as to minimize the line-insulation costs.

Because reactive-power compensation influences the power-transmission capacity of the connected line, controlled compensation can be used to improve the system stability (by changing the maximum power-transmission capacity), as well as to provide it with positive damping. Like other system components, reactive-power compensators are dimensioned, and their types are selected on the basis of both their technical and cost effectiveness.

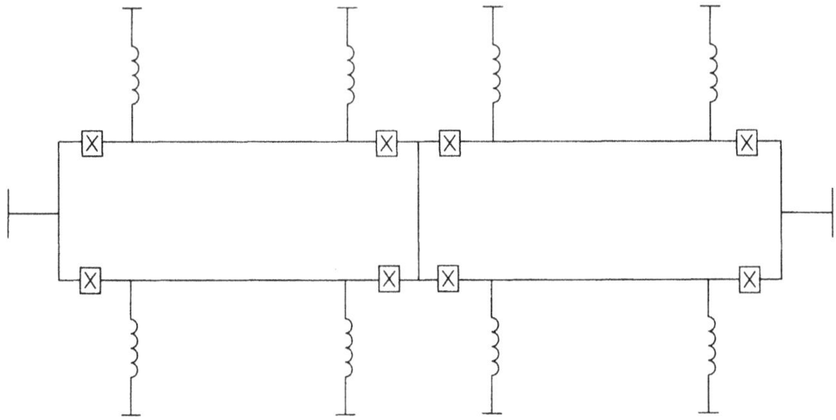

Figure 2.12 The two sections of a double-circuit high-voltage ac line for long-distance transmission.

2.3.1 Shunt Compensation

Passive reactive-power compensators include series capacitors and shunt-connected inductors and capacitors. Shunt devices may be connected permanently or through a switch. Shunt reactors compensate for the line capacitance, and because they control overvoltages at no loads and light loads, they are often connected permanently to the line, not to the bus. Figure 2.12 shows the arrangements of shunt reactors on a long-distance, high-voltage ac line. Many power utilities connect shunt reactors via breakers, thereby acquiring the flexibility to turn them off under heavier load conditions. Shunt reactors are generally gapped-core reactors and, sometimes, air-cored.

Shunt capacitors are used to increase the power-transfer capacity and to compensate for the reactive-voltage drop in the line. The application of shunt capacitors requires careful system design. The circuit breakers connecting shunt capacitors should withstand high-charging in-rush currents and also, upon disconnection, should withstand more than 2-pu voltages, because the capacitors are then left charged for a significant period until they are discharged through a large time-constant discharge circuit. Also, the addition of shunt capacitors creates higher-frequency–resonant circuits and can therefore lead to harmonic overvoltages on some system buses.

2.3.2 Series Compensation

Series capacitors are used to partially offset the effects of the series inductances of lines. Series compensation results in the improvement of the maximum power-transmission capacity of the line. The net effect is a lower load angle for a given power-transmission level and, therefore, a higher-stability margin. The reactive-power absorption of a line depends on the transmission current, so when series capacitors are employed, automatically the resulting reactive-power compensation is adjusted proportionately. Also, because the series compensation effectively reduces the overall line reactance, it is expected that the net line-voltage drop would become less susceptible to the loading conditions.

In an interconnected network of power lines that provides several parallel paths, for power flow between two locations, it is the series compensation of a selected line that makes it the principal power carrier. Series compensation is defined by the degree of compensation; for example, a 1-pu compensation means that the effective series reactance of a line will be zero. A practical upper limit of series compensation, on the other hand, may be as high as 0.75 pu.

One impact of the passive compensation of lines is that whereas the shunt-inductive compensation makes the line electrically resonant at a supersynchronous frequency, the series compensation makes the line resonant at a subsynchronous frequency. The subsynchronous resonance (SSR) can lead to problematic situations for steam turbine–driven generators connected to a series-compensated transmission line. These generators employ multiple turbines connected on a common shaft with the generator. This arrangement constitutes an elastically coupled multimass mechanical system that exhibits several modes of low-frequency torsional resonances, none of which should be excited as a result of the subsynchronous-resonant electrical transmission system.

The application of series compensation requires several other careful considerations. The application of series capacitors in a long line constitutes placing a lumped impedance at a point. Therefore, the following factors need careful evaluation:

1. The voltage magnitude across the capacitor banks (insulation);
2. The fault currents at the terminals of a capacitor bank;
3. The placement of shunt reactors in relation to the series capacitors (resonant overvoltages); and
4. The number of capacitor banks and their location on a long line (voltage profile).

2.3.3 Effect on Power-Transfer Capacity

The consideration of series compensation invariably raises the issue of its comparison with shunt compensation. A simple system analysis can be performed to develop a basic understanding of the effect of shunt and series compensation on power-transmission capacity.

Consider a short, symmetrical electrical line as shown in Fig. 2.13. For an uncompensated line, and assuming $V_s = V_r = V$, the power equation (2.9) becomes

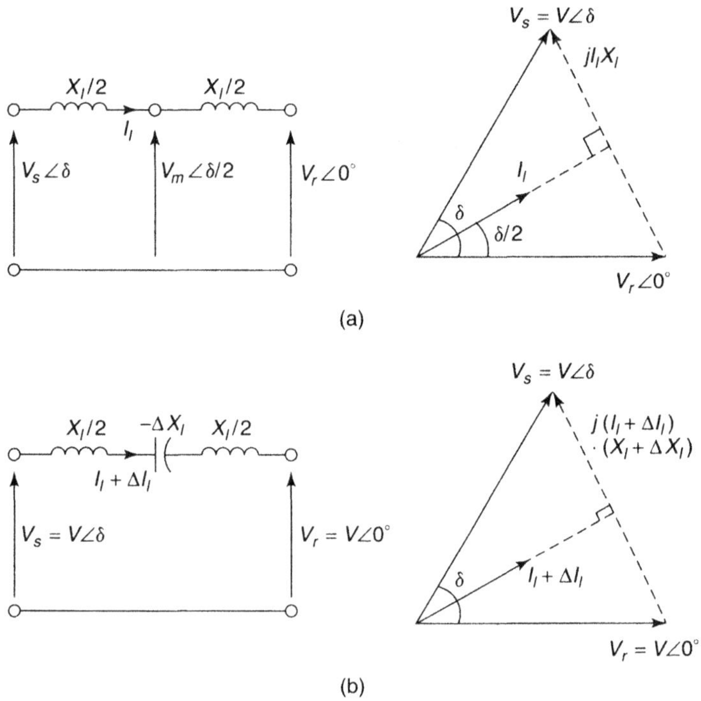

Figure 2.13 The series compensation of a short, symmetrical transmission line.

$$P = \frac{V^2}{X_l} \sin \delta = \frac{V^2}{X_l} 2 \sin \frac{\delta}{2} \cos \frac{\delta}{2} \tag{2.21}$$

From the voltage-phasor equations and the phasor diagram in Fig. 2.13(a),

$$I_l = \frac{2V}{X_l} \sin \frac{\delta}{2} \tag{2.22}$$

2.3.3.1 Series Compensation If the effective reactance of a line is controlled by inserting a series capacitor, and if the line terminal voltages are held unchanged, then a ΔX_l change in the line reactance will result in a ΔI_l change in the current, where

$$\Delta I_l = -\frac{2V}{X_l^2} \sin \frac{\delta}{2} X_l = -I_l \frac{\Delta X_l}{X_l} \tag{2.23}$$

Therefore, from Eq. (2.21), the corresponding change in the power transfer will be

$$\Delta P = -\frac{V^2}{X_l^2} 2 \sin \frac{\delta}{2} \cos \frac{\delta}{2} \Delta X_l \tag{2.24}$$

Using Eqs. (2.22) and (2.23), Eq. (2.24) may be written as

$$\Delta P = \frac{1}{2 \tan \frac{\delta}{2}} (-\Delta X_l I_l^2)$$

As $-\Delta X_l$ is the reactance added by series capacitors, $\Delta X_l I_l^2 = \Delta Q_{se}$ represents the incremental var rating of the series capacitor. Therefore

$$\frac{\Delta P}{\Delta Q_{se}} = \frac{1}{2 \tan \frac{\delta}{2}} \tag{2.25}$$

2.3.3.2 Shunt Compensation Reconsider the short, symmetrical line described in Fig. 2.13(a). Apply a shunt capacitor at the midpoint of the line so that a shunt susceptance is incrementally added (ΔB_c), as shown in Fig. 2.14. For the system in this figure, the power transfer in terms of the midpoint voltage on the line is

$$P = \frac{V V_m}{\frac{X_l}{2}} \sin \frac{\delta}{2} \tag{2.26}$$

The differential change in power, ΔP, as a result of a differential change, ΔV_m, is given as

$$\Delta P = \frac{2V}{X_l} \sin \frac{\delta}{2} \Delta V_m \tag{2.27}$$

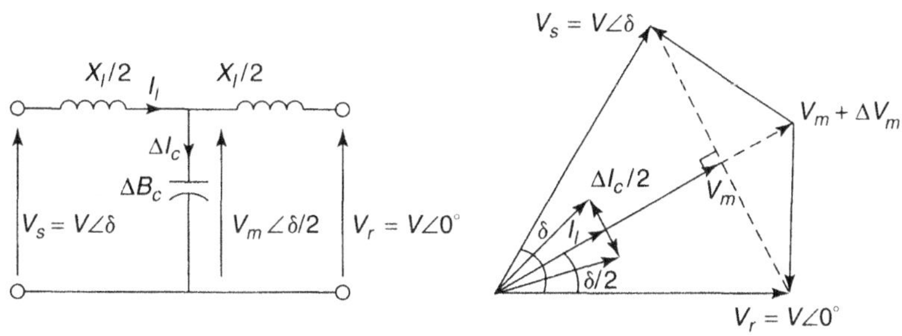

Figure 2.14 The midpoint-capacitor compensation of a short, symmetrical line.

Also as shown in Fig. 2.14,

$$\Delta I_c = V_m \Delta B_c$$

The current ΔI_c in the midline shunt capacitor modifies the line currents in the sending and receiving ends of the line to the following:

$$I_{ls} = I_l - \frac{\Delta I_c}{2} \quad \text{and} \quad I_{lr} = I_l + \frac{\Delta I_c}{2}$$

As $V_m = V_r + jI_{lr}X_l/2$,

$$\Delta V_m = \frac{\Delta I_c X_l}{4} = \frac{V_m X_l}{4} \Delta B_c \tag{2.28}$$

Substituting the results of Eq. (2.28) in Eq. (2.27), we get

$$\Delta P = \frac{V V_m}{2} \sin \frac{\delta}{2} \Delta B_c$$

If the midpoint voltage of the line is approximately equal to $V \cos \delta/2$, then the incremental rating of the shunt-capacitor compensation will be $\Delta Q_{\text{sh}} = V_m^2 \Delta B_c$, or

$$\frac{\Delta P}{\Delta Q_{\text{sh}}} = \frac{1}{2} \tan \frac{\delta}{2} \tag{2.29}$$

By comparing Eqs. (2.25) and Eqs. (2.29), we deduce that for an equivalent power transfer on a short electrical line,

$$\frac{\Delta Q_{\text{se}}}{\Delta Q_{\text{sh}}} = \left(\tan \frac{\delta}{2} \right)^2 \tag{2.30}$$

Assuming an operating load angle $\delta = 30°$, we get the ratio of the ratings of series (ΔQ_{se}) to shunt (ΔQ_{sh}) compensators to be 0.072, or 7.2%.

From the foregoing discussion, it is clear that the var net rating of the series compensator is only 7.2% of that required of a shunt compensator for the same change in power transfer. Therefore, one concludes that the series-capacitive compensation is not only achieved with a smaller MVAR rating, but also that it is automatically adjusted for the entire range of the line loading. However, the cost of the compensator is not directly related only to the MVAR-rating series-capacitor costs increase because they carry full line current and also both their ends must be insulated for the line voltage.

Practical application of series capacitors requires isolation and bypass arrangements as well as protection and monitoring arrangements. For a complete discussion of series compensation, it is recommended that readers consult refs. [2] and [3].

2.4 SUMMARY

This chapter elucidated the concepts of reactive power and presented the theoretical bases of reactive-power compensation in electrical transmission systems. A detailed case study was presented in which the principles of shunt-reactive-power compensation were illustrated, and a comparative analysis of both series and shunt compensation was included as well.

REFERENCES

[1] L. Gyugyi, "Fundamentals of Thyristor-Controlled Static Var Compensators in Electric Power System Applications," IEEE Special Publication 87TH0187-5-PWR, *Application of Static Var Systems for System Dynamic Performance*, 1987, pp. 8–27.

[2] P. M. Anderson and R. G. Farmer, *Series Compensation of Power Systems*, PBLSH! Inc., Encinitas, CA, 1996.

[3] T. J. E. Miller, Ed., *Reactive Power Control in Electric Systems*, John Wiley and Sons, New York, 1982.

CHAPTER 3

Principles of Conventional Reactive-Power Compensators

3.1 INTRODUCTION

This chapter presents power-circuit topologies, fundamental operating principles, and reactive-power control processes of different shunt-connected reactive-power control devices. For years, these devices have been addressed by different names in the literature—static var generators (SVGs), static var compensators (SVCs), static compensators, and static var systems (SVSs), for example. However, both the CIGRE [1] and the Institute of Electrical and Electronics Engineers (IEEE) [2] have established the following definitions to ensure consistency.

An SVG is a static electrical device, system, or piece of equipment that is capable of drawing a controlled capacitive or inductive current from an electrical power system, thereby generating or absorbing reactive power. An SVC is a shunt-connected static generator and/or absorber of reactive power in which the output is varied to maintain or control specific parameters of an electrical power system. An SVG is an integral part of an SVC. An SVS is a combination of different static and mechanically switched var compensators (capacitors and/or reactors) in which the outputs are coordinated. A var compensating system (VCS) is a combination of SVSs and rotating var compensators in which the outputs are coordinated. The general characteristics of SVCs are given in the list that follows. (See refs. [1] and [3]–[10].)

1. The lowering of maintenance requirements from the absence of rotating parts.
2. The very fast control-response time.
3. The feasibility of individual phase control.
4. The diminished losses.
5. The high reliability.
6. The lack of contribution to system short-circuit capacity.
7. The generation of harmonics by SVCs except thyristor-switched capacitors (TSCs).

8. The variation of SVC reactive-power generation as the square of terminal voltage when it is operating outside the linear controllable range, leading to a substantial reduction in reactive-power support at lower voltages.

Characteristics (1)–(5) of the foregoing list constitute the prominent advantages of SVCs, whereas characteristics (7) and (8) constitute disadvantages; however, characteristic (6) may be considered a merit or a demerit, depending on the application. Those SVCs using higher-power–rated thyristors have been identified as key elements in flexible ac transmission systems (FACTSs), which are either controlled individually or coordinated with other dynamic devices in power systems, such as power-system stabilizers (PSS) and HVDC controllers. The ultimate control objectives may be to extend the power loading of existing transmission lines, as well as to regulate voltage and provide additional system damping [8]–[11]. Although SVCs are used extensively at the present time for reactive-power compensation, one should note that rotating synchronous condensers have been the industry workhorses for reactive compensation for decades and still find some important applications, as described in the following section.

3.2 SYNCHRONOUS CONDENSERS

Until the mid-1970s, synchronous condensers or synchronous compensators were the only fully controllable reactive-power devices available for power systems.

3.2.1 Configuration

A synchronous condenser is a synchronous machine, the reactive-power output of which can be continuously controlled by varying its excitation current, as shown by the *V*-curves and performance characteristics of the machine in Fig. 3.1. When the synchronous machine is connected to the ac system and is underexcited, it behaves like an inductor, absorbing reactive power from the ac system. However, when it is overexcited, it functions like a capacitor, injecting reactive power into the ac system. The machine is normally excited at the base current when its generated voltage equals the system voltage; it thus floats without exchanging reactive power with the system. The broken-line characteristic curve corresponds to loading beyond the machine's rated stator current. A synchronous condenser is usually connected to the EHV ac system through a coupling transformer. For voltage-control applications, the desired slope in the steady-state voltage–current characteristics (see Section 5.2) is implemented through the reactance of the coupling transformer. The magnitude of the slope can be adjusted by excitation control.

Large, synchronous condensers are usually hydrogen-cooled. Ratings of up to 345 MVA have been reported in commercial use [12]. Various aspects of the design, operation, and starting methods of synchronous compensators are explained in ref. [3]. Synchronous compensators are characterized by relatively

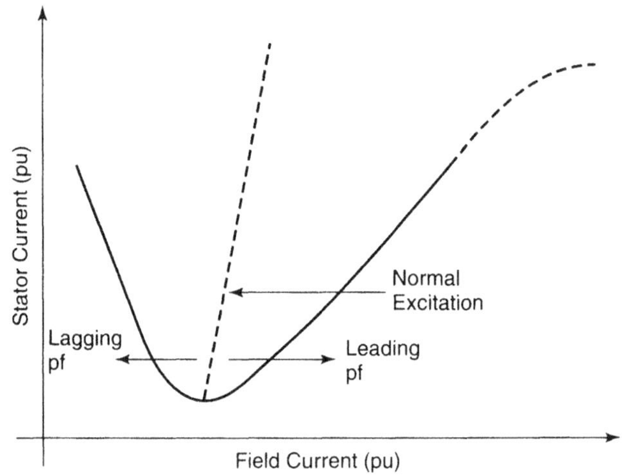

Figure 3.1 The relation between stator current and field current in the synchronous condenser.

slow control responses (100–500 ms) because of their large field-time constants. As they are rotating devices, they require regular maintenance and become more expensive than equivalent-rating static compensators.

3.2.2 Applications

Synchronous condensers are currently used for the following main applications:

1. control of large-voltage excursions, and
2. dynamic reactive-power support at HVDC terminals.

3.2.2.1 Control of Large-Voltage Excursions Under sudden terminal-voltage changes, a synchronous condenser's operation switches sequentially to subtransient and transient modes, thus absorbing significant amounts of reactive power. For example, under severe overvoltage conditions, intrinsically a synchronous condenser absorbs a substantial amount of reactive power even when the field current remains unchanged. (In fact, the field control provides a slower follow-up control.) As the thermal time constant of the condenser is usually large, it can be safely overloaded for a short time by field control as well [3]. For example, should the terminal voltage degrade to 0.8 pu, the condenser may be used to supply 1.5-pu reactive power for about one minute to correct the voltage decline. The extent of capacitive overloading depends on the margin of the exciter-ceiling voltage.

3.2.2.2 Dynamic Reactive-Power Support at HVDC Terminals Synchronous condensers are connected at the inverter end of an HVDC line to

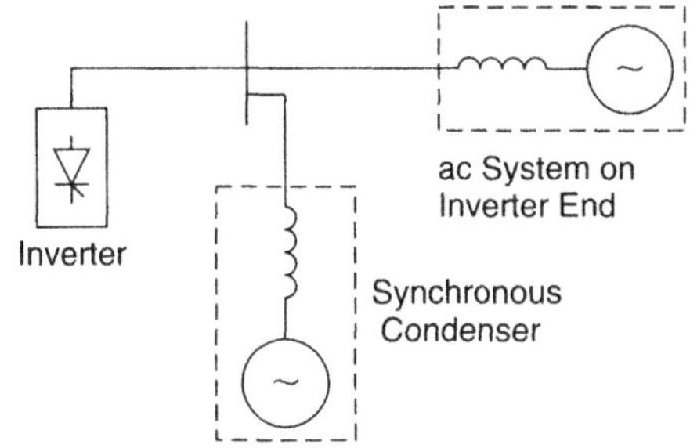

Figure 3.2 An equivalent circuit of a synchronous condenser connected at the inverter end of a dc link.

provide the controllable part of the reactive-power requirement of the inverter station and also to help regulate the inverter ac voltage by increasing the short-circuit capacity of the ac system. More simply stated, a synchronous condenser can be modeled as a controlled voltage source behind a reactance, as shown in Fig. 3.2. When it is viewed from the inverter terminals, the synchronous condenser reactance appears in parallel with the equivalent reactance of the ac system. The HVDC links are often connected to weak ac systems at the receiving end, which are susceptible to commutation failure if adequate control measures are not taken. Hence the compensation of ac system strength becomes an important consideration for utilities to adopt synchronous condensers instead of the much faster-acting SVCs, which do not contribute to the ac system fault level [13], [14].

3.3 THE SATURATED REACTOR (SR)

3.3.1 Configuration

This compensator comprises a polyphase, harmonic-compensated self-saturating reactor in shunt with a switched capacitor. The saturated reactor (SR) provides the control of reactive power, whereas the capacitor gives the bias in the leading power-factor range. A simple saturated iron-core reactor cannot be used in this compensator, as it would result in highly distorted voltage and current waveforms [3]. The harmonics are minimized by employing a specially designed multicoupled, core treble-tripler reactor. This reactor, shown in Fig. 3.3, constitutes nine equally displaced limbs, of which only one is unsaturated at a given instant. Furthermore, each limb saturates alternately in either a positive or a negative direction, resulting in a total of 18 distinct unsaturations in a cycle. This activity leads to the generation of characteristic harmonics of the order $18k \pm 1$, where $k = 1, 2, 3, \ldots$, that is, 17, 19, 35, 37, and so on. Additional internal compensation attenuates the level of these harmonics to less than 2%, thereby reducing the need for external filters.

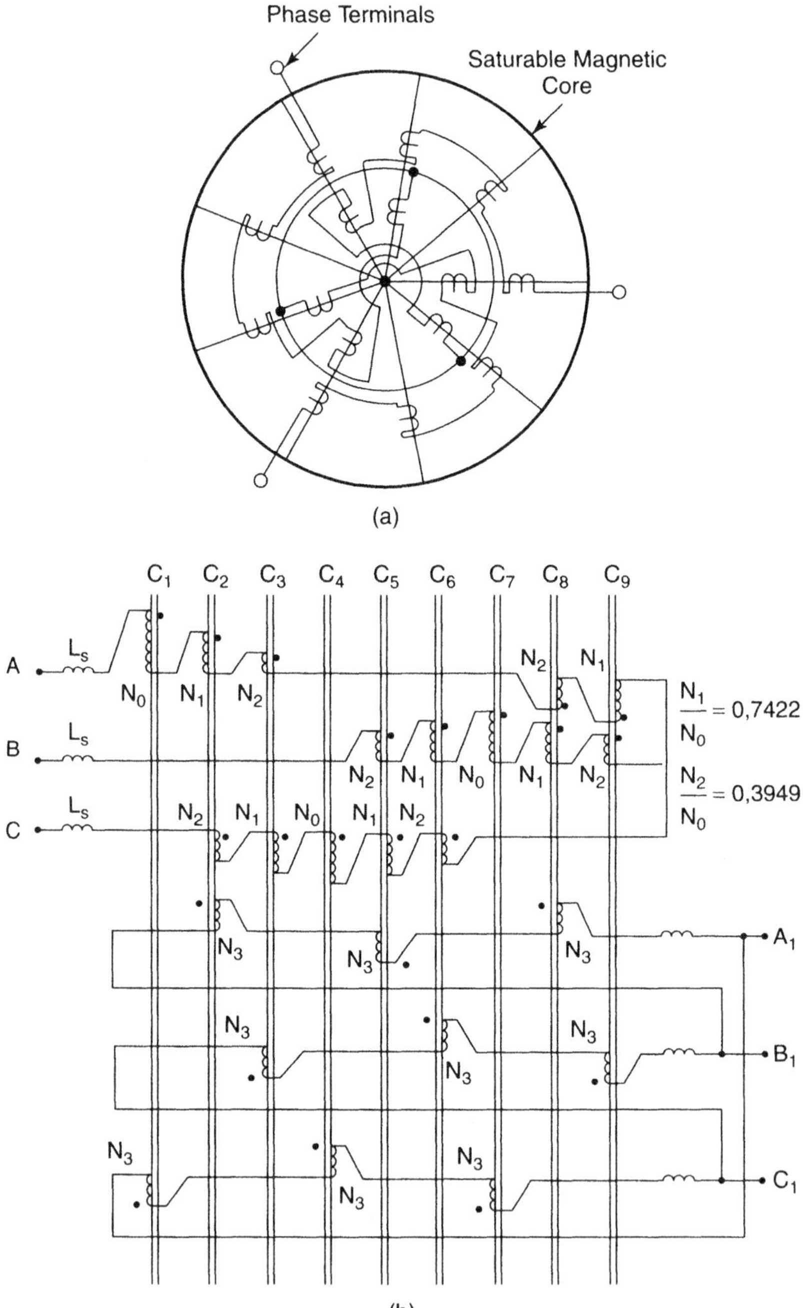

Figure 3.3 A polyphase treble-tripler SR: (a) magnetic-core layout and (b) winding connections. (*Courtesy:* The Canadian Electrical Association [CEA].)

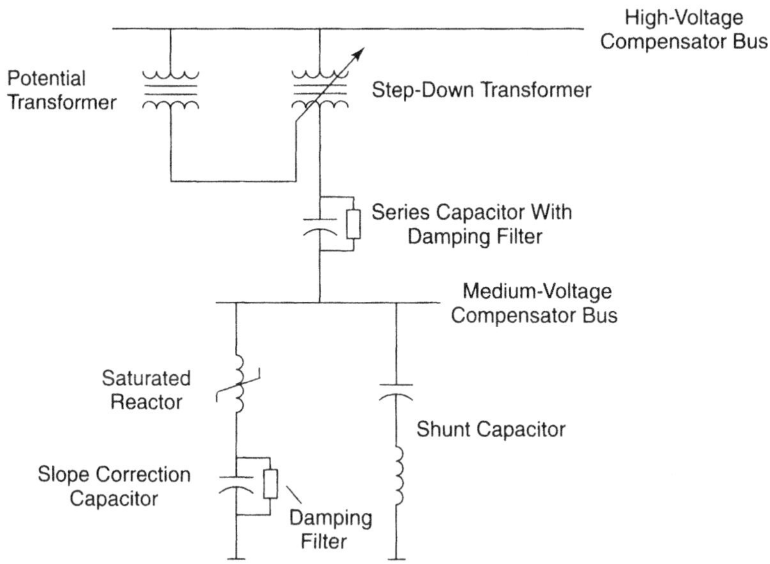

Figure 3.4 An SR compensator.

3.3.2 Operating Characteristics

The SR compensator has an inherent voltage-control capability. It directly responds to the variations in terminal voltage and does not use any thyristor switches or external control for voltage regulation. In EHV applications (usually those above 132 kV), SR compensators are connected to transmission system buses by means of a coupling transformer. A typical SR compensator is depicted in Fig. 3.4; its steady-state *V-I* characteristic is illustrated in Fig. 3.5. The effective reactance corresponding to the slope of saturation characteristic of the SR varies from 8 to 15% on its own rating, which is attributed to the residual inductance of saturated iron. For voltage regulation, however, this slope needs to be reduced to 3–5%. Because of the expense in achieving this slope reduction by improved reactor design, a slope-correcting capacitor is installed in series with a saturated reactor. To prevent the occurrence of subharmonic oscillations from the interaction of the slope-correcting capacitor, the saturated reactor, and the network reactance—especially in weak ac systems—invariably a damping filter is provided across the slope-correcting capacitor. Occasionally, a capacitor with an associated filter may be installed in series with the coupling transformer to offset its reactance and, consequently, improve the voltage regulation at the HV bus.

The connection of the shunt capacitor extends the range of continuously controllable vars to the leading power-factor range. The biasing shunt capacitor is equipped with small tuning inductors to provide filtering of the remnant harmonics generated by the SR. These inductors are designed to preclude the pos-

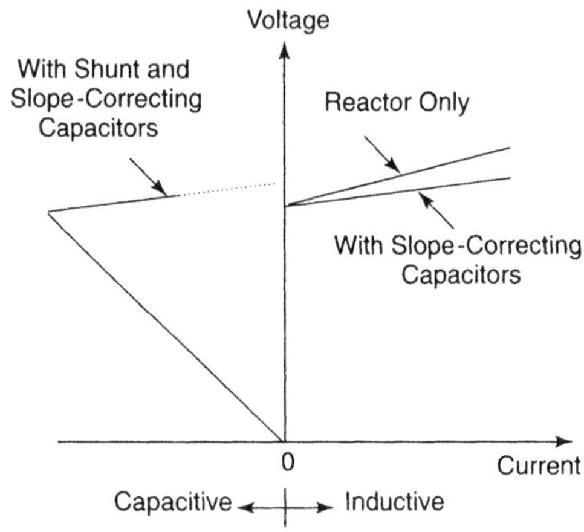

Figure 3.5 The operating characteristics of an SR compensator.

sibility of resonance with system impedance. The voltage reference, or knee point, of the *V-I* characteristic can be varied by adjusting the taps of the coupling transformer.

Because of its iron core, the SR compensator possesses an inherent current-overloading capability of 3–4 pu, which makes it very suitable for controlling temporary overvoltages. The overloading capability of the SR may become restricted from the slope-correcting capacitors, although it may, however, be restored by bypassing the slope-correcting capacitors with the use of spark gaps during severe overvoltage conditions, albeit at the expense of voltage regulation.

The SR in the absence of slope-correcting capacitor is the fastest of all commercially available SVCs. The slope-correcting capacitors reduce the response time to about one-and-a-half to two cycles, which is comparable to a thyristor-controlled reactor (TCR). The SR is more lossy (0.7–1% of its MVA rating) compared to TCR. Its high magnetostrictive noise usually forces its installation in thick enclosures. The SR is a very reliable device, except for spark-gap protection and load tap-changer components, and it is generally employed for

1. the control of large-voltage excursions,
2. the alleviation of voltage flicker, and
3. the reactive compensation at a HVDC terminal.

The SR compensators are not amenable to external controls, which deprives the SR of its capability to introduce the much-needed damping in the ac system for stability enhancement. Commercial sizes of the SR compensator extend up to 270 MVAR.

3.4 THE THYRISTOR-CONTROLLED REACTOR (TCR)

A TCR is one of the most important building blocks of thyristor-based SVCs. Although it can be used alone, it is more often employed in conjunction with fixed or thyristor-switched capacitors to provide rapid, continuous control of reactive power over the entire selected lagging-to-leading range.

3.4.1 The Single-Phase TCR

A basic single-phase TCR comprises an anti-parallel–connected pair of thyristor valves, T_1 and T_2, in series with a linear air-core reactor, as illustrated in Fig. 3.6. The anti-parallel–connected thyristor pair acts like a bidirectional switch, with thyristor valve T_1 conducting in positive half-cycles and thyristor valve T_2 conducting in negative half-cycles of the supply voltage. The firing angle of the thyristors is measured from the zero crossing of the voltage appearing across its terminals.

The controllable range of the TCR firing angle, α, extends from 90° to 180°. A firing angle of 90° results in full thyristor conduction with a continuous sinusoidal current flow in the TCR. As the firing angle is varied from 90° to close to 180°, the current flows in the form of discontinuous pulses symmetrically located in the positive and negative half-cycles, as displayed in Fig. 3.7. Once the thyristor valves are fired, the cessation of current occurs at its natural zero crossing, a process known as the *line commutation*. The current reduces to zero for a firing angle of 180°. Thyristor firing at angles below 90° introduces dc components in the current, disturbing the symmetrical operation of the two antiparallel valve branches. A characteristic of the line-commutation process with which the TCR operates is that once the valve conduction has commenced, any change in the firing angle can only be implemented in the next half-cycle, leading to the so-called thyristor deadtime.

Let the source voltage be expressed as

$$v_s(t) = V \sin \omega t \tag{3.1}$$

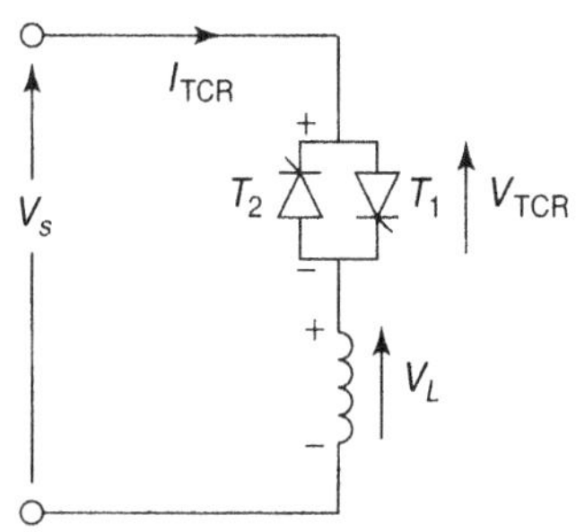

Figure 3.6 A TCR.

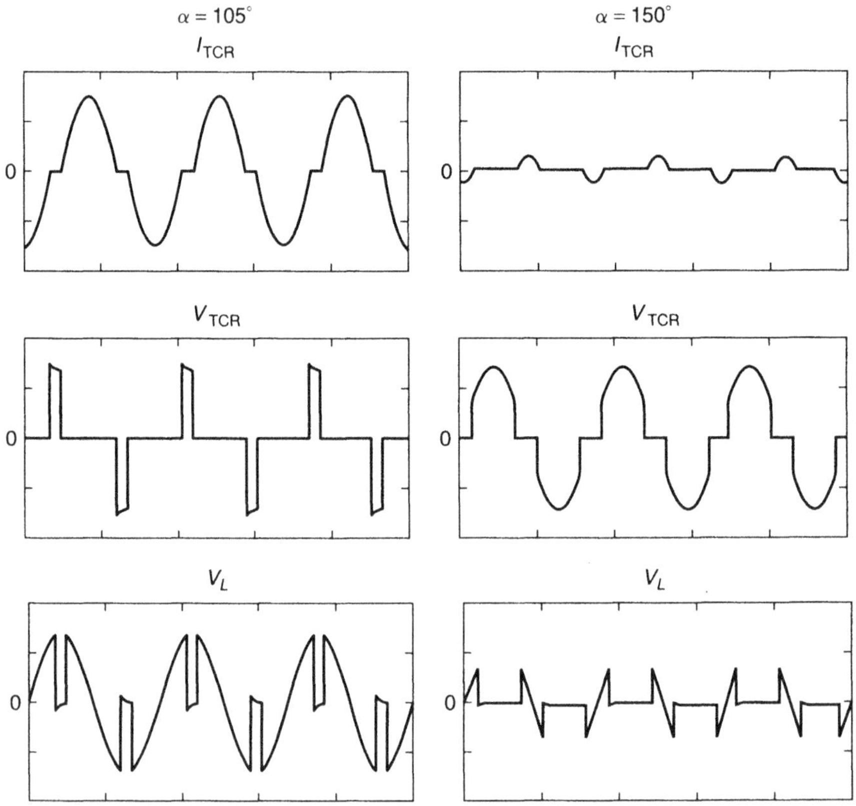

Figure 3.7 Current and voltages for different α in a TCR.

where V = the peak value of the applied voltage and ω = the angular frequency of supply voltage. The TCR current is then given by the following differential equation:

$$L \frac{di}{dt} - v_s(t) = 0 \tag{3.2}$$

where L is the inductance of the TCR. Integrating Eq. (3.2), we get

$$i(t) = \frac{1}{L} \int v_s(t)\, dt + C \tag{3.3}$$

where C is the constant.

Alternatively,

$$i(t) = -\frac{V}{\omega L} \cos \omega t + C \tag{3.4}$$

For the boundary condition, $i(\omega t = \alpha) = 0$,

$$i(t) = -\frac{V}{\omega L}(\cos\alpha - \cos\omega t) \tag{3.5}$$

where α = the firing angle measured from positive going zero crossing of the applied voltage. Fourier analysis is used to derive the fundamental component of the TCR current $I_1(\alpha)$, which, in general, is given as

$$I_1(\alpha) = a_1 \cos\omega t + b_1 \sin\omega t \tag{3.6}$$

where $b_1 = 0$ because of the odd-wave symmetry, that is, $f(x) = f(-x)$. Also, no even harmonics are generated because of the half-wave symmetry, that is, $f(x + T/2) = -f(x)$. The coefficient a_1 is given by

$$a_1 = \frac{4}{T}\int_0^{T/2} f(x)\cos\frac{2\pi x}{T}\,dx \tag{3.7}$$

Solving,

$$I_1(\alpha) = \frac{V}{\omega L}\left(1 - \frac{2\alpha}{\pi} - \frac{1}{\pi}\sin 2\alpha\right) \tag{3.8}$$

Equation (3.8) can also be rewritten as

$$I_1(\alpha) = VB_{\mathrm{TCR}}(\alpha) \tag{3.9}$$

where

$$B_{\mathrm{TCR}}(\alpha) = B_{\max}\left(1 - \frac{2\alpha}{\pi} - \frac{1}{\pi}\sin 2\alpha\right) \tag{3.10}$$

$$B_{\max} = \frac{1}{\omega L} \tag{3.11}$$

The firing angle α is related to the conduction angle σ, as follows:

$$\alpha + \frac{\sigma}{2} = \pi \tag{3.12}$$

Substituting Eq. (3.12) in Eq. (3.8) gives the alternative expression of the fundamental component of the TCR current:

$$I_1(\sigma) = VB_{\max}\left(\frac{\sigma - \sin\sigma}{\pi}\right) \tag{3.13}$$

or

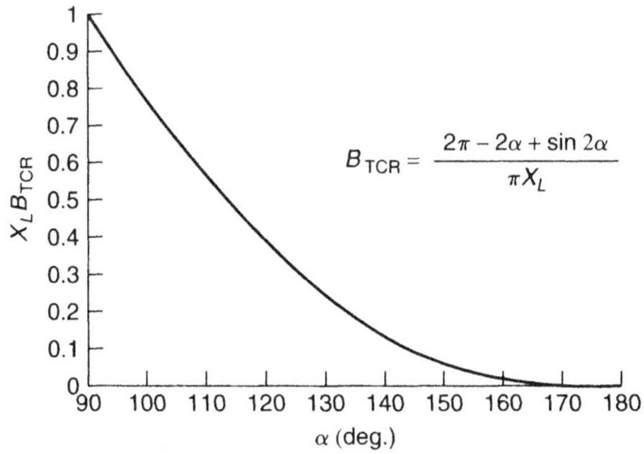

Figure 3.8 Control characteristics of the TCR susceptance, B_{TCR}.

$$I_1(\sigma) = VB_{TCR}(\sigma) \tag{3.14}$$

where

$$B_{TCR}(\sigma) = B_{max}\left(\frac{\sigma - \sin\sigma}{\pi}\right) \tag{3.15}$$

The variation of per-unit value of B_{TCR} with firing angle α is depicted in Fig. 3.8. The per-unit value of B_{TCR} is obtained with respect to its maximum value B_{max} as the base quantity.

The TCR thus acts like a variable susceptance. Variation of the firing angle changes the susceptance and, consequently, the fundamental-current component, which leads to a variation of reactive power absorbed by the reactor because the applied ac voltage is constant. However, as the firing angle is increased beyond 90°, the current becomes nonsinusoidal, and harmonics are generated. If the two thyristors are fired symmetrically in the positive and negative half-cycles, then only odd-order harmonics are produced. The harmonics can be deduced through a Fourier analysis of higher-frequency components. The rms value of the nth-order harmonic is expressed as a function of α in the following equation:

$$\begin{aligned} I_n(\alpha) &= \frac{V}{\omega L}\frac{2}{\pi}\left[-2\frac{\cos\alpha}{n}\sin n\alpha + \frac{\sin(n-1)\alpha}{n-1} + \frac{\sin(n+1)\alpha}{n+1}\right] \\ &= \frac{V}{\omega L}\frac{4}{\pi}\left[\frac{\sin\alpha\cos(n\alpha) - n\cos\alpha\sin(n\alpha)}{n(n^2-1)}\right] \end{aligned} \tag{3.16}$$

where $n = 2k + 1$ and $k = 1, 2, 3, \ldots$

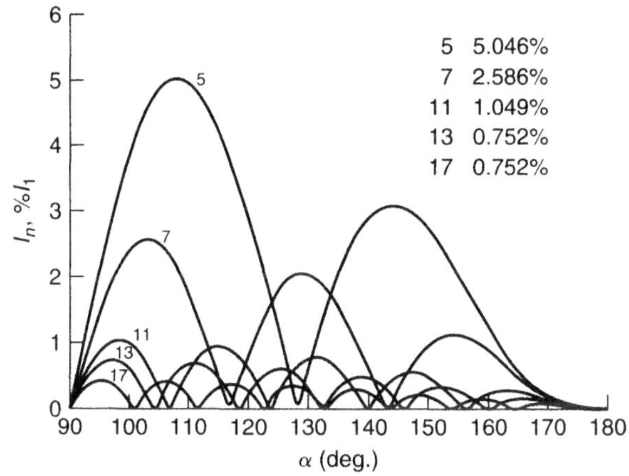

Figure 3.9 Harmonics in a TCR current.

The variation of the amplitude of different harmonics is shown in Fig. 3.9, whereas the same for the total harmonic-current content is displayed in Fig. 3.10. It is seen that all the harmonics do not peak at the same firing angle. The maximum values of various harmonic currents, each expressed as a percentage of the fundamental component, are listed in Fig. 3.9.

It should be noted that a thryistor valve usually comprises many parallel-connected strings, each constituting many serially connected thyristors. The series

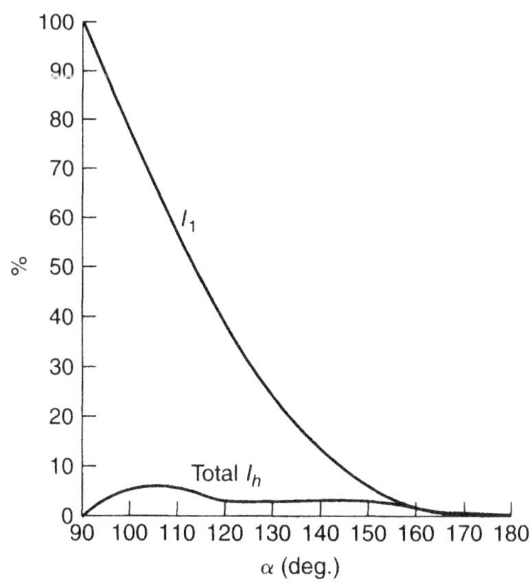

Figure 3.10 The I_1 and total I_h.

connection enhances the voltage-blocking capability of the valve to correspond to the secondary voltage of the coupling transformer. On the other hand, the parallel connection of strings extends the current capability of the valve. The exact number of thyristors in series and parallel is determined from an optimization process that depends on the rating of individual valves and the coupling transformer.

3.4.2 The 3-Phase TCR

A 3-phase, 6-pulse TCR comprises three single-phase TCRs connected in delta, as shown in Fig. 3.11. The inductor in each phase is split into two halves, as shown in Fig. 3.12, one on each side of the anti-parallel–connected thyristor pair, to prevent the full ac voltage appearing across the thyristor valves and damaging them if a short-circuit fault occurs across the reactor's two end terminals. The phase- and line-current waveforms are also displayed in Fig. 3.11. If the 3-phase supply voltages are balanced, if the three reactor units are identical, and also if all the thyristors are fired symmetrically—with equal firing angles in each phase—then the symmetric current pulses result in both positive and negative half-cycles and the generating of only odd harmonics. The percentage values of harmonic currents with respect to fundamental—both in the phases and in the lines—are the same.

The delta connection of the three single-phase TCRs prevents the triplen (i.e., multiples of third) harmonics from percolating into the transmission lines. The cancellation of its 3rd and multiple harmonics can be explained as follows: Let i_{ABn}, i_{BCn}, and i_{CAn} be the nth-order harmonic-phase currents in the respective delta branches, and let i_{An}, i_{Bn}, and i_{Cn} be the currents in the respective lines connected to the delta-configured TCR. Then, the 3rd harmonic currents are expressed as

$$
\begin{aligned}
i_{AB3} &= a_3 \cos(3\omega t + \phi_3) \\
i_{BC3} &= a_3 \cos\left(3\omega t + \phi_3 - 3\,\frac{2\pi}{3}\right) \\
&= a_3 \cos(3\omega t + \phi_3 - 2\pi) \\
i_{CA3} &= a_3 \cos\left(3\omega t + \phi_3 - 3\,\frac{4\pi}{3}\right) \\
&= a_3 \cos(3\omega t + \phi_3 - 4\pi)
\end{aligned} \tag{3.17}
$$

Thus

$$i_{AB3} = i_{BC3} = i_{CA3} \tag{3.18}$$

All three currents are in phase and circulate in the thyristor delta, forming a

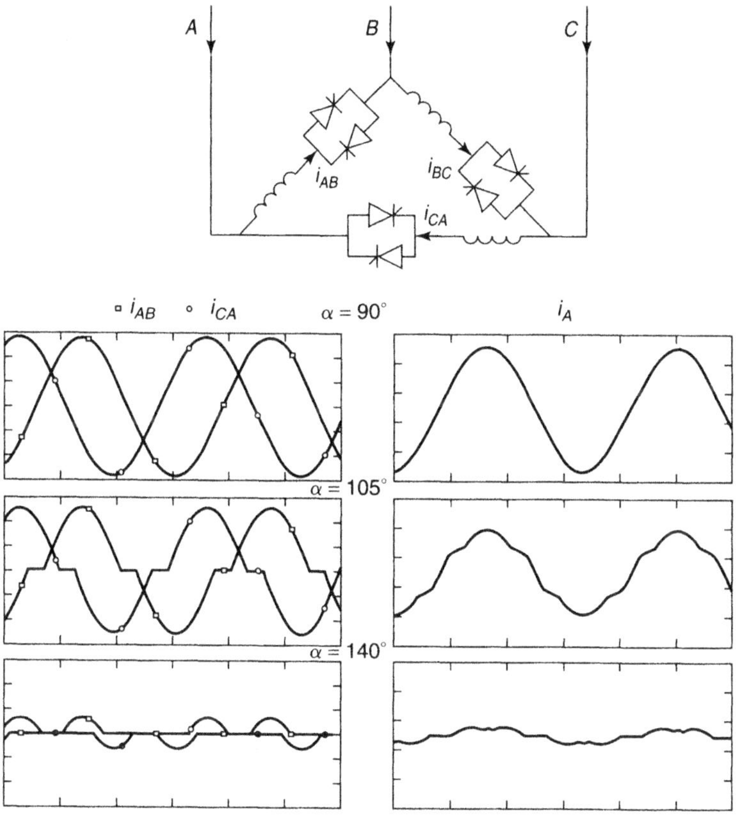

Figure 3.11 A delta-connected TCR and its phase and line currents for different α.

zero-sequence system. It follows that the 3rd harmonic line currents reduce to zero, as follows:

$$i_{A3} = i_{AB3} - i_{CA3} = 0 \tag{3.19}$$

Likewise

$$i_{B3} = 0, \quad i_{C3} = 0 \tag{3.20}$$

A closer analysis reveals that not only the 3rd harmonic but also all triplen harmonics get canceled out. Therefore, all harmonic components of the order $3p + 3$, where $p = 0, 1, 2, 3, \ldots$ (3, 9, 15, 21, 27, etc.) cannot flow in the lines during balanced operation.

A similar analysis can be done for the 5th and 7th harmonic current, as follows:

For the 5th harmonic, let

$$\begin{aligned} i_{AB5} &= a_5 \cos(5\omega t + \phi_5) \\ i_{BC5} &= a_5 \cos\left(5\omega t + \phi_5 - 5\,\frac{2\pi}{3}\right) \\ &= a_5 \cos\left(5\omega t + \phi_5 - \frac{4\pi}{3}\right) \\ i_{CA5} &= a_5 \cos\left(5\omega t + \phi_5 - 5\,\frac{4\pi}{3}\right) \\ &= a_5 \cos\left(5\omega t + \phi_5 - \frac{2\pi}{3}\right) \end{aligned} \tag{3.21}$$

Phase-shift angles for the three delta currents clearly indicate that the 5th harmonic represents a negative-sequence system of currents. The same applies for harmonics of the order $6p + 5$, where $p = 0, 1, 2, 3, \ldots$ (5, 11, 17, etc.). Furthermore, these harmonics do not cancel out in the line.

For the 7th harmonic, let

$$\begin{aligned} i_{AB7} &= a_7 \cos(7\omega t + \phi_7) \\ i_{BC7} &= a_7 \cos\left(7\omega t + \phi_7 - \frac{2\pi}{3}\right) \\ i_{CA7} &= a_7 \cos\left(7\omega t + \phi_7 - \frac{4\pi}{3}\right) \end{aligned} \tag{3.22}$$

The phase-shift angles reveal that the 7th harmonic and all other harmonics of the order $6p + 1$, where $p = 0, 1, 2, 3, \ldots$ (7, 13, 19, etc.), constitute a positive-sequence system and moreover, that they *do* flow in the lines connecting to the 6-pulse TCR.

The stringent conditions mentioned previously for balanced operation may not be fully satisfied in real life. For instance, the reactors may not be identical in all three phases or the supply voltages may not be balanced. This imbalance results in the emission of noncharacteristic harmonics, including triplen harmonics, into the line [4]. The magnitude of these noncharacteristic harmonics is insignificant under normal circumstances, but during severe disturbances, unequal firing of thyristors in positive and negative cycles may occur. This condition results in a dc component, which is sufficient to cause saturation of the coupling transformer, creating increased harmonic emanation.

In addition to the harmonics, a small, in-phase component of current (0.5–2%) of fundamental frequency also flows in the TCR, which represents the resistive losses in the TCR windings. Typical values of TCR quality factor ($QF = \omega L/R$) accounting for these losses range from 40 to 100.

Filters are usually provided in shunt with the TCR, which are either of series

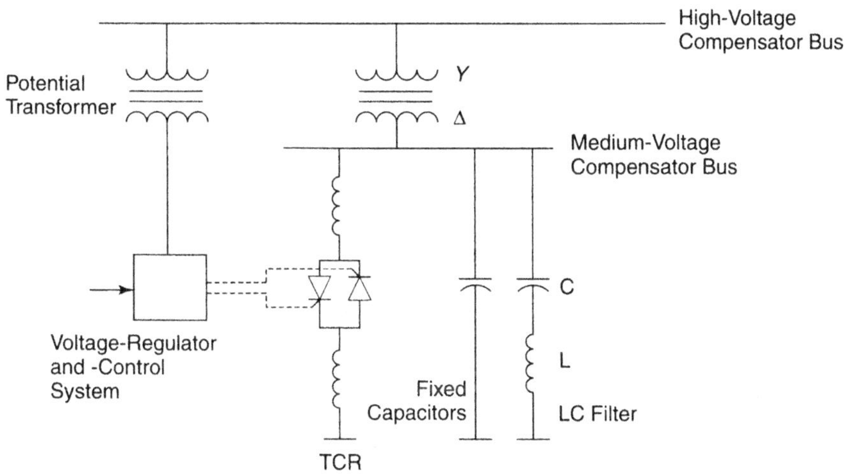

Figure 3.12 A 1-line diagram of a TCR compensator with fixed-shunt capacitors.

LC or LCR configuration. These filters are tuned to the dominant 5th and 7th harmonic frequencies. Sometimes, specific filters for 11th and 13th harmonics or a simple high-pass filter are also installed. If an individual phase control of the TCR is envisaged, or if the network resonance conditions so necessitate (see Section 5.5), a 3rd harmonic filter needs to be installed in parallel with the TCR.

The schematic diagram of a 6-pulse TCR with filters is depicted in Fig. 3.12. As it is desirable in power-system applications to have controllable capacitive-reactive power, a capacitor is connected in shunt with the TCR. This capacitor may be fixed, or it may be switchable by means of mechanical or thyristor switches. The main advantages of the TCR are flexibility of control and ease in uprating. Different control strategies can be easily implemented, especially those involving external supplementary signals to achieve significant improvements in system performance. The voltage reference and current slope can be controlled in a simple manner. Modular in nature, a TCR SVC can have its rating extended by the addition of more TCR banks, as long as the coupling-transformer rating is not exceeded.

The TCRs do not possess high overload capability because the air-core design of their reactors. If the TCRs are expected to transiently withstand high overvoltages, a short-term overload capacity must be built into the TCR by design, or additional thyristor-switched overload reactors may need to be installed.

The TCR responds rapidly, typically in duration of one-and-a-half to three cycles. The actual response time is a function of measurement delays, TCR-controller parameters, and system strength, which is discussed in Chapter 5. Those TCRs having a 300-MVA rating are already installed in power systems.

3.4.3 The Thyristor-Switched Reactor (TSR)

The TSR is a special case of a TCR in which the variable firing-angle control option is not exercised. Instead, the device is operated in two states only: either fully on or fully off.

If the thyristor valves are fired exactly at the voltage peaks corresponding to $\alpha = 90°$ for the forward-thyristor valve T_1 and $\alpha = 270°$ $(90° + 180°)$ for the reverse-thyristor valve T_2, as depicted in Fig. 3.6, full conduction results. The maximum inductive current flows in the TCR as if the thyristor switches were replaced by short circuits. However, if no firing pulses are issued to the thyristors, the TSR will remain in a blocked-off state, and no current can flow.

The TSR ensures a very rapid availability of rated inductive reactive power to the system. When a large magnitude of controlled reactive power, Q, is required, a part of Q is usually assigned to a small TSR of rating, say, $Q/2$; the rest is realized by means of a TCR also of a reduced rating $Q/2$. This arrangement results in substantially decreased losses and harmonic content as compared to a single TCR of rating Q.

3.4.4 The Segmented TCR

One method of reducing harmonic generation is to segment the main TCR into n $(n \geq 2)$ parallel-connected TCRs, each having a reactive-power rating $1/n$ of the total TCR. Of these n TCRs, the firing angle of only one unit is controlled, whereas the others are either switched on or off to absorb the desired amount of reactive power. As the size of the controllable reactor is decreased with a consequent increase in its inductance by a factor of n, the magnitude of each generated harmonic—as given by Eq. (3.16)—also attenuates by a factor of n compared to the rated fundamental component of the current. Correspondingly, the size and rating of required harmonic filters also attenuates.

The harmonic reduction achieved with this SVC configuration is accompanied by increased costs because of the greater number of involved thyristors. Thus a segmented TCR with TSRs emerges as a more expensive option compared to a nonsegmented TCR if the number of TCR segments is large.

3.4.5 The 12-Pulse TCR

As in HVDC systems, a substantial harmonic reduction is achieved by employing a 12-pulse–configured TCR. In this arrangement, two 6-pulse TCRs are supplied by three phase voltages, which are phase-displaced by 30°. The 12-pulse TCR necessitates either a customized 3-winding transformer with two secondaries or two separate transformers with identically connected primaries. In both cases, one of the secondaries is star-connected; the other, delta-connected.

The concept behind a 12-pulse TCR is explained in Fig. 3.13 by phasor diagrams of two delta-connected TCRs, each energized by differently configured ideal transformers. Figure 3.13(a) displays on its left side a TCR circuit with a

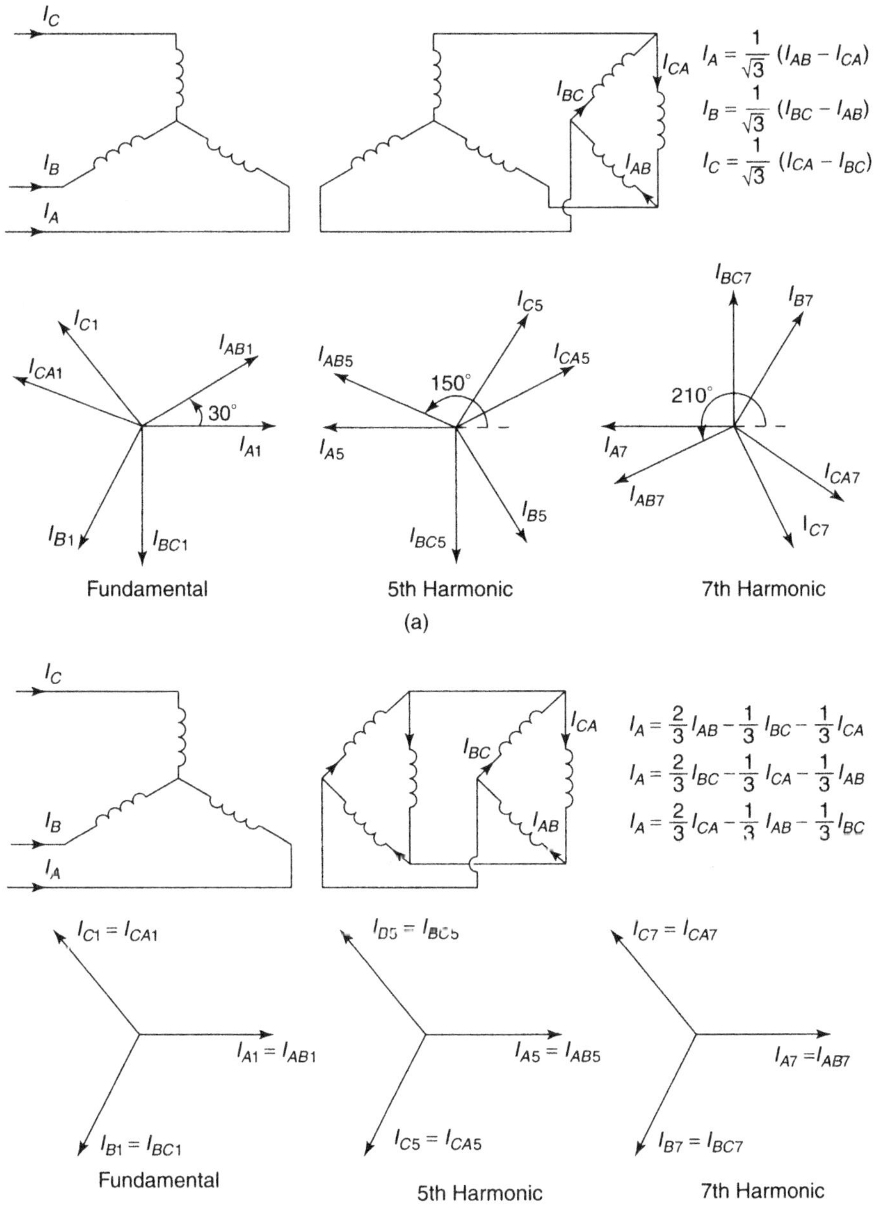

Figure 3.13 The phasor relationship in a 12-pulse TCR: (a) a star–star coupling transformer and (b) a star–delta coupling transformer.

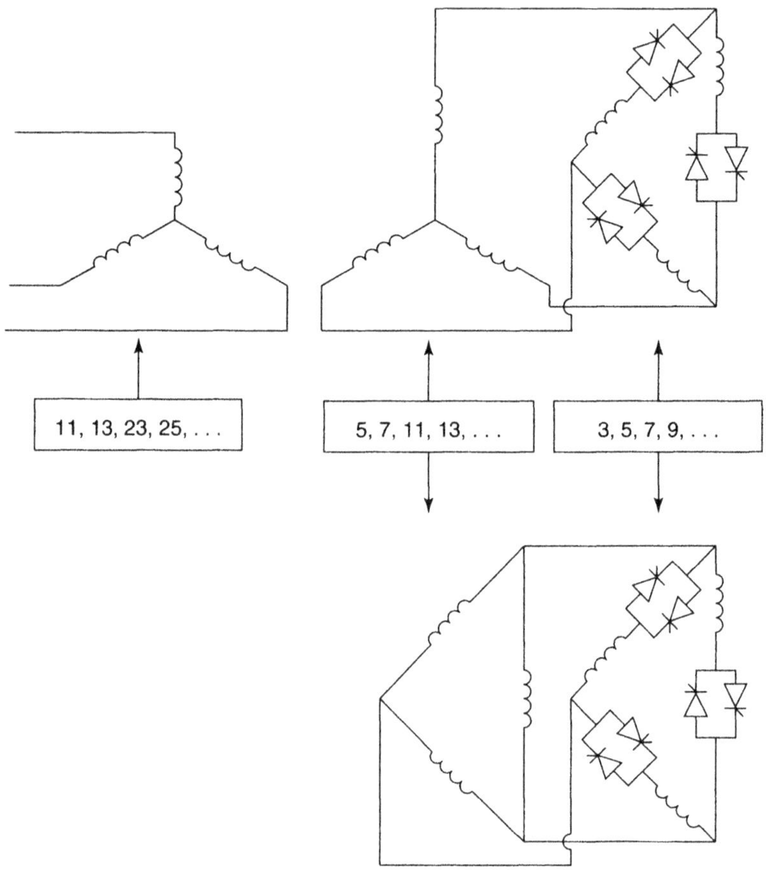

Figure 3.14 A TCR in a 12-pulse arrangement.

star-star–connected transformer and the phasor diagrams for fundamental, the 5th and 7th harmonic phases, and the primary line current. The corresponding circuit for the star-delta–connected transformer is shown in Fig. 3.13(b). Note that the turns ratio of both the transformers are chosen so that they receive the same magnitude of the primary line currents as in the valves.

The two sets of phasor diagrams are directly comparable because the phasor reference on both sides is the primary fundamental current in phase A. It is seen that both alternatives generate the same fundamental currents (phase and magnitude) on the primary side of the transformer. The 3rd harmonic is not existent in the transformer primary currents because it gets canceled by the delta arrangement of the TCRs. Both the 5th and 7th harmonics show the same magnitude but opposite phase relations in the two alternatives. The primary star windings of the two systems may be combined, resulting in a 3-winding transformer arrangement, as shown in Fig. 3.14. The currents of the two 6-pulse systems add on the system side for fundamental frequency, but they subtract and cancel for 5th and 7th—and for all higher frequencies of $[6(2n + 1) \pm 1, n = 0, 1, 2, \ldots]$—order.

Figure 3.14 shows the harmonic-current content in a 12-pulse scheme at various locations. The large reduction in harmonic content achieved with the 12-pulse TCR greatly alleviates the requirements for harmonic filters. Instead of employing tuned 5th and 7th harmonic filters, as in the 6-pulse TCR, only high-pass filters may prove adequate. Once again, the harmonic reduction is associated with enhanced costs because of the increased number of thyristors and the presence of a special double-secondary transformer as well as a complex firing sequence. An additional advantage that occurs with the 12-pulse TCR is increased reliabiilty. Should one of the 6-pulse TCR units fail, the other TCR unit continues to operate, although with half the reactive-power rating. The 12-pulse TCRs have higher overload capabilities than the 6-pulse ones.

Those TCRs with pulse numbers higher than 12 are not used in practice, although their use would ensure a drastic reduction in harmonics. The reason for this lack of use is that they have become too complex and expensive; for instance, a 3-secondary-winding transformer would be needed for an 18-pulse TCR. Furthermore, the precision required in firing control to ensure the symmetrical firing of the thyristors is not easily attainable.

3.4.6 Operating Characteristics of a TCR

3.4.6.1 Operating Characteristics Without Voltage Control The simplest SVC configuration consists of a TCR connected to the power system as shown in Fig. 3.15. In the analysis of compensator performance, the fundamental frequency behavior is generally considered. In practice, harmonics are filtered and reduced to very low values. The approach shown in Fig. 3.15 is very convenient for the performance analysis—the whole TCR branch is replaced by an equivalent continuously variable reactor. The sinusoidal current flowing in this reactor is equal to the fundamental component of the nonsinusoidal current

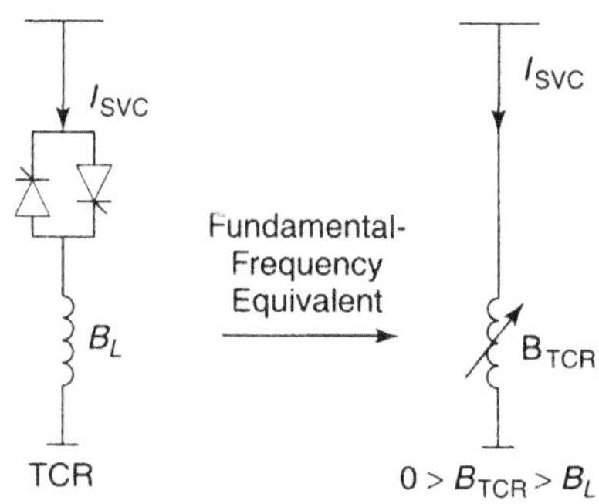

Figure 3.15 A simple SVC circuit using a TCR.

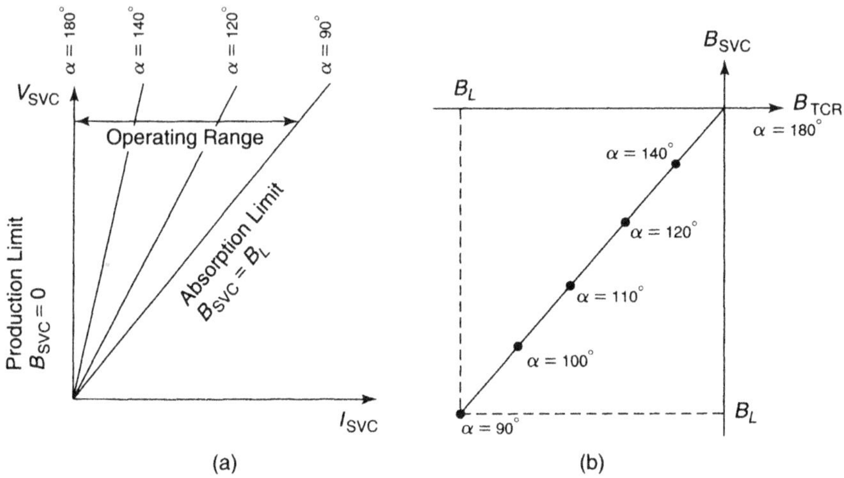

Figure 3.16 Different characteristics of an SVC: (a) the voltage–current characteristic and (b) the SVC TCR susceptance characteristic.

flowing in the TCR. Note that B_{TCR} is the variable susceptance in the foregoing equivalent of the TCR; it is given by Eq. (3.15).

For a general SVC, which can be considered as a black box with an unknown but purely reactive circuit inside, the overall compensator susceptance B_{SVC} can be defined with the following equation:

$$\bar{I}_{SVC} = \bar{V} j B_{SVC} \tag{3.23}$$

In the simple case of a TCR, the compensator susceptance is

$$B_{SVC} = B_{TCR} \tag{3.24}$$

Usually, three kinds of characteristics are of interest while analyzing an SVC, as described in the paragraphs that follow.

Voltage–Current Characteristic or Operating Characteristic This shows the SVC current as a function of the system voltage for different firing angles, as depicted in Fig. 3.16(a). This *V-I* characteristic is given in a very general sense. No control system is assumed to vary the firing angle, and any operating point within the two limits is possible depending on the system voltage and the setting of the firing angle (other currents and voltages may be shown, too). This characteristic clearly illustrates the limits of the operating range, and it may include the steady-state characteristics of the various possible controls. This characteristic is the usual way in which the system engineers prefer to look at the compensator, because the characteristic shows the steady-state performance of the SVC plant.

SVC TCR Susceptance Characteristics These illustrate the change of the total SVC susceptance when the TCR susceptance is varied, as shown in Fig. 3.16(b). The susceptance characteristic for this case is very simple because $B_{SVC} = B_{TCR}$. Note that the TCR susceptance is negative, indicating that the TCR is an absorbing reactive component. These characteristics are of most interest to control-system analysis—the controls affect the TCR firing angle, whereas the total susceptance B_{SVC} influences the power system.

Current Characteristics For more complex SVC arrangements, especially those including TSCs, it is not easy to see how the various branches contribute to the total SVC current. Therefore, these characteristics show the branch currents as a function of the total SVC current, and they are important for determining steady-state current ratings of the various components. (An example of such a characteristic is given in Fig. 3.38, shown later in this chapter.)

3.4.6.2 Operating Characteristic With Voltage Control The operating range of Fig. 3.16(a) can be reduced to a single characteristic of operating points if the effect of the voltage control is incorporated. Let us assume that the compensator is equipped with the voltage control shown in Fig. 3.17(a). The system voltage is measured, and the feedback system varies B_{TCR} to maintain V_{ref} on the system. This control action is represented in the operating characteristic in Figure 3.17(b) by the horizontal branch marked as *control range*. This characteristic shows the hard-voltage control of the compensator, which stabilizes the system voltage exactly to the set point V_{ref}.

Two system characteristics—system 1 and system 2—are depicted in Fig. 3.17(b) that illustrate the decline in system node voltage when the node is

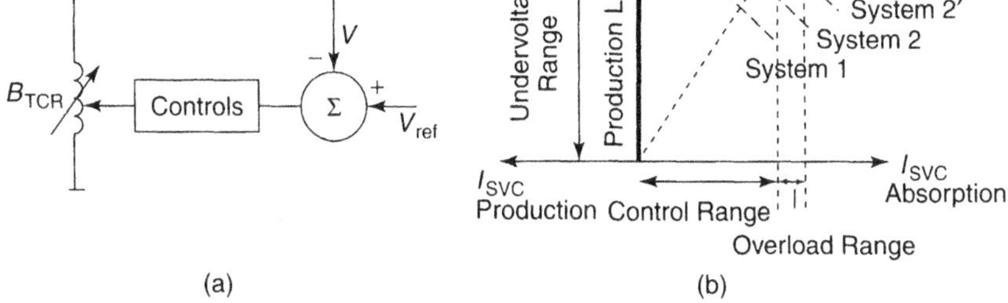

Figure 3.17 The operating characteristics of a TCR with voltage control: (a) an SVC control system and (b) the *V-I* characteristic.

loaded inductively and reactive power is absorbed. The corresponding operating points for the two system conditions are A_1 and A_2. If the system voltage of system 2 rises, a new characteristic—system 2′—results. Operating point A then moves to the right and reaches the absorption limit of the compensator. Any further increase in system voltage cannot be compensated for by the control system, because the TCR reactor is already fully conducting. The operating point A_2 will, therefore, move upward on the characteristic, corresponding to the fully on reactor connected to the system ($\alpha = 90°$). The compensator then operates in the *overload range*, beyond which a current limit is imposed by the firing control to prevent damage to the thyristor valve from an overcurrent. At the left-hand side, the compensator will reach the *production limit* if the system voltage drops excessively; the operating point will then lie on the characteristic of the *undervoltage range*.

3.5 THE THYRISTOR-CONTROLLED TRANSFORMER (TCT)

The TCT [1]–[5] is a special kind of transformer designed with a high value of leakage reactance (typically 100%). The secondary windings of this transformer are closed through anti-parallel–connected pairs of thyristors, as shown in Fig. 3.18. Firing-angle control of these thyristors provides a continuous variation of reactive power. The TCT SVC offers a reduction in the cost of the transformer and reactor by integrating the two devices into one. Furthermore, the iron core of the TCR allows a greatly enhanced var-absorption overload capability that is not attainable with conventional air-core TCRs.

The operation of the TCT can be understood from Fig. 3.19, which shows the secondary side-phase voltage v and phase current i [4]. The corresponding primary side-phase voltage and current signals are similar. The effective reactance of the TCT is varied by short-circuiting the secondary delta-connected winding by the firing-angle control of the associated thyristor-valve pair. The valve firing is synchronized with the zero crossing of the primary phase voltage. At any instant of time, the current in a given phase winding is the sum of the current i_1 in that phase from its associated thyristor firing and the current i_2 flowing in the other two phase windings, as influenced by their associated valve firings. The current component i_1 flows through the reactance of one phase winding, whereas the component i_2 traverses through two phase windings.

For firing angles between 150° and 180°, no overlap between the conducting intervals of two valves exists, whereas the same occurs if the firing angle is in the range of 120° to 150°. Two sets of current waveforms result, as depicted in Fig. 3.19. Full sinusoidal conduction is achieved for a firing angle of 120°, as opposed to 90° for a TCR. Thus the total controllable range for a TCT is $120° \leq \alpha \leq 180°$.

Only odd harmonics are generated in a TCT. The triplen harmonics are eliminated by the special connection of the transformer-secondary windings and the thyristor valves. Overall harmonic generation is the same as in a TCR.

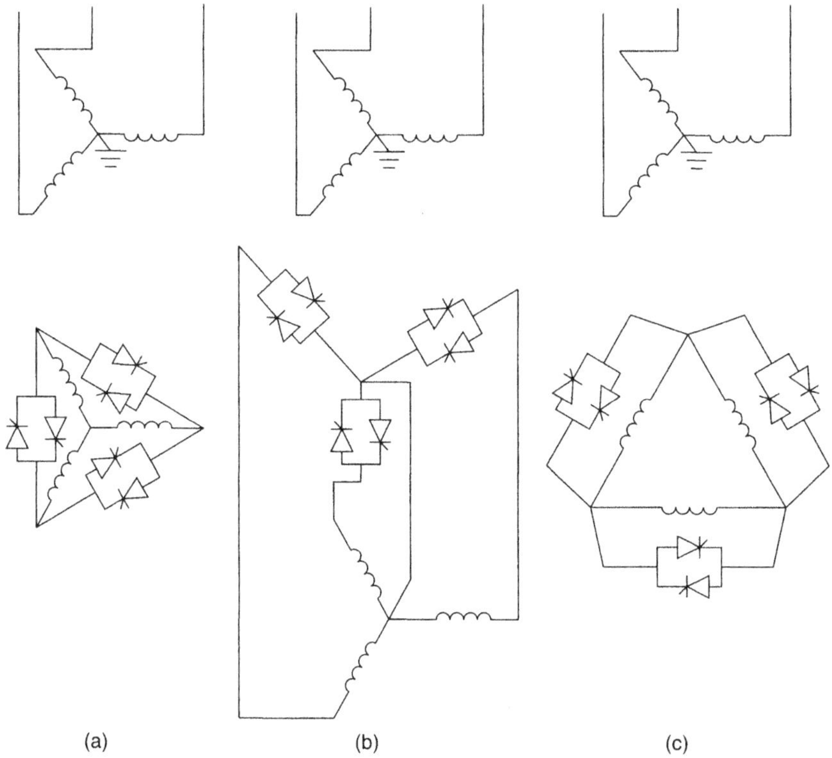

Figure 3.18 Various arrangements of a TCT compensator: (a) star-connected reactors and a delta-connected thyristor controller; (b) star-connected reactors in series with a thyristor controller; and (c) delta-connected reactors and a thyristor controller.

The main disadvantage of the TCT SVC is the high capital cost involved in the fabrication of a special transformer. Moreover, as no low-voltage secondary winding is available, the capacitors and filters must be installed on the high-voltage bus, necessitating a higher voltage rating of the passive components and, consequently, an increased cost. One of the largest TCT static var compensators is installed in the Hydro-Quebec System [15], which has a rating of +350 MVA, −100 MVA at 735 kV.

3.6 THE FIXED-CAPACITOR–THYRISTOR-CONTROLLED REACTOR (FC–TCR)

3.6.1 Configuration

The TCR provides continuously controllable reactive power only in the lagging power-factor range. To extend the dynamic controllable range to the leading power-factor domain, a fixed-capacitor bank is connected in shunt with the

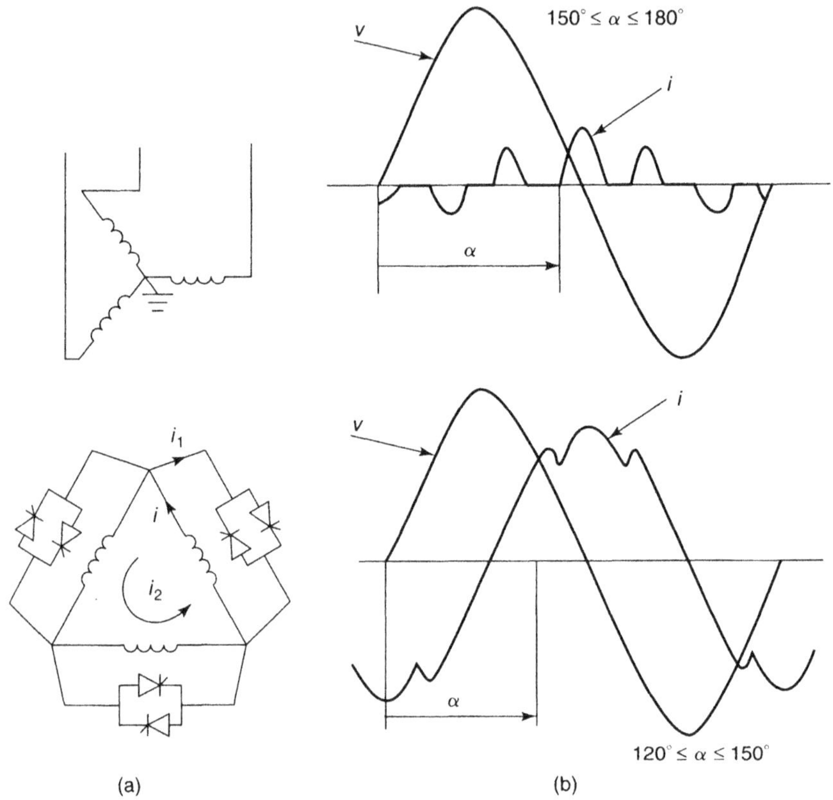

Figure 3.19 The reactor-transformer compensator: (a) the configuration and (b) the voltage and currents for two values of α.

TCR. The TCR MVA is rated larger than the fixed capacitor to compensate (cancel) the capacitive MVA and provide net inductive-reactive power should a lagging power-factor operation be desired. The fixed-capacitor banks, usually connected in a star configuration, are split into more than one 3-phase group. Each capacitor contains a small tuning inductor that is connected in series and tunes the branch to act as a filter for a specific harmonic order. For instance, one capacitor group is tuned to the 5th harmonic and another to the 7th, whereas yet another is designed to act as a high-pass filter. At fundamental frequency, the tuning reactors slightly reduce the net MVA rating of the fixed capacitors. An FC–TCR compensator is shown in Fig. 3.20.

3.6.2 Operating Characteristic

3.6.2.1 Without the Step-Down Transformer The operating *V-I* characteristic of an FC–TCR compensator is illustrated in Fig. 3.21. The fixed capacitor extends the operating-control range of the SVC to the leading side as com-

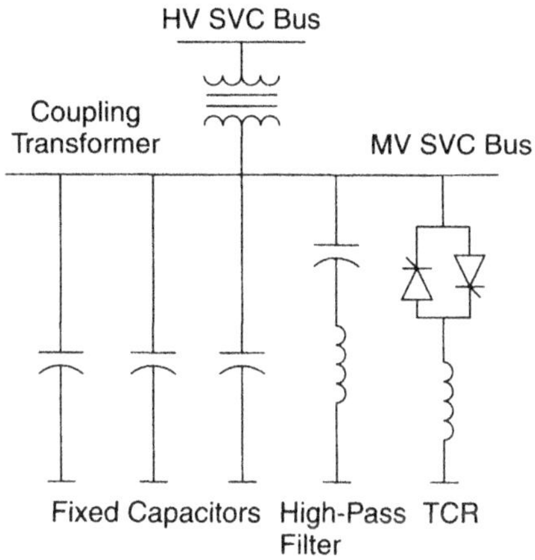

Figure 3.20 An FC–TCR SVC.

pared to that shown in Fig. 3.17. The SVC current, I_{SVC}, can be expressed as a function of system voltage, V, and compensator susceptance, B_{SVC}, as follows:

$$\bar{I}_{\text{SVC}} = \bar{V} j B_{\text{SVC}} \tag{3.25}$$

where

$$B_{\text{SVC}} = B_C + B_{\text{TCR}} \quad \text{and} \quad B_C = \omega C \tag{3.26}$$

Figures 3.21(b) and (c) show the operating characteristic and the susceptance of this type of compensator, respectively, and both also show that var production as well as var absorption is possible. By dimensioning the ratings of the TCR and the capacitor, respectively, the production and absorption ranges can be selected according to the system requirements.

3.6.2.2 With the Step-Down Transformer An FC–TCR SVC is usually connected to the high-voltage power system by means of a step-down coupling transformer, as shown in Fig. 3.22.

The Compensator Susceptance The compensator susceptance, B_{SVC}, is given by

$$B_{\text{SVC}} = \frac{B_\sigma (B_C + B_{\text{TCR}})}{B_\sigma + B_C + B_{\text{TCR}}} = \frac{1}{1 + \dfrac{B_C + B_{\text{TCR}}}{B_\sigma}} (B_C + B_{\text{TCR}}) \tag{3.27}$$

where B_σ is the susceptance of the transformer and B_{TCR} is variable from 0 to B_L, according to the firing angles from 180° to 90°.

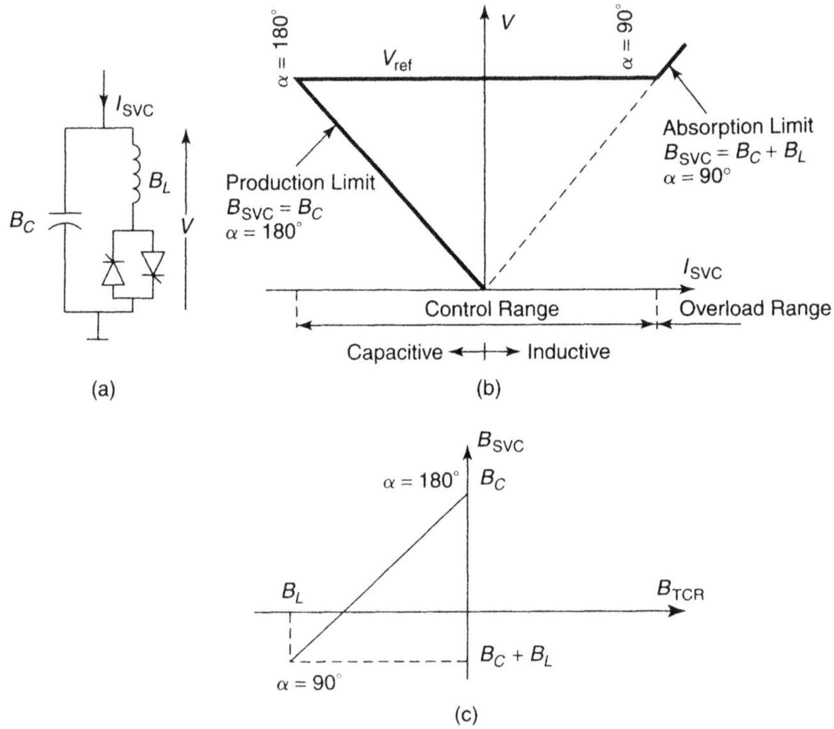

Figure 3.21 The operating characteristics of an FC–TCR without a coupling transformer.

From Eq. (3.27), the susceptance limits can be calculated. Susceptance at the production (capacitive) limit, that is, with $B_{\mathrm{TCR}} = 0$ at $\alpha = 180^\circ$, is expressed as

$$B_{\mathrm{SVC\ max}} = \frac{B_\sigma B_C}{B_\sigma + B_C} \tag{3.28}$$

Susceptance at the absorption (inductive) limit, that is, with $B_{\mathrm{TCR}} = B_L$ at $\alpha = 90^\circ$, is given by

$$B_{\mathrm{SVC\ min}} = \frac{B_\sigma (B_C + B_L)}{B_\sigma + B_C + B_L} \tag{3.29}$$

It must be noted that B_L is a negative quantity. An analysis of Eq. (3.27) shows that the total susceptance B_{SVC} of the static var compensator does not change linearly with B_{TCR}. However, if $(B_C/B_\sigma) \ll 1$ and $(B_L/B_\sigma) \ll 1$, which is usually the case, the nonlinearity is relatively small. This assumption implies that the reactance of the coupling transformer is greatly smaller than the reac-

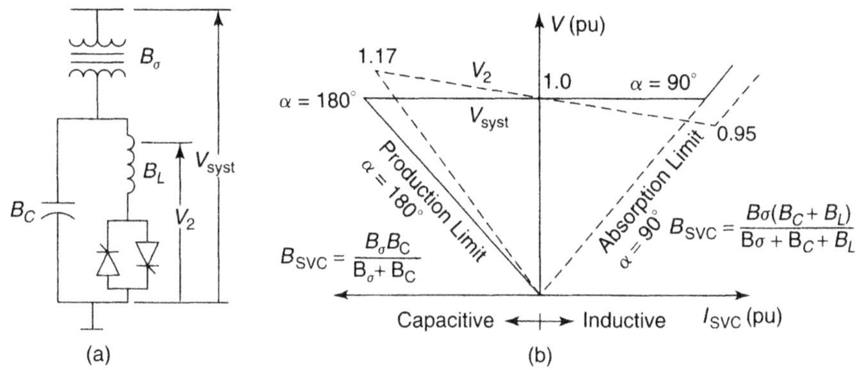

Figure 3.22 An FC–TCR with a step-down transformer and its *V-I* characteristics.

tance of either the fixed capacitor or TCR. Equation (3.27) can then be approximated by a linear relation, as follows:

$$B_{SVC} = \left(1 - \frac{B_C}{B_\sigma}\right) B_C + \left(1 - \frac{2B_C + B_L}{B_\sigma}\right) B_{TCR} \tag{3.30}$$

The susceptance limits based on the linearized equation (3.30) are

$$B_{SVC\,max} = \left(1 - \frac{B_C}{B_\sigma}\right) B_C$$

$$B_{SVC\,min} = \left(1 - \frac{B_C + B_L}{B_\sigma}\right) (B_C + B_L) \tag{3.31}$$

The Transformer-Secondary Voltage The voltage at the secondary of the transformer is

$$\overline{V}_2 = \overline{I}_{SVC} \frac{1}{j(B_C + B_{TCR})} \tag{3.32}$$

The $\overline{I}_{SVC}$ can be expressed as a function of the system voltage $\overline{V}$ and the total compensator susceptance B_{SVC} using Eqs. (3.25) and (3.27), as follows. (Note that the secondary voltage is in phase with the system voltage.)

$$V_2 = V \frac{B_\sigma}{B_\sigma + B_C + B_{TCR}} \tag{3.33}$$

At the operating limits of the compensator, the following secondary voltages are found:

$$V_{2\,\text{cap}} = V \frac{B_\sigma}{B_\sigma + B_C} \simeq V\left(1 - \frac{B_C}{B_\sigma}\right) \tag{3.34}$$

$$V_{2\,\text{ind}} = V \frac{B_\sigma}{B_\sigma + B_C + B_L} \simeq V\left(1 - \frac{B_C + B_L}{B_\sigma}\right) \tag{3.35}$$

A Practical Example Figure 3.22 gives typical data for an FC–TCR SVC scheme. The characteristic data at the boundaries of the operating range can be calculated from the previously deduced formulas. Note that for the calculation, a per-unit system is used with the following base quantities:

$V_{\text{base}} = V_{\text{nom}}$ (the nominal system voltage)
$Q_{\text{base}} = Q_{\text{SVC cap}}$ (the reactive power at full production at V_{nom})

The SVC current and the transformer-secondary voltage are calculated with the assumption that the system voltage is 1 pu.

Production Limit	Absorption Limit
$B_{\text{SVC}} = 1$ pu	$B_{\text{SVC}} = -0.3$ pu
$I_{\text{SVC}} = -1$ pu	$I_{\text{SVC}} = 0.3$ pu
$V_2 = 1.167$ pu	$V_2 = 0.95$ pu

Figure 3.22 shows the operating characteristic of the SVC with the data of the preceding table. The transformer-leakage reactance is in the range of 15%, and the production side of the compensator is about three times the rating of the absorption side, which can be considered typical. The figure also depicts the transformer-secondary voltage as a function of the total SVC current. It is interesting to note that the voltage has a negative slope if the SVC steady-state characteristic is considered flat. Figure 3.23 illustrates the susceptance characteristic. Both the exact characteristic from Eq. (3.27) and the linearized characteristic from Eq. (3.30) are displayed. The errors from linearization are clearly visible.

Losses A drawback of the FC–TCR SVC is the circulation of large currents in the FC–TCR loop needed for cancellation of capacitive vars. This results in high steady-state losses, even when the SVC is not exchanging any reactive power with the power system, as shown in Fig. 3.24. Typical losses in an FC–TCR scheme vary from 0.5% to 0.7% of the MVA rating. However, these losses can be min-

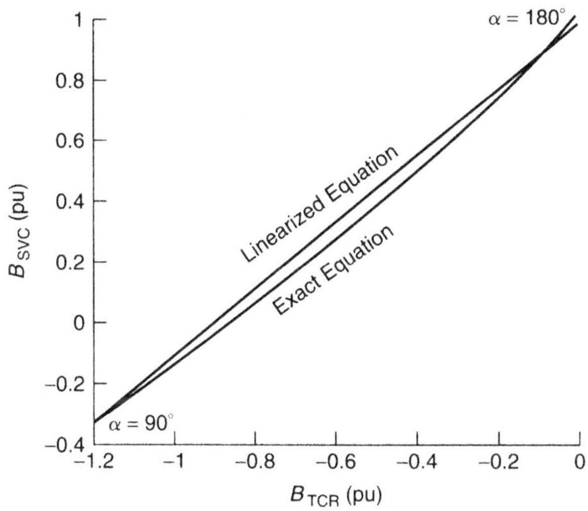

Figure 3.23 Susceptance characteristics of an FC–TCR SVC with a step-down transformer.

imized by switching the fixed capacitors through mechanical breakers, ensuring that the capacitors are inserted in the compensator circuit only when leading vars are needed. Thus a smaller-size–interpolating TCR can be used, and consequently, the steady-state operating losses can be reduced. Those FC–TCRs having a 300-MVA inductive rating have been already installed in the field.

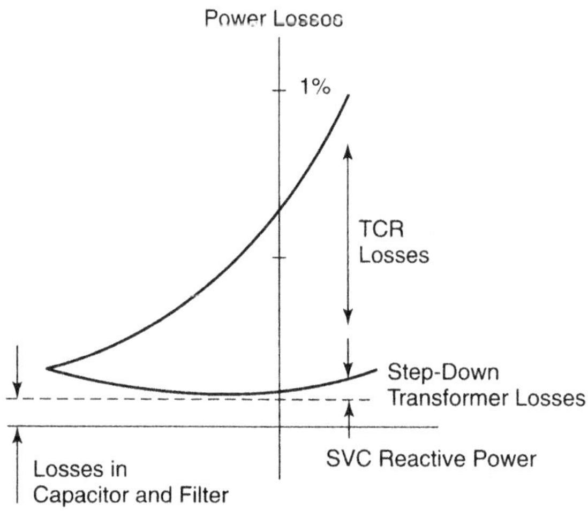

Figure 3.24 Losses in an FC–TCR. (*Courtesy:* The CIGRE.)

3.7 THE MECHANICALLY SWITCHED CAPACITOR–THYRISTOR-CONTROLLED REACTOR (MSC–TCR)

In certain applications, especially those involving few capacitor switchings, an MSC–TCR has been shown to offer acceptable performance at much lower compensating system costs than a TSC–TCR.

The different MSC–TCR circuit configurations are shown in Fig. 3.25. The mechanically switched capacitor can be located at the high-voltage bus; however, in such a case, fixed-harmonic filters must be installed in shunt with the TCR on the transformer secondary to reduce the harmonic loading of the transformer.

One advantage of the MSC–TCR scheme lies in its lower capital cost from the elimination of the thyristor switches in the capacitor branches; another advantage, in the reduced operating costs in terms of losses. The disadvantage of the MSC–TCR is a slower speed of response. The mechanical switches can close in two cycles and open in about eight, compared to one-half to one cycle with thyristor switches. Some studies show that to compensate for the slower speed and to achieve a level of transient stability similar to a thyristor-switched capacitor–thyristor-controlled reactor (TSC–TCR; see Section 3.9), a 25% higher–rated MSC–TCR SVC may be needed.

Another problem with the MSC–TCR relates to the trapped charge that is invariably left on the capacitor following deenergization. The residual charge on the capacitors is usually dissipated in about five minutes through discharge resistors built into the capacitor units. If the capacitor is switched on within five minutes after deenergization, the trapped charge may lead to increased switching transients. The MSCs can be switched in only when the capacitors are discharged. One solution [11] to the problem of trapped charges is to connect a small magnetic transformer, such

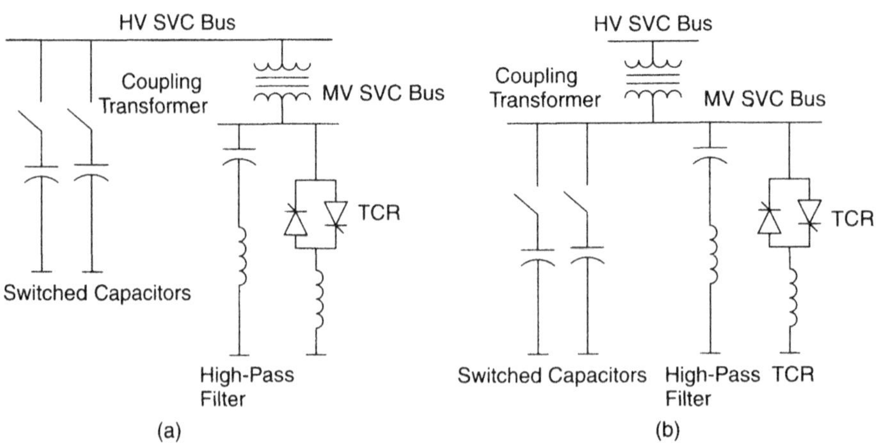

Figure 3.25 Different configurations of an MSC–TCR compensator.

as a potential transformer, in parallel with each phase of the capacitor bank. Doing so aids in dissipating the trapped charge within 0.15 s.

The TCR in an MSC–TCR is designed to have a lower inductance as compared to a TCR in a TSC–TCR SVC of similar rating. This design is to permit the increase of its overload capacity to transiently balance the capacitive-reactive power output. A lower inductance TCR produces an increased level of harmonics and thus needs a more elaborate filtering than a TSC–TCR.

The mechanical switches also possess a finite life, typically 2000–5000 operations, compared to the infinite switching life of thyristors. The MSCs are usually switched two to four times a day; they are connected during heavy-load conditions and removed under light-load conditions. An MSC–TCR may not be very suitable for voltage-control applications in a system experiencing frequent disturbances; nevertheless, a study [11] has shown that MSC–TCRs provide performance comparable to a TSC–TCR in damping-power swings between two areas and also in alleviating severe voltage depression from system faults, all at a much lower installed capital cost than an equivalent TSC–TCR. Losses with mechanically switched capacitors are fairly low; they lie in the range of 0.02–0.05%. Because capacitors are very sensitive to overvoltages and overcurrents, appropriate protection strategies need to be employed.

3.8 THE THYRISTOR-SWITCHED CAPACITOR (TSC)

Before describing the configuration and operating characteristics of a TSC, certain concepts are explained.

3.8.1 Switching a Capacitor to a Voltage Source

The circuit shown in Fig. 3.26 consists of a capacitor in series with a bidirectional thyristor switch. It is supplied from an ideal ac voltage source with neither resistance nor reactance present in the circuit. The analysis of the current transients after closing the switch brings forth two cases:

1. The capacitor voltage *is not equal* to the supply voltage when the thyristors are fired. Immediately after closing the switch, a current of infinite magnitude flows and charges the capacitor to the supply voltage in an infinitely short time. The switch realized by thyristors cannot withstand this stress and would fail.
2. The capacitor voltage *is equal* to the supply voltage when the thyristors are fired, as illustrated in Fig. 3.26(b). The analysis shows that the current will jump immediately to the value of the steady-state current. The steady-state condition is reached in an infinitely short time. Although the magnitude of the current does not exceed the steady-state values, the thyristors have an upper limit of di/dt values that they can withstand during the

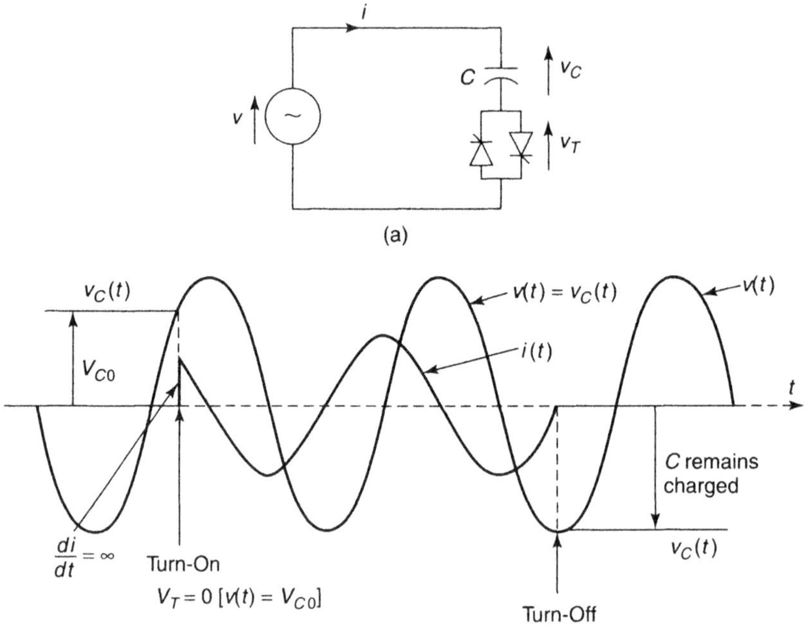

Figure 3.26 Switching of a capacitor at a voltage source: (a) a circuit diagram and (b) the current and voltage waveforms.

firing process. Here, di/dt is infinite, and the thyristor switch will again fail.

It can therefore be concluded that this simple circuit of a TSC branch is not suitable.

3.8.2 Switching a Series Connection of a Capacitor and Reactor

To overcome the problems discussed in the preceding list, a small damping reactor is added in series with the capacitor, as depicted in Fig. 3.27. Let the source voltage be

$$v(t) = V \sin \omega_0 t \tag{3.36}$$

where ω_0 is the system nominal frequency. The analysis of the current after closing the thyristor switch at $t = 0$ leads to the following result.

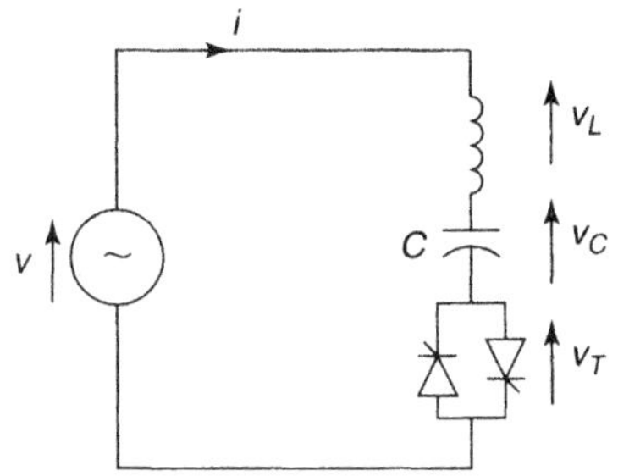

Figure 3.27 A TSC with a series reactor.

$$i(t) = I_{AC}\cos(\omega_0 t + \alpha) - nB_C\left(V_{C0} - \frac{n^2}{n^2-1}V\sin\alpha\right)$$

$$\cdot \sin\omega_n t - I_{AC}\cos\alpha\cos\omega_n t \tag{3.37}$$

where the natural frequency is

$$\omega_n = n\omega_0 = \frac{1}{\sqrt{LC}} \tag{3.38}$$

and the per-unit natural frequency is

$$\omega_n = \sqrt{\frac{|X_C|}{|X_L|}} \tag{3.39}$$

$$I_{AC} = V\,\frac{B_C B_L}{B_C + B_L} \tag{3.40}$$

Here, V_{C0} is initial capacitor voltage at $t = 0$. It is well-worth discussing this result in some detail. Note that no damping is considered in the circuit.

3.8.2.1 The Term Involving Fundamental Frequency, ω_0 This term represents the steady-state solution. As expected, the current leads the voltage by 90°. The current magnitude, I_{AC}, as obtained in the foregoing equation, can be alternatively expressed as

$$I_{AC} = VB_C\,\frac{n^2}{n^2-1} \tag{3.41}$$

A magnification in current by a factor of $n^2/(n^2 - 1)$ is seen as compared to the case without reactor. The same magnification factor is also inherent in the magnitude of the capacitor voltage.

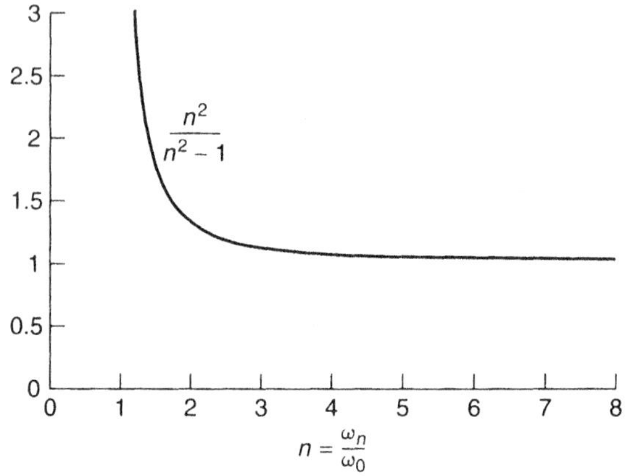

Figure 3.28 The magnification factor for the fundamental-frequency quantities in a TSC.

$$V_C = IX_C = V \frac{n^2}{n^2 - 1} \tag{3.42}$$

It is interesting to study this magnification as a function of the tuning of the TSC branch. The result is given in Fig. 3.28. For LC circuits tuned to resonance frequencies of three times the supply frequency and higher, the magnification factor is close to 1.0; for tuning below $3\omega_0$, the magnification factor increases very rapidly. For practical schemes, therefore, n should be chosen higher than 3 (typically, between the 4th and 5th harmonic).

3.8.2.2 The Terms Involving Natural Resonance Frequency, ω_n These terms constitute the oscillatory transients. Let us first discuss if these terms can be made equal to zero. From Eq. (3.27), it is seen that the following two conditions need to be fulfilled simultaneously to avoid the transients:

$$\cos \alpha = 0 \Rightarrow \sin \alpha = \pm 1 \tag{3.43}$$

$$V_{C0} = \pm V \frac{n^2}{n^2 - 1} = I_{AC} X_C \tag{3.44}$$

The first condition, Eq. (3.43), expresses that to avoid transients, the switch must be closed either at the positive or the negative peak of the supply-voltage sine wave. The second condition, Eq. (3.44), shows that the capacitor must be charged to a predetermined value. In practice, there are many problems in the realization of a firing strategy that avoids the transients, including the following:

1. At places where SVCs are installed, usually the voltages are not purely

sinusoidal and constant, which makes the decision of when to switch on less predictive than the ideal condition. A certain amount of transients is to be expected, even with very precise firing strategies.

2. Keeping the capacitor charged at $Vn^2/(n^2 - 1)$ asks for extra charging equipment. It is easier to keep a capacitor charged at V, but then a certain amount of transients will occur anyway, for the supply voltage V, line inductance X, and, consequently, n can change randomly during system operation.
3. Large ac capacitors are not designed to withstand prolonged dc stress of precharging.

If losses are considered in the analysis of the circuit, the oscillatory terms will decay, but the principal conclusions made in the preceding list will still remain valid.

3.8.2.3 Practical Switching Strategies The switching strategies presented here limit the transients to acceptable limits and are based on very simple processes to decide when the thyristors should be fired. The control hardware is simple and the performance is very robust. The following two simple firing schemes are the basis for the switching strategies:

- Firing when the initial capacitor voltage is equal to the supply voltage; hence $V_{C0} = v(t)$. Firing, therefore, takes place when the valve voltage has a zero crossing:

$$V \sin \alpha = V_{C0} \tag{3.45}$$

and the firing angle is

$$\alpha = \sin^{-1} \frac{V_{C0}}{V} \tag{3.46}$$

Equation (3.37) with α from Eqs. (3.45) and (3.46) gives the magnitude of the oscillatory term, I_{osc}, as follows:

$$\frac{I_{osc}}{I_{AC}} = \sqrt{1 - \left(\frac{V_{C0}}{V}\right)^2 \left(1 - \frac{1}{n^2}\right)} \tag{3.47}$$

Figure 3.29(a) shows this relation for different tuning of the TSC branch and for different precharge conditions of the capacitor. It can be noted that the magnitude of the oscillatory term never exceeds the magnitude of the steady-state current. With a higher precharge on the capacitor, the transients become smaller, especially for branches that are tuned to higher-resonance frequencies. The highest transients occur if the capacitor is completely dis-

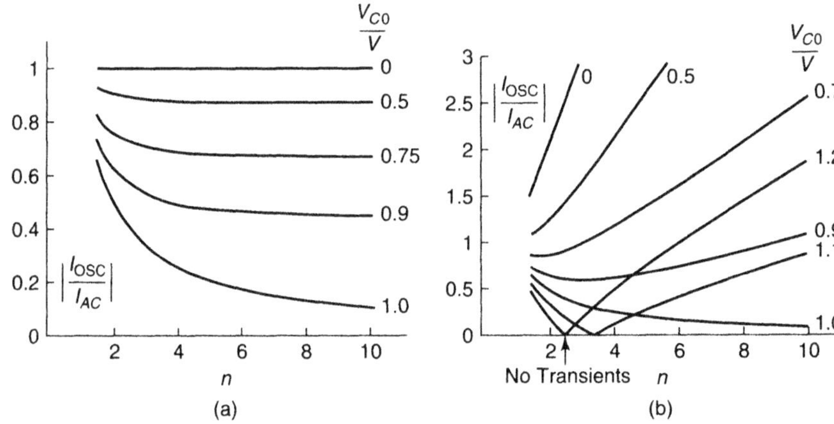

Figure 3.29 The magnitude of the oscillatory component for different tuning of the LC branch and different precharging on C: (a) firing criterion, $v(t) = V_{C0}$ ($V_T = 0$) and (b) firing criterion, crest of vt.

charged. This firing strategy cannot be used if the capacitor is overcharged to $V_{C0} > V$.

- Firing at the crest of the supply-voltage sine wave; hence $\cos\alpha = 0$. Again, Eq. (3.37) gives the magnitude of the oscillatory term as follows:

$$\frac{I_{osc}}{I_{AC}} = n\left(\frac{n^2 - 1}{n^2}\frac{V_{C0}}{V} - 1\right) \tag{3.48}$$

Figure 3.29(b) demonstrates this result. Note that this scheme is applicable even if the capacitor is overcharged. However, it gives worse transients if the capacitor initial voltage is below the crest of the supply voltage (e.g., $V_{C0}/V = 0.75$). The fewest transients are expected if tuning is in the range $n = 2$ to 5. For an initial capacitor voltage of 1.0, the two firing schemes are identical, which can be seen by comparing the characteristics for $V_{C0}/V = 1.0$ in Figs. 3.29(a) and (b), respectively.

Based on these firing schemes, the following two firing strategies are used in practice:

Strategy A

- The capacitors discharge if they are not connected to the system. Therefore, any initial voltage across the capacitor is possible.
- If $V_{C0} < V$ at the time of demand for the capacitor, it is switched on according to the first explained firing scheme, that is, as soon as the voltage across the valve reaches zero and the capacitor voltage is equal to the supply voltage.

- If the capacitor is overcharged ($V_{C0} > V$) at the time of demand, it is switched on according to the second firing scheme, that is, when the supply voltage reaches the crest and the voltage across the valve is minimal. This scheme is also called a *forced switch-on*.

This firing strategy uses both of the two aforementioned firing schemes to minimize the current transients. It requires no special capacitor-charging strategy and can operate with conventional ac power capacitors. It is widely used in TSC plants for transmission systems. The strategy is presented in Fig. 3.30 in a simplified manner for four cases.

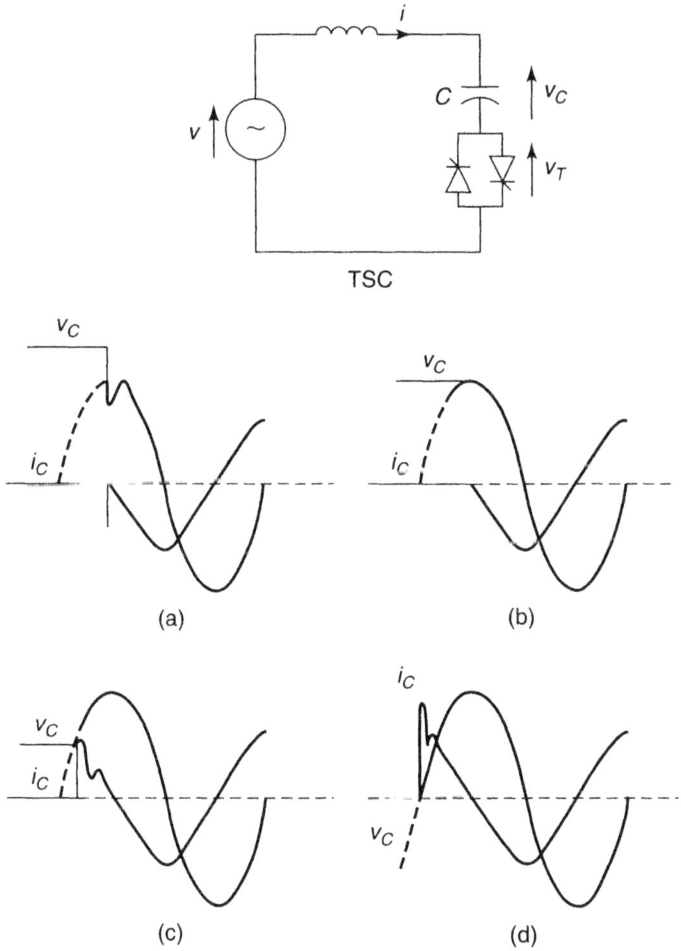

Figure 3.30 The switching strategy for a TSC: (a) firing at the minimum valve voltage, $v_C > v$; (b) firing at the zero valve voltage, $v_C = v$; (c) firing at the zero valve voltage, $v_C < v$; and (d) firing at the zero valve voltage, $v_C = 0$.

Strategy B

- The capacitors are charged to the crest of the supply voltage by firing just one of the two thyristors. It is noted that they are not charged to the optimal voltage $Vn^2/(n^2 - 1)$.
- Firing always takes place at the crest of the supply voltage, where the valve voltage is minimal.

This firing strategy is based on the second firing scheme. It makes use of the relatively small transients if the capacitor is charged to 1 pu voltage, and it can be found in some industrial TSC applications. There are, however, many problems related to its use. The problem of the prolonged dc stress on the capacitor has already been mentioned. To overcome this problem, the capacitor is periodically charged to opposite polarity, which eventually becomes tolerable in industrial plants but, because of the high current when the polarity is reversed, is hardly acceptable for transmission compensation. Furthermore, the capacitors are not properly isolated from the system when they are turned off, and interactions with the power system can still occur. Because of the voltage transients, it is also possble for the capacitor to get charged to voltages above the ideal value, which is undesirable.

3.8.3 Turning-Off of the TSC Valve

Thyristors turn off when the current crosses zero, as shown in Fig. 3.31. A zero crossing in the capacitor current occurs simultaneously with a crest in the capacitor voltage. If the current is interrupted by the thyristors, the capacitor will remain charged. The valve, therefore, has to block the voltage difference between the supply voltage and the capacitor voltage, which becomes higher than the supply voltage at the following supply-voltage peak.

In Fig. 3.31, the supply voltage $v(t)$, the capacitor voltage $v_C(t)$ and the valve-blocking voltage $v_T(t)$ are shown for an ideal case without capacitor discharging and a more realistic case where discharging of the capacitor is assumed. The initial step in the valve-blocking voltage immediately after turning off the valve is caused by the inductance in the circuit. It can be concluded that the blocking voltage across the thyristor valve is quite high—in general, $V_T > 2V$. In combined schemes with TSCs and TCRs at the same bus, the number of thyristors in the TSC valves is typically about two times the number of thyristors in the TCR valves. The TSC valve must further withstand a dc voltage stress for some time after switching off.

3.8.4 The TSC Configuration

A basic single-phase TSC consists of an anti-parallel–connected thyristor-valve pair that acts as a bidirectional switch in series with a capacitor and a current-limiting small reactor, as shown in Fig. 3.32(a). The thyristor switch allows

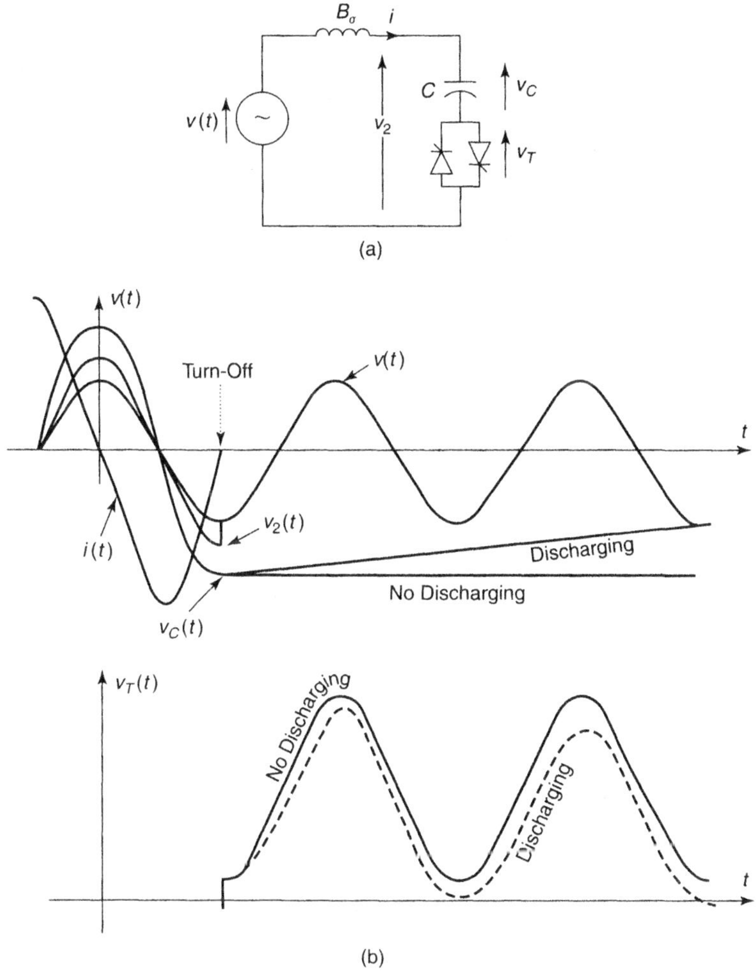

Figure 3.31 Voltages after turn-off to the TSC: (a) a circuit diagram and (b) the voltage–current waveforms.

the conduction for integral number of half-cycles. The capacitor is not phase-controlled, as is a TCR.

As discussed in previous sections of this chapter, the thyristor valves are turned on at an instant when minimum voltage is sensed across the valves to minimize the switching transients. Barring these initial transients, the TSC current is sinusoidal and free from harmonics, thus obviating the need for any filters.

The small-series inductor is installed to limit current transients during overvoltage conditions and planned switching operations, as well as when switching at incorrect instants or at the inappropriate voltage polarity. The inductor magnitude is chosen to give a natural resonant frequency of four to five times the

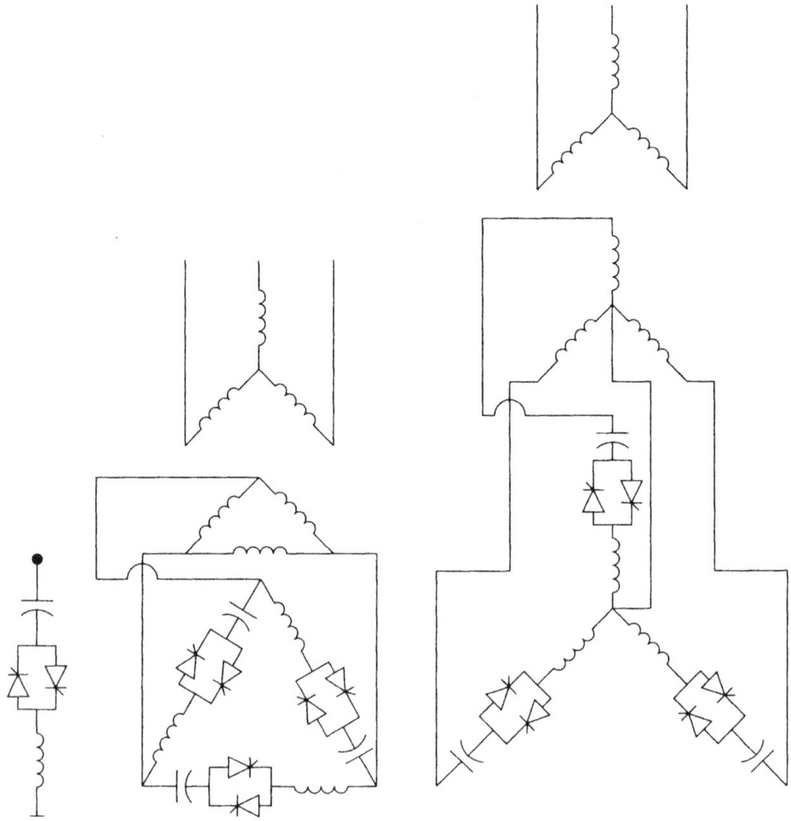

Figure 3.32 Different TSC configurations: (a) a single-phase TSC branch; (b) a 3-phase delta-connected TSC; and (c) a 3-phase Y-connected transformer secondary with neutrals connected.

system nominal frequency, which ensures that the inductance neither creates a harmonic-resonant circuit with the network nor hampers the TSC control system. Another function of this series inductor is to act in combination with the capacitor as a filter for harmonics generated by the associated TCR. In some cases, discharge circuits are provided with the capacitors to rapidly dissipate the remnant charge on the capacitor after a switch-off.

A 3-phase TSC unit comprises three single-phase TSCs connected in a delta, which are usually supplied by the delta secondary winding of a step-down transformer, as depicted in Fig. 3.32(b). An alternative 3-phase, 4-wire star-connected TSC configuration is shown in Fig. 3.32(c).

A practical TSC compensator involves n 3-phase TSC banks of equal rating connected in shunt. The overall TSC susceptance at any given instant is the sum of conducting TSC. In some cases, the ratings of different constituent TSC steps

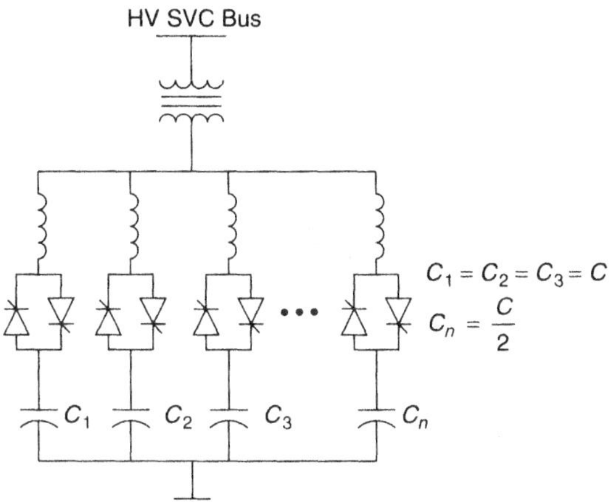

Figure 3.33 A general TSC scheme.

may be chosen based on a binary system. In this scheme, $n - 1$ capacitors are rated for susceptance B and one capacitor is rated for susceptance $B/2$. Thus the total number of possible TSC steps get extended to $2n$. An example of such a TSC is shown in Fig. 3.33.

The TSC provides a fast response—typically between one-half to one cycle. However, this response time may be extended because of any delays in the measurement and control systems. The TSCs provide virtually unlimited switching operations, in stark contrast to MSCs.

3.8.5 Operating Characteristic

The TSC has a discrete voltage–current operating characteristic [1], as shown in Fig. 3.34. The shape of this characteristic is a function of the number of TSC

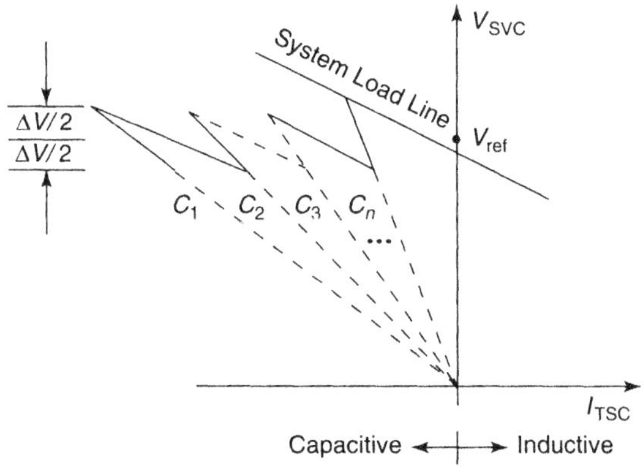

Figure 3.34 The operating characteristic of a TSC.

units, their individual ratings, and a hysteresis voltage ΔV, which is built in to avoid undesirable frequent switchings of capacitors. In a closed-loop voltage-control operation, the TSC regulates the bus voltage within the range $V_{ref} \pm \Delta V/2$.

3.9 THE THYRISTOR-SWITCHED CAPACITOR–THYRISTOR-CONTROLLED REACTOR (TSC–TCR)

3.9.1 Configuration

The TSC–TCR compensator shown in Fig. 3.35 usually comprises n TSC banks and a single TCR that are connected in parallel. The rating of the TCR is chosen to be $1/n$ of the total SVC rating. The capacitors can be switched in discrete steps, whereas continuous control within the reactive-power span of each step is provided by the TCR. Thus the maximum inductive range of the SVC corresponds to the rating of the relatively small interpolating TCR.

As the size of TCR is small, the harmonic generation is also substantially reduced. The TSC branches are tuned with the series reactor to different dominant harmonic frequencies. To avoid a situation in which all TSCs and, consequently, the associated filters are switched off (with only the TCR in operation), an additional nonswitchable capacitive-filter branch is provided.

The main motivations in developing TSC–TCRs was for enhancing the operational flexibility of the compensator during large disturbances and for reducing the steady-state losses. A fixed capacitor–thyristor-controlled reactor (FC–TCR) behaves like a parallel LC circuit that tends to set up a resonance with the ac system impedance during large disturbances. What particularly aggravates the problem is when severe voltage swings are experienced and followed by load rejection. In this event, a TSC–TCR can quickly operate to disconnect all the capacitors from the compensator, precluding the resonant oscillations. This feature of disconnecting the capacitor in exigencies is not available with FC–TCRs.

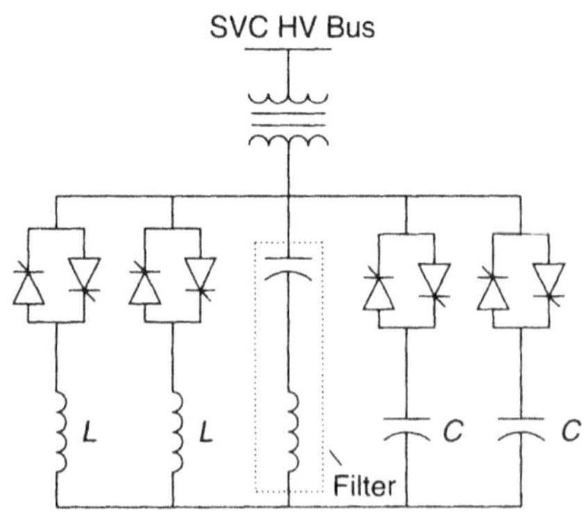

Figure 3.35 A general TSC–TCR SVC.

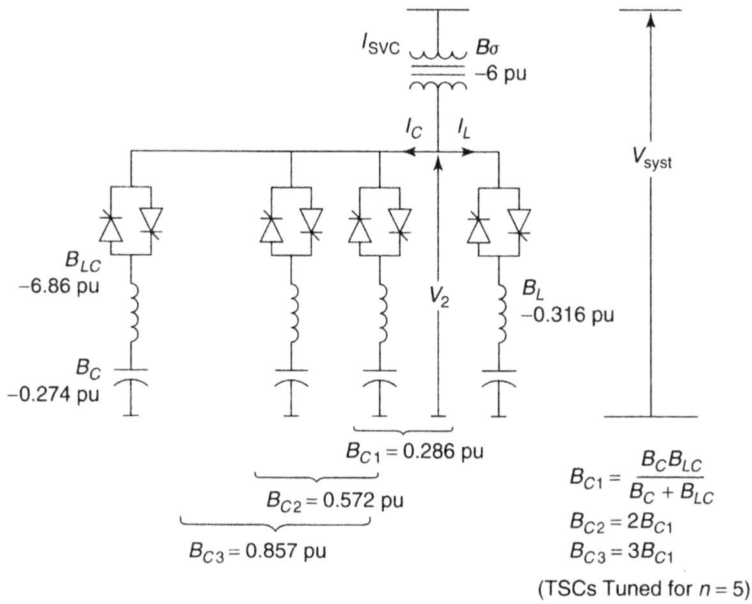

Figure 3.36 A practical example of a 6-pulse TCR–TSC compensator with three TSCs.

3.9.2 Operating Characteristic

3.9.2.1 A Practical Example A TSC–TCR scheme can be considered as a fixed capacitor TCR scheme, where the capacitor may have a number of different values. Thus an understanding of the TSC–TCR can be obtained by applying the theory of an FC–TCR SVC developed in Section 3.6 to a practical example of TSC–TCR.

An example configuration is given in Fig. 3.36. The SVC consists of three TSCs and one TCR in a 6-pulse arrangement and has the same ratings as the example for a fixed-capacitor scheme in Fig. 3.22. It is assumed that the TSC branches are tuned to $n = 5$ and, furthermore, that all TSC branches are identical.

Calculation of the Operating-Range Limits Equations (3.28) and (3.29) can be used to determine the susceptance of the SVC at the limits of the operating range. In Eq. (3.28), B_C is replaced by B_{C3} of the TSC scheme to get the susceptance at the production limit. Figure 3.36 defines B_{C3} as the susceptance of all three TSC branches in parallel and considers the influence of the damping reactors. In Eq. (3.29), B_C is considered zero for the TSC scheme at the absorption limit, for all capacitors are switched off. With the given data, the susceptance at production limit is

$$B_{\text{SVC max}} = \frac{B_\sigma B_{C3}}{B_\sigma + B_{C3}} = 1 \text{ pu} \tag{3.49}$$

and the susceptance at the absorption limit is

$$B_{\text{SVC min}} = \frac{B_\sigma B_{C3}}{B_\sigma + B_{C3}} = -0.3 \text{ pu} \tag{3.50}$$

This shows that the rating of the TSC scheme discussed here is the same as that for the previously discussed FC–TCR scheme. (See Section 3.6.2.)

The overall operating characteristic of the TSC–TCR scheme can be assembled by applying the equations for FC–TCR schemes, assuming that there is one, two, three, or no capacitors connected. Here, the total operating range consists of four subranges. Equations (3.25), (3.27), and (3.29) are used to determine the characteristics.

Subrange With Three TSCs Turned On The total compensator current is

$$I_{\text{SVC}} = -V \frac{B_\sigma(B_{C3} + B_{\text{TCR}})}{B_\sigma + B_{C3} + B_{\text{TCR}}} \tag{3.51}$$

The negative sign indicates a capacitive current. The currents at the two limits are

$$I_{\text{SVC cap}} = -V \frac{B_\sigma B_{C3}}{B_\sigma + B_{C3}} = -1 \text{ pu} \tag{3.52}$$

and

$$I_{\text{SVC ind}} = -V \frac{B_\sigma(B_{C3} + B_L)}{B_\sigma + B_{C3} + B_L} = -0.595 \text{ pu} \tag{3.53}$$

The current at the inductive limit of the subrange with three TSCs is still capacitive.

Other Subranges The subranges with one, two, or no TSCs are determined in an analogous manner as the subrange with three TSCs. Figure 3.37 gives the results; it can be seen that the subranges overlap, which is required for continuous, stable control. The figure also shows the transformer-secondary voltage calculated with Eq. (3.33) and the capacitor voltage calculated with Eq. (3.42).

3.9.3 Current Characteristic

It is interesting to visualize how the TSCs and the TCR share the current and contribute to the total SVC current. This is, however, difficult to observe from

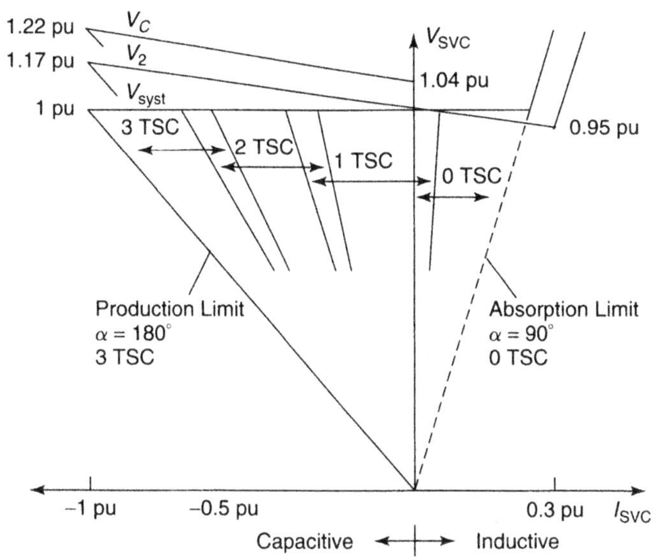

Figure 3.37 The voltage–current characteristic of the example TSC–TCR SVC.

the operating characteristics derived above. Generally, the TSC and TCR currents add to give the total current as follows:

$$I_{SVC} = I_C + I_L \tag{3.54}$$

where

$$I_C = -V_{2n}B_{Cn}, \qquad I_L = V_{2n}B_{TCR} \tag{3.55}$$

The subscript n indicates the number of TSC branches turned on and, therefore, can have values 0, 1, 2, . . . , n_{max}, where n_{max} is the total number of TSCs—in this case, three. In the preceding two equations, V_{2n} can be substituted from Eq. (3.33) to give

$$I_C = -V\frac{B_\sigma B_{Cn}}{B_\sigma + B_{Cn} + B_{TCR}} \tag{3.56}$$

$$I_L = -V\frac{B_\sigma B_{TCR}}{B_\sigma + B_{Cn} + B_{TCR}} \tag{3.57}$$

where $n = 1, 2, \ldots$ is the number of TSC circuits in operation and B_{Cn} is the total susceptance of n TSC branches. Substituting $B_{TCR} = 0$ and $B_{TCR} = B_L$ in the foregoing equations, respectively, results in the currents at the absorption limit and at the production limit of each subrange for different numbers of TSCs. The results for the scheme analyzed here are presented in Fig. 3.38. It is

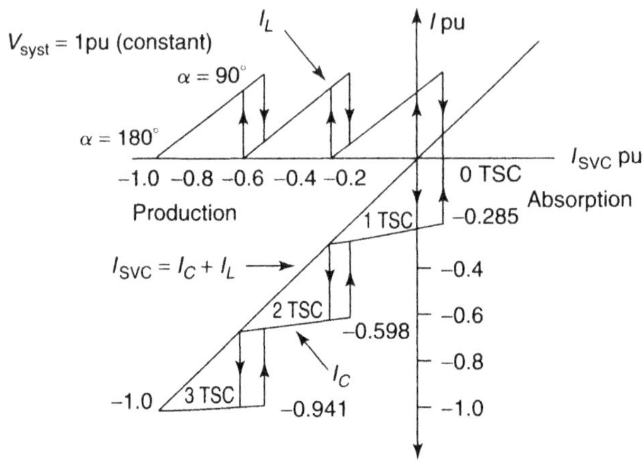

Figure 3.38 The current characteristic of the example TSC–TCR SVC.

emphasized that the variation of the transformer-secondary voltage influences these characteristics. The highest steady-state current in the TCR is with three connected TSCs, because the transformer-secondary voltage is high when three TSCs are on (assuming the system voltage is constant over the entire range).

3.9.4 Susceptance Characteristic

Equation (3.27) can be used to calculate the SVC susceptance in the TSC–TCR scheme as follows:

$$B_{SVC} = \frac{B_\sigma(B_{Cn} + B_{TCR})}{B_\sigma + B_{Cn} + B_{TCR}} \tag{3.58}$$

where $n = 1, 2, \ldots$ is the number of TSC branches in operation and B_{Cn} is the total susceptance of n TSC branches. With a linear approximation, Eq. (3.58) reduces to

$$B_{SVC} = \left(1 - \frac{B_{Cn}}{B_\sigma}\right) B_{Cn} + \left(1 - \frac{2B_{Cn} + B_L}{B_\sigma}\right) B_{TCR} \tag{3.59}$$

Figure 3.39 gives the total susceptance B_{SVC} as a function of the susceptance of the controlled reactor B_{TCR} for the example data. These characteristics are of importance for control design, for the controls vary B_{TCR} and the effect on the system is caused by B_{SVC}.

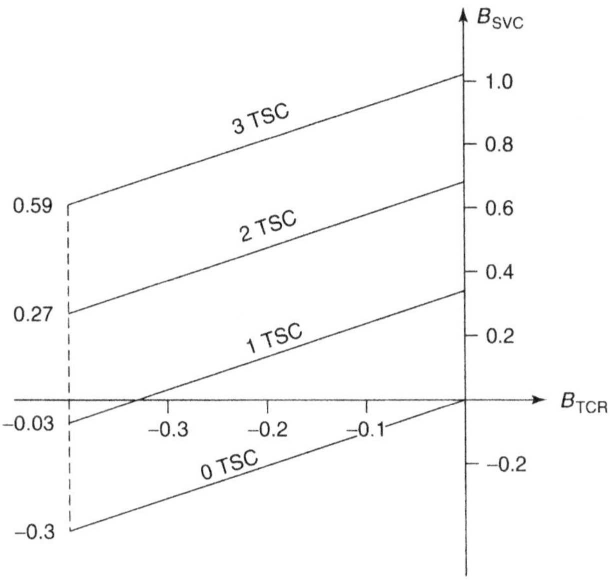

Figure 3.39 The susceptance characteristic of the example TSC–TCR SVC.

3.9.5 Mismatched TSC–TCR

A TSC–TCR scheme in which the TCR cannot compensate for one TSC step because its rating is too small, is shown in Fig. 3.40. Each subrange shows the typical operating characteristic of an FC scheme. However, instead of overlap areas, there are now gaps, and no operation is possible within these dead zones. To illustrate the problems arising from such a scheme, a typical system characteristic is given in the Fig. 3.40. The compensator cannot stabilize the system voltage to 1 pu. The operating points with voltages closest to nominal voltage are *A* or *B*. Operating point *A* is with two TSCs switched on and results in some overvoltage, whereas operating point *B* is with only one TSC but shows some undervoltage. This is a discontinuous system and creates problems in control.

The standard design for TSC–TCR schemes is therefore to include a finite overlap. However, discontinuous operation may sometimes occur under degraded operating conditions; for example, in a 12-pulse scheme, one TCR is shut down and the remaining 6-pulse TCR is too small to compensate for one TSC.

The *V-I* characteristic of TSC–TCR SVC shown in Fig. 3.40 is a simple case when the bus voltage is regulated strictly at the specified reference level. Practical SVCs incorporate a slight slope in the *V-I* characteristics to obtain certain advantages, as discussed in Chapter 5 (Section 5.1.3). The problem of coordination between the TSC and TCR for an SVC having a droop in the operating characteristics is illustrated in Fig. 3.41. A careful examination reveals that to obtain a smooth characteristic, it is necessary to vary slightly the TCR knee

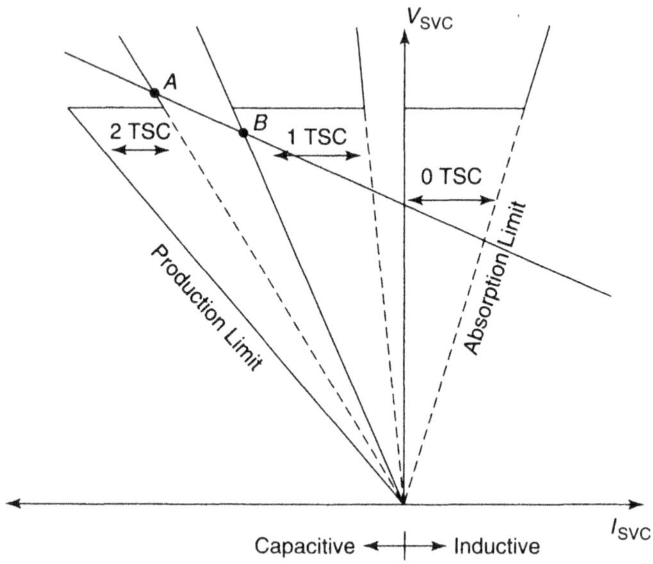

Figure 3.40 The operating characteristic of a TSC–TCR SVC with mismatched ratings of TSCs and TCRs.

voltage and current slope at every TSC switching [3]. An overlap or hysteresis, as discussed previously, is always desirable to avoid chattering of TSCs. In this scheme, a TSC is turned on at a lower system voltage but turned off at a higher voltage level.

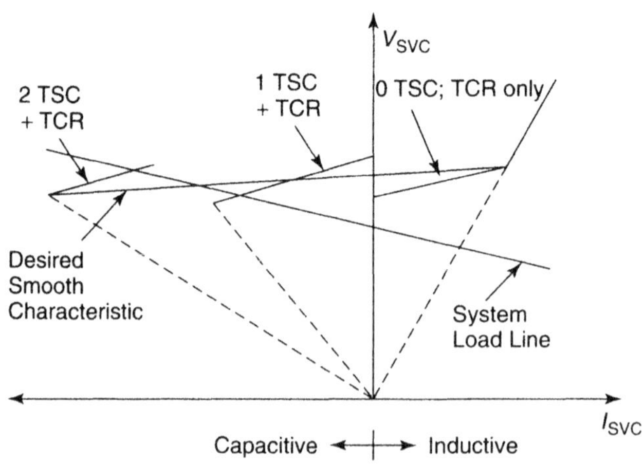

Figure 3.41 Coordination issues in a TSC–TCR SVC.

3.10 A COMPARISON OF DIFFERENT SVCS

3.10.1 Losses

The real-power losses are an important factor during the selection of a specific SVC configuration. Because these losses are constantly present, the capitalized cost of these losses keep accruing to significantly high levels.

A comparison of losses of three main SVC configurations, namely, FC–TCR, TSC–TCR, and MSC–TCR, is presented in Fig. 3.42. In each of these cases, the losses associated with the step-down transformer are neglected; however, the losses are expressed as the percent of rated MVA of SVC. The following losses are contributed by different components:

1. Small, resistive losses are in the permanently connected filter branches in the TSC–TCR and MSC–TCR.
2. Losses in the main capacitors in all three SVCs.
3. Valve-conduction losses and switching losses in the thyristor power circuit.
4. Resistive losses in the inductor of the TCR, which increases substantially with the TCR current.

Maximum losses occur with FC–TCR in the floating state, that is, when the SVC is not exchanging any reactive power with the power system. This condition is a prime disadvantage with this type of SVC, as for an extended period of time, the SVC largely remains in the floating state with its reactive power margins in standby to meet any system exigency.

The TSC–TCR losses are the least in the floating state, just as with MSC–TCR. However, these losses increase as more TSCs are switched into service. The TCR losses in a TSC–TCR are lower because of the smaller reac-

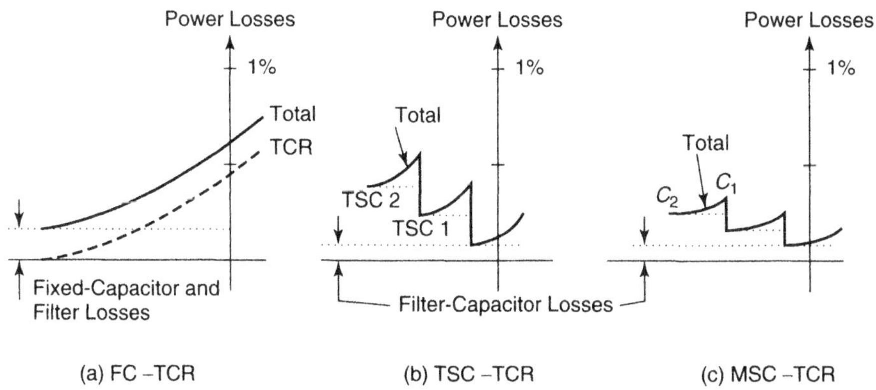

Figure 3.42 A comparison of losses for different SVC configurations.

TABLE 3.1 Comparison of Different Reactive Compensators

SVC No.	Feature	Synchronous Condenser	SR/FC	FC–TCR/FC–TCT	TSC	TSC–TCR	MSC–TCR
1	Control range	Inductive and capacitive	Inductive and capacitive (with FC)	Inductive and capacitive	Capacitive	Inductive and capacitive	Inductive and capacitive
2	Nature of control	Continuous, active	Continuous, inherent	Continuous, active	Discrete, active	Continuous, active	Continuous, active
3	Response time	Slow	Fast: system-, slope-, correction-capacitor-, and filter-dependent	Fast: system- and control dependent	Fast: control-dependent	Fast: system-dependent	Medium: system-dependent
4	Control capability:						
	Voltage control	Good	Limited	Good	Limited	Good	Good
	Auxiliary-stabilizing signals	Limited	No	Good	No	Good	Good
	Individual-phase control	Limited	Limited	Good	Limited	Good	Good
5	Harmonic generation	None	Very low: lowest harmonic (17th)	Low: filters needed	None	Low: filters needed	Low: filters needed
6	Overvoltage limitation; overload capability	Very good	Very good: limited by slope-correction capacitor	Moderate	None	Limited	Limited
7	Rotating inertia	Yes	No	No	No	No	No
8	Sensitive to frequency deviations	Yes	No	No	No	No	No
9	Losses	Moderate; rotating losses as well	Moderate: increase with lagging current	Medium: increase with lagging current	Small: increase with leading current	Small: configuration-dependent	Small
10	Direct EHV connection	No	SR: no; FC: yes	TCR: no; TCT, FC: yes	No	No	TCR: no; MSC: yes
11	Energization	Slow	Fast and direct; some transients	Fast with control action; minimal transients	Fast with control action; some transients	Fast with control action; some transients	Fast with control action; some transients

tor rating. This SVC generates higher losses only for short duration when it is involved in system stabilization in case of a contingency. The MSC–TCR losses show a similar trend as those of the TSC–TCR, but they have a much lower magnitude in comparison, for the losses associated with thyristor valves are absent. Losses in each of the three SVCs are dependent on the operating point, so the preference of one SVC configuration over another must be based on the operating point at which the SVC will generally reside for most of the time in a given application.

3.10.2 Performance

A performance comparison of different SVC configurations is presented in Table 3.1 [1]. It is evident that there is no single SVC that acts as a panacea for all reactive-power compensation requirements. The choice of a specific SVC is based on several considerations—the application requirement, speed of response, frequency of operation, losses, capital cost, and so forth. Nevertheless, the TSC–TCR is by far the most versatile among all SVC configurations, athough with a cost premium.

3.11 SUMMARY

This chapter presented in depth the fundamental operating principles and characteristics of the various conventional reactive-power compensators, including the time-tested synchronous condenser and the SRC, as well as several thyristor-based compensators that constitute the initial series of FACTS controllers. Practical examples were incorporated to illustrate the principles of operation of some main compensators. Also, a comparative performance analysis was provided for the different compensators.

REFERENCES

[1] I. A. Erinmez, Ed., *Static Var Compensators*, CIGRE Working Group 38-01, Task Force No. 2 on SVC, Paris, 1986.

[2] A. Edris et al., "Proposed Terms and Definitions for Flexible AC Transmission System (FACTS)," *IEEE Transactions on Power Delivery*, Vol. 12, No. 4, October 1997, pp. 1848–1853.

[3] T. J. E. Miller, Ed., *Reactive Power Control in Electric Power Systems*, John Wiley and Sons, New York, 1982.

[4] R. M. Mathur, Ed., *Static Compensators for Reactive Power Control*, Canadian Electrical Association (CEA), Cantext Publications, Winnipeg, Manitoba, 1984.

[5] L. Gyugyi, "Fundamentals of Thyristor-Controlled Static Var Compensators in Electric Power System Applications," IEEE Special Publication 87TH0187-5-PWR, *Application of Static Var Systems for System Dynamic Performance*, 1987, pp. 8–27.

[6] CIGRE Task Force 38.05.04, "Analysis and Optimizaiton of SVC Use in Transmission Systems," CIGRE Technical Brochure No. 77, 1993.

[7] IEEE Power Engineering Society/CIGRE, *FACTS Overview*, Publication 95TP108, IEEE Press, New York, 1995.

[8] N. G. Hingorani and L. Gyugyi, *Understanding FACTS*, IEEE Press, New York, 1999.

[9] Y. H. Song and A. T. Johns, Eds., *Flexible AC Transmission Systems (FACTS)*, IEEE Press, U.K., 1999.

[10] IEEE Power Engineering Society, *FACTS Applications*, Publication 96TP116-0, IEEE Press, New York, 1996.

[11] The Electric Power Research Institute (EPRI) Report TR-100696, "Improved Static Var Compensator Control," Final Report of Project 2707-01, Prepared by General Electric (GE) Company, Schenectady, NY, June 1992.

[12] J. A. Oliver et al., "345 kV MVA Fully Water-Cooled Synchronous Condenser for Dumont Station," *IEEE Transactions on Power Apparatus and Systems*, Vol. 90, November/December 1971, pp. 2758–2777.

[13] C. V. Theo and J. B. Davies, "New Synchronous Compensators for the Nelson River HVDC System—Planning Requirement and Specification," *IEEE Transactions on Power Delivery*, Vol. 6, No. 2, April 1991, pp. 922–928.

[14] C. A. Gama, E. H. Ellery, D. C. Azevedo, and J. R. R. Poute, "Static Var Compensators (SVC) Versus Synchronous Condensers (SC) for Inverter Stations' Compensation—Technical and Economic Aspects in Electronorte Studies," CIGRE Paper 14-103, Paris, 1992.

[15] H. E. Schweickardt, W. M. Pfyl, and G. Romegialli, "Laurentides—The First 735 kV Static Var System, Description, and First Operational Results," *International Symposium on Controlled Reactive Compensation*, IREQ, Varennes, Quebec, 1979.

CHAPTER 4

SVC Control Components and Models

4.1 INTRODUCTION

This chapter describes the various components of a general SVC control system, such as a measurement system, a voltage regulator, a gate-pulse generator, a synchronizing system, and supplementary control and protection functions [1]–[16]. The demodulation effect of the measurement system [7], [17], [18], which causes various kinds of control instabilities, is discussed in depth, and different implementations of the voltage regulator [11] are presented. The genesis of control-system delays is included. The SVC features are described in detail; each SVC is provided with many additional control and protection features to ensure a fast yet secure operation in the wake of severe system disturbances. For the various control-system components, mathematical models are presented that vary in their complexity to accommodate the nature of the study to be performed for an SVC-compensated power system [1]–[16], [19]–[27].

The block diagram of a general TSC–TCR type of SVC control system is depicted in Fig. 4.1. This control system incorporates features of both simple voltage control and auxiliary control.

4.2 MEASUREMENT SYSTEMS

These systems provide the necessary inputs to the SVC controller for performing its control operations. The different inputs required by an SVC depend on the function that the SVC controller is intended to perform. The required inputs that correspond to three basic modes of SVC control are as follows:

1. Voltage control based on balanced control of the three phases of SVC:
   a. 3-phase–rectified root mean square/average voltage
   b. Positive-sequence voltage
   c. 3-phase average/root mean square currents
   d. Squared voltage, V^2
2. Individual-phase voltage/reactive-power control:
   a. Individual-phase voltages

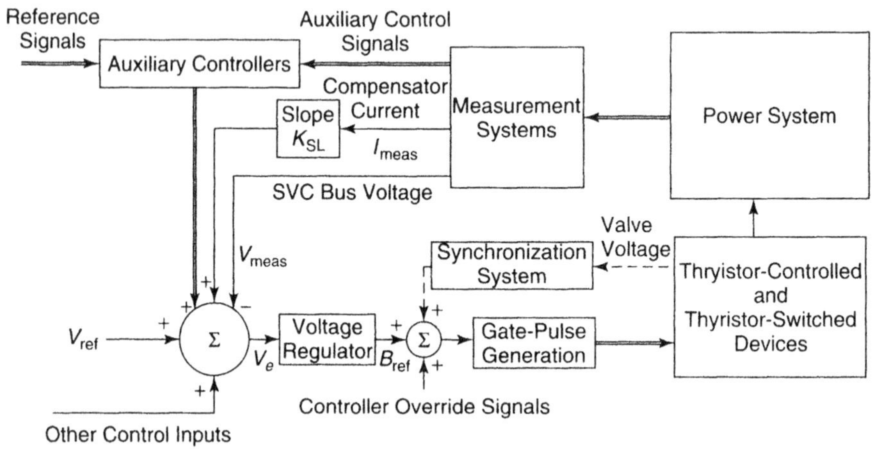

Figure 4.1 A general schematic diagram of an SVC control system.

   b. Positive- and negative-sequence voltages
   c. Squared voltage, V^2
   d. Individual-phase currents
   e. Individual-phase reactive power

3. Auxiliary control for electrical damping enhancement using the following major supplementary signals:
   a. Transmission-line current
   b. Transmission-line active/reactive power
   c. Bus angle
   d. Bus frequency
   e. Angular velocity or accelerating power of a synchronous generator

Some signals of the preceding list can be measured directly, whereas others may be derived within the control system [3] using the basic voltage and current signals. The block-schematic diagram of a general measurement system incorporating voltage and current measurement, together with reactive-power computation, is given in Fig. 4.2.

4.2.1 Voltage Measurement

The purpose of the voltage-measurement system [2], [3], [17] is to generate a dc signal proportional to the root mean square (rms) value of the 3-phase–balanced voltages at fundamental frequency. The SVC bus voltage is stepped down using a potential transformer (PT), after which one of the following techniques is applied:

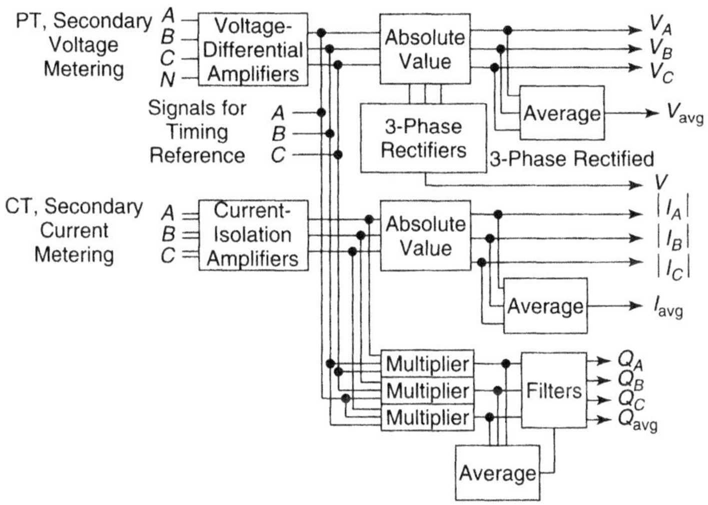

Figure 4.2 The measurement circuits for a general SVC control module.

1. ac-to-dc rectification,
2. coordinate transformation, or
3. Fourier analysis.

4.2.1.1 ac/dc Rectification This commonly employed method involves full-wave rectification of the 3-phase ac voltages using a 6-pulse diode bridge or two 6-pulse diode bridges fed by two sets of 30° phase–displaced ac voltages, resulting in a 12-pulse mode of rectification. A $6n$-pulse diode bridge, when $n = 1, 2, \ldots,$ produces harmonics of the order $6nf_0$ on the dc side, corresponding to a system frequency f_0 Hz. A low-pass filter is usually installed on the dc side to eliminate this high-frequency ripple.

4.2.1.2 Coordinate Transformation This method is based on a 3-phase (a, b, c)–2-phase (α, β) scalar transformation, defined as

$$\begin{bmatrix} v_\alpha(t) \\ v_\beta(t) \end{bmatrix} = \frac{1}{3} \begin{bmatrix} 2 & -1 & -1 \\ 0 & \sqrt{3} & -\sqrt{3} \end{bmatrix} \begin{bmatrix} v_a(t) \\ v_b(t) \\ v_c(t) \end{bmatrix} \tag{4.1}$$

where the subscripts a, b, c, α, and β denote the instantaneous values of the respective axis-voltage components. The rms value of the bus voltage V_{rms} is then computed as

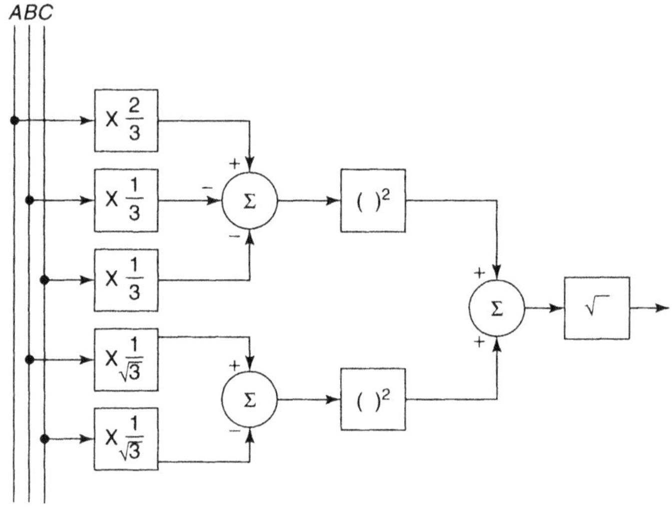

Figure 4.3 The coordinate transformation method for the instantaneous measurement of 3-phase quantities.

$$V_{\text{rms}} = \sqrt{v_\alpha^2 + v_\beta^2} \tag{4.2}$$

A graphical illustration [4] of this measurement technique is presented in Fig. 4.3. It is assumed that no zero-sequence components exist in the 3-phase voltages.

4.2.1.3 Fourier Analysis Another method for calculating the magnitude of fundamental voltage is based on the Fourier analysis technique [17]. This method is insensitive to harmonics and can be implemented with ease in digital-control systems. The computation principle for a single-phase measurement system is depicted in Fig. 4.4(a).

The magnitude of the fundamental-voltage wave $V_{\text{meas}}(\tau)$ at a time instant τ is given as

$$V_{\text{meas}}(\tau) = \sqrt{a^2 + b^2} \tag{4.3}$$

where a and b are the first-order coefficients of the Fourier series and are expressed as

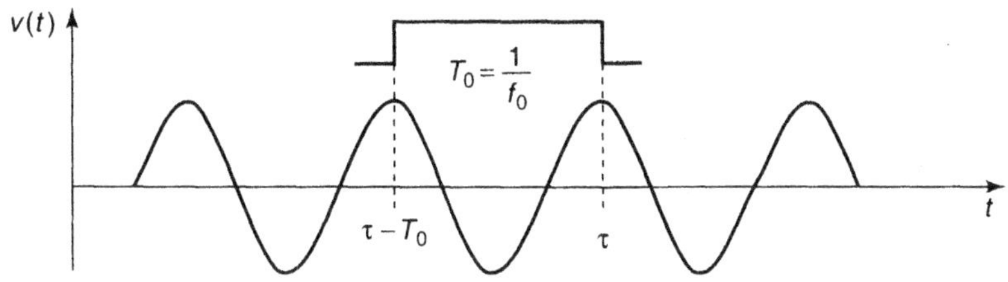

(a)

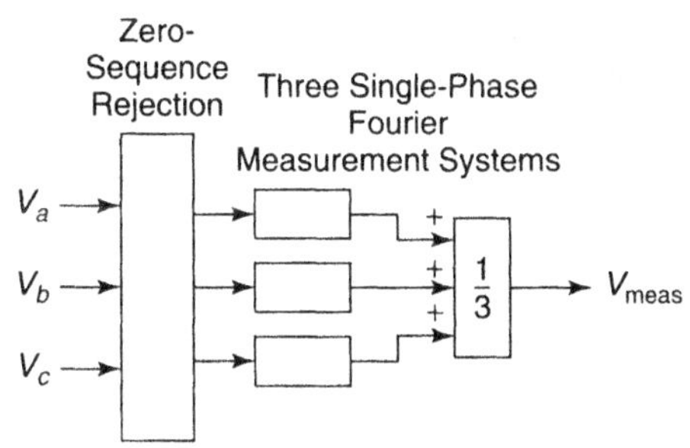

(b)

Figure 4.4 (a) The principle of operation of a single-phase Fourier measurement system and (b) a 3-phase Fourier measurement system.

$$a(\tau) = \frac{2}{T_0} \int_{\tau - T_0}^{\tau} v(t) \sin(2\pi f_0 t)\, dt \tag{4.4}$$

$$b(\tau) = \frac{2}{T_0} \int_{\tau - T_0}^{\tau} v(t) \cos(2\pi f_0 t)\, dt \tag{4.5}$$

A 3-phase measurement system is realized by interconnecting three single-phase measurement systems, as illustrated in Fig. 4.4(b). The output of the 3-phase measurement system depends on the phase relationship of the fundamental components of the three single-phase voltages. Generally, the voltage in each phase is composed of all three—the positive-, negative-, and zero-sequence components. The zero-sequence component is usually eliminated in measurement systems either through filtering or by choosing line-to-line voltage instead of the line-to-neutral voltages as inputs.

4.2.1.4 The Measurement of Squared Voltage In very special cases, to prevent instability of the 3rd harmonic (see Section 5.6), it may be advantageous to generate a signal at the output of the measurement system that is proportional to the square-of-the-bus voltage [28]. Such a measurement is recommended if independent phase control is employed using a phase-locked oscillator in either a 6-pulse or 12-pulse scheme and when the SVC capacitors are resonant with the ac system near the 3rd harmonic. The measurement system in this case comprises a multiplier circuit instead of a rectifier circuit. For a normal bus voltage of $\sqrt{2}V_1 \sin 2\pi f_0 t$, the output of the multiplier is $2V_1^2 \sin^2 2\pi f_0 t$, which is equivalent to

$$2V_1^2 \sin^2 2\pi f_0 t = V_1^2(1 - \cos 4\pi f_0 t) \tag{4.6}$$

If a phase-locked loop (PLL) voltage regulator (see Section 4.3.2) is used, then the PLL will ignore the 2nd harmonic components and the controller will respond to V_1^2. Should there be a 3rd harmonic component in the bus voltage, the dc output of the measurement system (multiplier) will be proportional to $(V_1^2 + V_3^2)$ in each phase. Thus, this feedback signal is unaffected by the phases of 3rd harmonic components relative to the fundamental-frequency signals. The three different TCR phases respond symmetrically to the total rms voltage and do not augment the preexisting 3rd harmonic currents in the system, therefore obviating 3rd harmonic instability.

With squared voltage, or $(\text{voltage})^2$, feedback, the reference voltage input could also be squared to achieve proportionality between the bus voltage and reference voltage.

4.2.2 The Demodulation Effect of SVC Voltage-Measurement System

The SVC voltage-measurement systems have an inherent demodulation effect that plays a critical role in the excitation of different types of harmonic instabilities [7], [17], [18], [29], [30]. These instabilities are caused by the adverse interaction between the SVC control system and the network resonances, transformer saturation, or series capacitors, as discussed in Chapter 5.

To understand the demodulation effect, we need to review some basic concepts [17]. There are two ways in which nonfundamental frequencies caused by system disturbances get integrated with the fundamental voltage and current waveforms. These additional excited frequencies either add to or modulate the fundamental-voltage wave.

4.2.2.1 Addition The phenomenon of addition of subsynchronous components to the fundamental frequency is explained here in the context of a series-compensated network, as depicted in Fig. 4.5. The R_s and L_s represent the line resistance and inductance, respectively, whereas C_s represents the series capacitance. The line is also shunt-compensated by an inductor having parameters R_p and L_p. A disturbance is simulated by closing the switch, thereby short-circuiting the shunt admittance. The equivalent impedance of the network, as viewed from the switch, exhibits a zero (minimum) impedance at frequency f_s:

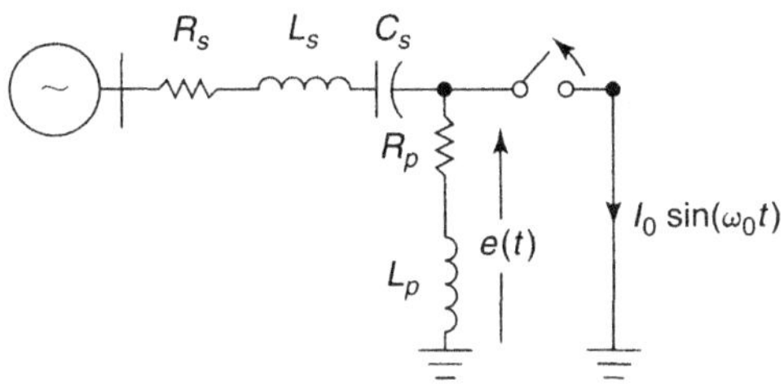

Figure 4.5 Fault clearing in a series-compensated network with a shunt reactor.

$$f_s = \frac{1}{2\pi\sqrt{L_s C_s}} \tag{4.7}$$

The network impedance also shows a pole (peak impedance) at frequency f_1

$$f_1 = \frac{1}{2\pi\sqrt{(L_s + L_p)C_s}} \tag{4.8}$$

As the shunt inductance L_p is usually very large compared to the series inductance L_s, the frequency f_1 can be approximated by

$$f_1 \approx \frac{1}{2\pi\sqrt{L_p C_s}} \tag{4.9}$$

If the applied fault is subsequently cleared the bus voltage, $e(t)$ is given by

$$e(t) = E_0 \cos(2\pi f_0 t) + E_1 e^{-(t/\tau_1)} \cos(2\pi f_1 t) \tag{4.10}$$

where

$$E_0 = X_0 I_0 \frac{\left(1 - \frac{f_s}{f_0}\right)^2}{\left(1 - \frac{f_1}{f_0}\right)^2}$$ = the magnitude of the fundamental-voltage component

$$E_1 = X_0 I_0 \frac{\left(\frac{f_s}{f_1}\right)^2 - 1}{\left(\frac{f_0}{f_1}\right)^2 - 1} = \text{the magnitude of subsynchronous-voltage}$$

component

$$X_0 = \frac{L_s L_p}{L_s + L_p}(2\pi f_0) = \text{the equivalent parallel reactance of } L_s \text{ and } L_p \text{ at}$$

fundamental frequency

$$\tau_1 = 2\frac{L_s + L_p}{R_s + R_p} = \text{the decay-time constant of the subsynchronous component}$$

An estimate of the magnitude of different quantities can be obtained from the following realistic example [17]: Let the system have a short-circuit level of 10,000 MVA at 735 kV when it is series-compensated by 30%, and furthermore, let the X/R ratio of the system reactance be 30 and that of the 330-MVAR shunt inductor be 400 at 60 Hz. Using the preceding equations, it can be computed that a subsynchronous-frequency component magnitude $E_1 = 0.4$ pu and frequency $f_1 = 7$ Hz, with a decay-time constant $\tau_1 = 1.37$ s added to the fundamental component. The current and voltage waveforms corresponding to this case are depicted in Fig. 4.6. It is seen that the short-circuit current is composed of a 32.8-Hz component corresponding to the series-resonance mode frequency, f_s, that is given by Eq. (4.7). Similarly, voltages and currents of different frequencies that correspond to various other excited modes also get added to the fundamental.

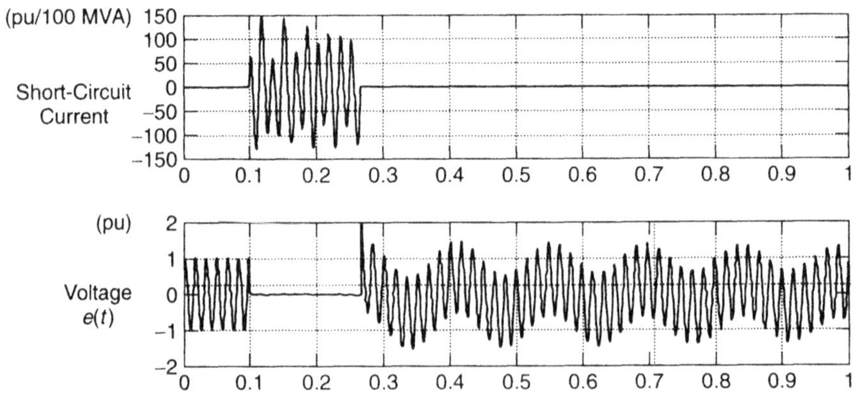

Figure 4.6 Current and voltage waveforms at fault application and clearing on the circuit.

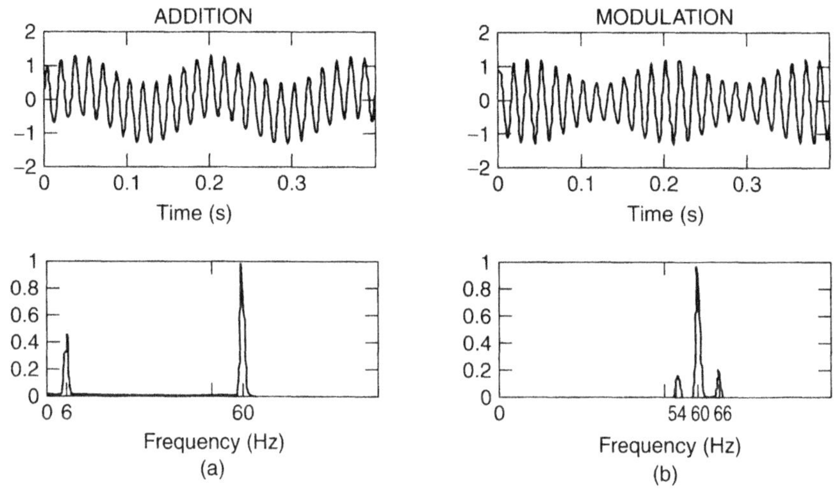

Figure 4.7 Time- and frequency-domain representation of a voltage: (a) 0.5 pu, 6 Hz added to a fundamental of 1 pu, 60 Hz and (b) 1-pu fundamental modulated by a 0.5-pu, 6-Hz voltage.

4.2.2.2 Modulation The phenomenon of modulation occurs under different circumstances, such as during a power swing when the magnitude of the fundamental voltage varies with the frequency of the modulating signal. A fundamental-frequency voltage undergoing modulation can be expressed as

$$v(t) = [V_0 + V_1 \sin(2\pi f_1 t + \varphi_1)] \sin 2\pi f_0 t \tag{4.11}$$

Equation (4.11) can be rewritten as

$$v(t) = V_0 \sin 2\pi f_0 t + \frac{V_1}{2} \cos[2\pi(f_0 - f_1)t - \varphi_1] - \frac{V_1}{2} \cos[2\pi(f_0 + f_1)t + \varphi_1] \tag{4.12}$$

It is seen that modulation results in two sideband frequencies, $f_0 - f_1$ and $f_0 + f_1$ besides the fundamental frequency f_0.

A comparative study of the addition and modulation phenomenon—together with their corresponding frequency spectra, as obtained from a fast Fourier transform analysis—is presented in Fig. 4.7.

4.2.2.3 The Fourier Analysis–Based Measurement System Let the voltage to be measured be given by

$$v(t) = V_0 \sin(2\pi f_0 t + \varphi_0) + V_1 \sin(2\pi f_1 t + \varphi_1) \tag{4.13}$$

A single-phase Fourier analysis–based measurement system will generate an output

$$V_{\text{meas}}(t) = V_0 + V_d \cos(2\pi f_d t + \alpha - \varphi_0) + V_s \cos(-2\pi f_s t - \alpha - \varphi_0) \quad (4.14)$$

where

$$V_d = \frac{V_1 \sin\left(\frac{\pi f_1}{f_0}\right)}{\pi\left(1 - \frac{f_1}{f_0}\right)} \quad (4.15)$$

$$V_s = \frac{V_1 \sin\left(\frac{\pi f_1}{f_0}\right)}{\pi\left(1 + \frac{f_1}{f_0}\right)} \quad (4.16)$$

$$\alpha = \varphi_1 + \pi\left(1 - \frac{f_1}{f_0}\right) \quad (4.17)$$

$$f_d = f_0 - f_1 \quad (4.18)$$

$$f_s = f_0 + f_1 \quad (4.19)$$

It is seen from these equations that when a frequency component f_1 is added to the fundamental frequency f_0, the output comprises the following components:

1. A constant dc voltage V_0 representing the magnitude of the fundamental component.
2. A sinuosidal time-varying voltage V_d of frequency $f_d = f_0 - f_1$.
3. A sinusoidal time-varying voltage V_s of frequency $f_s = f_0 + f_1$.

To exemplify the preceding list, if a 0.25-pu, 6-Hz voltage is added to a 1-pu, 60-Hz fundamental, the output is composed of the following components:

1. A dc voltage $V_0 = 1$ pu.
2. An ac voltage $V_d = 0.027$ pu of frequency $f_d = 60 - 6 = 54$ Hz.
3. An ac voltage $V_s = 0.022$ pu of frequency $f_s = 60 + 6 = 66$ Hz.

The input and output signals of the single-phase measurement system, along with their frequency spectra, are shown in Fig. 4.8.

In a 3-phase Fourier measurement system, if the 3-phase voltages are balanced the output is obtained as a vectorial addition of the three input voltages, as follows:

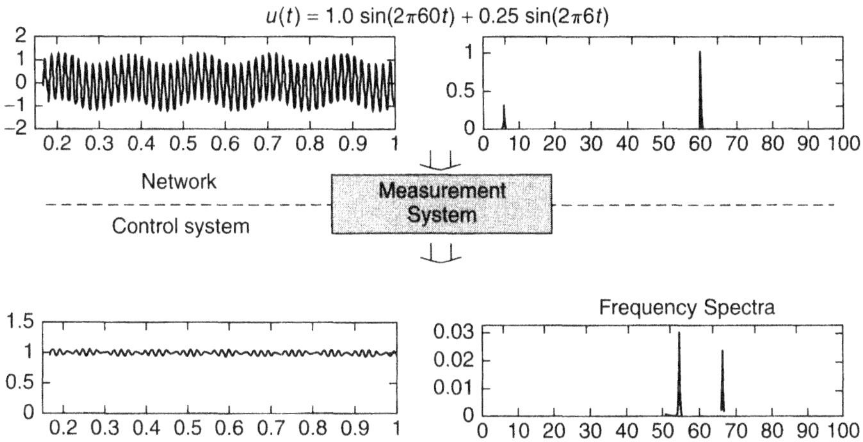

Figure 4.8 The measurement technique and frequency spectrum obtained by using Fourier analysis.

$$V_{\text{meas}} = \tfrac{1}{3}(V_{a\,\text{meas}} + V_{b\,\text{meas}} + V_{c\,\text{meas}}) \tag{4.20}$$

The frequency components in V_{meas} are determined by the phase sequence of the added frequency component f_1 Eq. (4.13), as given in the following table:

Sequence of Added Frequency, f_1	Generated Frequency	Output, V_{meas}
Positive	$f_d = f_0 - f_1$	$V_0 + V_d \sin(\omega_d t + \varphi_d)$
Negative	$f_s = f_0 + f_1$	$V_0 + V_s \sin(\omega_s t + \varphi_s)$
Zero	—	V_0

where V_d and V_s are given by Eqs. (4.15) and (4.16), and the phase shifts φ_d and φ_s are, respectively, obtained as

$$\varphi_d = \varphi_1 - \varphi_0 + \pi\left(\frac{3}{2} - \frac{f_1}{f_0}\right) \tag{4.21}$$

$$\varphi_s = -\varphi_1 - \varphi_0 - \pi\left(\frac{1}{2} - \frac{f_1}{f_0}\right) \tag{4.22}$$

which shows that if the frequency f_1 in Eq. (4.13) has a positive-phase sequence, only the sideband $f_0 - f_1$ is produced. On the other hand, if f_1 has a negative-phase sequence, then only the sideband $f_0 + f_1$ will only be generated. Usually both positive- and negative-sequence components of frequency f_1 are present at the input; therefore, a linear combination of both sideband frequencies $f_0 - f_1$ and $f_0 + f_1$ will be produced at the output.

4.2.2.4 Coordinate Transformation–Based Measurement Systems Measurement systems such as those based on coordinate transformation, which compute the rms value by squaring the input voltage, also produce additional frequencies, as described here. Let the input voltage be expressed as

$$v(t) = V_0 \sin 2\pi f_0 t + V_1 \sin 2\pi f_1 t \tag{4.23}$$

Then

$$\begin{aligned} v^2(t) &= V_0^2 \sin^2 2\pi f_0 t + V_1^2 \sin^2 2\pi f_1 t + 2V_0 V_1 \sin 2\pi f_0 t \sin 2\pi f_1 t \\ &= \frac{V_0^2}{2}(1 - \cos 4\pi f_0 t) + \frac{V_1^2}{2}(1 - \cos 4\pi f_1 t) \\ &\quad + V_0 V_1 [\cos 2\pi (f_0 - f_1)t - \cos 2\pi (f_0 + f_1)t] \end{aligned} \tag{4.24}$$

If f_0 and f_1 are 60 Hz and 6 Hz, respectively, then the output of the measurement system will contain, in addition to dc, the frequency components 12 Hz ($2f_1$), 54 Hz ($f_0 - f_1$), 66 Hz ($f_0 + f_1$), and 120 Hz ($2f_0$). These measurement systems therefore produce not only the sideband frequencies $f_0 - f_1$ and $f_0 + f_1$ but also the frequencies $2f_1$ and $2f_2$.

4.2.2.5 ac/dc Rectification–Based Measurement Systems Measurement systems involving ac/dc rectification are nonlinear in nature and hence produce the frequencies $2f_1$ in addition to the sideband frequencies $f_0 + f_1$ and $f_0 - f_1$. A comparison of the Fourier analysis–based and rectification-based measurement systems is presented in Fig. 4.9. The rectification system is also shown to introduce an error in the computation of the mean value of the input voltage.

4.2.2.6 Filtering Requirement Any disturbance in the power system excites all the natural modes of oscillation in the network. Corresponding to each single resonant frequency f_1, the measurement systems will generally produce sideband frequencies $f_0 + f_1$ and $f_0 - f_1$ on the dc side. However, measurement systems based on rms voltage computation or rectification will generate two additional frequency components: $2f_1$ and $2f_0$. The frequencies that lie within or close to the bandwidth of the voltage regulator (typically 30 Hz) appear as superimposed oscillations in the susceptance-order signal B_{ref} and modulate the TCR currents. If these frequency components are not filtered, they tend to reinforce the natural oscillation modes, resulting in different forms of resonance and instabilities (see Chapter 5).

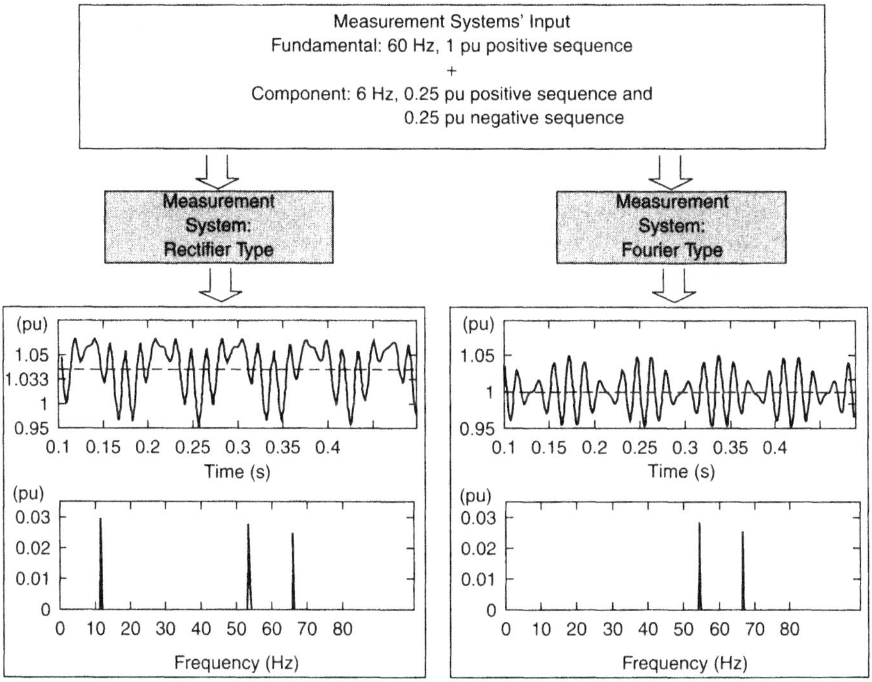

Figure 4.9 The comparison between two types of measurement systems: rectifier and Fourier.

For instance, in systems without series compensation, a network-resonant pole, typically at 85 Hz, excites an 85-Hz oscillation that is superimposed on the fundamental 60-Hz voltage wave. The measurement system then generates the frequencies 25 Hz ($f_1 - f_0$), 120 Hz ($2f_0$), 145 Hz ($f_0 + f_1$), and 170 Hz ($2f_1$) in addition to dc at its output. The frequencies of 145 Hz and 170 Hz lie well outside the bandwidth of the voltage regulator and hence are not significant, but the 25-Hz frequency component may lead to harmonic instability. This instability can be avoided by installing band-reject (i.e., notch) filters on the ac side of the voltage transducer. These filters are tuned to the critical network-resonant modes, which typically lie within 80–100 Hz.

The choice of installing the filter on the ac side or dc side of the measurement system is based on its effect on the phase margin of the system. It is found that a band-reject filter on the ac side reduces the phase margin to a much lesser extent than a corresponding filter on the dc side and is therefore used widely. The filter is designed to cause a minimum phase lag at 60 Hz to not affect the normal SVC response. For example, in the Hydro-Quebec James Bay system, a combination of notch filters (80 Hz and 96 Hz) are provided [18]. If in any ac system the first network-resonant frequency lies beyond $2f_0$ (120 Hz), these ac notch filters are not required.

Because of the saturation of the SVC coupling transformer, sometimes 2nd harmonic currents are injected into the system, causing the inception of 2nd harmonic voltage oscillations. This 120-Hz (f_1) voltage component gets translated to 60 Hz ($f_1 - f_0$), 120 Hz ($2f_0$), 180 Hz ($f_0 + f_1$), and 240 Hz ($2f_1$) on the dc side of the voltage transducer. The 60-Hz frequency component in the control loop tends to augment the original 120-Hz oscillation in the system, leading to 2nd harmonic instability (see Section 5.4). To mitigate this effect, a notch filter tuned to 60 Hz is connected to the dc side.

Provided on the dc side of the voltage transducer are two additional notch filters: one tuned to 120 Hz, the other to 360 Hz. It may be recalled that the 360-Hz ripple is caused by the 6-pulse rectification process. There is a need to filter the 120-Hz frequency component; otherwise, it may generate a 3rd harmonic sideband-frequency component in the SVC bus voltage. The other high-frequency components, arising on the dc side and caused by the demodulation effect of the voltage transducers, are eliminated through a low-pass filter having a time constant in the range of 1–8 ms.

Furthermore, in series-compensated ac networks, the interaction between the series capacitors and shunt reactors results in the excitation of the shunt-reactor mode having a frequency in the range of several to 20 Hz. This mode is decisive in determining the controller stability. A much-improved SVC response is obtained by filtering this mode and is achieved by placing a high-pass filter with a break frequency of about 25 Hz on the ac side.

The total arrangement of filters in the voltage measurement system and the typical characteristics of the filters within the context of the Hydro-Quebec James Bay system SVC [18] are depicted in Fig. 4.10.

4.2.3 Current Measurement

Current measurement is needed for implementing different features, including the following:

1. the slope in the SVC steady-state characteristic;
2. the current limit;
3. any protection strategies; and
4. any balancing schemes.

The SVC current is stepped down and processed using transducers and filters similar to those of voltage-measurement systems.

Although a normal ac current transformer (CT) might seem to be the appropriate choice for stepping down the current and also for providing isolation between the power and control systems, this is indeed not so. The reason is that it tends to saturate quickly when the thyristor valves fire asymmetrically in the event of major system disturbances. One option is to use a zero-flux CT [31], although this is quite expensive.

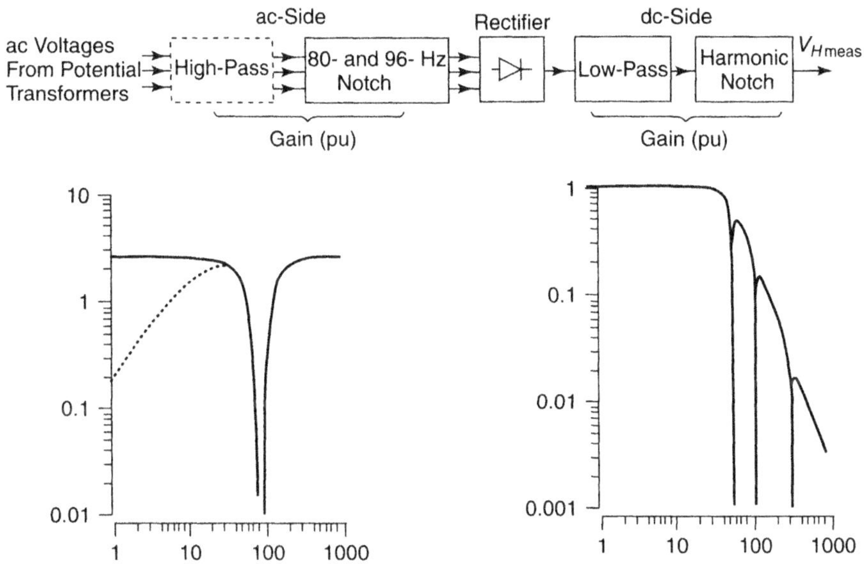

Figure 4.10 The filtering of measurement voltages (solid lines represent existing filters; dashed lines represent additional filters needed for series-compensated ac lines).

A better option is to use a differentiating-current transformer (DCT) [28], which is an air-gapped-current transformer that supplies a voltage output with an extremely high burden resistance. An application of DCT is illustrated in Fig. 4.11, in which the di/dt signal is integrated to reconstruct the current signal and the integrator is reset at every firing instant when the thyristor valves are just commencing to conduct. An advantageous feature of this application is that any dc component in the actual current can be faithfully generated in the reconstructed signal. To prevent amplification of the fast interference pulses, a small despiking-lag circuit of time constant 200 μs and a metal-oxide varistor (MOV) are connected at the output of DCT.

In many control applications, such as load compensation and voltage balancing, the control strategies require the phasor quantities of positive- and negative-sequence currents and voltages. These quantities can be obtained by suitably transforming the instantaneous values of the currents or voltages. A general procedure [32] for deriving the phasor components of the kth harmonic is described in the following text.

Let a sinusoidal signal $x(t)$ be represented by Fourier series as

$$x(t) = \frac{C_0}{2} + \sum_{n=1}^{\infty} C_n \cos(2\pi n f_0 t + \varphi_n) \tag{4.25}$$

The kth harmonic phasor is then described by

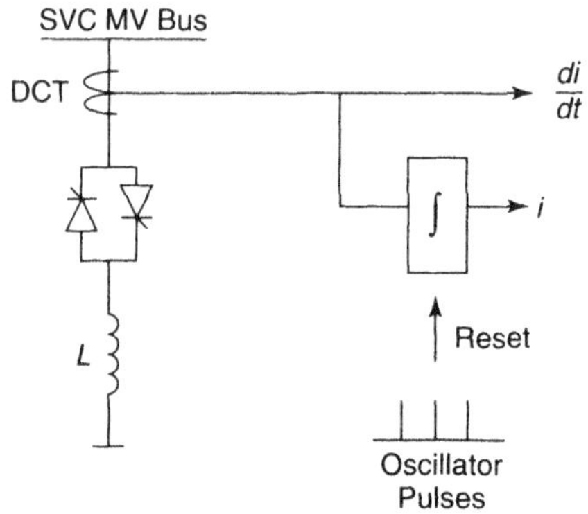

Figure 4.11 The TCR current measurement using a differentiating-current transformer (DCT).

$$C_k = c_k \angle \varphi_k = c_k(\cos \varphi_k + \sin \varphi_k) \tag{4.26}$$

Multiplying $x(t)$ by a sinusoidal signal $\cos 2\pi f_r t$, where $f_r = k f_0$ and $k \in n$, results in

$$x(t)\cos(2\pi f_r t) = \frac{c_k}{2}\cos(\varphi_k) + \sum_{m=1, m\neq k}^{\infty} d_m \cos(2\pi m f_0 t + \gamma_m) \tag{4.27}$$

The first term on the right-hand side of Eq. (4.27) represents a dc term, whereas the second term corresponds to a series of sinusoids of frequency $m f_0 = (n \pm k) f_0$. The magnitude and phase of these sinusoids are denoted by d_m and γ_m, respectively. It follows from Eq. (4.22) that

$$2x(t)\cos(2\pi f_r t) = \text{Re}\{C_k\} + \text{sinusoids} \tag{4.28}$$

Also

$$2x(t)\sin(2\pi f_r t) = -\text{Im}\{C_k\} + \text{sinusoids} \tag{4.29}$$

where $\text{Re}\{C_k\}$ and $\text{Im}\{C_k\}$ represent the real and imaginary components of the kth harmonic phasor of signal $x(t)$. The sinusoids in Eqs. (4.28) and (4.29) can be eliminated by an averaging filter, which averages over one time period of the smallest $(n \pm k) f_0$ frequency.

If $x(t)$ corresponds to the α-axis current, $i_\alpha(t)$, then the kth harmonic phasor, I_α, is given by

$$\begin{aligned} I_\alpha &= \text{Re}[I_\alpha] + j\,\text{Im}[I_\alpha] \\ &= 2i_\alpha(t)\cos(2\pi f_r t) - j2i_\alpha(t)\sin(2\pi f_r t) \end{aligned} \tag{4.30}$$

Equation (4.30) is strictly valid only after filtering out the sinusoids of Eqs. (4.28) and (4.29). A similar expression, however, can be obtained for the kth harmonic phasor, I_β, that corresponds to the β-axis current, $i_\beta(t)$, as discussed in the text that follows.

The symmetrical component phasors I_1 and I_2 are derived using the following phasor transformation:

$$\begin{bmatrix} I_1 \\ I_2 \end{bmatrix} = \frac{1}{2} \begin{bmatrix} 1 & -j \\ 1 & j \end{bmatrix} \begin{bmatrix} I_\alpha \\ I_\beta \end{bmatrix} \tag{4.31}$$

The real and imaginary components of the kth harmonic positive-sequence phasor are then given by

$$\begin{bmatrix} \mathrm{Re}\{I_1\} \\ \mathrm{Im}\{I_1\} \end{bmatrix} = \begin{bmatrix} \cos(2\pi f_r t) & \sin(2\pi f_r t) \\ -\sin(2\pi f_r t) & \cos(2\pi f_r t) \end{bmatrix} \begin{bmatrix} i_\alpha(t) \\ i_\beta(t) \end{bmatrix} \tag{4.32}$$

The corresponding negative-sequence phasor quantities are

$$\begin{bmatrix} \mathrm{Re}\{I_2\} \\ \mathrm{Im}\{I_2\} \end{bmatrix} = \begin{bmatrix} \cos(2\pi f_r t) & -\sin(2\pi f_r t) \\ -\sin(2\pi f_r t) & -\cos(2\pi f_r t) \end{bmatrix} \begin{bmatrix} i_\alpha(t) \\ i_\beta(t) \end{bmatrix} \tag{4.33}$$

The α, β components $i_\alpha(t)$, $i_\beta(t)$ are obtained from the 3-phase quantities $i_\alpha(t)$, $i_b(t)$, and $i_c(t)$ using the transformation given by Eq. (4.1). Implementation of this measurement scheme is described in ref. [32].

4.2.4 Power Measurement

Conventional PTs and CTs are used to obtain the line-voltage and -current signals, respectively. These are subsequently multiplied using multiplier circuits to compute the active and reactive power (see Fig. 4.2). One of the commonly employed methods for measuring reactive power Q_a in phase a is based on

$$Q_a = i_a(v_b - v_c) \tag{4.34}$$

where i_a = the current in phase a and v_b, v_c = the voltages across phases b and c, respectively.

An analog pulse ratio–modulated (PRM) scheme for measuring real and reactive power, which is employed by the Bonneville Power Administration (BPA) [16], [33], is depicted in Fig. 4.12. The width of the voltage-input wave, V_{in}, modulates a sequence of square pulses using a triangular-wave generator. This pulse width–modulated (PWM) signal is subsequently amplitude-modulated (AM) by the current signal, I_{in}. The combined process of PWM followed

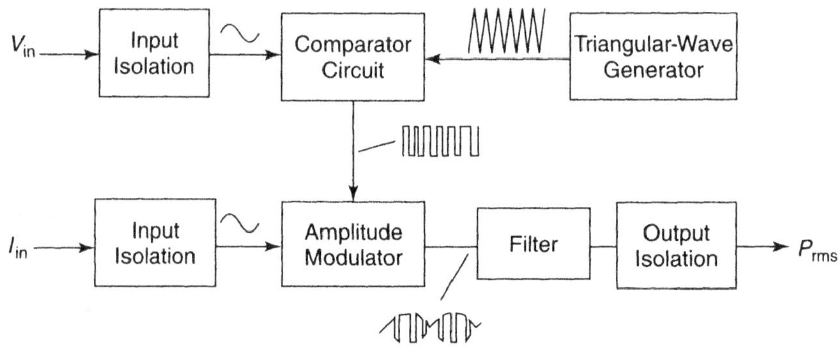

Figure 4.12 The general architecture for a pulse-ratio–modulated (PRM) megawatt transducer.

by AM effectively results in analog multiplication and generates a PRM signal that is filtered to give the rms power, P_{rms}. In fact, the PRM signals of the three phases are added together before the filtering stage, which minimizes the filtering requirements.

In digital-control schemes—fast becoming integral components of all modern SVC installations—the analog signals available at the output of instrument transformers are converted to digital signals using A/D (analog-to-digital) converters for subsequent processing. The necessary filtering is accomplished by digital filters in software.

4.2.5 The Requirements of Measurement Systems

The SVC controller performance and its robustness largely depend on the authenticity of the measured signals. However, the measurement units, together with their transducers, operate in environments that cannot be predicted from simulation studies of power-system component models but can only be witnessed in the field. The voltage, current, power, and frequency transducers can potentially produce spurious outputs from the effects of extraneous dynamics present in the input signals. (These likely dynamics are listed in Table 4.1 and discussed in refs. [16] and [34].) The transducers thus have the onerous task of retrieving desired information from a combination of the following amplitude- and frequency-modulated signals:

1. modulated harmonics of 50/60 Hz;
2. extraneous-current frequencies that may not be of 50/60–Hz harmonics;
3. modulated extraneous-carrier frequencies; and
4. any added transients.

A graphical illustration of the different possible inputs to an rms (root mean

TABLE 4.1 Extraneous Dynamics in the Measurement Systems

Dynamic Activity	Frequency Range (Hz)
Torsional oscillation	5–120
Transient torque	5–50
Turbine-blade vibration	80–250
Fast-bus transfer	1–1000
Controller interaction	10–30
Harmonic interaction and resonance	60–600
Ferroresonance	1–1000
Network resonance	10–300

square) transducer is presented in Fig. 4.13 [16]. Another aspect to be considered is that the 3-phase, 50/60-Hz carrier-frequency signals may not be balanced. Hence the sequence components of these signals may need to be evaluated and adequately filtered according to the expectations of the particular application. The rms transducers also require to be designed with appropriate bandwidths to eliminate the undesirable frequency signals. They must possess high resolution, accuracy, and reliability, yet they should be cost-effective. The following list gives the other desirable features sought in the next generation of transducers/measurement systems:

1. programmable outputs for increased versatility;
2. effective networking options with transducers of other controllers in the system, both at local-area and wide-area level; and
3. possible synchronization of the measurements with a precise global external reference.

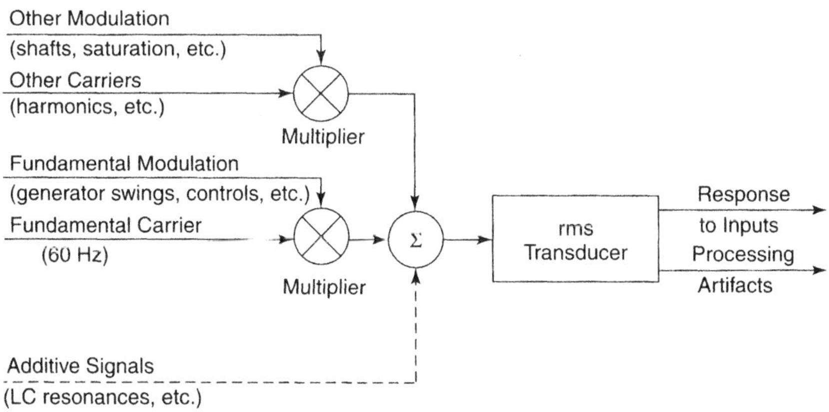

Figure 4.13 The input signals of an rms transducer in a real-power system environment.

It is anticipated that the future technology of measurement systems will be mostly digital in nature, as it is quite capable of meeting the aforementioned requirements.

4.2.5.1 Phasor Transducers Although the various measurement systems described in previous sections are based on analog techniques, measurement systems using digital transducers based on phasor computations are fast emerging and have already been employed in field. Extensive experience with such phasor transducers has been acquired by the BPA in its Wide-Area Measurement Systems Project [35]. The phasor transducers essentially determine the projections of voltage, $v(t)$, and current, $i(t)$, along two reference waveforms—sine reference and cosine reference, as indicated in Fig. 4.14 [16]. These projections constitute the voltage and current phasors that are used for further computation of different variables, such as real powers, reactive powers, and frequency. The digital phasor–based measurement systems allow networking with other measurement units in the system and can produce measurements synchronized with a global-external reference [35], [36]. The advantage of this synchronization scheme is that all phasors in the network provide consistent angle information.

4.2.5.2 Optical Sensors Presently used measurement units are based on traditional instrument transformers, such as PTs, CTs, and constant-voltage transformers (CVTs). Recently, however, optical sensors have become available [37], [38], offering the following advantages:

1. reduced size and weight;
2. electrical isolation;
3. the avoidance of difficulties associated with oil-filled transformers;
4. a lack of need for secondary electrical wirings in substations;
5. the elimination of the instrumentation's transformer-burden requirement;
6. a wide bandwidth and dynamic range; and
7. high accuracy and compatibility with digital-measurement systems.

The optical-sensors technology is promising, and it is expected to find wide use in future applications.

4.3 THE VOLTAGE REGULATOR

4.3.1 The Basic Regulator

The SVC voltage regulator processes the measured system variables and generates an output signal that is proportional to the desired reactive-power compensation. Different control variables and transfer functions of the voltage

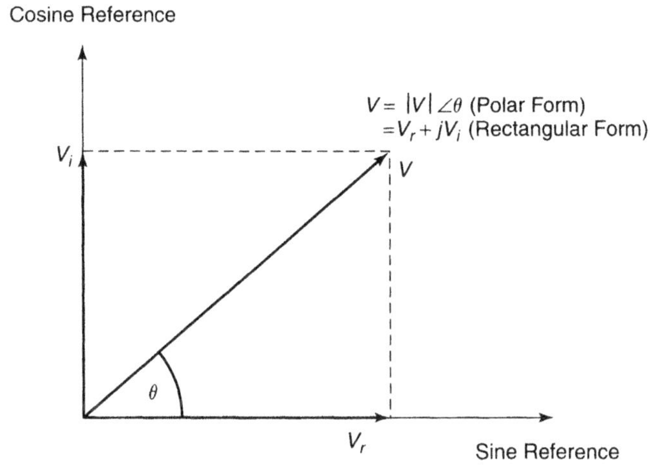

Figure 4.14 The phasor determination via projections.

regulator are used, depending on the specific SVC application. The measured control variables are compared with a reference signal, usually V_{ref}, and an error signal is input to the controller transfer function. The output of the controller is a per-unit susceptance signal B_{ref}, which is generated to reduce the error signal to zero in the steady state. The susceptance signal is subsequently transmitted to the gate pulse–generation circuit.

A small slope or droop (3–5%) is typically incorporated into the steady-state characteristics of SVCs to achieve specific advantages, as discussed in Section 5.2.3. Alternative implementations of this slope in the voltage regulator model are illustrated in Fig. 4.15; the current droop-feedback arrangement is depicted in Fig. 4.15(a). The SVC current is explicitly measured and multiplied by a factor K_{SL} representing current droop before feeding as a signal V_{SL} to the summing junction. The sign of V_{SL} is such that it corresponds to an increase of reference voltage for inductive SVC currents and a decrease of the reference voltage for capacitive SVC currents. Simple integral control finds most common usage in voltage controllers. R_R is termed the *response rate*, which is indicative of the time taken by an SVC to move across its entire reactive-power range, that is, from a fully capacitive to a fully inductive state, in response to a large (1-pu) voltage error.

In certain cases, it may be difficult to faithfully obtain the current signal. This occurs when the SVC is operating close to its floating state, that is, zero MVA reactive power. The current signal then comprises a predominant harmonic component and a fundamental resistive component corresponding to the real losses in SVC. To overcome this problem, in certain SVC controllers the reactive power is computed and fed back instead of using the SVC current. The reactive-power signal is calculated by multiplying the phase currents in SVC by a fundamental-frequency voltage lagging behind the actual phase voltage by 90°.

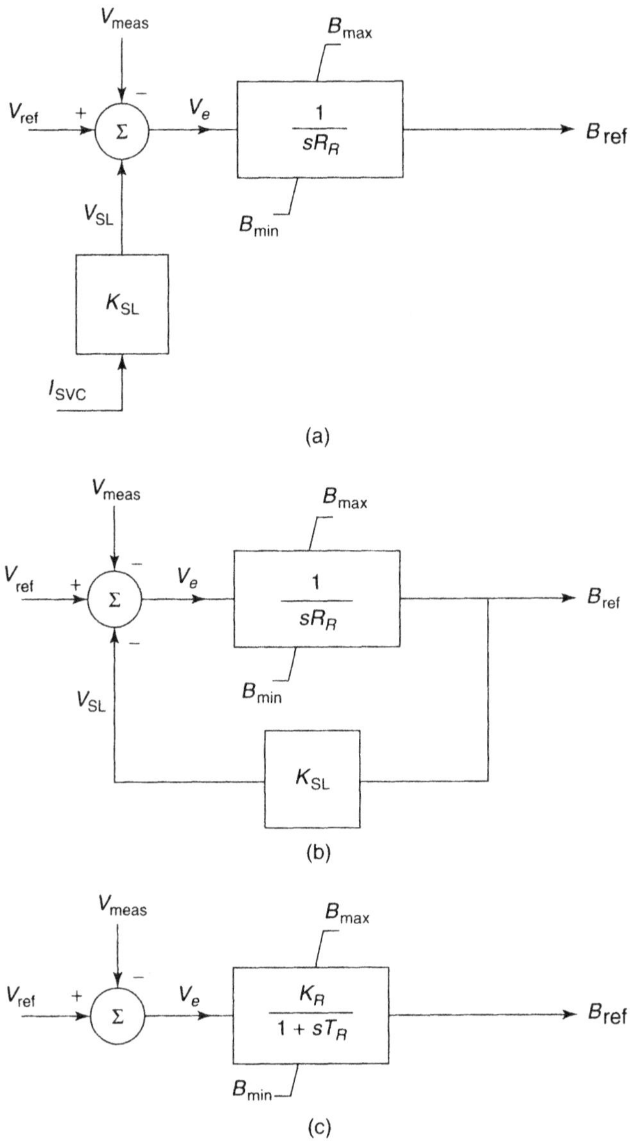

Figure 4.15 Alternative implementations of current droop (slope) in the voltage regulator: (a) an integrator with current-droop feedback; (b) an integrator with susceptance-droop feedback; and (c) a gain–time constant.

The other easily realizable option is the susceptance-droop feedback demonstrated in Fig. 4.15(b). It is implicitly assumed that the SVC bus voltage remains close to 1 pu; thus the SVC current that is strictly equal to $B_{ref} \cdot V_{SVC}$ can be expressed as simply B_{ref}. The closed-loop control system in Fig. 4.15(b) can be simplified to the gain–time-constant form of the controller depicted in Fig. 4.15(c). The gain K_R is termed the *static gain*, which is defined as the inverse of the current droop.

$$K_R = \frac{1}{K_{\text{SL}}} \tag{4.35}$$

Also

$$T_R = \frac{R_R}{K_{\text{SL}}} \tag{4.36}$$

In the foregoing context, the term *transient gain*, K_T, which is representative of the dynamic nature of the voltage regulator, is defined as

$$K_T = \frac{K_R}{T_R} \tag{4.37}$$

The current-droop form of the voltage regulator ensures linearity between SVC terminal voltage and current in the SVC control range. Meanwhile, the susceptance-droop feedback provides linearity between SVC susceptance and voltage that strictly translates to a slightly nonlinear voltage–current relationship in the event of voltage variations. However, the discrepancy between the two representations is not significant, as the droop is usually very small and the SVC terminal voltage variations are generally not large. The dynamic performance of the SVC can thus be modeled by either of the two representations.

The advantage of the current-droop model is that the steady-state characteristic (described by K_R) and the dynamic characteristic (implied by R_R) can be independently specified. In the gain–time-constant model, the transient and steady-state characteristics are interlinked through the parameters K_R and T_R. The gain–time-constant model has been widely used in SVC modeling for SVC studies, although the integrator with the current-droop model represents the physical realization of most installed SVCs. Table 4.2 [7] presents certain example cases demonstrating the relationship between the different parameters mentioned in the preceding text and their typical range of magnitudes.

TABLE 4.2 Typical SVC Voltage-Regulator Parameters

	Gain–Time-Constant Format			Integrator–Droop Format	
Regulation and Speed	K_R (pu)	T_R (s)	K_T (pu/s)	D_R (%)	R_R (ms/pu)
Tight and slow	100	2.0	50	1	20
Tight and fast	100	0.2	500	1	2
Medium and slow	20	0.4	50	5	20
Medium and fast	20	0.04	500	5	2

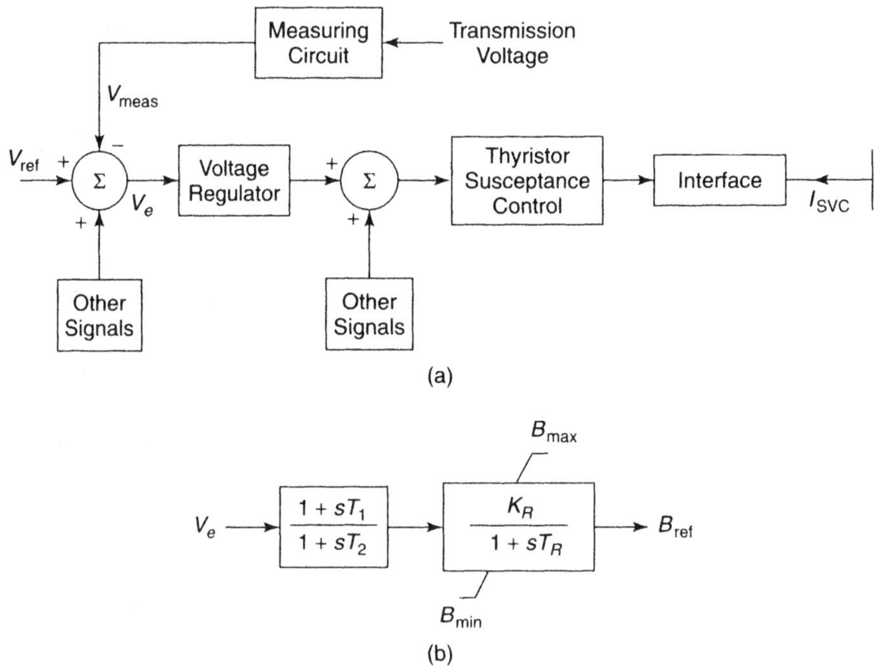

Figure 4.16 (a) The IEEE Basic Model 1 for the SVC control system and (b) the voltage-regulator model.

The IEEE [11] has proposed two basic models for SVC: the IEEE Basic SVC Model 1, which corresponds to the gain–time-constant format, and the IEEE Basic SVC Model 2, which relates to the integrator with current-droop format. Model 1 and its voltage-regulator model are shown in Figs. 4.16(a) and (b), respectively. The gain K_R (inverse of the current slope) is typically between 20 pu (5% slope) and 100 pu (1% slope) on the base of the SVC-rated reactive power. The time constant T_R usually lies between 20 and 150 ms, and the time constants T_1 and T_2 are zero in most cases.

A phase lead can, however, be provided to enhance the damping contribution of the SVC, although this may have a slightly detrimental effect on the synchronizing torque. The lead-lag time constants can also be used to generate sufficient phase and gain margins when large values of static gain are employed. The maximum and minimum limits on the susceptance output B_{ref} are given by B_{max} and B_{min}, respectively. The transfer function of voltage regulator in Model 1 is given by

$$G(s) = \frac{K_R}{1 + sT_R}\left(\frac{1 + sT_1}{1 + sT_2}\right) \tag{4.38}$$

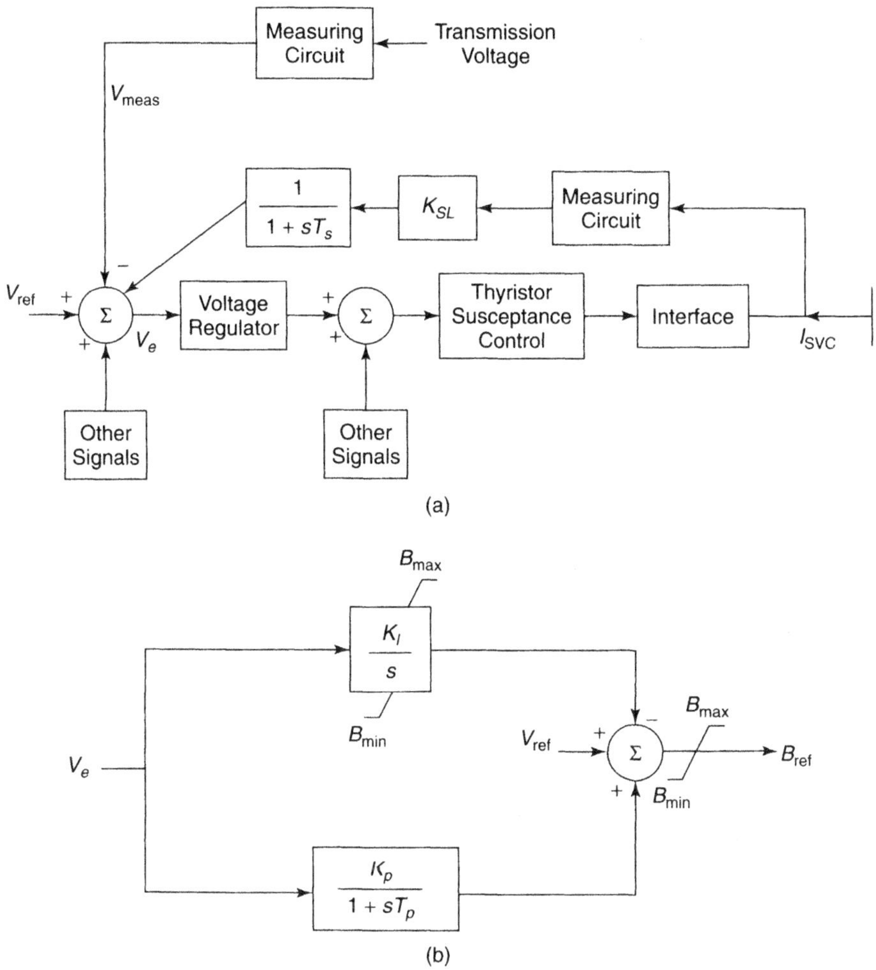

Figure 4.17 (a) The IEEE Basic Model 2 for the SVC control system and (b) the voltage-regulator model.

Model 2 and its corresponding voltage-regulator model are shown in Figs. 4.17(a) and (b), respectively. A proportional gain K_p is employed to increase the speed of response. The voltage regulator can be equivalently expressed as

$$G(s) = \frac{K_I}{s}\left(\frac{1 + sT_Q}{1 + sT_p}\right) \tag{4.39}$$

where

$$T_Q = T_p + \frac{K_p}{K_I} \tag{4.40}$$

Generally, T_P is kept at zero, thereby rendering the controller to be of the simple proportional–integral (PI) type. The integrator in both Models 1 and 2 are of the non-windup type [2].

4.3.2 The Phase-Locked Oscillator (PLO) Voltage Regulator

Phase-locked oscillators (PLOs) are extensively employed in the control of HVDC converters [39], [40]. A similar closed-loop control of SVCs using PLOs was proposed by Ainsworth [28] in 1988 and is now used in many applications. The PLO-based control does not involve any control filter or other lead-lag circuits and provides immunity from instabilities at harmonic frequencies in addition to generating a fast response. The main features of this control are as follows:

1. The feedback signal must possess a mean dc component proportional to the quantity to be controlled.
2. The PLL incorporates the integral control. The TCR firing angle and, consequently, the TCR current are so controlled that in the steady state, the error between the controlled quantity and its reference value becomes zero.
3. The control reacts only to the mean value of the measured signal averaged over half-cycles. It does not respond to harmonics or transient impulses, thereby imparting a strong immunity from the instability at harmonic and other supersynchronous frequencies. However, special measures need to be taken to preclude the 2nd and 3rd harmonic instabilities discussed in Chapter 5.
4. The PLO does not also respond to spurious harmonic components having frequencies that are integral multiples of (system frequency $f_0 \cdot$ pulse number p). Thus a single-phase (2-pulse) PLO will ignore the frequencies $2f_0$, $4f_0$, $6f_0$, and so on, in the control signal. Similarly, a 3-phase, 6-pulse or 3-phase, 12-pulse PLO will ignore harmonic components of frequencies $6pf_0$ or $12pf_0$, respectively, where $p = 1, 2, 3, \ldots$ This special characteristic of the PLO obviates the need for installing any delay-introducing filters into the input-measurement system.
5. The process of integration extends in time to the very firing instant in each half-cycle. The control is thus of a "zero-delay" type, with the only unavoidable delay caused by the half-cycle integrations. A rapid response, combined with a good stability characteristic, is the key feature of the PLO.

4.3.2.1 The Basic Single-Phase Oscillator The principle of operation

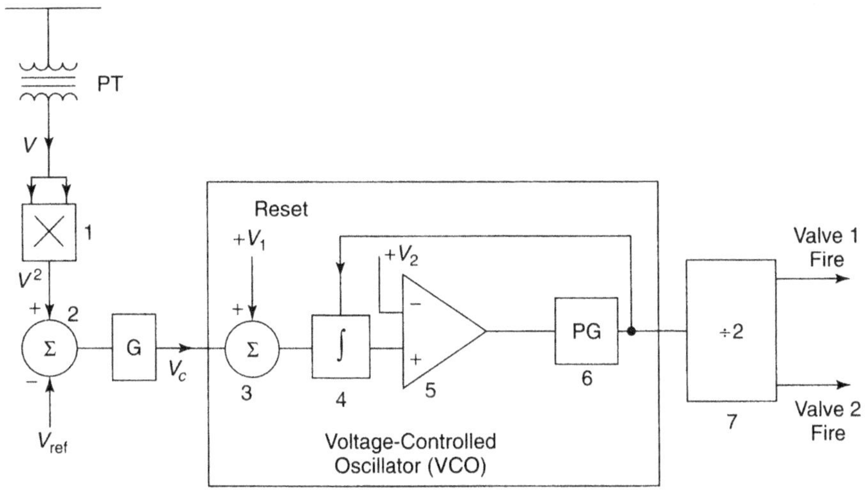

Figure 4.18 Phase-locked oscillator control for a single-phase TCR with (voltage)^2 feedback.

of a PLO in a 3-phase system can be explained based on its control performance in a single-phase (2-pulse) TCR. The control system for a single-phase TCR involving (voltage)^2 feedback is illustrated in Fig. 4.18. The output of this squarer transducer contains a dc component proportional to square of the voltage (in addition to a 2nd harmonic component that is ignored by the PLO). This signal is compared with a reference signal V_{ref}, and the error signal V_c is added to a fixed-bias voltage V_1 in the summer 3 of the figure. The combined signal is integrated, and the output is compared with a constant-voltage V_2 in a comparator. As soon as the integrator output becomes equal to V_2, a short-duration output pulse activates the pulse generator (PG). The PG, acting through the divide-by-two counter, releases two firing pulses spaced 180° apart that are appropriately amplified and transmitted to the anti-parallel–connected set of thyristors. The PLO thus effectively operates at twice the system frequency. The relevant waveforms are depicted in Fig. 4.19. The positive- and negative-cycle TCR-current pulses are denoted by i_1 and i_2; the conduction period is denoted by 2β.

For an integrator-transfer function of K/s, the magnitudes of V_1 and V_2 are chosen to reduce the error voltage V_c to zero in the steady state. The required magnitudes of V_1 and V_2 are obtained from the relation [28]:

$$K_1 V_1 = 2 f_0 V_2 \tag{4.41}$$

where K_1 is the constant.

The PLOs for multiphase SVC systems can be realized in two configurations: individual-phase control and equidistant-pulse control (also known as common-oscillator control).

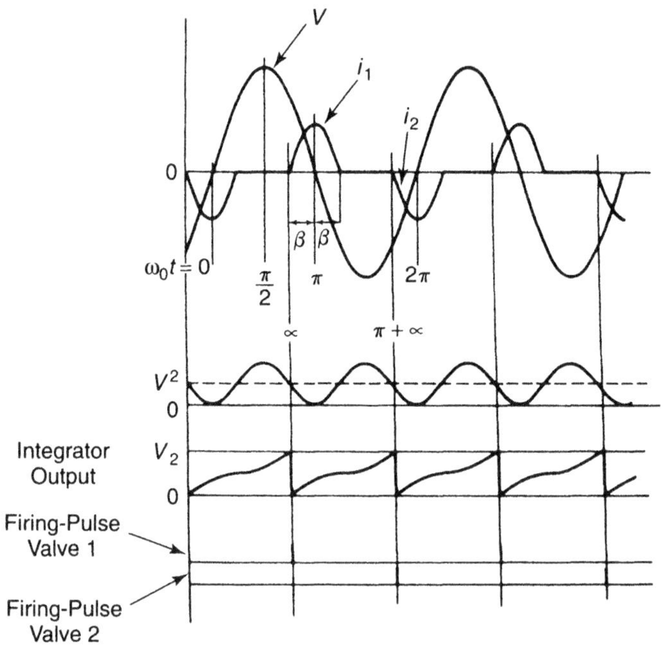

Figure 4.19 Waveforms for a single-phase TCR.

4.3.2.2 The 3-Phase Oscillator

Individual-Phase Control (6-Pulse Operation) This requires three sets of 2-pulse oscillators, one for each phase. Each set operates on the basis of the corresponding phase voltage. In case of load compensation or voltage balancing (where such a scheme is most commonly employed), unequal currents flow in the three TCR phases, resulting in the generation of noncharacteristic harmonics—among which a predominant 3rd harmonic component exists for which adequate filtering must be provided.

Equidistant-Pulse Control (Common-Oscillator Control) Only one common PLO is needed in this configuration. The layout is the same as shown in Fig. 4.18, except that a divide-by-6 or -12 ring counter is employed for the 6- or 12-pulse operation, respectively, instead of a divide-by-2 ring counter. The 6-pulse PLO is impervious to harmonics of order 6, 12, 18, and so on; hence, even if the feedback signal is derived from the output of a 6-pulse rectifier, instead of a (voltage)^2 measurement there is no need to filter out the 6-pulse ripple.

Other closed-loop control applications and features of PLOs are described in ref. [28].

4.3.3 The Digital Implementation of the Voltage Regulator

The dramatic evolution of microprocessor technology in recent times, together with the availability of powerful software tools, has opened up new dimensions in the digital control of power-system controllers, such as FACTS [16], [33]. The digital technology offers the following advantages:

1. Enhanced flexibility and amenability to different system requirements.
2. Control functions synthesized in software. This hardware can implement different control schemes; for instance, controllers with variable parameters, as well as adaptive and nonlinear controller. Controller parameters are not subject to variation with age and environmental conditions, and their performance is completely repeatable.
3. Graphical, friendly, and interactive user interface. Operator interface can be both local and remote.
4. Improved alarm and protection functions.
5. Advanced autodiagnostics, internal supervision, and debugging facilities.
6. Data communication with supervisory controls.
7. On-line monitoring and recording of variables affected by transient phenomena.
8. Ease in testing and commissioning.
9. High resolution and accuracy.
10. Adequate bandwidth for FACTS controller applications, even though it is somewhat limited in extremely fast, complex control loops.
11. Steadily declining costs of digital controllers.

Most modern FACTS installations are equipped with digital controllers. The details of the application of digital control to SVCs in practical installations are provided in refs. [16] and [33]. The SVC controller is implemented on a MACH 2 computer. The high-speed functions, such as the regulator and firing controls, are realized in digital signal processors (DSPs), whereas the other regular features, such as the operator interface, are provided in a main computer's central processing unit (CPU). The other control features of SVCs that are implemented on the main computer are

1. closed-loop PI voltage regulator with variable gain, which responds to the positive-sequence component of the measured voltage;
2. voltage measurement;
3. gain supervisor;
4. gate-pulse generation;
5. power oscillation damper;
6. control of external reactive-power sources;
7. undervoltage-control strategy;

8. sequence control of breakers; and
9. TCR direct-current control.

These functions are explained later in this chapter, as well as in Chapter 5.

4.3.3.1 Digital Control The concept of digital control of SVCs is based on the Z-transform technique [19], [20]. In a digital controller, a new transfer function, $G(z)$, is derived that corresponds to the continuous-transfer function, $G(s)$, which provides the same output at discrete time instants in response to sampled values of the input. The Z transform is subsequently expressed as a difference equation of the general form:

$$Y_n = \sum_{k=0}^{M} A_k X_{n-k} - \sum_{k=1}^{L} B_k Y_{n-k} \tag{4.42}$$

where Y_n = the present output values
Y_{n-1}, Y_{n-2} = the past output values
X_n, X_{n-1}, X_{n-2} = the present and past input values
A_k, B_k = the constant coefficients that are dependent on the poles and zeroes of the continuous-transfer function $G(s)$ and the sampling frequency
L, M = the positive integers that depend on $G(s)$

For instance, let a pure integral SVC controller $G(s)$ be expressed as

$$G(s) = \frac{K}{s} \tag{4.43}$$

The corresponding Z transform is given by

$$G(z) = \frac{KTz^{-1}}{1 - z^{-1}} \tag{4.44}$$

where K = the controller gain
T = the sample period
z^{-1} = a delay of T seconds

The transform $G(z)$ is step-invariant, that is, it has a similar response as $G(s)$ corresponding to a step change in input. Then, $G(z)$ is expressed as a difference equation:

$$Y_n = KTX_{n-1} + Y_{n-1} \tag{4.45}$$

where Y_n = the present output at t seconds
Y_{n-1} = the past output at $t - T$ seconds
X_{n-1} = the past input at $t - T$ seconds

For an SVC controller, X and Y correspond to the error signal V_e and desired susceptance B_{ref}, respectively. Thus, Eq. (4.45) can be rewritten as

$$B_{\text{ref}(n)} = KTV_{e(n-1)} + B_{\text{ref}(n-1)} \tag{4.46}$$

A general difference equation corresponding to a third-order $G(s)$ having three poles is expressed as

$$\begin{aligned} B_{\text{ref}(n)} = {} & K[A_1V_{e(n-1)} + A_2V_{e(n-2)} + A_3V_{e(n-3)}] \\ & + B_1B_{\text{ref}(n-1)} + B_2B_{\text{ref}(n-2)} + B_3B_{\text{ref}(n-3)} \end{aligned} \tag{4.47}$$

4.4 GATE-PULSE GENERATION

The susceptance reference output from the voltage regulator is transmitted to the gate pulse–generation (GPG) unit, which produces appropriate firing pulses for all the thyristor-controlled and thyristor-switched devices of the SVC so that the desired susceptance is effectively made available at the SVC bus to achieve the specified control objectives. In a general TSC–TCR configuration of SVC, the GPG unit performs the following functions [5]:

1. It ascertains the number of TSC branches to be switched in to meet the capacitive susceptance demand and also to allow an excess capacitive susceptance to appear in the SVC.
2. It calculates the magnitude of TCR-inductive susceptance to offset the surplus capacitive susceptance.
3. It determines the sequence in which the TSC connections should be actuated, depending on the existing polarity of charges on the different capacitors, and thereby ensures transient-free capacitor switching.
4. It computes the firing angle for TCR thyristors for implementing the desired TCR-inductive susceptance at the SVC terminals.

Function 1 is achieved by dividing the SVC susceptance reference output from the voltage regulator, B_{ref}, by the susceptance of one capacitor bank, B_C. The quotient rounded off to the next integer corresponds to the number of capacitor banks required, say, n_C. The difference between the total capacitive susceptance n_CB_C and B_{ref} provides the inductive susceptance to be realized by TCR through firing control.

The TSC switching is implemented through a logic circuit [5] that follows two rules for transient-free capacitor switching: that the capacitor bank should be switched on when the voltage across the TSC valves is either zero or at a minimum. The firing pulses for TSC are produced in a similar manner as the TCR, except that they are applied incessantly to the TSC thyristors to maintain continuous conduction. A hysteresis is incorporated in the switching scheme of TSCs to prevent rapid switching (chattering) from the small variations of system voltage around the threshold.

The TCR firing angle α computation corresponding to its calculated susceptance is done using a susceptance-to-firing (B-to-α) angle converter or through a lookup table residing in a microprocessor.

4.4.1 The Linearizing Function

Let us assume that the desired susceptance output, B_{ref}, of the voltage regulator is to be entirely implemented through the TCR, that is, there are no fixed or switchable capacitors. Then the implementation of the voltage-regulator output B_{ref} as an actual installed susceptance B_{SVC} takes place through an intermediate stage of the firing-angle calculation, as shown in Fig. 4.20. Because the relationship between the firing angle α and the susceptance B_{SVC}—expressed as $F_1(\alpha)$—is nonlinear, it necessitates the inclusion of a linearizing function $F_2(\alpha)$ to ensure that

$$F_2(\alpha)F_1(\alpha) = 1 \tag{4.48}$$

where

$$F_2(\alpha) = [F_1(\alpha)]^{-1} \tag{4.49}$$

For the case of a single TCR alone, $F_1(\alpha)$ is expressed as

$$F_1(\alpha) = B_{SVC} = \frac{2\pi - 2\alpha + \sin 2\alpha}{\pi} \tag{4.50}$$

Thus

$$F_2(\alpha) = \frac{\pi}{2\pi - 2\alpha + \sin 2\alpha} \tag{4.51}$$

In physical terms, the function $F_2(\alpha)$ represents the calculation of the firing angle, corresponding to B_{ref} if all of the B_{ref} is to be implemented on the TCR.

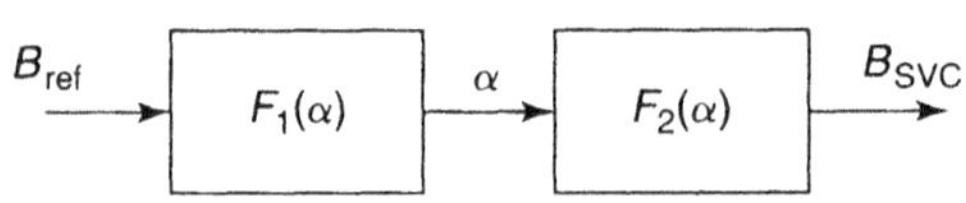

Figure 4.20 The linearization function.

Subsequently, when the TCR is fired with an angle α, the net susceptance B_{SVC} connected to the SVC bus will become equal to B_{ref}.

In a realistic case, however, the SVC will not be a simple TCR but a combination of TCR and either fixed capacitors or switched capacitors on the secondary of the coupling transformer. Furthermore, the number of switched capacitors also varies with time; thus the nonlinearity of function $F_1(\alpha)$ gets compounded. The linearizing function $F_2(\alpha)$ must therefore accommodate all the nonlinearities of the function $F_1(\alpha)$ in addition to any other nonlinearity in the voltage regulator, especially the one caused by susceptance feedback instead of the actual SVC current for implementing the current droop. Sometimes, the linearizing function is inherently embedded in the scheme used for firing-pulse generation, as in the cosine-wave–zero-crossing method [2].

In analog implementations of the GPG, the linearizing function is modeled by a piecewise linear circuit that closely resembles the linearizing function. In digital implementations, it is often realized through a lookup table, as mentioned previously.

The firing pulses are produced in the GPG unit at a low voltage level with respect to the ground potential, but they need to be transmitted to thyristors operating at very high potential levels. The present technology, which ensures isolation between the two widely different potential levels, is to transmit the firing-pulse information through fiber-optic cables.

4.4.2 Delays in the Firing System

4.4.2.1 Thyristor Deadtime The concept of thyristor deadtime in SVC controls [3], [4], [13], [41] can be explained on the basis of a single-phase, 2-pulse TCR in which the firing angle can change from 90° to 180°. In such a TCR, the firing angle can only be varied once in each half-cycle; when the conduction starts in either half-cycle, any change in the firing angle of the same thyristor will not have any effect, impling that the desired firing-angle signal, α_{order}, may be sampled just twice in each cycle just before the positive- and negative-voltage peaks. The sampling frequency thus needs to be no faster than twice the fundamental frequency. If the α_{order} changes to a new value just before the positive-voltage peak, it can be implemented instantaneously with a zero delay in the forward thyristor of the anti-parallel–connected pair. However, if the update in α_{order} occurs, in the worst case—just after the positive peak of the voltage across TCR—there will be a time delay of a half-cycle until this update is picked up at the next sampling just before the negative-voltage peak. This delay, caused by the thyristors inability to respond to changes in α_{order} at any arbitrary time instant, is termed *thyristor deadtime*. For a 2-pulse TCR, the deadtime is a random quantity varying from 0 to $T/2$, where T is the time period of the voltage wave. It is therefore assigned an average value of $T/4$.

Extending the preceding argument to a 6-pulse TCR, it can be seen that the frequency of sampling the signal α_{order} needs to be only six times the fundamental frequency. The sampling instants occur shortly before the six voltage peaks corresponding to the 3-phase voltages in one time period. Thus similar to the 2-pulse TCR the average thyristor deadtime is set at half of $T/6$ or one-twelfth cycle time, which is 4.17 ms for 60-Hz systems. In general, for a p pulse TCR, the thyristor deadtime is given by $T/(2p)$, where T is the time period of the fundamental voltage. Sampling at a frequency $6f_0$ has the additional advantage of filtering out the 6th harmonic ripple in the controller output introduced by the rectifier in measurement circuit.

4.4.2.2 *Thyristor Firing-Delay Time* There is another kind of delay caused by the magnitude of the firing angle at which the TCR is operating. It is attributed to the sequential nature of firing of a 3-phase TCR and corresponds to the delay between the sampling instant and the instant when the thyristors commence conduction. If a change occurs in α_{order} shortly before a voltage peak (i.e., at one of the sampling instants), the TCR current will achieve its steady state after the new firing angle is sampled by all the three phases. The TCR firing angle varies from 90° to 180°, implying a delay that varies between zero and one-fourth of a cycle time, as measured from the peak of voltage in one phase. The firing angle in the forward thyristors of other phases will be delayed by 60° and 120°, respectively. The transfer function of the firing delay is therefore expressed as follows [41]:

$$G_Y(s) = \frac{1}{3}(e^{-sT_1} + e^{-sT_2} + e^{-sT_3}) \tag{4.52}$$

When α is close to 90°, the time constants are

$$\begin{aligned} 0 < T_1 &< T/6 \\ T/6 < T_2 &< 2T/6 \\ 2T/6 < T_3 &< 3T/6 \end{aligned} \tag{4.53}$$

and when α is close to 180°, these time constants are

$$\begin{aligned} T/4 < T_1 &< 5T/12 \\ 5T/12 < T_2 &< 7T/12 \\ 7T/12 < T_3 &< 9T/12 \end{aligned} \tag{4.54}$$

where

$$T = \frac{1}{f_0} \tag{4.55}$$

The following time constants can be used as a rough approximation:

$$\begin{aligned} T_1 &= T/3 \\ T_2 &= T/2 \\ T_3 &= 2T/3 \end{aligned} \tag{4.56}$$

In most stability studies, the thyristor deadtime, T_d, and the thyristor firing-delay time, T_Y, are together represented by the following transfer function [4], [11], [13], [14]:

$$G_Y(s) = \frac{e^{-sT_d}}{1 + sT_Y} \tag{4.57}$$

where

$$T_d = \frac{T}{2p} \tag{4.58}$$

and

$$T_Y = \frac{T}{4} \tag{4.59}$$

corresponding to a maximum delay of 180°. As T_d is small, $G_Y(s)$ can also be approximated by

$$G_Y(s) = \frac{1}{(1 + sT_d)(1 + sT_Y)} \tag{4.60}$$

4.5 THE SYNCHRONIZING SYSTEM

The purpose of the synchronizing system is to generate reference pulses in synchronism with the fundamental component of system voltage [3], [5], [16], [42]. These pulses are then used by the GPG unit to time the firing pulses to the TCR and TSCs. The synchronizing system must possess the following attributes:

1. insensitivity to distortions in supply voltage;
2. minimal generation of noncharacteristic harmonics;
3. unhindered operation during severe system faults;
4. accurate tracking of system frequency and phase angle; and
5. quick resynchronizing capability on reappearance of the system voltage subsequent to fault clearing.

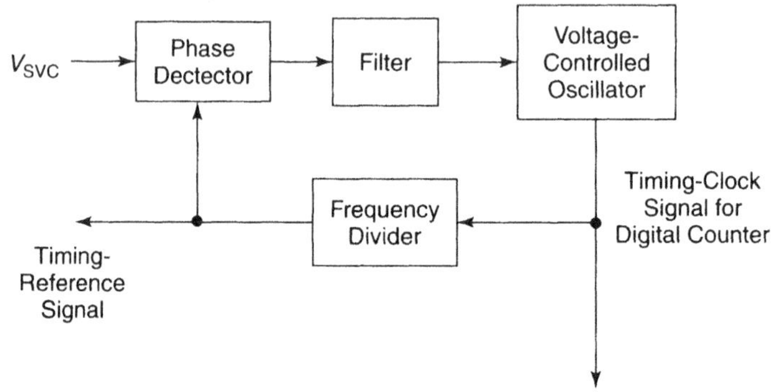

Figure 4.21 A general PLL model for digital-control system representation.

A very commonly employed synchronizing system satisfying the requirements in the foregoing list is based on the PLL shown in Fig. 4.21. The PPL not only provides a signal at the zero-crossing instant of the fundamental voltage, but it also generates the necessary timing-clock signals that are phase-locked to the fundamental frequency for the digital counters that count the firing angle.

Another synchronizing technique employs actual thyristor-valve voltages for producing the firing-angle reference ramps with which the firing angles are synchronized [42]. This scheme strongly satisfies criteria 4 and 5 of the foregoing list, but not criterion 1—especially during faults and system disturbances when the voltages undergo significant variations and distortions. This problem may, however, be overcome by resorting to balancing techniques for equalizing the firing angles in the three phases and equalizing the positive- and negative-current pulses.

4.6 ADDITIONAL CONTROL AND PROTECTION FUNCTIONS

The basic voltage regulator is usually augmented by certain special control and protection functions [3]–[9], [13]–[15] described in this section. Modeling of these functions is dependent on the nature of study, such as fundamental-frequency stability simulation or electromagnetic-transient program (EMTP) simulation studies.

4.6.1 The Damping of Electromechanical Oscillations

Auxiliary control of SVC based on supplementary control signals, such as bus frequency, line current, line active/reactive power, and generator-rotor velocity, is often employed to enhance the damping contributed by SVC to electromechanical or power oscillations, thereby attaining much higher levels of stable power transfer. Details of these controls are presented in Chapter 6.

The relevant auxiliary signal is measured, computed, and processed through a controller that is similar in configuration to that of a power-system stabilizer. The auxiliary controller typically includes one or two lead-lag stages, a gain, and a washout-transfer function. The output of the auxiliary controller may be applied at the reference-summing junction, as shown in Fig. 4.16(a) for IEEE Basic SVC Model 1 at the location labeled "Other Signals." The auxiliary controller modifies the voltage reference only when a change occurs in the auxiliary control signal. In the steady state, this contribution is zero.

In another variant of the supplementary controller, the output of the auxiliary controller is applied to the voltage-regulator output, thereby bypassing the voltage regulator itself. This application is either to expedite the SVC response by avoiding the time constants inherent in the voltage regulator or to transform the controller for the exclusive purpose of damping the electromechanical oscillations. If the power oscillations exceed a threshold, then a discontinuous damping signal is applied that provides a "bang-bang" kind of control. The voltage-regulator output is disabled in this case.

The strategy of applying the auxiliary-control signal to the output of voltage regulator is also adopted in some schemes for damping the subsynchronous oscillations in series-compensated lines.

4.6.2 The Susceptance (Reactive-Power) Regulator

The SVC requires a substantial reactive-power reserve capacity to improve system stability. In the event of a disturbance, the fast-voltage regulator control uses a significant part of the reactive-power range of the SVC to maintain a prespecified terminal voltage. If the SVC continues to be in this state, not enough reactive-power capacity may be available for it to respond effectively to a subsequent disturbance. A slow susceptance (or var) regulator is provided in the control system that changes the voltage reference to return the SVC to a preset value of reactive-power output, which is usually quite small. Other neighboring compensating devices, such as mechanically switched capacitors or inductors, can then be employed to take up the required steady-state reactive-power loading.

The susceptance regulator shown in Fig. 4.22 [11] compares the voltage-regulator output susceptance B_{ref} with a setpoint B_{set} [8]. The B_{set} is usually chosen to be the fundamental-frequency reactive-power contribution of the permanently connected harmonic filters of the SVC. In these cases, the B_{set} corresponds to a near-floating-state operation of the SVC. The error signal is transmitted through an integral control, and a corrective voltage contribution is applied to the voltage-reference junction. This control operates in a time period of tens of seconds or minutes; hence it does not interfere with the fast voltage regulation or auxiliary control of the SVC. The susceptance regulator may be inhibited during large disturbances to permit full voltage control for system restoration.

The operation of the susceptance regulator is illustrated in Fig. 4.23 [5], [13]. Let the SVC initially be at the steady-state operating point 1, which marks the intersection of the system load line and SVC *V-I* characteristics. If a sudden

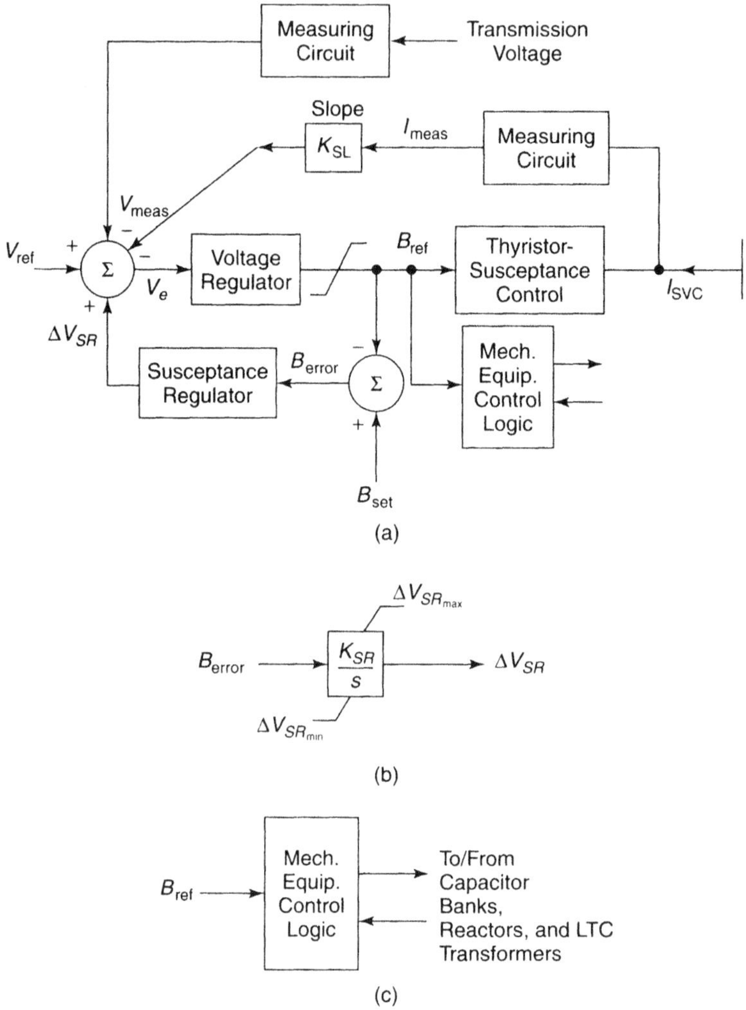

Figure 4.22 The susceptance regulator and mechanical-equipment switching: (a) the general structure; (b) the susceptance regulator; and (c) the mechanical equipment switching.

disturbance occurs in the system, reducing the SVC bus voltage by ΔV_T, the SVC moves rapidly to operating point 2 by the action of the voltage regulator. This operating point is described by the intersection of the new system load line and the SVC *V-I* characteristic.

If this decrease in voltage persists for some time, the susceptance regulator, through its slow integrator action, will modify the SVC reference voltage by ΔV_{SR} and bring the SVC steady-state operating point to 3. Now, although the SVC voltage has been reduced below the desired reference, the SVC reactive-

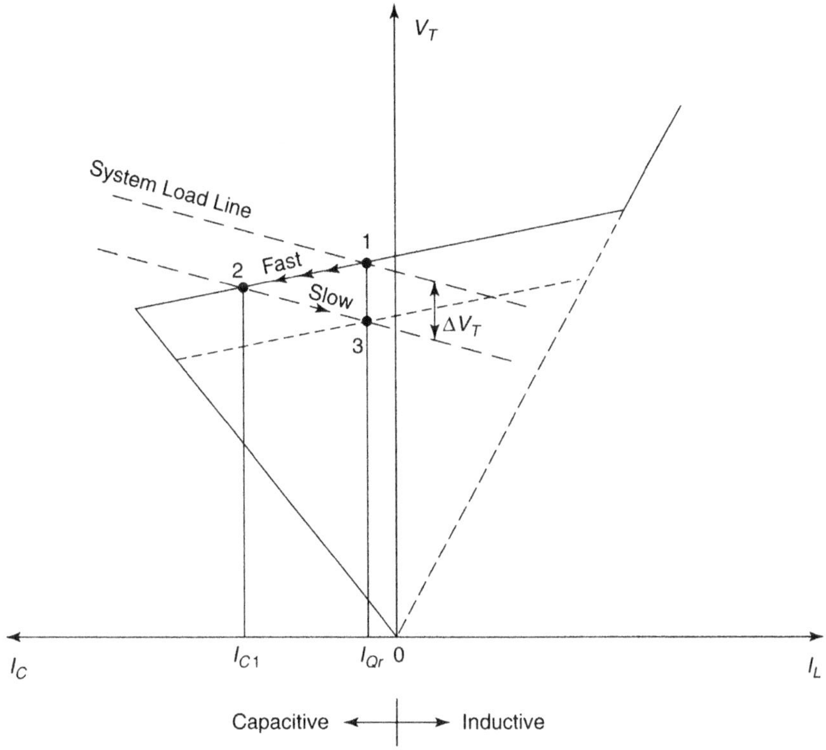

Figure 4.23 Operation of the SVC susceptance regulator.

power range is still available for coping with any system contingency. The neighboring var sources may then be switched to raise the SVC voltage to the desired value.

4.6.3 The Control of Neighboring Var Devices

Mechanically switched capacitors (MSCs), mechanically switched reactors (MSRs), and LTC transformers are some of the major neighboring reactive-power devices that constitute the overall static var system and thus need to be controlled appropriately. Some of the switching strategies are:

1. switching MSCs subsequent to sensing of the short circuit;
2. switching MSRs with overload capability for voltage regulation;
3. switching MSCs and MSRs to bring back the SVC to nearly its reactive-power setpoint; and
4. controlling the tap changing on LTC transformers.

All these switchings are slow and may be deliberately delayed for allow-

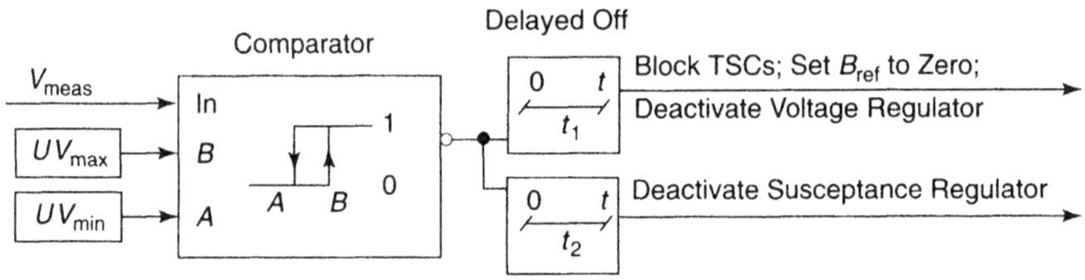

$UV_{max} = 0.69$ pu, $t_1 = 30$ ms

$UV_{min} = 0.60$ pu, $t_2 = 200$ ms

Figure 4.24 The undervoltage strategies.

ing fast control devices, such as an SVC, to act. The process of mechanical equipment switching is illustrated conceptually in Fig. 4.22(c) [11].

4.6.4 Undervoltage Strategies

A logic is incorporated in the SVC control to block the SVC during instances of severe undervoltage, such as faults. If the SVC continues to operate under these conditions, the voltage regulator will act to make the SVC output immensely capacitive, which may lead to excessive overvoltages as soon as the fault is cleared. One such model of undervoltage strategies [8] is illustrated in Fig. 4.24. There, V_{meas} represents the SVC bus voltage on HV side in parts per unit, and the limits UV_{min} and UV_{max} specify the permissible maximum and minimum levels of the undervoltage.

When a 3-phase–rectified voltage drops below a threshold UV_{min} (typically 0.6 pu), in case of a 3-phase fault, the following actions are initiated:

1. the voltage regulator is disabled (B_{ref} is clamped to zero),
2. all the TSCs are blocked, and
3. the susceptance regulator is inhibited.

After a prespecified time (about 30 ms) of ac voltage recovery to a second setpoint UV_{max}, typically 0.69 pu (higher than the first setpoint UV_{min}), the TSCs are permitted to deblock, and the B_{ref} clamp is withdrawn. After an additional prespecified time (about 170 ms), the susceptance regulator is reinstated. The undervoltage strategy is designed to become effective mainly during 3-phase faults, not during single line-to-ground faults. The discrimination between the two types of faults is implemented through the choice of threshold voltage.

4.6.5 The Secondary-Overvoltage Limiter

The control function provided by the secondary-overvoltage limiter ensures that the secondary (low-voltage) side of the SVC coupling transformer does not

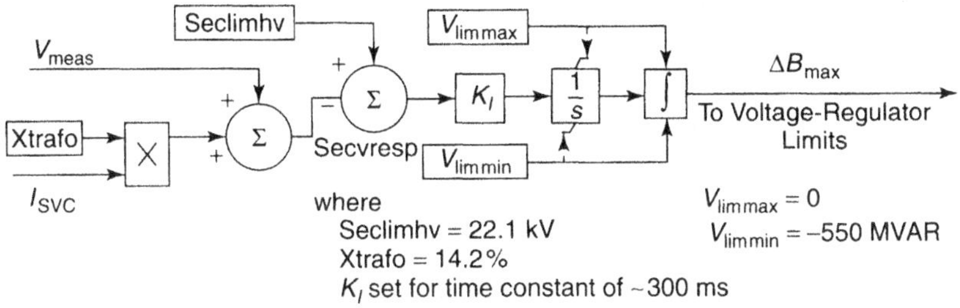

Figure 4.25 The secondary-overvoltage limiter.

exceed the design limits during an abnormal situation when the SVC output is highly capacitive and the voltage regulator is inoperative.

The secondary-overvoltage limiter [8] shown in Fig. 4.25 comprises a closed-loop integral control, the output of which modifies the susceptance-output limits B_{max} and B_{min} of the main voltage regulator shown in Figs. 4.15 and 4.16 [11]. The time constant of this slow-acting limiter typically is chosen as 300 ms, which is much longer than that incorporated in the main voltage control. The operation of this limiter therefore does not adversely affect the fast performance of the main voltage controller.

4.6.6 The TCR Overcurrent Limiter

The special control function provided by the TCR overcurrent limiter restricts the TCR current during periods of high voltage, such as those caused by load rejections, to prevent any damage to the thyristors. This function is also closed loop integral in nature, and it alters the firing angle to restrain the current to a preset level. The time constant of this controller is chosen so that it does not interact with other fast-acting controls. A value of 100 ms is typical. This overcurrent limiter can be modeled by the time-varying magnitude of B_{min} in Figs. 4.16 and 4.17.

4.6.7 TCR Balance Control

The TCR balance-control function monitors and restricts the magnitude of dc flow in the TCR reactors during situations of large 2nd harmonic ac voltage distortion. This control function essentially acts through the TCR firing angle; it is discussed in Section 5.4.3.

4.6.8 The Nonlinear Gain and the Gain Supervisor

A nonlinear gain is incorporated into the SVC voltage regulator to speed up the response during large disturbances. In many cases, a gain supervisor is installed to ensure a fast, stable response during changing values of system strength (short-circuit levels) [3], [15], [41]. (Both features are discussed in Chapter 5.) The nonlinear gain is modeled in transient-stability simulations, but the need for an appropriate model of a gain supervisor may arise only during the electromagnetic-transient program (EMTP) simulations.

4.7 MODELING OF THE SVC FOR POWER-SYSTEM STUDIES

The extent of modeling of the SVC and the power system is dependent on the nature of the power-system studies to be performed. In this section, the basic SVC modeling concepts involved in different power-system studies are presented.

4.7.1 Modeling for Load-Flow Studies

The SVC models in these studies should represent the fundamental-frequency, steady-state, and balanced performance of the SVC [3], [11], [14], [25], [26]. It may be necessary to model the SVC in terms of its three individual phases when an unbalanced operation of the SVC is considered, such as during load compensation or voltage balancing. The features of conventional load-flow programs are described in the text that follows.

4.7.1.1 SVC Operation Within the Control Range If the slope of the SVC is neglected, then the SVC is modeled as a *PV* bus, with $P = 0$ and $V = V_{ref}$. However, if the slope is considered (as in the analysis of weak ac systems), the same is modeled by connecting the high-voltage side of the SVC bus to a fictitious auxiliary bus by means of a reactance equal to the slope expressed in per units on the SVC base. Such a model is shown in Fig. 4.26(a).

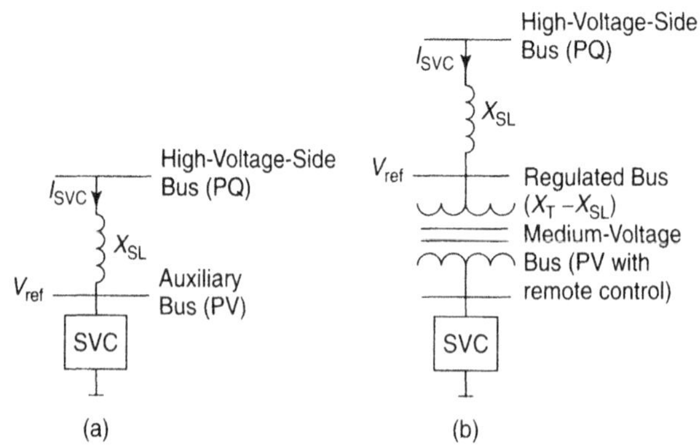

Figure 4.26 The SVC models with slope representation using conventional power-flow *PV* buses: (a) without a coupling transformer and (b) with a coupling transformer.

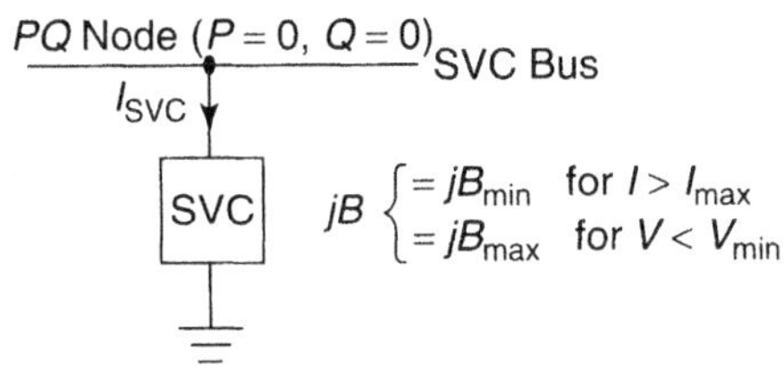

Figure 4.27 The SVC model with slope for operation outside the control range.

It may become necessary to model the coupling transformer should the SVC be connected to the tertiary winding. When the transformer is represented explicitly, the susceptance range of the SVC must be appropriately adjusted to represent the correct reactive-power rating as seen at the high-voltage bus [11]. The corresponding load-flow model is illustrated in Fig. 4.26(b).

4.7.1.2 SVC Operation Outside the Control Range The SVC is represented as an appropriate shunt admittance, depending on which limit is violated [3].

If $I_{SVC} > I_{max}$ (the inductive-limit violation), then

$$B = B_{min} = -\frac{Q_{max}}{V_{max}^2} \tag{4.61}$$

If $V < V_{min}$ (the capacitive-limit violation), then

$$B = B_{max} = -\frac{Q_{min}}{V_{min}^2} \tag{4.62}$$

where Q_{max} = the maximum inductive-reactive-power rating at $V_{SVC} = V_{max}$

Q_{min} = the maximum capacitive-reactive-power rating at $V_{SVC} = V_{min}$

It may be noted that Q_{max} is a positive quantity, whereas Q_{min} is a negative quantity.

The SVC model in this case is shown in Fig. 4.27. Modeling of the SVC as a *PV* node with *Q* limits is not appropriate for representing the SVC under overload conditions. This characterizes an incorrect behavior outside the control range, as depicted in Fig. 4.28.

In advanced load-flow programs, a comprehensive SVC model is used that models not only the slope in the controlled range but also the *Q* limits in the uncontrolled domain [3], [11].

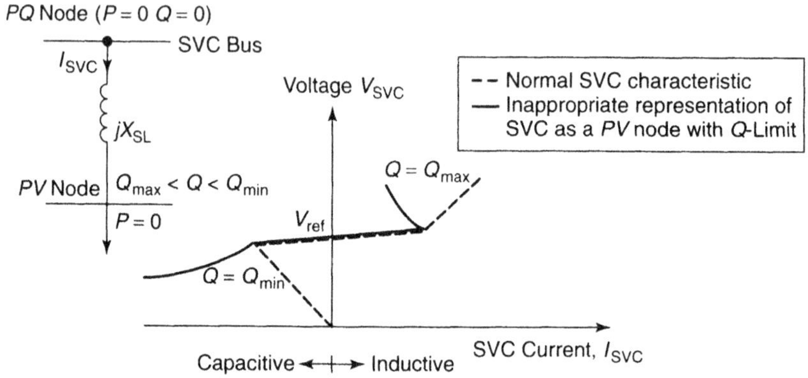

Figure 4.28 Inappropriate SVC modeling outside the control range for load-flow studies.

4.7.2 Modeling for Small- and Large-Disturbance Studies

In these studies, only the positive-sequence behavior of the SVC-compensated system is modeled. The electromagnetic transients in the SVC (TSC, TCR) and the network can be neglected if the objective is to investigate the stability related to electromechanical oscillations.

The models corresponding to the different components of the control system have already been presented. Because the bandwidth of stability programs is limited, the thyristor firing control is not explicitly represented. In such a case, a suitable interface model is needed to convert the output signal B_{ref} into a controlled network component connected to the SVC bus. This can be implemented using any of the two models shown in Fig. 4.29 and listed as follows:

1. The *variable-susceptance model*, in which the SVC current in response to the susceptance output is given by

$$I_{SVC} = B_{ref} V_{SVC} \tag{4.63}$$

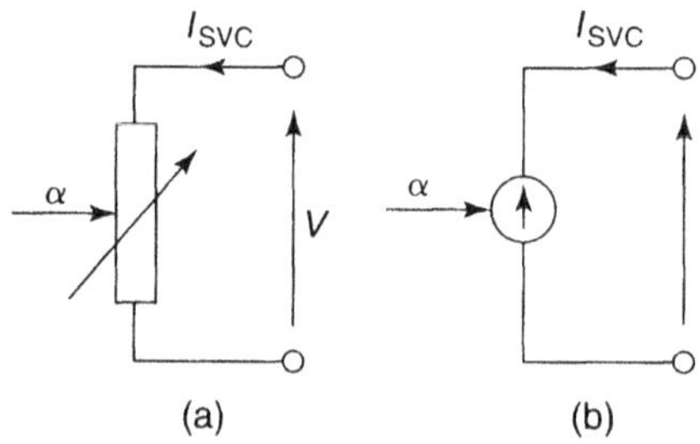

Figure 4.29 Basic static var generator models for small- and large-disturbance studies: (a) the susceptance model and (b) the current-source model.

2. The *controlled-current-source model*, in which the SVC current is again given as

$$I_{SVC} = B_{ref} V_{SVC} \tag{4.64}$$

Although both models in the preceding list are equivalent, a significance difference exists in their implementation. In case of the variable-susceptance model, the system-admittance matrix $\boldsymbol{B}$ must be updated in case any changes occur in B_{ref}, whereas a constant $\boldsymbol{B}$ matrix can continue to be used in case of the controlled-current-source model.

4.7.3 Modeling for Subsynchronous Resonance (SSR) Studies

In subsynchronous resonance (SSR) studies [43], a wide bandwidth of electromechanical frequencies is considered, so a need exists to model the network transients, as well as the thyristor-controlled and thyristor-switched elements. Additional filtering in the measurement systems may be required to eliminate network-resonant frequencies close to fundamental for the satisfactory, stable operation of the SVC control system. These aspects are discussed in detail in Chapter 5.

4.7.4 Modeling for Electromagnetic-Transient Studies

The requirements for modeling the SVC in a general electromagnetic-transient study [3], [13], [14], [21]–[24] are

1. representation of three phases;
2. accuracy over a wide frequency range;
3. representation of all system nonlinearities, as well as different controls and protection functions;
4. time-domain simulation;
5. detailed modeling of the static var system components (e.g., reactors, capacitors, and transformer models—including their saturation characteristics);
6. representation of the GPG system and synchronizing system for studying the instant-to-instant behavior of SVCs; and
7. modeling of additional control and protection functions.

4.7.5 Modeling for Harmonic-Performance Studies

The basic SVC model required for conducting harmonic-performance studies [2], [3] comprises an ideal harmonic-current source (single-phase or 3-phase symmetrical) with a specified current spectrum. The magnitudes of the nth-order harmonic current, I_n, are dependent on the SVC configuration and its operating condition. A harmonic-voltage source model in series with a variable source admittance is proposed in ref. [27]. Harmonic-performance studies are based on certain assumptions, as follows:

1. The effect of multiple harmonic sources can be investigated by applying the superposition principle.
2. The SVC harmonic generation can be modeled by positive-, negative-, and zero-sequence harmonic sources.
3. The system can be represented by linear models at each harmonic frequency.
4. The precise evaluation of harmonic distortion must have accurate load modeling.

4.8 SUMMARY

This chapter described the various measurement systems employed in the SVC control system. The demodulation effect of the measurement systems was discussed in detail. The different components of the basic SVC voltage control system were described, together with the accompanying protection functions embedded in the SVC control schemes, and the basic features of SVC modeling for various studies were highlighted.

REFERENCES

[1] T. J. E. Miller, Ed., *Reactive Power Control in Electric Power Systems*, John Wiley and Sons, New York, 1982.

[2] R. M. Mathur, Ed., *Static Compensators for Reactive Power Control*, The Canadian Electrical Association, Cantext Publications, Winnipeg, Manitoba, 1984.

[3] I. A. Erinmez, Ed., *Static Var Compensators*, CIGRE Working Group 38-01, Task Force No. 2 on SVCs, Paris, 1986.

[4] H. E. Schweickardt, G. Romegialli, and K. Reichert, "Closed Loop Control of Static Var Sources (SVS) on EHV Transmission Lines," Paper A 78 135-6, Presented at the IEEE/PES 1978 Winter Meeting, New York, January 29–February 3, 1978.

[5] L. Gyugyi, "Fundamentals of Thyristor-Controlled Static Var Compensators in Electric Power System Applications," IEEE Special Publication 87TH0187-5-PWR, *Application of Static Var Systems for System Dynamic Performance*, 1987, pp. 8–27.

[6] IEEE Power System Engineering Committee, "Application of SVC for System Dynamic Performance," IEEE Special Publication 87TH0187-5-PWR, 1987.

[7] The Electric Power Research Institute (EPRI) Report TR-100696, "Improved Static Var Compensator Control," Final Report of Project 2707-01, Prepared by General Electric (GE) Company, Schenectady, NY, June 1992.

[8] D. Dickmander, B. Thorvaldsson, G. Stromberg, and D. Osborn, "Control System Design and Performance Verification for the Chester, Maine, Static Var Compensator," *IEEE Transactions on Power Delivery*, Vol. 7, No. 3, July 1992, pp. 1492–1503.

[9] CIGRE Task Force 38.05.04, "Analysis and Optimization of SVC Use in Transmission Systems," CIGRE Technical Brochure No. 77, 1993.

[10] CIGRE Working Group 31-01, "Modelling of Static Shunt Var Systems (SVS) for System Analysis," *ELECTRA*, Vol. 51, March 1977.

[11] IEEE Special Stability Controls Working Group, "Static Var Compensator Models for Power Flow and Dynamic Performance Simulation," *IEEE Transactions on Power Systems*, Vol. 9, No. 1, February 1994, pp. 229–239.

[12] IEEE Power Engineering Society/CIGRE, *FACTS Overview*, Publication 95TP108, IEEE Press, New York, 1995.

[13] N. G. Hingorani and L. Gyugyi, *Understanding FACTS*, IEEE Press, New York, 1999.

[14] Y. H. Song and A. T. Johns, Eds., *Flexible AC Transmission Systems (FACTS)*, IEE Press, London, 1999.

[15] CIGRE Working Group 14.29, "Coordination of Controls of Multiple FACTS/HVDC Links in the Same System," CIGRE Technical Brochure No. 149, Paris, December 1999.

[16] CIGRE Task Force 38.02.17, "Advanced Angle Stability Controls," CIGRE Technical Brochure No. 155, Paris, April 2000.

[17] G. Sybille, L. Gerin-Lajoie, and P. Giroux, "Interactions Between Static Var Compensators and Series Compensation on Hydro-Quebec 735 kV Network," *CEA Transactions of Engineering and Operating Division*, Paper 92-SP-181, Vol. 31, Pt. 4, 1992.

[18] L. Gerin-Lajoie, G. Scott, S. Breault, E. V. Larsen, D. H. Baker, and A. F. Imece, "Hydro-Quebec Multiple SVC Application Control Stability Study," *IEEE Transactions on Power Delivery*, Vol. 5, No. 3, July 1990, pp. 1543–1551.

[19] M. F. McGranaghan, J. G. Koepfinger, R. G. Rocamora and L. W. Esker, "Design of a Generalized Static-Var-System Model for TNA Simulation," *IEEE Transactions on Power Apparatus and Systems*, Vol. PAS–101, No. 9, September 1982, pp. 3413–3420.

[20] M. F. McGranaghan, R. G. Rocamora, and J. G. Koepfinger, "Simulation of Static Var System (SVS) Performance on a Transient Network Analyzer," *IEEE Transactions on Power Apparatus and Systems*, Vol. PAS–101, No. 9, September 1982, pp. 3373–3378.

[21] R. H. Lasseter and S. Y. Lee, "Digital Simulation of Static Var System Tranients," *IEEE Transactions on Power Apparatus and Systems*, Vol. PAS–101, No. 10, October 1982, pp. 4171–4177.

[22] A. N. Vasconcelos, A. J. P. Ramos, J. S. Monteiro, M. V. B. C. Lima, H. D. Silva, and L. R. Lins, "Detailed Modeling of an Actual Static Var Compensator for Electromagnetic Transient Studies," *IEEE Transactions on Power Systems*, Vol. 7, No. 1, February 1992, pp. 11-19.

[23] S. Y. Lee, S. Bhattacharya, T. Lejonberg, A. E. Hammad, and S. Lefebvre, "Detailed Modeling of Static Var Compensators Using the Electromagnetic Trans-

ients Program (EMTP)," *IEEE Transactions on Power Delivery*, Vol. 7, No. 2, April 1992, pp. 836–847.

[24] S. Lefebvre and L. Gerin-Lajoie, "A Static Compensator Model for the EMTP," *IEEE Transactions on Power Systems*, Vol. 7, No. 2, 1992, pp. 477–484.

[25] S. Arabi and P. Kundur, "A Versatile FACTS Device Model for Power Flow and Stability Simulations," *IEEE Transactions on Power Systems*, Vol. 11, No. 4, November 1996, pp. 1944–1950.

[26] G. Therond et al., "Modelling of Power Electronics Equipment (FACTS) in Load Flow and Stability Programs," CIGRE Task Force 38-01-08, Report, 1998.

[27] L. J. Bohmann and R. H. Lasseter "Equivalent Circuit for Frequency Response of a Static Var Compensator," *IEEE Transactions on Power Systems*, Vol. 1, No. 4, Nov. 1986, pp. 68–74.

[28] J. D. Ainsworth, "Phase-Locked Oscillator Control System for Thyristor-Controlled Reactors," *IEE Proceedings*, Vol. 135, Pt. C., No. 2, March 1988, pp. 146–156.

[29] P. Czech, S. Y. M. Hung, N. H. Huynh, and G. Scott, "TNA Study of Static Compensator Performance on the 1982–83 James Bay System, Results and Analysis," *Proceedings of the International Symposium on Controlled Reactive Compensation*, Varennes, Quebec, 1979.

[30] J. Belanger, L. E. N. Dias, S. A. Moraes, and S. O. Frontin, "Application of a Static Var System on the Furnas (Brazil) 138-kV Transmission System," *Proceedings of the International Symposium on Controlled Reactive Compensation*, Varennes, Quebec, 1979.

[31] J. Lisser and A. J. van de Water, "Zero Flux Current Transformer for Wideband Precision Measurement in ac and dc HV Systems," CIGRE Paper 34-03, Paris, 1986.

[32] R. W. Menzies and G. B. Mazur, "Advances in the Determination of Control Parameters for Static Compensators," *IEEE Transactions on Power Delivery*, Vol. 4, No. 4, October 1989, pp. 2012–2017.

[33] ABB Power Systems, "MACH 2 Description," Technical Report RU 8037, AU, 1999.

[34] M. R. Iravani et al., "Modelling and Analysis Guidelines for Slow Transients: Part I (Torsional Oscillations; Transient Torques; Turbine Blade Vibrations; Fast Bus Transfer)," *IEEE Transactions on Power Delivery*, Vol. 10, No. 4, October 1995, pp. 1950–1955.

[35] J. F. Hauer, W. A. Mittelstadt, W. H. Litzenberger, C. Clemans, D. Hamai, and P. Overholt, "Wide-Area Measurements for Real-Time Control and Operation of Large Electric Power Systems: Evaluation and Demonstration of Technology for the New System," Report prepared for the U.S. Department of Energy by the Bonneville Power Administration (BPA) and the Western Area Power Administration, Portland, OR, April 1999. (This report and its attachments are available from the BPA on CD-ROM).

[36] A. G. Phadke, "Synchronized Phasor Measurements in Power Systems," *IEEE Computer Applications in Power Systems*, April 1993, pp. 10–15.

[37] J. Tillet, J. Pease, J. Hall, and D. Bradley, "Experience with Optical PTs and CTs for Relaying and Metering," *Proceedings of the Western Protective Relay Confer-*

ence, Spokane, WA, October 23–25.

[38] D. Chatrefau, *Application of Optical Sensors in Extra High Voltage Substations*, General Electric (GE) Company Alsthom T&D Review, January 1997, pp. 17–24.

[39] J. D. Ainsworth, "The Phase Locked Oscillator—A New Control System for Controlled Static Converters," *IEEE Transactions on Power Apparatus and Systems*, Vol. PAS–87, pp. 859–865.

[40] J. D. Ainsworth, "Developments in the Phase-Locked Oscillator Control System for HVDC and other Large Converters," *IEE Conference Publication 255 on AC and DC Power Transmission*, September 1985, pp. 98–103.

[41] J. Belanger, G. Scott, T. Anderson, and S. Torseng, "Gain Supervisor for Thyristor-Controlled Shunt Compensators," CIGRE Paper 38-01, Paris, 1984.

[42] R. W. Lye and G. M. McAllister, "The Rimouski Static Compensator—A Pioneer on the Hydro-Quebec System," *Proceedings of the International Symposium on Controlled Reactive Compensation*, Varennes, Quebec, September 1979.

[43] IEEE Committee Report, "Third Supplement to a Bibliography for the Study of Subsynchronous Resonance Between Rotating Machines and Power Systems," *IEEE Transactions on Power Systems*, Vol. 6, No. 2, May 1991, pp. 830–834.

CHAPTER 5

Concepts of SVC Voltage Control

5.1 INTRODUCTION

Static var compensators (SVCs) are used primarily in power systems for voltage control as either an end in itself or a means of achieving other objectives, such as system stabilization [1]–[8]. This chapter presents a detailed overview of the voltage-control characteristics of SVC and the principles of design of the SVC voltage regulator. The performance of SVC voltage control is critically dependent on several factors, including the influence of network resonances, transformer saturation, geomagnetic effects, and voltage distortion. When SVCs are applied in series-compensated networks, a different kind of resonance between series capacitors and shunt inductors becomes decisive in the selection of control parameters and filters used in measurement circuits.

Various considerations affecting the design of the SVC voltage regulator are discussed in this chapter as well.

5.2 VOLTAGE CONTROL

5.2.1 *V-I* Characteristics of the SVC

The steady-state and dynamic characteristics of SVCs describe the variation of SVC bus voltage with SVC current or reactive power. Two alternative representations of these characteristics are shown in Fig. 5.1: part (a) illustrates the terminal voltage–SVC current characteristic and part (b) depicts the terminal voltage–SVC reactive-power relationship. The dynamic *V-I* characteristics of SVCs are described in the following text.

5.2.1.1 *Dynamic Characteristics*

Reference Voltage, V_{ref} This is the voltage at the terminals of the SVC during the floating condition, that is, when the SVC is neither absorbing nor generating any reactive power. The reference voltage can be varied between the maximum and minimum limits—$V_{ref\ max}$ and $V_{ref\ min}$—either by the SVC control system, in case of thyristor-controlled compensators, or by the taps of the

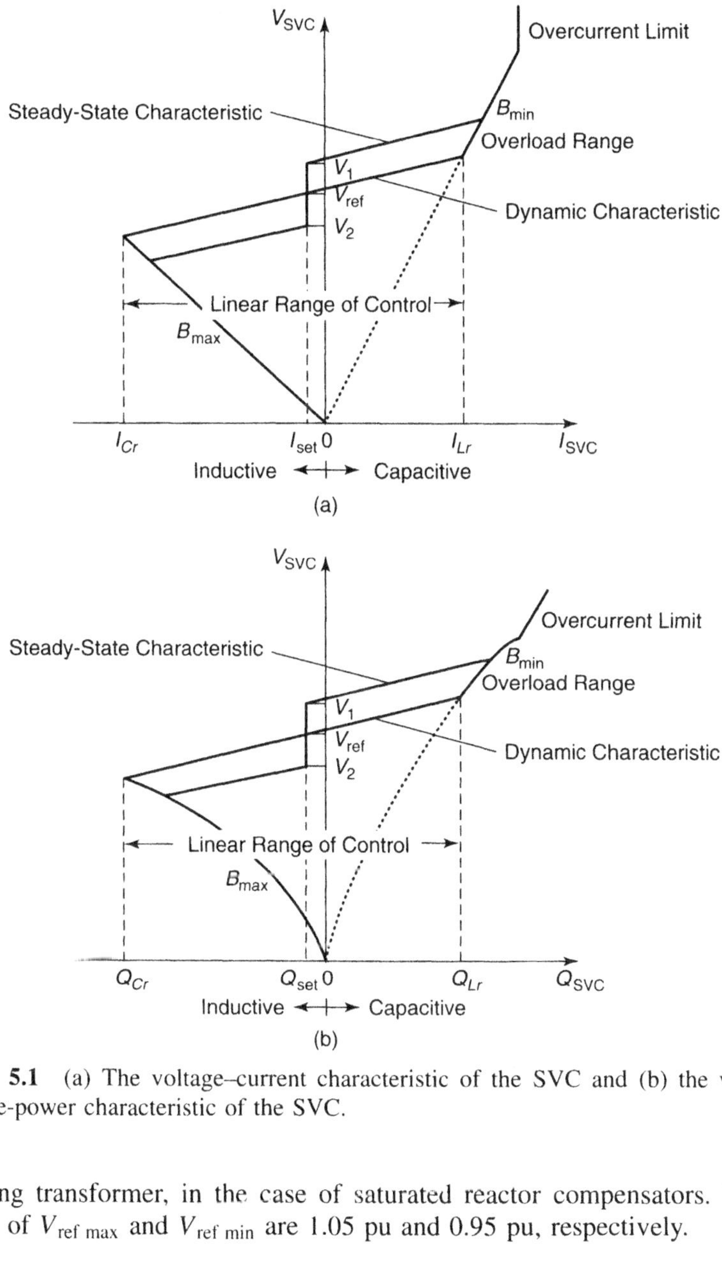

Figure 5.1 (a) The voltage–current characteristic of the SVC and (b) the voltage–reactive-power characteristic of the SVC.

coupling transformer, in the case of saturated reactor compensators. Typical values of $V_{ref\ max}$ and $V_{ref\ min}$ are 1.05 pu and 0.95 pu, respectively.

Linear Range of SVC Control This is the control range over which SVC terminal voltage varies linearly with SVC current or reactive power, as the latter is varied over its entire capacitive-to-inductive range.

Slope or Current Droop The slope or droop of the *V-I* characteristic is defined as the ratio of voltage-magnitude change to current-magnitude change over the linear-controlled range of the compensator. Thus slope K_{SL} is given by

$$K_{SL} = \frac{\Delta V}{\Delta I} \; \Omega \tag{5.1}$$

where ΔV = the change in voltage magnitude (V)
ΔI = the change in current magnitude (A)

The per-unit value of the slope is obtained as

$$K_{SL} = \frac{\Delta V / V_r}{\Delta I / I_r} \; \text{pu} \tag{5.2}$$

where V_r and I_r represent the rated values of SVC voltage and current, respectively.

For $\Delta I = I_r$,

$$K_{SL} = \frac{\Delta V(\text{at } I_r \text{ or } Q_r)}{V_r} \; \text{pu}$$

$$= \frac{\Delta V(\text{at } I_r \text{ or } Q_r)}{V_r} \cdot 100\% \tag{5.3}$$

where Q_r represents the rated reactive power of SVC.

Thus the slope can be defined alternatively as the voltage change in percent of the rated voltage measured at the larger of the two—maximum inductive- or maximum capacitive-reactive-power outputs, as the larger output usually corresponds to the base reactive power of the SVC. In some literature, the reactive-power rating of the SVC is defined as the sum of its inductive and capacitive rating. The slope is often expressed as an equivalent reactance:

$$X_{SL} = K_{SL} \qquad \text{in pu} \tag{5.4}$$

The slope can be changed by the control system in thyristor-controlled compensators, whereas in the case of saturated reactor compensators, the slope is adjusted by the series slope-correction capacitors. The slope is usually kept within 1–10%, with a typical value of 3–5%. Although the SVC is expected to regulate bus voltage, that is, maintain a flat voltage-current profile with a zero slope, it becomes desirable to incorporate a finite slope in the *V-I* characteristics for reasons described in Section 5.2.3.

Overload Range When the SVC traverses outside the linear-controllable range on the inductive side, the SVC enters the overload zone, where it behaves like a fixed inductor.

Overcurrent Limit To prevent the thyristor valves from being subjected to excessive thermal stresses, the maximum inductive current in the overload range is constrained to a constant value by an additional control action.

5.2.1.2 Steady-State Characteristic The steady-state *V-I* characteristic of the SVC is very similar to the dynamic *V-I* characteristic except for a deadband in voltage, as depicted in Figs. 5.1(a) and (b). In the absence of this deadband, in the steady state the SVC will tend to drift toward its reactive-power limits to provide voltage regulation. It is not desirable to leave the SVC with very little reactive-power margin for future voltage control or stabilization excursions in the event of a system disturbance. To prevent this drift, a deadband about V_{ref} holds the I_{SVC} at or near zero value, depending on the location of the deadband. Thus the reactive power is kept constant at a setpoint, typically equal to the MVA output of the filters. This output is quite small; hence the total operating losses are minimized [9], [10]. A slow susceptance regulator (see Section 4.6.2) is employed to implement the voltage deadband, which has a time constant of several minutes. Hence the susceptance regulator is rendered virtually ineffective during fast transient phenomena, and it does not interfere with the operation of the voltage controller.

5.2.2 Voltage Control by the SVC

The voltage-control action of the SVC can be explained through a simplified block representation of the SVC and power system, as shown in Fig. 5.2. The power system is modeled as an equivalent voltage source, V_s, behind an equivalent system impedance, X_s, as viewed from the SVC terminals. The system impedance X_s indeed corresponds to the short circuit MVA at the SVC bus and is obtained as

$$X_s = \frac{V_b^2}{S_c} \cdot \mathrm{MVA}_b \qquad \text{in pu} \tag{5.5}$$

where S_c = the 3-phase short circuit MVA at the SVC bus
V_b = the base line-to-line voltage
MVA_b = the base MVA of the system

If the SVC draws a reactive current I_{SVC}, then in the absence of the SVC voltage regulator, the SVC bus voltage is given by

$$V_s = V_{SVC} + I_{SVC}X_s \tag{5.6a}$$

or

$$V_s = V_{SVC}\angle 0^\circ + I_{SVC}\angle{-90^\circ}X_s\angle 90^\circ) \tag{5.6b}$$

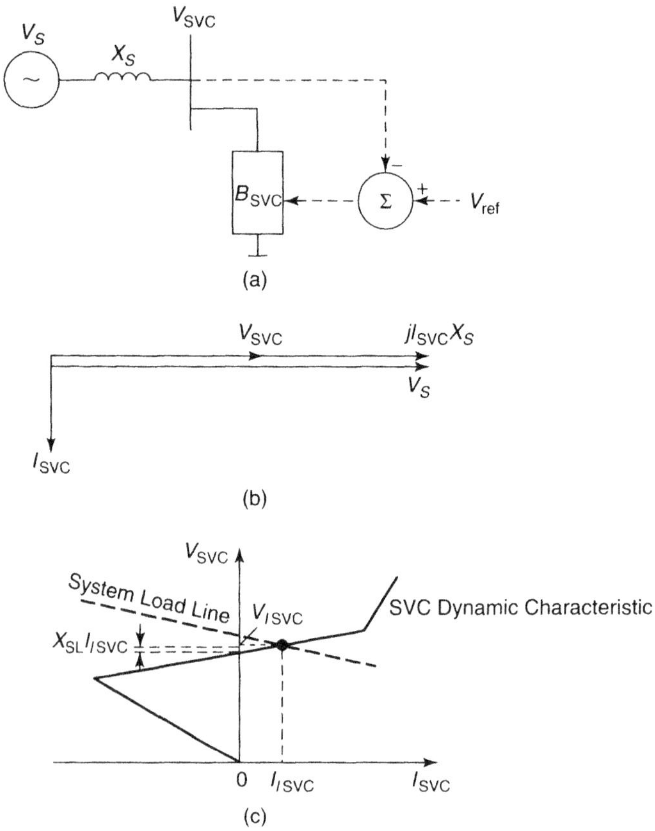

Figure 5.2 (a) A simplified block diagram of the power system and SVC control system; (b) a phasor diagram of the ac system for the inductive SVC current; and (c) characteristics of the simplified power system and the SVC.

or

$$V_s = (V_{\text{SVC}} + I_{\text{SVC}}X_s)\angle 0^\circ \tag{5.6c}$$

or

$$V_s = V_{\text{SVC}} + I_{\text{SVC}}X_s \tag{5.6d}$$

The SVC current thus results in a voltage drop of $I_{\text{SVC}}X_s$ in phase with the system voltage V_s. The SVC bus voltage decreases with the inductive SVC current and increases with the capacitive current. Equation (5.6d) represents the power-system characteristic or the system load line. An implication of Eq. (5.6d) is that the SVC is more effective in controlling voltage in weak ac systems (high X_s) and less effective in strong ac systems (low X_s).

The dynamic characteristic of the SVC depicted in Fig. 5.1 describes the reactive-power compensation provided by the SVC in response to a variation in SVC terminal voltage. The intersection of the SVC dynamic characteristic and the system load line provides the quiescent operating point of the SVC, as illustrated in Fig. 5.2(c).

The voltage-control action in the linear range is described as

$$V_{\mathrm{SVC}} = V_{\mathrm{ref}} + X_{\mathrm{SL}} I_{\mathrm{SVC}} \tag{5.7}$$

where I_{SVC} is positive if inductive, negative if capacitive.

It is emphasized that the *V-I* characteristics described here relate SVC current/reactive power to the voltage on the high-voltage side of the coupling transformer. An example characteristic describing the relationship between the voltage on the low-voltage side of the coupling transformer, and the SVC current has been derived in Chapter 3.

5.2.3 Advantages of the Slope in the SVC Dynamic Characteristic

Although the SVC is a controller for voltage regulation, that is, for maintaining constant voltage at a bus, a finite slope is incorporated in the SVC's dynamic characteristic and provides the following advantages despite a slight deregulation of the bus voltage. The SVC slope

1. substantially reduces the reactive-power rating of the SVC for achieving nearly the same control objectives;
2. prevents the SVC from reaching its reactive-power limits too frequently; and
3. facilitates the sharing of reactive power among multiple compensators operating in parallel.

5.2.3.1 Reduction of the SVC Rating Figure 5.3 illustrates two dynamic *V-I* characteristics of an SVC. Characteristic $OA'B'C'$ incorporates a finite slope, whereas characteristic $OABC$ does not. The slope has been deliberately exaggerated to demonstrate its effect. Assuming that the system load line varies between L_1 and L_2, the reactive-power rating of the SVC needed for providing flat voltage regulation is Q_{Cm} capacitive to Q_{Lm} inductive, as determined from the characteristic $OABC$. However, if a small deregulation in the SVC bus voltage is considered acceptable (as demonstrated by the characteristic $OA'B'C'$), the maximum reactive-power rating of the SVC required for performing the voltage control corresponding to the same variation in the system load line is Q'_{Cm} capacitive to Q'_{Lm} inductive. Evidently, $Q'_{Cm} < Q_{Cm}$ and $Q'_{Lm} < Q_{Lm}$. Thus a much lower SVC reactive-power rating and, hence, a much lower cost is required for nearly the same control objective.

It has been shown for an example system [2] that the SVC rating can be

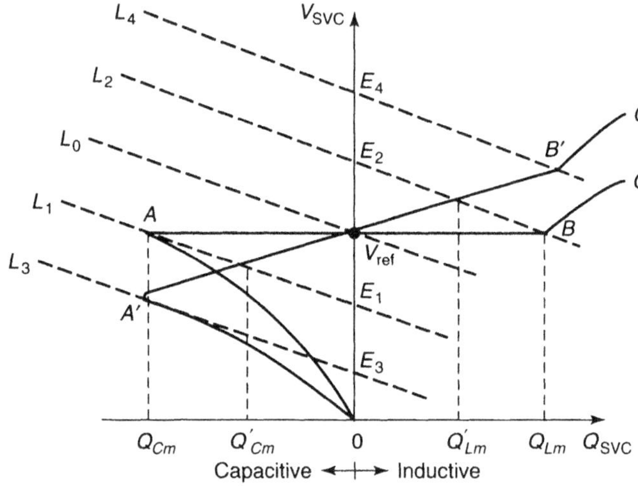

Figure 5.3 Reduction in the SVC reactive-power rating by the current slope.

reduced to half, with a 5% slope in the *V-I* characteristic. The resulting trade-off is a 2.5% voltage excursion.

5.2.3.2 Prevention of Frequent Operation at Reactive-Power Limits

Figure 5.3 also shows that if there is no slope in the dynamic characteristic, even a small change in the system load line (from a small variation, $E_2 - E_1$, in the no-load equivalent system voltage, as viewed from the SVC bus) may cause the SVC to traverse from one end of the reactive-power range to the other end to maintain constant voltage. The reactive-power limits of the SVC are reached more frequently if the ac system tends to be strong, that is, when the slope of the system load line is quite small. The effectiveness of the SVC as a voltage-control device therefore becomes limited.

With a finite slope in the *V-I* characteristic, the SVC continues to operate in the linear-controllable range for a much larger variation in the load line of the external ac system. For instance, the SVC can exercise voltage control for a significantly larger variation, $E_4 - E_3$, in the equivalent ac system no-load voltage.

When the external ac system is subjected to a disturbance, both the slope of the load line (indicative of the equivalent system reactance) and the system no-load voltage are influenced. However, the feature discussed here has been explained in terms of changes in the no-load voltage only.

5.2.3.3 Load Sharing Between Parallel-Connected SVCs For reliability via redundancy, and also for minimizing the net harmonic generation, it is not uncommon to divide the net-required SVC range into several equal-sized compensators. When more than one compensator is used at one location, the control action must be coordinated. This section discusses such coordination.

Consider two SVCs, SVC_1 and SVC_2, connected at a system bus as depicted in Fig. 5.4(a). The two SVCs have the same ratings but the reference voltages, V_{ref}, of the two control characteristics differ by a small amount, ε. In practice, ε is small, although it is not zero. Two cases are examined: one in which both the SVCs have a zero slope, as shown in Fig. 5.4(b), and the other in which the two SVCs have a finite slope, as illustrated in Fig. 5.4(c). The composite *V-I* control characteristic of the two SVCs is derived by summing up the individual currents of both SVCs for the same bus-voltage magnitude—procedure that is repeated over the entire range of SVC bus voltage. The composite characteristic is indicated by the thicker line.

In the case of zero current slope, the composite operating characteristic is beset with a discontinuity around point *A*. When the system load line intersects the *V-I* characteristic at *A*, a quiescent operating point results that corresponds to full reactive-power production on SVC_1 (point *B*) and full inductive-reactive-power absorption on SVC_2 (point *C*). Thus one SVC partially compensates the output of the other, which is uneconomical because the losses are high. On the left of point *A*, SVC_2 controls the bus voltage, whereas SVC_1 remains at full production. However, on the right of point *A*, it is SVC_1 that controls the bus voltage and SVC_2 that is at full absorption. This operation clearly demonstrates that the two SVCs are not well coordinated.

The current droop ensures that the composite *V-I* control characteristic of both SVCs is continuous despite the difference in the voltage-reference set-points. If the two SVCs and the power system achieve a stable-operating point at *A*, SVC_1 operates at *B* and SVC_2 at *C*. The reactive-load sharing of the two compensators is improved, and the losses are minimized. The zones where only one compensator controls the voltage while the other is already at a limit reduce to small portions at both ends of the control range.

In practice, the error ε is much smaller than depicted in Figs. 5.4(b) and (c), and the achievable accuracy of load sharing is often acceptable without any further controls. With additional balancing controls, exact load sharing can be attained.

5.2.4 Influence of the SVC on the System Voltage

5.2.4.1 Coupling Transformer Ignored The SVC behaves like a controlled susceptance, and its effectiveness in regulating the system voltage is dependent on the relative strength of the connected ac system. The system strength or equivalent system impedance, as seen from the SVC bus, primarily determines the magnitude of voltage variation caused by the change in the SVC reactive current. This can be understood [5] from the simplistic representation of the power system and SVC shown in Fig. 5.2(a). In this representation, the effect of the coupling transformer is ignored and the SVC is modeled as a variable susceptance at the high-voltage bus. The SVC is considered absorbing reactive power from the ac system while it operates in the inductive mode.

The SVC bus voltage, V_{SVC}, is given by Eq. (5.7). Linearizing Eq. (5.7)

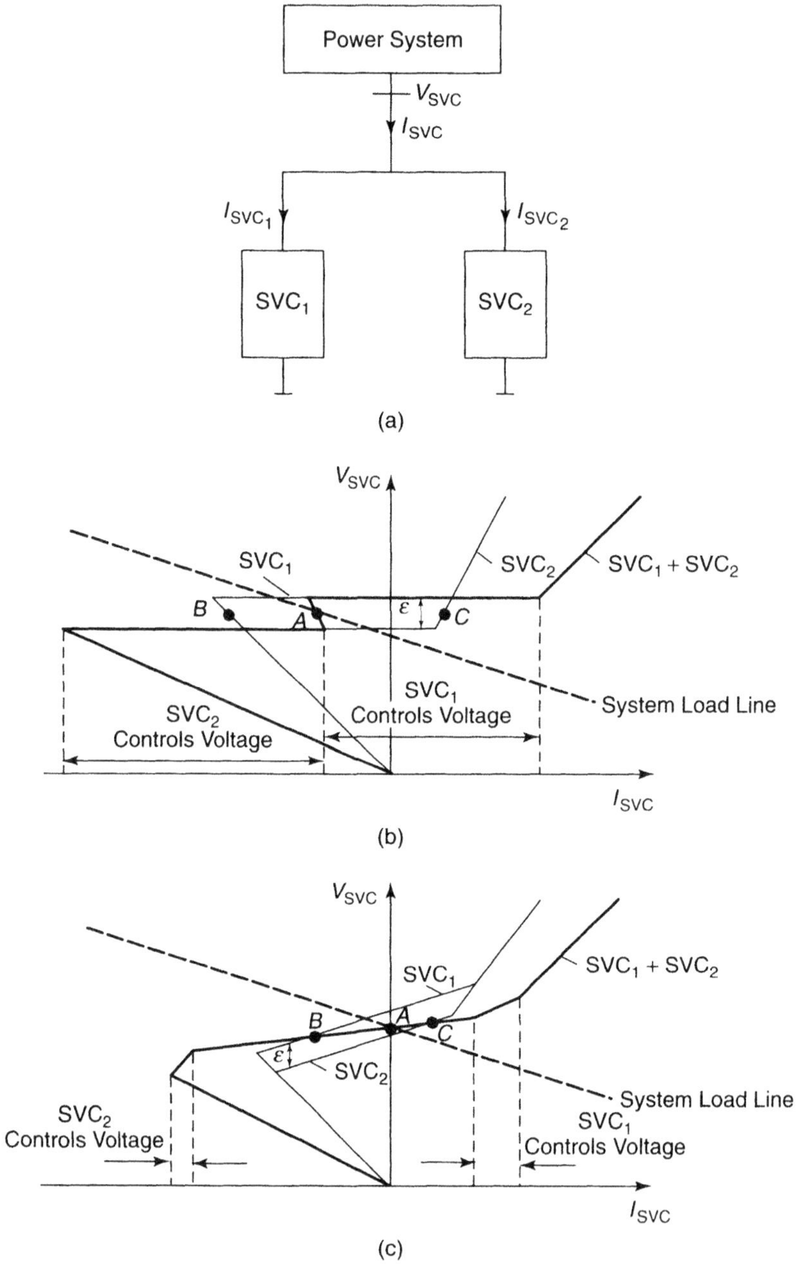

Figure 5.4 (a) Two parallel-connected SVCs at a system bus; (b) two SVCs in parallel with difference ε in the reference-voltage setpoints without current droop; and (c) two SVCs in parallel with current droop and with difference ε in the reference-voltage setpoints.

gives the variation in the V_{SVC} as a function of change in the SVC current, I_{SVC}. Thus for the constant-equivalent-source voltage V_s,

$$\Delta V_{SVC} = -X_s \Delta I_{SVC} \tag{5.8}$$

The V_{SVC} is also related to I_{SVC} through the SVC reactance, B_{SVC}, as follows:

$$I_{SVC} = B_{SVC} V_{SVC} \tag{5.9}$$

For incremental changes, Eq. (5.9) is linearized to give

$$\Delta I_{SVC} = B_{SVC_0} \Delta V_{SVC} + \Delta B_{SVC} V_{SVC_0} \tag{5.10}$$

Substituting ΔI_{SVC} from Eq. (5.10) in Eq. (5.8),

$$\frac{\Delta V_{SVC}}{\Delta B_{SVC}} = \frac{-V_{SVC_0}}{\text{ESCR} + B_{SVC_0}} \tag{5.11}$$

where the effective short-circuit ratio (ESCR) is defined as

$$\begin{aligned} \text{ESCR} &= \frac{1}{(-\Delta V_{SVC}/\Delta I_{SVC})} \\ &= \frac{1}{X_s} = B_s \end{aligned} \tag{5.12}$$

where B_s = the equivalent system susceptance

5.2.4.2 Coupling Transformer Considered As shown in Fig. 5.5, the representation of the SVC coupling transformer creates a low-voltage bus connected to the SVC and the transformer reactance X_T is separated from X_s. The high-voltage side, V_H, is then related to low-voltage side, V_{SVC}, as

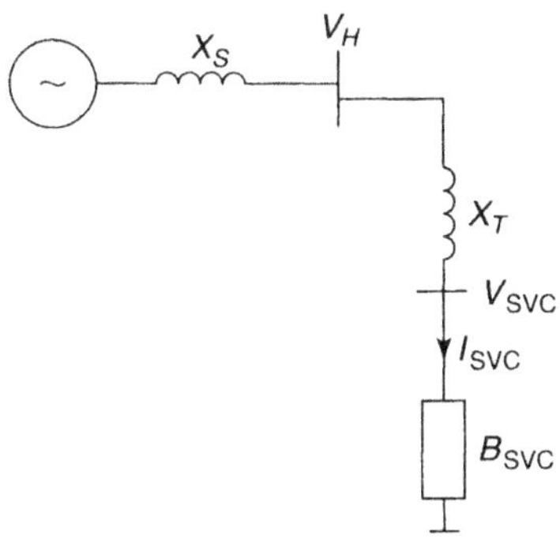

Figure 5.5 Representation of the power system and the SVC, including the coupling transformer.

$$\frac{V_{\mathrm{SVC}}}{V_H} = \frac{1}{1 + X_T B_{\mathrm{SVC}}} \tag{5.13}$$

Linearizing Eq. (5.13) gives

$$\Delta V_{\mathrm{SVC}}(1 + X_T B_{\mathrm{SVC}_0}) + V_{\mathrm{SVC}_0} X_T \Delta B_{\mathrm{SVC}} = \Delta V_H \tag{5.14}$$

Substituting Eq. (5.14) and the expression $V_{\mathrm{SVC}_0}/V_{H_0}$ from Eq. (5.13) in Eq. (5.11) results in

$$\frac{\Delta V_H}{\Delta B_{\mathrm{SVC}}} = \frac{-V_{H0}}{(\mathrm{ESCR} + B_{\mathrm{SVC}_0})} \left(\frac{1 - X_T \mathrm{ESCR}}{1 + X_T B_{\mathrm{SVC}_0}} \right) \tag{5.15}$$

5.2.4.3 The System Gain The effect of the SVC on system voltage can also be evaluated in an alternate [11] (though approximate) manner, as follows in Eq. (5.16). Equations (5.7) and (5.9) can be combined to give

$$\begin{aligned} V_{\mathrm{SVC}} &= \frac{V_s(1/B_{\mathrm{SVC}})}{(X_s + 1/B_{\mathrm{SVC}})} \\ &= \frac{V_s}{(1 + B_{\mathrm{SVC}}/\mathrm{ESCR})} \end{aligned} \tag{5.16}$$

where X_s is the equivalent short-circuit impedance of the system in shunt with the capacitive reactance of SVC.

For ac systems, generally ESCR $\gg B_{\mathrm{SVC}}$ (or effectively, $X_s \ll 1/B_{\mathrm{SVC}}$). Thus Eq. (5.16) can be expanded as

$$V_{\mathrm{SVC}} \cong V_S \left(1 - \frac{B_{\mathrm{SVC}}}{\mathrm{ESCR}} \right) \tag{5.17}$$

The change in SVC bus voltage, ΔV, is then given by

$$\Delta V = V_S - V_{\mathrm{SVC}} \tag{5.18}$$

or

$$\Delta V = \frac{V_S B_{\mathrm{SVC}}}{\mathrm{ESCR}} \tag{5.19}$$

or

$$\Delta V = K_N B_{\mathrm{SVC}} \tag{5.20}$$

where K_N is defined as the "system gain" and is expressed as

$$K_N = \frac{V_S}{\text{ESCR}} = \frac{V_S}{B_S} \tag{5.21}$$

The system gain, K_N, thus relates the deviation in SVC bus voltage to SVC susceptance. An increase in inductive susceptance, B_{SVC}, causes ΔV to become more positive, thereby leading to a drop in the SVC bus voltage. In fact, Eq. (5.20) can be obtained from Eq. (5.11) if it is assumed that

$$V_{\text{SVC}_0} \cong V_S \quad \text{and} \quad \text{ESCR} \gg B_{\text{SVC}_0} \tag{5.22}$$

and understanding that

$$\Delta V = -\Delta V_{\text{SVC}} \tag{5.23}$$

The foregoing derived expression for system gain can be used to arrive at a preliminary design of an SVC voltage regulator. However, it may, be noted that the system gain K_N depends on the equivalent system voltage V_S and equivalent impedance X_S—both of which are subject to change with the dynamically varying power-system configuration. The gain K_N is thus not a constant and, in fact, varies in a certain range. A weak ac system would correspond to a high system gain; a strong ac system would result in a relatively lower system gain.

Equation (5.21) was derived based on the absolute values of the various parameters involved. For control studies, it is desirable to derive a corresponding equation based on per-unit values of different variables. Let the base voltage, V_b, and base susceptance, B_b, be chosen as

$$V_b = V_{\text{nominal}} \tag{5.24}$$

where V_{nominal} = the rated bus voltage

$$B_b = B_{\text{max}} - B_{\text{min}} \tag{5.25}$$

where B_{max} = the maximum susceptance of the SVC (fully capacitive)

B_{min} = the minimum susceptance of the SVC (fully inductive)

The per-unit system gain is therefore given from Eq. (5.21) as

$$K_N = \frac{V_S}{V_b} \frac{B_b}{B_S} \tag{5.26}$$

Multiplying and dividing Eq. (5.20) by V_b^2,

$$K_N = \frac{V_S}{V_b} \frac{B_b}{B_S} \frac{V_b^2}{V_b^2}$$

$$= \frac{V_S}{V_b} \frac{Q_{SVC}}{B_S V_S V_b \dfrac{V_b}{V_S}} \tag{5.27}$$

or

$$K_N = \frac{V_S Q_{SVC}}{V_b S_c \dfrac{V_b}{V_S}} \tag{5.28}$$

where S_c = the short-circuit power
= the base voltage · the short-circuit current
= $V_b \cdot (B_S V_S)$

Assuming V_S/V_b is close to unity, which is usually the case in power systems, the per-unit system gain is expressed as

$$K_N = \frac{\Delta V_{SVC}}{B_{SVC}} = \frac{Q_{SVC}}{S_c} \text{ pu} \tag{5.29}$$

It should be noted that the system gain will change with variations in network configuration, line switchings, and any event that may change the system short-circuit level at the SVC bus.

5.2.5 Design of the SVC Voltage Regulator

The principles of design of the SVC voltage regulator are now presented. The basic block schematic of a voltage-control system for a thyristor-switched capacitor–thyristor-controlled reactor (TSC–TCR) SVC is depicted in Fig. 5.6. The voltage regulator is expressed in the gain–time-constant format. As discussed in Chapter 4, two alternative ways of modeling the voltage regulator exist: the gain–time-constant form and the integrator–current-droop form. The design methodology is based on these models, for which reason they are repeated in Fig. 5.7.

In the gain–time-constant representation, the voltage regulator is expressed by the following transfer function:

$$G_R(s) = \frac{K_R}{1 + sT_R} \tag{5.30}$$

where K_R = the static gain = $1/K_{SL}$ (pu)

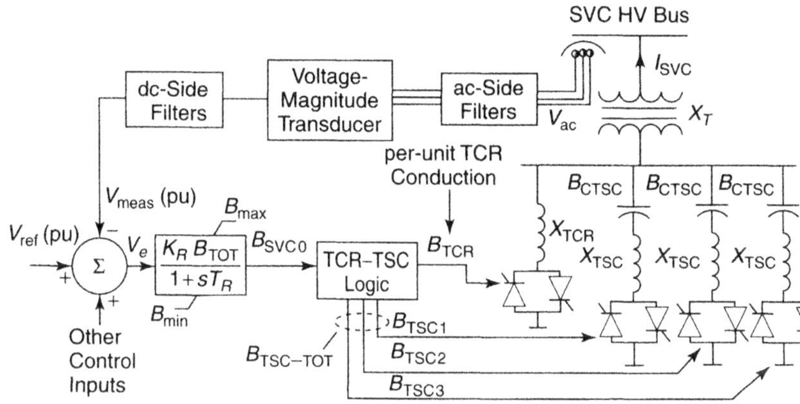

where

$B_{min} = B_{SVC}$ at the TCR only
$B_{max} = B_{SVC}$ at all TSCs only
$B_{TOT} = B_{max} - B_{min}$

K_R = the static gain (full range for the voltage change of $1/K_R$)
T_R = the regulator time constant
B_{SVC0} = the net susceptance at the SVC HV bus

Figure 5.6 Basic elements of SVC voltage-regulation control with TSC.

K_{SL} = the current droop (pu)
T_R = the regulator time constant (s)

An additional term-transient gain, K_T, is defined as

$$K_T = \frac{K_R}{T_R} \tag{5.31}$$

In the integrator–current-droop model, the voltage regulator is described as an integrator $G'_R(s)$, with the following explicit current-feedback loop:

$$G'_R(s) = \frac{1}{sR_R} \tag{5.32}$$

where R_R = the response rate (ms/pu)

The regulator time constant, T_R, and the response rate, R_R, are related as

$$T_R = \frac{R_R}{K_{SL}} \tag{5.33}$$

5.2.5.1 Simplistic Design Based On System Gain This design methodology, proposed in ref. [11], considers a proportional–integral (PI) controller with an explicit current feedback similar to that described in the previous section. This method is called *simplistic* because it models the power system by a

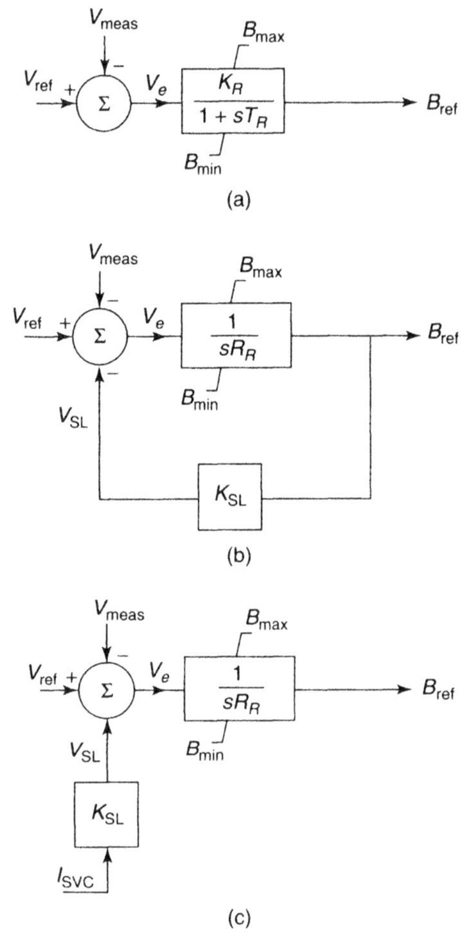

Figure 5.7 Alternate voltage-regulator forms: (a) a gain–time constant; (b) an integrator with susceptance-droop feedback; and (c) an integrator with current-droop feedback.

pure reactance or equivalent gain and does not account for the effect of generator dynamics on controller performance. This method is, however, highly important, as it explains very clearly the influence of system strength and current-droop on the speed of controller response.

The block diagram of an SVC-compensated power system is shown in Fig. 5.8(a). It is assumed that

1. the change in system voltage ΔV caused by the SVC is small;
2. the SVC bus voltage is very close to the nominal-rated voltage, that is, $V_{SVC} \cong 1$ pu; and
3. the variations in the SVC reference voltage are also quite small.

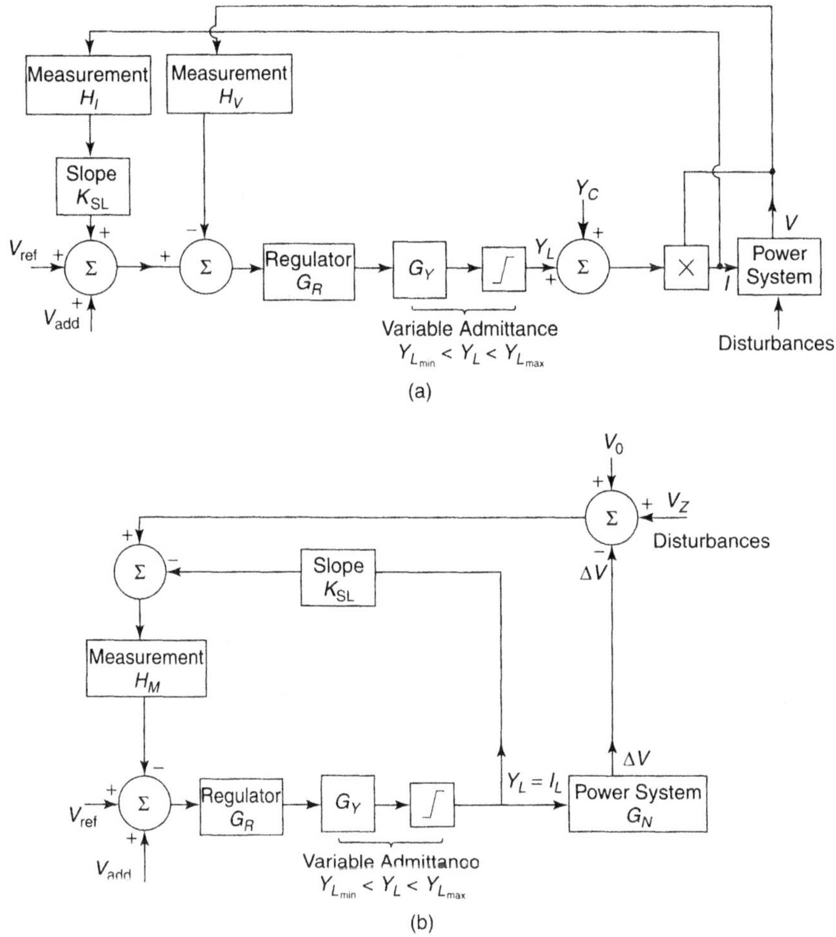

Figure 5.8 (a) A block diagram of the system voltage controller incorporating an SVC and (b) a simplified block diagram of the system voltage controller for $V \simeq V_{\text{rated}}$.

Then, the currents I_L, in the TCR, and I_C, in the TSC, are given by

$$I_L = Y_L V_{\text{SVC}} \approx Y_L \qquad \text{in pu} \tag{5.34}$$

$$I_C = Y_C V_{\text{SVC}} \approx Y_C \qquad \text{in pu} \tag{5.35}$$

An equivalent circuit of the SVC-compensated system is depicted in Fig. 5.8(b). The following simplifications are made:

1. The voltage- and current-measurement systems are considered identical.
2. The TSC switchings are ignored, and the droop effect of the capacitive current is merged with V_{ref}.

3. The only variable considered is the inductive current I_L, which reduces the system bus voltage by ΔV as given by Eqs. (5.18)–(5.20). The effect of constant-capacitive SVC current on the SVC bus voltage is incorporated in V_0. The influence of any power-system disturbance, V_z, is neglected.

The thyristor phase control is denoted by G_Y, as already explained in Chapter 4, and is given by

$$G_Y(s) = \frac{e^{-sT_d}}{1 + sT_Y} \tag{5.36}$$

where T_d = the thyristor dead time (= one-twelfth cycle time)
T_Y = the thyristor firing-delay time caused by the sequential switching of thyristors ($\cong$ one-quarter cycle time)

A PI controller that gives the fastest stable response for the weakest system configuration having gain $K_{N\,\max}$ is determined [11] as

$$G_R(s) = K_p\left\{1 + \frac{1}{sT_Y}\right\} = -\frac{1}{2(K_{\text{SL}} + K_{N\,\max})}\left\{1 + \frac{1}{sT_Y}\right\} \tag{5.37}$$

The overall closed-loop transfer function, $G_W(s)$, of the control system for incremental variations is given by

$$\begin{aligned} G_W(s) &= \frac{\Delta V(s)}{\Delta V_{\text{ref}}(s)} \\ &= \frac{K_N G_R G_Y}{1 + (K_{\text{SL}} + K_N) G_R G_Y H_M} \end{aligned} \tag{5.38}$$

where H_M = the transfer function of the measurement system

To obtain the response time of the control system only, the largest time constants are considered. Thus

$$G_W(s) \cong \frac{K_N}{(K_{\text{SL}} + K_N)} \frac{sT_W}{1 + sT_W} \tag{5.39}$$

where

$$T_W = 2\left(\frac{K_{\text{SL}} + K_{N\,\max}}{K_{\text{SL}} + K_N}\right) T_Y \tag{5.40}$$

Hence for a change in the reference voltage ΔV_{ref}, the SVC bus voltage varies as

$$\Delta V = \frac{K_N}{K_{SL} + K_N} \Delta V_{ref}(1 - e^{-t/T_W}) \tag{5.41}$$

The following conclusions are drawn from the foregoing analysis:

1. A change in the ΔV_{ref} causes the SVC bus voltage to vary by $k \cdot V_{ref}$ because of the current droop, where $k = K_N/(K_{SL} + K_N)$.
2. The fastest response is obtained when $K_N = K_{N\,max}$ (i.e., for the weakest system state). In this case, the response time constant, $T_{W\,opt}$, is obtained from Eq. (5.40) as $T_{W\,opt} = 2T_y = 8.33$ ms for a 60-Hz system frequency. The settling time (to 95% of the step change) $\cong 3T_{W\,opt} = 25$ ms $= 1.5$ cycle time.

The voltage controller is expected to maintain stability under all reasonably realizable network configurations and to ensure a rapid response under those conditions. Hence the voltage controller is optimized for the lowest system short-circuit level.

Equation (5.36) can be rewritten as

$$\frac{T_W}{2T_Y} = \frac{T_W}{T_{W\,opt}} = \frac{K_{SL} + K_{N\,max}}{K_{SL} + K_N} \tag{5.42}$$

Dividing the numerator and denominator on the right-hand side by $K_{N\,max}$, we get

$$\frac{T_W}{T_{W\,opt}} = \frac{\dfrac{K_{SL}}{K_{N\,max}} + 1}{\dfrac{K_{SL}}{K_{N\,max}} + \dfrac{K_N}{K_{N\,max}}} \tag{5.43}$$

From Eq. (5.29), it follows that

$$K_{N\,max} = \frac{Q_{SVC}}{S_{c\,min}} \tag{5.44}$$

Substituting Eqs. (5.29) and (5.44) in Eq. (5.43) gives

$$\frac{T_W}{T_{W\,opt}} = \frac{\dfrac{K_{SL}}{K_{N\,max}} + 1}{\dfrac{K_{SL}}{K_{N\,max}} + \dfrac{S_{c\,min}}{S_c}} \tag{5.45}$$

A plot of the normalized-response time constant, $T_W/T_{W\,opt}$, as a function of the normalized-system short-circuit level, $S_c/S_{c\,min}$, is shown in Fig. 5.9. This

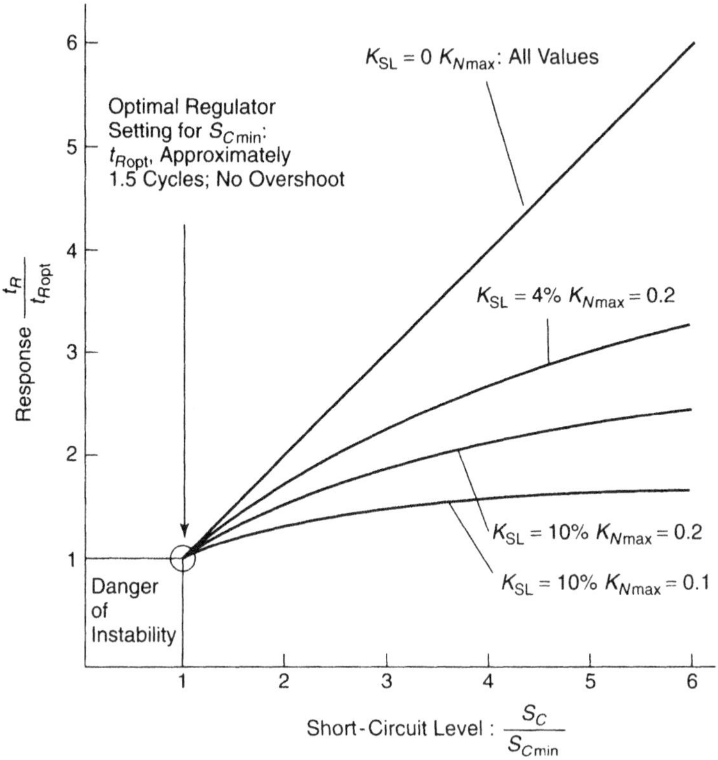

Figure 5.9 Effect of the system short-circuit level on the SVC response time.

plot shows the influence of the network short-circuit level and current droop on the SVC response time.

For a given short-circuit level, the control response time increases (becomes slow) with decreasing values of the current slope. The response time is also enhanced as the system strength increases relative to the weakest network state. If a larger current slope is chosen (of course, at the expense of voltage regulation), the increase in response time with system strength is comparatively lower. If the weakest network states are chosen to be those with very low short-circuit levels, then the response time exhibits a much sharper increase with increasing system strength.

On the basis of the preceding observations, it is concluded that a large value of slope should be chosen, preferably in the 3–5% range. Furthermore, the weakest network should also be identified judiciously. If an unduly weak system, especially that encountered while rebuilding a network after a major system collapse, is chosen as the weakest network state, the response time will be unusually longer when realistic system states are existent.

Example 5.1 An SVC connected to a 735-kV system has a reactive-power range of 350 MVAR production to 100 MVA absorption. The droop is set to 4%. The system short-circuit level is specified as follows:

- The maximum short-circuit current: 50 kA.
- The minimum short-circuit current under normal operating conditions: 5 kA.
- The minimum short-circuit current during system restoration after loss of a transmission line: 500 A.

From these specifications,

1. Determine the per-unit regulator gain that ensures stable operation from 5 kA to 50 kA system short-circuit current.
2. Show the change of voltage-control response for the system variation and regulator setting in item 1 of this sublist.
3. Determine the per-unit regulator gain for stable operation in the system short-circuit-level range 500 A to 50 kA.
4. Show the change of the voltage-control response for item 1.

Solution 5.1

Item 1 The system gain is given from Eq. (5.29) as

$$K_N = \frac{Q_{\text{SVC}}}{S_c} \text{ pu} \tag{5.46}$$

where $S_c = \sqrt{3} V_b I_c$

$$S_{c\ \min} = \sqrt{3} \cdot 735 \cdot 10^3 \cdot 5 \cdot 10^3 = 6365 \text{ MVA} \tag{5.47}$$

$$S_{c\ \max} = \sqrt{3} \cdot 735 \cdot 10^3 \cdot 50 \cdot 10^3 = 63653 \text{ MVA} \tag{5.48}$$

$$K_{N\ \max} = \frac{Q_{\text{SVC}}}{S_{c\ \min}} = \frac{350 - (-100)}{6365} = 0.0707 \text{ pu} \tag{5.49}$$

$$K_{N\ \min} = \frac{Q_{\text{SVC}}}{S_{c\ \max}} = \frac{350 - (-100)}{63653} = 0.00707 \text{ pu} \tag{5.50}$$

The regulator is optimized for $K_{N\ \max}$. Thus the proportional gain of the voltage regulator, as given from Eq. (5.37), is

$$K_p = -\frac{1}{2(K_{\text{SL}} + K_{N\ \max})} = -\frac{1}{2(0.04 + 0.0707)} = -4.52 \tag{5.51}$$

Item 2 The variation in response speed is calculated as follows from Eq. (5.42):

$$\frac{T_W}{2T_Y} = \frac{K_{SL} + K_{N\,\max}}{K_{SL} + K_N} = \frac{0.04 + 0.0707}{0.04 + K_N} = \frac{0.111}{0.04 + K_N} \tag{5.52}$$

For

$$K_N = K_{N\,\max} = 0.0707, \qquad \frac{T_W}{2T_Y} = 1 \tag{5.53}$$

For

$$K_N = K_{N\,\min} = 0.00707, \qquad \frac{T_W}{2T_Y} = 2.36 \tag{5.54}$$

This demonstrates that the SVC response can slow down by a factor of 2.36 if the system short-circuit current changes from 5 kA to 50 kA.

Item 3 If the weakest network state is chosen to correspond to a short-circuit current of 500 A,

$$K_N \max = \frac{Q_{SVC}}{S_{c\,\min}} = \frac{350 - (-100)}{\sqrt{3} \cdot 735 \cdot 0.5} = 0.070 \text{ pu} \tag{5.55}$$

The proportional gain of the voltage regulator is

$$K_p = -\frac{1}{2(K_{SL} + K_{N\,\max})} = -\frac{1}{2(0.04 + 0.707)} = -0.669 \tag{5.56}$$

Item 4 The change of response speed is

$$\frac{T_W}{2T_Y} = \frac{K_{SL} + K_{N\,\max}}{K_{SL} + K_N} = \frac{0.04 + 0.707}{0.04 + K_N} = \frac{0.747}{0.04 + K_N} \tag{5.57}$$

For

$$K_N = K_{N\,\max} = 0.0707, \qquad \frac{T_W}{2T_Y} = \frac{0.747}{0.04 + 0.707} = 1 \tag{5.58}$$

For

$$K_N = K_{N\,\min} = 0.00707, \qquad \frac{T_W}{2T_Y} = \frac{0.747}{0.04 + 0.00707} = 15.86 \tag{5.59}$$

The system configuration with the 500-A short-circuit level is definitely abnormal and should not be chosen as the weakest network state for optimizing the regulator parameters. Otherwise, a very slow response (approximately 16 times the optimum) would result during the strong system condition.

5.2.5.2 Design That Considers Generator Dynamics A technique often resorted to for designing the voltage regulator of the SVC, by including the impact of generator dynamics, consists of the following steps:

1. Deriving a system model linearized around an operating point.
2. Obtaining the stability range of parameters of the preselected voltage-regulator structure from eigenvalue analysis.
3. Determining the best regulator parameters within the stability range that result in a fast response with acceptable overshoot.
4. Validating of the voltage-regulator performance through a detailed non-linear simulation of the system.

A case study of voltage-regulator design for an SVC located at the mid-line in a single-machine infinite-bus (SMIB) system [12], [13] is presented in Appendix A of this book.

5.3 EFFECT OF NETWORK RESONANCES ON THE CONTROLLER RESPONSE

The controller-design methodology discussed in Section 5.2.5 was based on the assumption that the network was inductive up to twice the nominal system frequency f_0, which is usually a characteristic of reasonably strong systems. However, in practical systems, the network impedance may exhibit a strong first resonance at a frequency much lower than $2f_0$, as shown in Fig. 5.10. This resonant frequency results from the interaction between the series inductances and shunt capacitances of the total length of the major transmission line. Although the subsequent resonant frequencies (poles) and antiresonant frequencies (zeroes) are highly sensitive to changes in the network state (e.g., the number of parallel lines and the number of shunt reactors), the variation in the first-resonant frequency is very slight (80–100 Hz). This first-resonant frequency plays a critical role in determining the SVC response and may lead to voltage-regulator instability.

The effect of the first-resonant frequency (e.g., 80 Hz) can be explained from Fig. 5.11. Any power-system disturbance will excite this resonant mode, causing a 80-Hz voltage component in the network. Because of the demodulation effect of the SVC voltage-measuring system (explained in Chapter 4), a 20-Hz (80 − 60 Hz) frequency component will be introduced in both the voltage-transducer output and the voltage-regulator susceptance output. Consequently, the TCR current will become amplitude-modulated, with a carrier frequency of

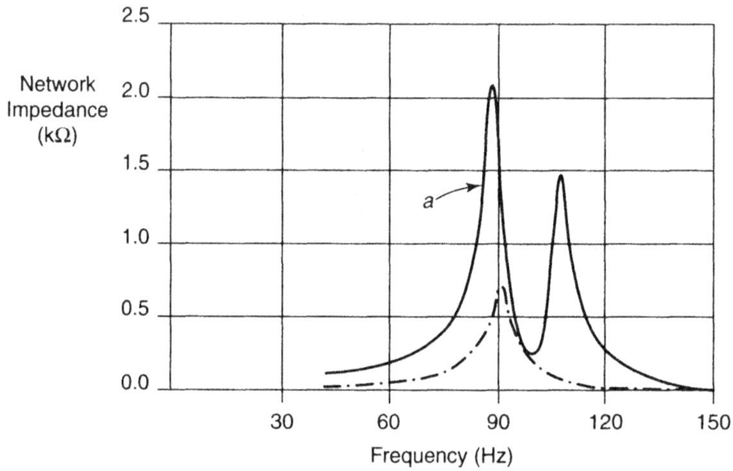

Figure 5.10 Network impedance as a function of frequency.

60 Hz and a modulating frequency of 20 Hz. When this TCR current is injected into the network, it will create sideband currents and voltages of frequencies 60 Hz and 80 Hz that again will be demodulated to a dc signal and 20-Hz ripple by the voltage transducer, thus augmenting the initially present 20-Hz oscillations. In the limiting case, the control system may become unstable.

The strength of the ac system that decides the magnitude of the ac bus voltage variation, as caused by a change in the SVC current, indeed contributes to the total gain around the voltage-control loop. A weak system having high impedance results in a higher loop gain and therefore a faster response, but it also results in an increased proximity to the stability limit.

The network resonance essentially causes a time delay between the SVC control action and the variation in system voltage. If this delay, together with the delays of the voltage regulator and thyristor-firing circuits, equals 180°, then the system may become unstable, showing increased oscillations if the loop

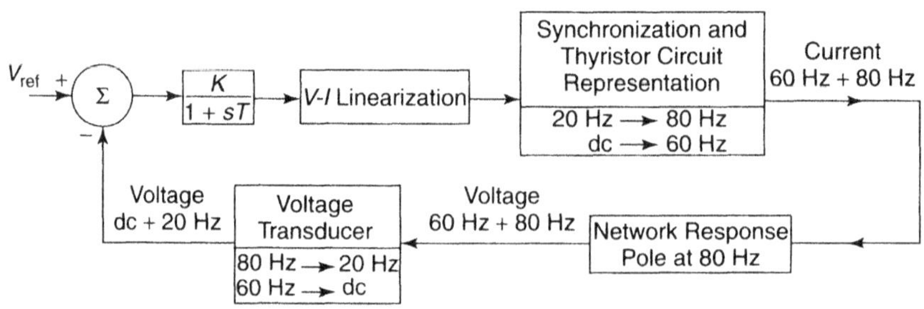

Figure 5.11 An 80-Hz instability in the SVC control loop.

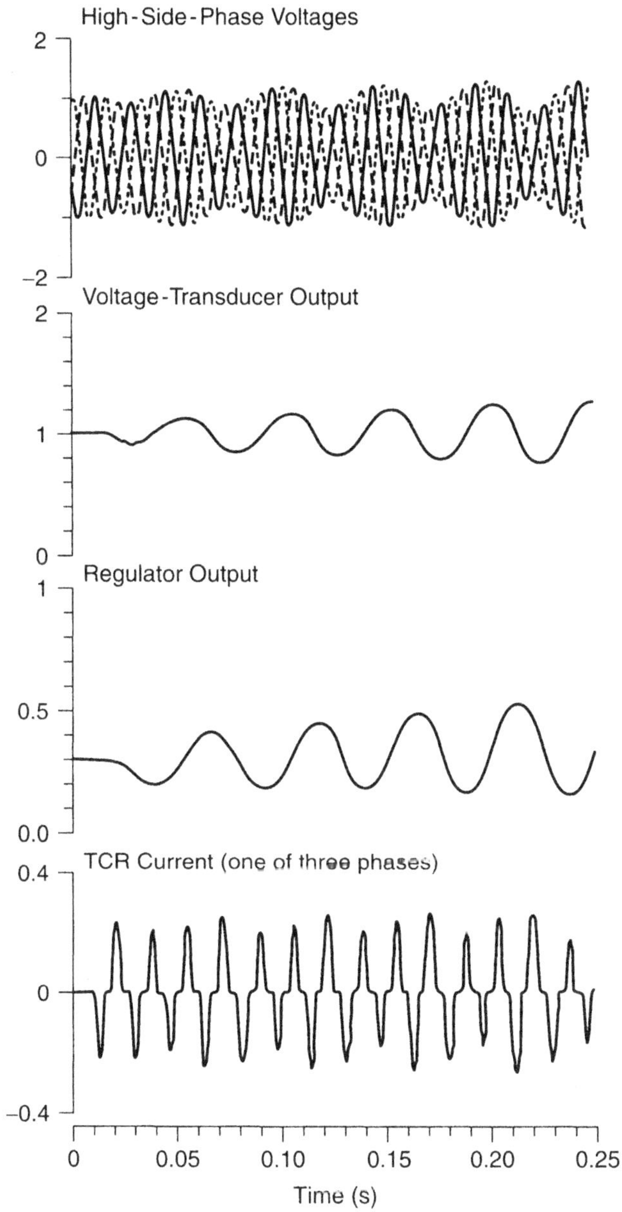

Figure 5.12 An example of SVC voltage-regulator instability.

gain is high. The different voltage and current waveforms are illustrated in Fig. 5.12 for a case when a total phase lag of 180° is achieved that corresponds to a frequency of 20 Hz [5]. Both the TCR current and the high-voltage-side SVC bus voltage are amplitude-modulated with a 20-Hz frequency.

5.3.1 Critical Power-System Parameters

The stability of the SVC voltage regulator can be explained on the basis of two power-system parameters: the effective short-circuit ratio (ESCR) and the first-resonant frequency of the ac system (F_{r0}) [4], [5]. Consider a typical power-transmission system with an SVC connected at midpoint and its corresponding impedance-versus-frequency characteristic depicted in Fig. 5.13.

The $ESCR_0$ relates to the total fundamental-frequency impedance of the power system as viewed from TCR terminals. The power system then includes all the filters, shunt capacitors, and the SVC coupling transformer. The $ESCR_0$ is usually expressed in per units on the basis of full-conduction admittance of the TCR. The F_{r0} is indicative of the natural transient response of the power system that is critical in determining the dynamic performance of the SVC voltage regulator; the Z_S is the equivalent impedance of the power system without the filters, shunt capacitors, and the SVC coupling transformer; and the Z_T is the total system-equivalent impedance when the aforementioned components are included.

It is evident from Fig. 5.13 that the addition of impedance of the filter capacitors and the SVC transformer to the ac system impedance Z_S has two effects:

1. It increases the effective system impedance, thus reducing the effective $ESCR_0$.
2. It lowers the first-resonant frequency, F_{r0}.

The system impedance as a function of frequency should be computed over the full range of system conditions for which SVC control stability must be ensured and the corresponding variation in F_{r0} and $ESCR_0$ must be determined.

An example [5] of a power system that is compensated at its midpoint by an SVC is shown in Fig. 5.14. Because all the lines connected to the SVC are approximately of the same length, they are represented by an equivalent line of surge-impedance loading (SIL) equal to the sum of SILs of all the lines. The B_C represents the shunt-capacitive susceptance of the SVC, including filters. It is observed that with decreasing system strength $ESCR_0$, the first-resonant frequency F_{r0} gets lowered. Also, for a given line length the F_{r0} decreases with increasing capacitive susceptance.

5.3.2 Sensitivity to Power-System Parameters

The influence of power-system parameters on the transient response of the SVC voltage regulator, with varying regulator gains K_T [4], [5], is depicted in Fig. 5.15. The SVC bus-voltage transducer output and the voltage-regulator output, B_{ref}, responses are presented for three kinds of power-systems: a very weak system ($ESCR_0 = 0.7$ pu, $F_{r0} = 80$ Hz); a moderately weak system ($ESCR_0 = 1.6$ pu, $F_{r0} = 110$ Hz); and a strong system ($ESCR_0 = 2.5$ pu, $F_{r0} = 180$ Hz). For each system configuration, the responses are presented for four widely different

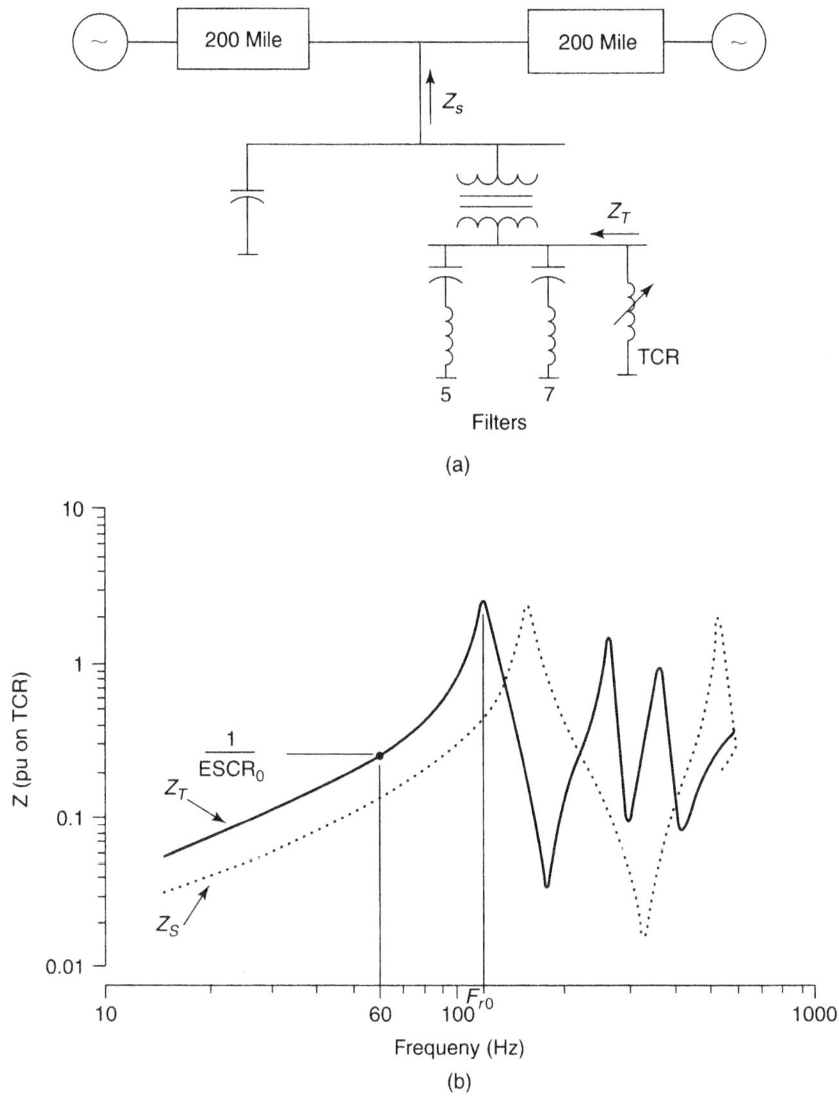

Figure 5.13 (a) A single-line diagram of an SVC-compensated system and (b) impedance-versus-frequency characteristics for an SVC-compensated system.

values of the SVC controller transient gain $K_T\{= 1/(K_{SL} \cdot T_R)\}$, that is, K_T = 30, 100, 300, and 1000 pu/s/pu. For each combination of system strength and controller-transient gain, the SVC voltage response is obtained for a 3-phase fault applied at 50 ms and cleared at 100 ms. All the foregoing SVC responses are determined for a TCR-average conduction (operating susceptance) B_{T0} of 0.3 pu.

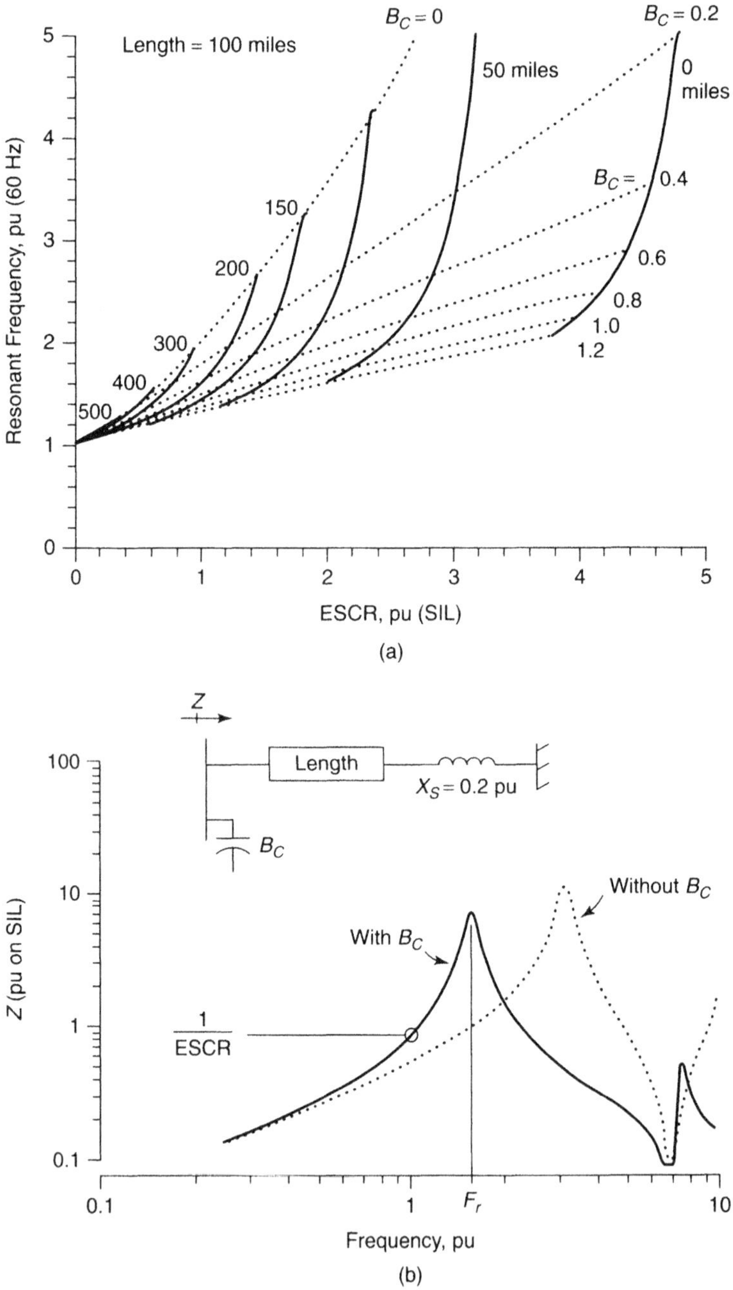

Figure 5.14 Plots illustrating the relationships of (a) ESCR and (b) F_{r0} to parameters of a system with long lines.

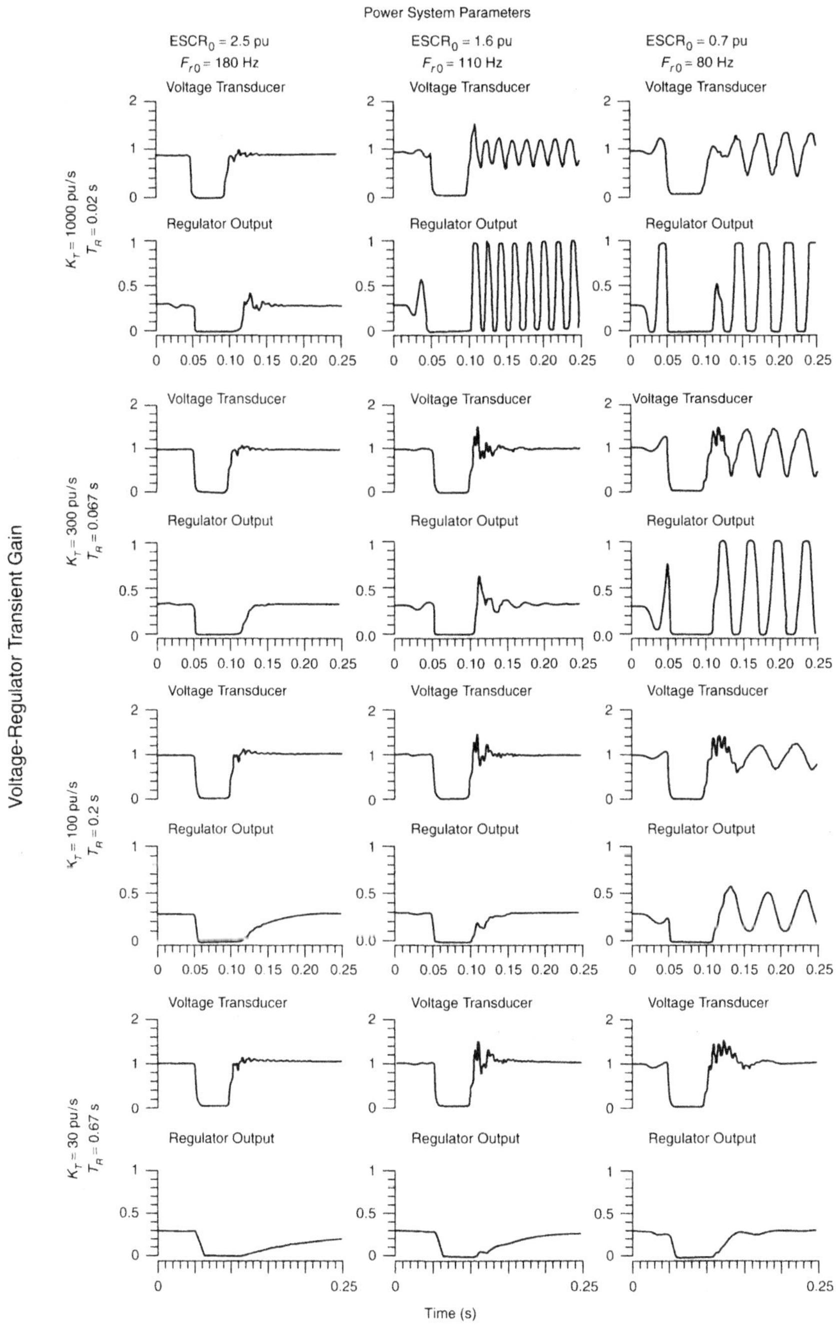

Figure 5.15 Effect of power-system characteristics on SVC transient response with $B_{T0} = 0.3$ pu.

5.3.2.1 Response Variation With Regulator Transient Gain, K_T For the strong system ($ESCR_0 = 2.5$ pu), the SVC response is very sluggish for $K_T = 30$ pu/s/pu, but quite fast for $K_T = 1000$ pu/s/pu. For the moderately weak system ($ESCR_0 = 1.6$ pu), the response for $K_T = 30$ pu/s/pu is less sluggish; however, as the gain is increased, the response becomes faster but also becomes oscillatory for $K_T = 300$ pu/s/pu and unstable for $K_T = 1000$ pu/s/pu. At this high value of K_T, the voltage-regulator susceptance output increases substantially and starts hitting the susceptance limits. For the very weak system ($ESCR_0 = 0.7$ pu), the SVC response is reasonably fast for the lowest selected gain, that is, $K_T = 30$ pu/s/pu; it becomes oscillatory at a slightly increased value of gain $K_T = 100$ pu/s/pu, unstable if the gain is increased further to $K_T = 300$ pu/s/pu and 1000 pu/s/pu. The stability limit of K_T is much lower with weak systems than it is with strong systems.

5.3.2.2 Response Variation With System Strength, $ESCR_0$ For the regulator gain $K_T = 30$ pu/s/pu, as the system strength $ESCR_0$ decreases from 2.5 pu to 0.7 pu, the SVC response tends to become faster. For $K_T = 100$ pu/s/pu, however, the sluggish response for the strong system ($ESCR_0 = 2.5$ pu) becomes quicker for the system with medium strength ($ESCR_0 = 1.6$ pu) and oscillatory for the weak system ($ESCR_0 = 0.7$ pu). With a high gain $K_T = 1000$ pu/s/pu, the SVC response is very fast, with acceptable overshoot for the strong system, but it becomes oscillatory and eventually unstable as the system strength declines.

From the foregoing studies, the following are illustrated

1. The SVC response, in general, becomes faster with the increase in transient gain of the voltage regulator.
2. If the gain is made very large, the SVC response may become oscillatory or even unstable.
3. A sluggish SVC response for a strong system becomes faster as the system strength deteriorates. If the regulator gain is optimized for a high system strength, the SVC response may become unstable for a weak system, implying that the SVC regulator gain should always be optimized for the weakest system state to ensure the stable response for any variation in system strength. The only repercussion of this strategy is that the SVC response will become slower as the system strength increases. However, this problem can be resolved through a variable gain strategy.

5.3.2.3 Voltage-Sensitivity Transfer Function The effect of the network resonance can be understood in terms of the voltage-sensitivity transfer function, which relates the SVC voltage-regulator output, ΔB_{CV}, to the output of the voltage-magnitude transducer, ΔV_{meas}. This transfer function essentially includes the characteristics of the network. The Bode plots for this transfer function corresponding to two power-system configurations having different resonant frequencies—80 Hz and 110 Hz—are illustrated in Fig. 5.16.

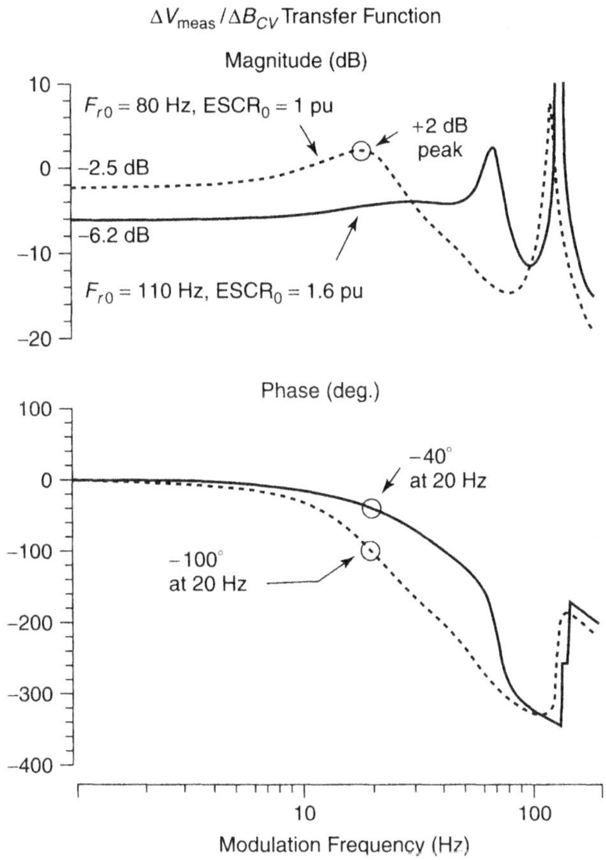

Figure 5.16 Effect of power-system response characteristics on voltage-sensitivity transfer function.

The two systems exhibit different steady-state gains for low-modulating frequencies. In case of the weak-system configuration ($ESCR_0 = 1.0$ pu), the effect of first-resonant frequency, $F_{r0} = 80$ Hz, is manifested as a peak in the gain at 20 Hz, which is the lower sideband (80 − 60 Hz) of the network resonance. At this modulating frequency, there occurs not only a large gain amplification of 45 dB over the steady-state gain, but also a significantly large phase lag of 100°, leading to a small stability margin. In comparison, the phase lag of the strong system at 20 Hz is just 40°. Thus for the weak system, if the phase lag associated with the voltage regulator and the various firing delays at 20 Hz is shown to be 80° (resulting in an overall phase lag of 180° in the voltage-control loop), an instability will occur.

Within the voltage-control loop, the frequency characteristics of the power system resemble that of a second-order control filter having a break frequency at the lower sideband of the network-resonant frequency. Since the phase lag of the network block at the break frequency (20 Hz) is fairly large, the stability margin of the voltage-controller loop deteriorates substantially.

5.3.3 Sensitivity to the TCR Operating Point

The operating-point average susceptance, B_{T0}, of the TCR is seen to have a positive influence on the voltage-controller stability [4], [5]. An increased value of stability margin is obtained for a higher average susceptance of the TCR. The reason is that the TCR presents itself as an effective inductance in shunt with the power system impedance at the fundamental frequency. Therefore, the system strength gets improved and the first-resonant frequency F_{r0} gets increased [4]. The modified values of system parameters are given by

$$\mathrm{ESCR}_M = \mathrm{ESCR}_0 + B_{T0} \qquad \text{in pu} \tag{5.60}$$

where ESCR_M = the ESCR that determines the system performance for small variations around B_{T0}

B_{T0} = the average TCR susceptance at the steady-state operating condition

The modified-resonant frequency, F_{rm}, with TCR conduction, B_{T0}, is given as [14]

$$F_{rm}^2 = \lfloor F_{r0}^2(\mathrm{ESCR}_0 + B_{T0}) + F_b^2 B_{T0} \rfloor / \mathrm{ESCR}_0 \tag{5.61}$$

where F_b = the system-base frequency (60 Hz)

Let the maximum modified value of the resonant frequency corresponding to full TCR conduction be denoted by F_{r1}. The variation of F_{r1} with F_{r0} with ESCR_0 as a parameter is illustrated in Fig. 5.17. A greater rise in resonant frequency is experienced if the system strength is lower, implying that weak ac systems are more sensitsive to TCR conduction.

The influence of increased operating-state TCR conduction can also be explained from the Bode plot of the voltage-sensitivity transfer function relating the measured voltage magnitude to the regulator output shown in Fig. 5.18. Comparing the plots depicted in Figs. 5.16 and 5.18 shows that a higher value of TCR conduction raises the resonant frequency to resemble that of a strong ac system.

The SVC transient responses with varying magnitudes of system strength corresponding to a TCR conduction $B_{T0} = 0.8$ pu are depicted in Fig. 5.19, which presents the same cases as shown in Fig. 5.17. The same general SVC behavior described in Section 5.3.2 for $B_{T0} = 0.3$ pu is also witnessed for $B_{T0} = 0.8$ pu. However, a comparison of SVC responses corresponding to TCR average conduction $B_{T0} = 0.3$ pu (shown in Fig. 5.17) with those corresponding to $B_{T0} = 0.8$ pu (shown in Fig. 5.19) reveals that increasing the TCR conduc-

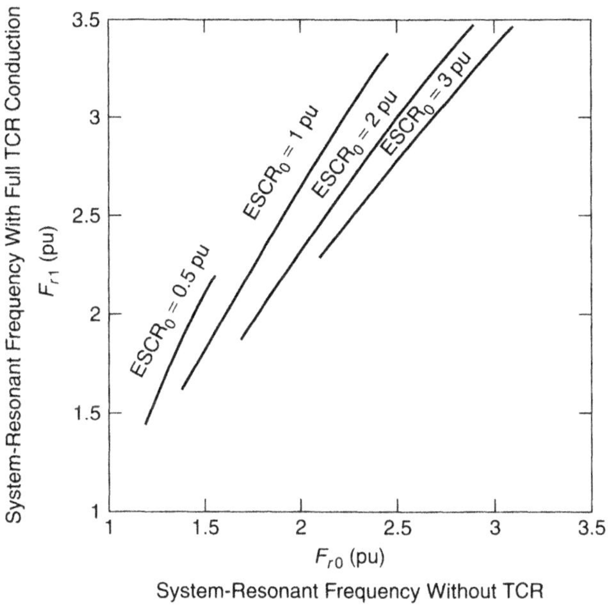

Figure 5.17 Effect of full TCR conduction on effective system-resonant frequency.

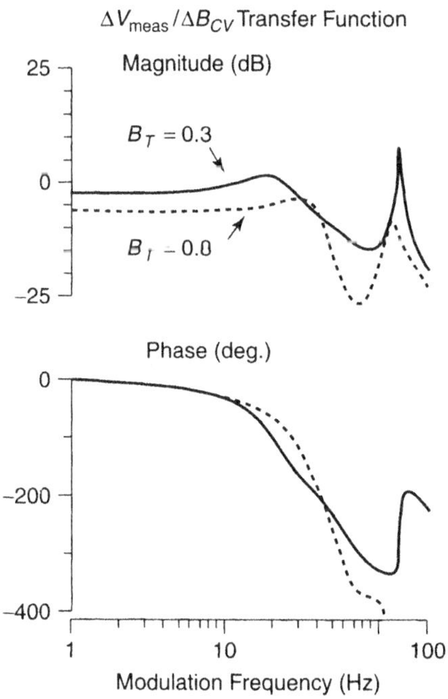

Figure 5.18 Effect of the TCR operating point on voltage-sensitivity transfer function.

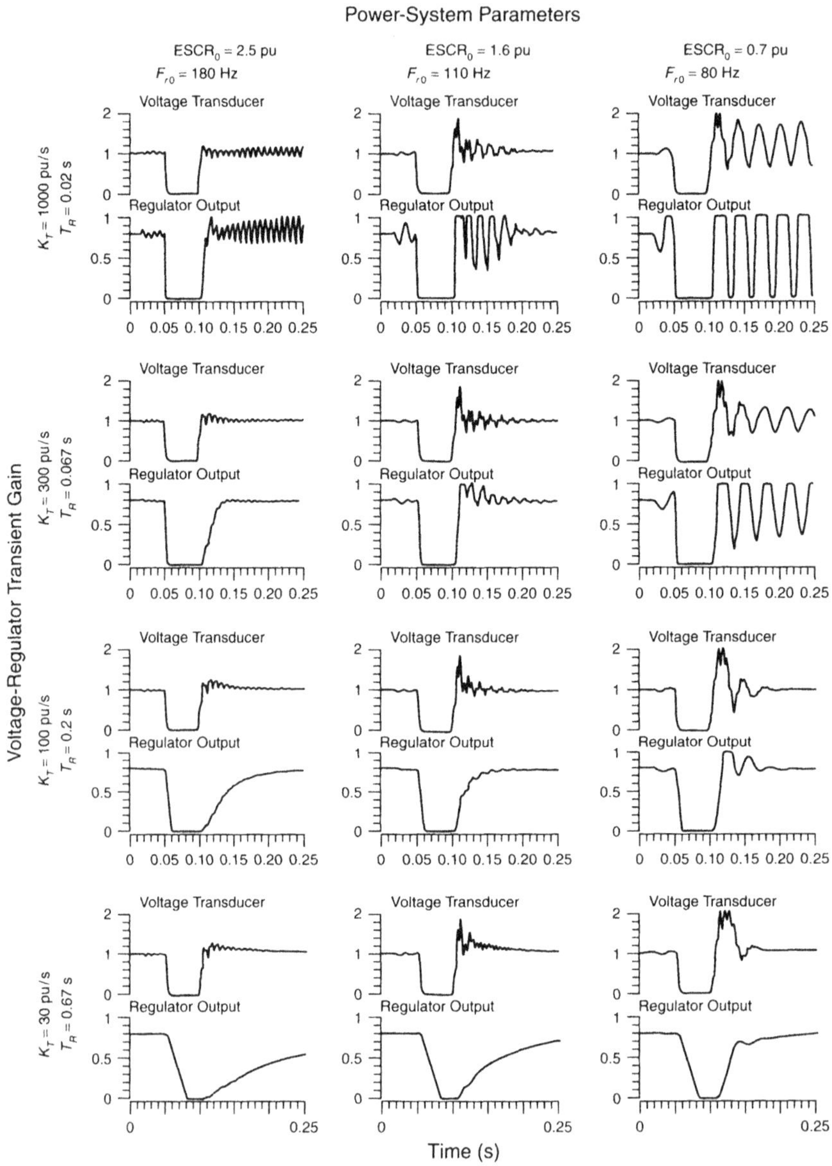

Figure 5.19 Effect of power-system characteristics on the SVC transient response with $B_{T0} = 0.8$ pu.

tion enhances the stability margin of the overall system. Cases that result in an unstable response with $B_{T0} = 0.3$ pu get stabilized with $B_{T0} = 0.8$ pu. Cases that generated a slow response with $B_{T0} = 0.3$ pu result in a faster response with $B_{T0} = 0.8$ pu. The only exception is the case corresponding to $\text{ESCR}_0 = 2.5$ pu and

$K_T = 1000$ pu/s/pu, which tends to become unstable with supersynchronous oscillations as the TCR conduction is increased because of a very high integral gain. However, this gain cannot be practically employed in systems subject to low-frequency resonance.

Certain SVC control systems are equipped with an inner loop where the TCR current is controlled directly [2] instead of the SVC susceptance. If this control loop is fast-acting, then the voltage-regulator stability margin becomes less sensitive to TCR conduction.

5.3.4 Choice of Transient Gain

The maximum values of transient gain K_{T1} at which instability sets in are obtained from eigenvalue analysis for a wide range of power-system parameters [4], [5] and are plotted in Fig. 5.20. These results correspond to a 60-Hz system with $B_{T0} = 0.3$ pu. For a 50-Hz system, both F_{r0} and K_{T1} need to be multiplied by a faster 50/60. It is evident from Fig. 5.20 that for the same ESCR_0, the transient-gain limit gets reduced as the resonant frequency decreases. For instance, as the resonant frequency decreases from 120 Hz to 80 Hz, the maximum transient gain gets lowered from 400 to 100 (i.e., four times).

The methodology usually adopted to obtain a preliminary estimate of the transient gain is as follows:

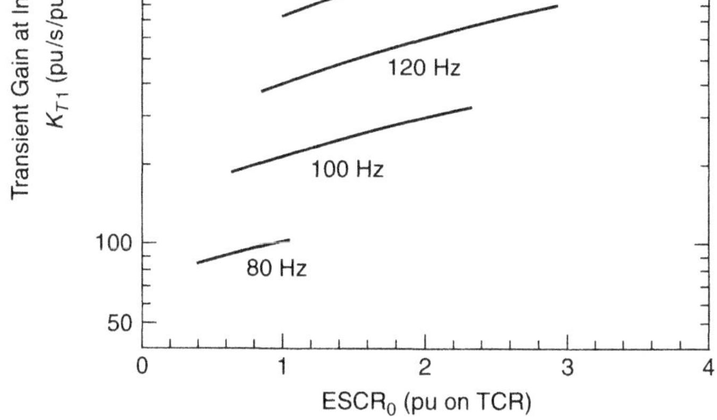

Figure 5.20 Transient gain at instability versus the power-system parameters with $B_{T0} = 0.3$ pu.

1. identify the weakest network state corresponding to the worst contingency;
2. determine the ESCR_0 and F_{r0} from impedance-versus-frequency studies;
3. calculate the transient-gain limit from eigenvalue studies; and
4. choose the regulator gain as half of the gain limit obtained previously.

Example 5.2 (Note: See also ref. [4].) Let the weakest network configuration be characterized by $\mathrm{ESCR}_0 = 1.5$ pu and $F_{r0} = 100$ Hz. The maximum value of the transient gain is obtained from Fig. 5.20 as 250 pu/s/pu. A tentative value of regulator gain is then calculated as half of 250, that is, 125 pu/s/pu. The steady-state gain, K_R, corresponding to a 5% current droop, is $1/0.05 = 20$ pu/pu. The regulator time constant is given by

$$T_R = \frac{K_R}{K_T} = \frac{20 \text{ pu/pu}}{125 \text{ pu/s/pu}} = 0.16 \text{ s} \tag{5.62}$$

It is emphasized that this method generates only a tentative estimate of the regulator gain that could be used in planning studies. The optimal value of regulator gain and other controller parameters must be determined at the time of detailed design of SVC equipment for the specific application.

5.3.5 Certain Features of the SVC Response

Figures 5.15 and 5.19 illustrate some interesting characteristics of the SVC response. As soon as a fault occurs, the response of the SVC is determined by its transient gain, not by the power-system parameters. Note that the initial decay of the SVC susceptance following the fault takes the same amount of time irrespective of the system strength. The rate of SVC-susceptance decay is equal to the transient gain.

When the fault is cleared, certain overvoltage is experienced. Although the SVC eventually controls this overvoltage, it has no influence on the first overvoltage peak, irrespective of how fast the control may be, because the initial overvoltage peak is determined by the resonance in the ac system. Even if the TCR is fired instantaneously, it will not affect the initial response of the ac resonant system.

The subsequent stabilization of the SVC voltage and TCR susceptance will, however, be dependent on the ac system characteristics, as discussed already in previous sections of this chapter. The SVC response is faster with weak ac systems and relatively slower with strong systems, as evident from Figs. 5.15 and 5.19. However, the SVC controller is more susceptible to instability with the weak systems.

5.3.6 Methods for Improving the Voltage-Controller Response

In SVC applications in which large variations in network short-circuit power are anticipated, the voltage-regulator gain is optimized for the weakest network state or the worst contingency to ensure a fast, stable response under such conditions. If this gain is maintained constant, even when the network assumes a normal configuration with a significantly enhanced short-circuit power, a much slower response will result. Nevertheless, it is, desirable for the SVC to have a fast response over the entire range of network configurations. The different methods available for improving the SVC response are discussed in the following text.

5.3.6.1 Manual Gain Switching This method involves predetermining the optimal regulator gains for different system-operating conditions and allowing the operating personnel to manually switch the gains according to the existing network states based on breaker-status signals. Higher gains can thus be used with stronger systems, resulting in a fast response. The drawback of this switching procedure is that the manual selection of gains cannot keep pace with the dynamically changing network conditions. In instances of large, sudden changes taking place within the network, this manual procedure may result in control instability. Even if the procedure were initiated automatically, it might not always be practical or possible to determine the optimum gains for numerous network-operating conditions. A block diagram of the control system incorporating gain switching is given in Fig. 5.21.

5.3.6.2 The Nonlinear Gain This gain function introduces an enhanced gain to provide a fast response in case of large voltage fluctuations. In this scheme, the

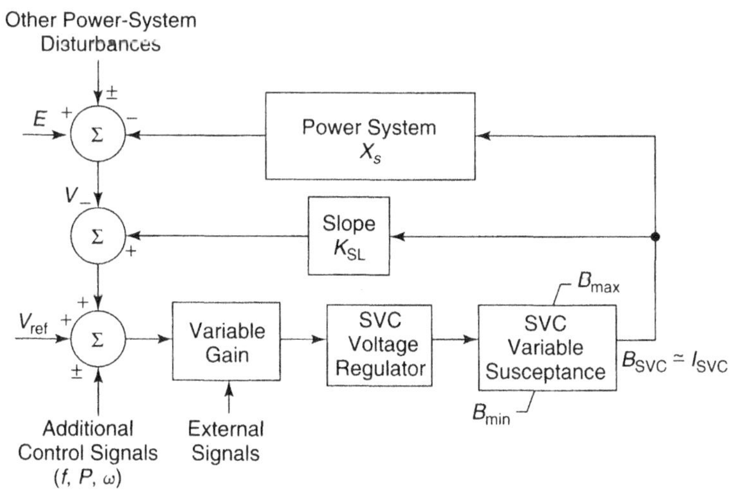

Figure 5.21 The SVC voltage controller with variable gain ($V \simeq V_{\text{rated}}$).

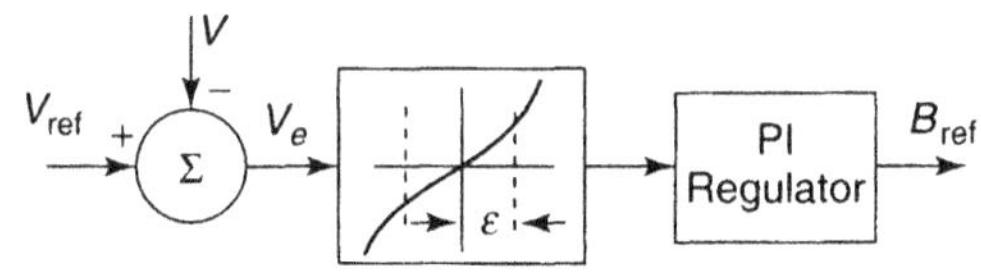

Figure 5.22 Nonlinear gain in the SVC voltage regulator.

variable gain block shown in Fig. 5.21 incorporates a nonlinear function, as depicted in Fig. 5.22. If the error voltage, ΔV, at the output of the referenced summing junction is less than a preset small value, ε, a unity gain is offered by the nonlinear function. Then, the response of the voltage regulator corresponds to small signals. If ΔV is large and exceeds ε, the gain increases continuously to give a fast response. If placing emphasis on a one-sided operation (overvoltage or undervoltage control) is desired, a larger gain on one side of the error signal can easily be built into the nonlinear characteristics. An example case demonstrating the effectiveness of nonlinear gain is given in Fig. 5.23.

5.3.6.3 Bang-Bang Control This is a limiting case of the nonlinear gain in which the TCR and/or the TSCs switch between their off and on states. This control is highly sensitive to the initiating disturbance and, hence, must be carefully preevaluated for all possible disturbances to avoid any instability.

5.3.6.4 The Gain Supervisor This [15], [16] is an automatic gain-control scheme that provides a continuous control of the regulator gain over a wide range of system-operating conditions, ensuring a consistently stable response in all situations. The gain supervisor implements an optimal gain based on normal system configuration and continuously monitors the output of the voltage regulator for any sustained or growing oscillations. Such oscillations result if any degradation in the system strength occurs while the regulator still operates with a high gain. The supervisor then repeatedly reduces the gain until the oscillations cease, thus maintaining control stability. As this gain super-

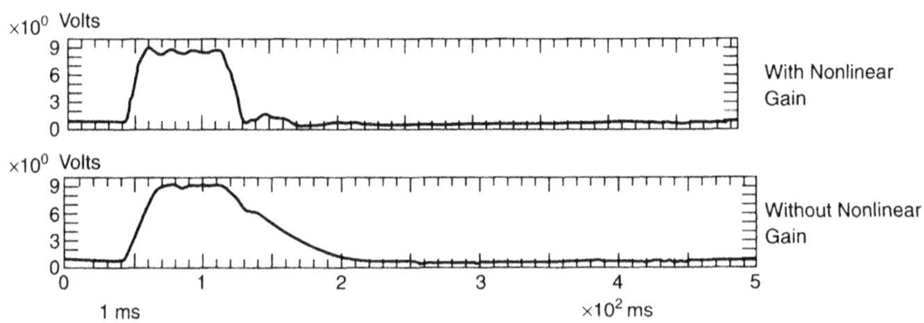

Figure 5.23 An example of the SVC response improvement with a nonlinear gain.

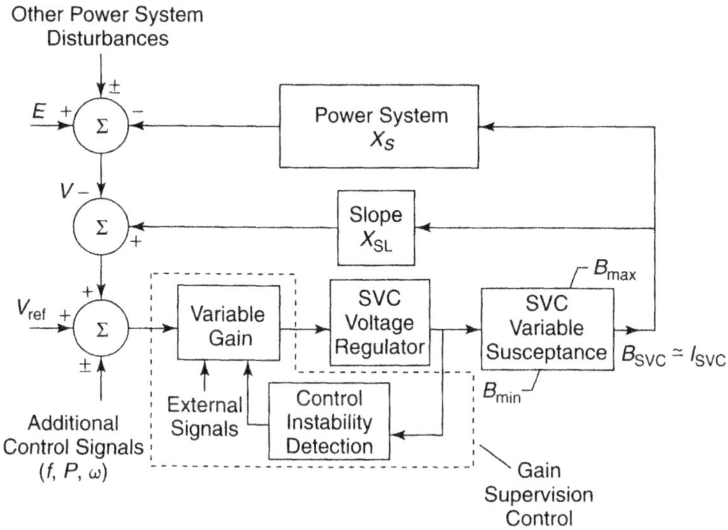

Figure 5.24 The connection of gain-supervisor control to the SVC voltage-control system ($V \simeq V_{\text{rated}}$).

visor is effective over a broad range of system-operating conditions, it has found widespread use in recent SVC installations.

The gain supervisor is connected in the SVC control system as shown in Fig. 5.24. It is a 3-terminal device, with one terminal used for instability detection and the other two used as the input and output, respectively, of the variable gain or voltage-control amplifier. The gain supervisor [15] comprises the following five main elements [as depicted in Fig. 5.25(a)]:

1. the input filter;
2. the level detector;
3. the pulse discriminator;
4. the integrator; and
5. the voltage-control amplifier (variable gain).

Input Filter This is a bandpass filter with its center frequency tuned to the frequency of the unstable controller mode. It thus allows the supervisor to respond only to the controller instability frequency, not to other system instabilities.

Level Detector This unit detects the presence of any oscillations. It compares the filtered voltage-regulator output with a preset level and generates pulses of duration equal to the time in which the input signals exceed the reference level. The magnitude of the preset level determines the sensitivity of the gain supervisor.

Pulse Discriminator The unit deletes certain erroneous pulses emitted by the level detector (such pulses do not imply an unstable operation). These unwanted pulses are generated when, for instance, there is a sudden change in the regulator output in response to a step change in the bus voltage. A fixed number of pulses are eliminated in a predefined time interval to avoid an unnecessary reduction in the regulator gain.

Integrating Unit This unit integrates the total number of pulses emitted by the pulse discriminator and maintains this output until such time that the integrator is reset. The integrator output constitutes a multiplication input to the voltage-controlled amplifier.

Figure 5.25(b) illustrates the various signals in the gain supervisor [15]. Note that the pulse discriminator deletes two pulses generated by the pulse detector. Figure 5.26 demonstrates how an unstable case with constant-regulator gain can be stabilized with a gain supervisor.

The gain supervisor is an effective device as long as the controller-instability frequency is far removed from the electromechanical-oscillation frequencies of the power system. It may be noted that the electromechanical-mode frequencies will not be given a path through the input filter and that the gain supervisor ensures control stability under degraded-system conditions but does not automatically guarantee optimal regulator gain during such conditions. It thus functions as a protection device. In system contingencies where optimal gains need to be selected to ensure a continually fast response, modifications in the gain supervisor or adoption of alternate methods are necessary.

One way of achieving optimal gains is through the use of integrator gain optimizers [16], [17]. Once a gain reduction has been initiated by the gain supervisor, these optimizers continuously adjust the gain to give the best response for those system conditions. They cause a small change in the voltage reference and examine the response of the voltage regulator. Based on the magnitude of the first undershoot, the gain is modified and another test is conducted. This process of adjusting the gain continues until either the original gain value has been restored or the gain has been optimized for the new prevailing system conditions. If the latter, the gain-optimization procedure is restarted after a certain delay period.

5.3.6.5 Series-Dynamic Compensation The voltage-regulator response can be improved by installing one or more dynamic compensator in series with the regulator [5], as shown in Fig. 5.27. This series compensator provides a phase lead to counteract the phase lags and delays that are inherent in the SVC voltage-control loop, including the ac system. As discussed previously, the delay introduced by the ac network is dependent on both the system strength and the SVC operating point. Figure 5.28 depicts the gain-and-phase characteristics of the open-loop controller-transfer function, B_{SVCR}/B_{SVC_0}, without the series-dynamic compensation. These plots are obtained for a specified set of voltage-regulator parameters for two power systems, one of which is moder-

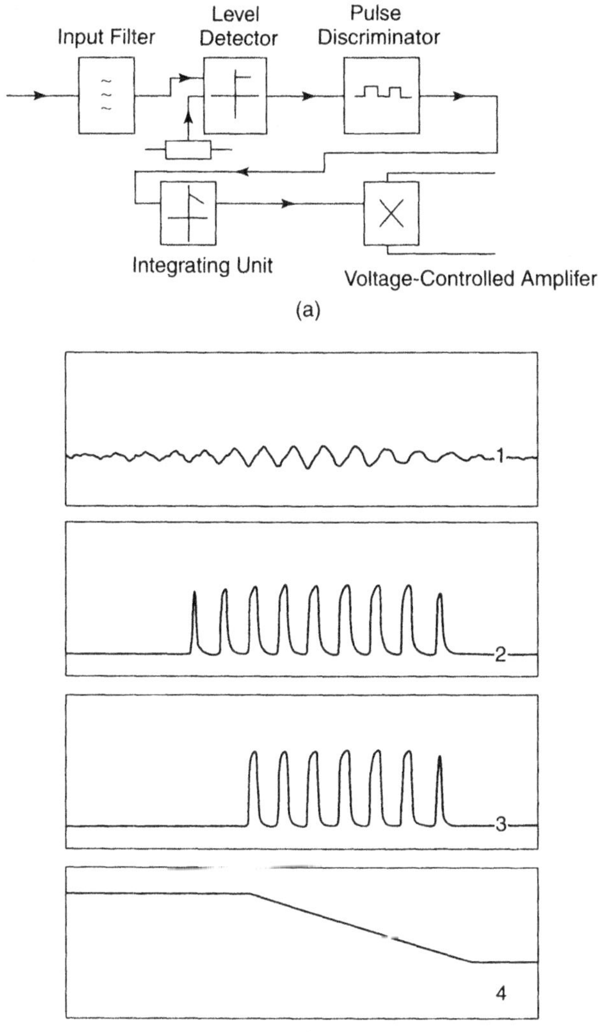

Figure 5.25 (a) A block diagram of the gain supervisor and (b) the gain-supervisor signals: (1) input signal; (2) output from the pulse detector; (3) output from the pulse discriminator (two pulses deleted); and (4) the signal proportional to the gain reduction.

ately weak with $\text{ESCR}_0 = 0.7$ pu and the other has $\text{ESCR}_0 = 1.6$ pu. For each power system, the transfer-function characteristics are plotted for two levels of SVC susceptance outputs, B_{SVC}. A low-average TCR conduction of $B_{\text{TCR}} = 0.3$ pu corresponds to B_{SVC} $(= 1 - B_{\text{TCR}}) = 0.7$ pu, which is depicted by the solid curve. On the other hand, a high-average TCR conduction of $B_{\text{TCR}} = 0.8$ pu corresponds to $B_{\text{SVC}} = 0.2$ pu, which is depicted by the dashed curve.

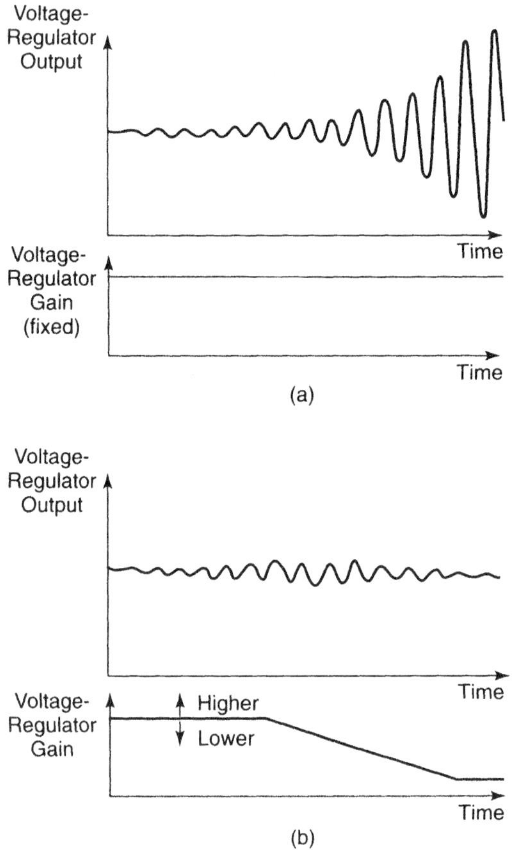

Figure 5.26 Behavior of the SVC voltage controller: (a) without gain-supervisor control and (b) with gain-supervisor control.

The low TCR conduction ($B_{SVC} = 0.7$ pu) results in a greater phase lag (lower phase margin) relative to the high-average TCR conduction ($B_{SVC} = 0.2$ pu). This effect is even greater if the power system is very weak ($ESCR_0 = 0.7$ pu). Thus, to maintain system stability, a greater need exists to provide phase compensation under such weak network configurations.

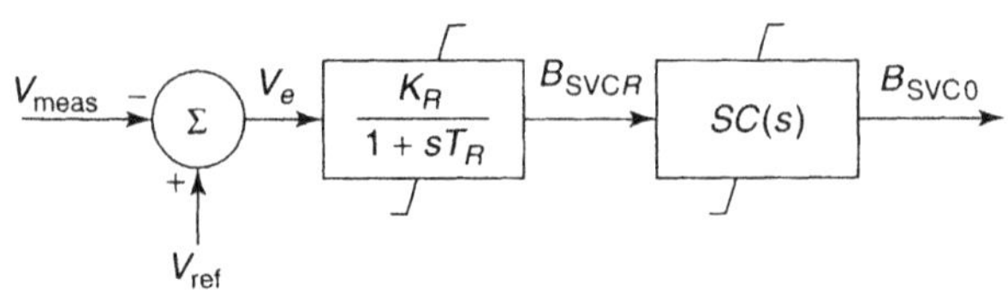

Figure 5.27 The SVC voltage regulator with series-dynamic compensation.

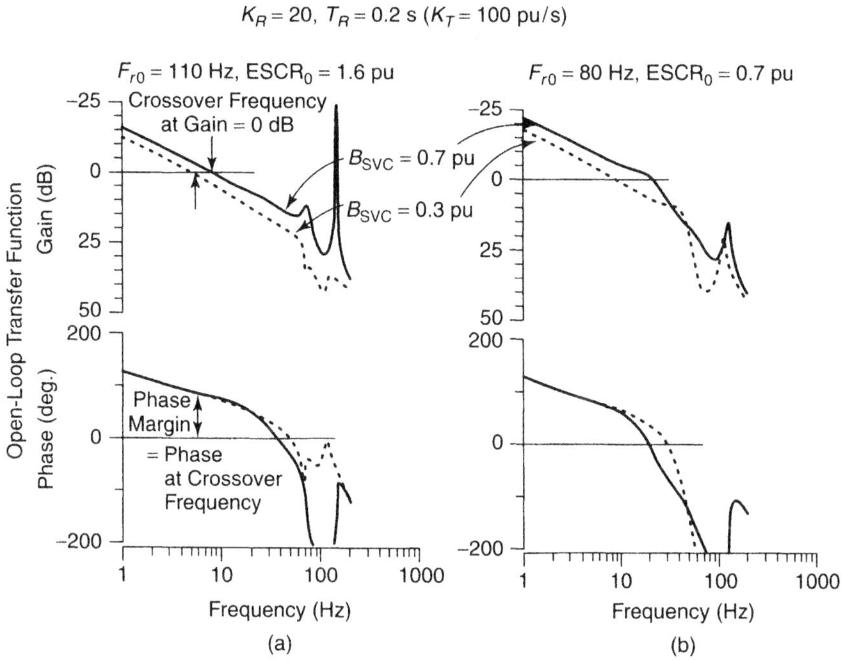

Figure 5.28 Gain-and-phase margins for a simple SVC voltage regulator.

The series-compensation device offers a phase lead in the range of gain crossover that occurs typically between frequencies of 10 to 20 Hz. As the phase lag offered by the ac network is not a constant but, instead, varies with TCR conduction, the series-compensating device is designed to be nonlinear for providing adequate compensation over the entire range of SVC operations. An example of a series compensator and its corresponding gain-and-phase characteristics is depicted in Fig. 5.29. It is seen that, corresponding to a lower TCR conduction (which implies higher B_{SVC}), the compensator provides an increased phase but with a diminished gain value.

The influence of the nonlinear series compensator on the transient performance of the system is demonstrated in Fig. 5.30. This figure demonstrates the SVC responses for the two levels of SVC susceptance outputs: $B_{SVC} = 0.2$ pu and $B_{SVC} = 0.7$ pu, corresponding to two settings of the voltage-regulator transient gain K_T. All the responses are depicted both with and without the series-dynamic compensation. It is seen that the compensator helps stabilize the low-TCR-conduction cases and provides enhanced damping with high-TCR-conduction cases. The series-dynamic compensator is more effective when the K_T is small.

5.3.6.6 ac-Side Control Filters Another approach that has been employed successfully in many SVC installations for counteracting the adverse interaction

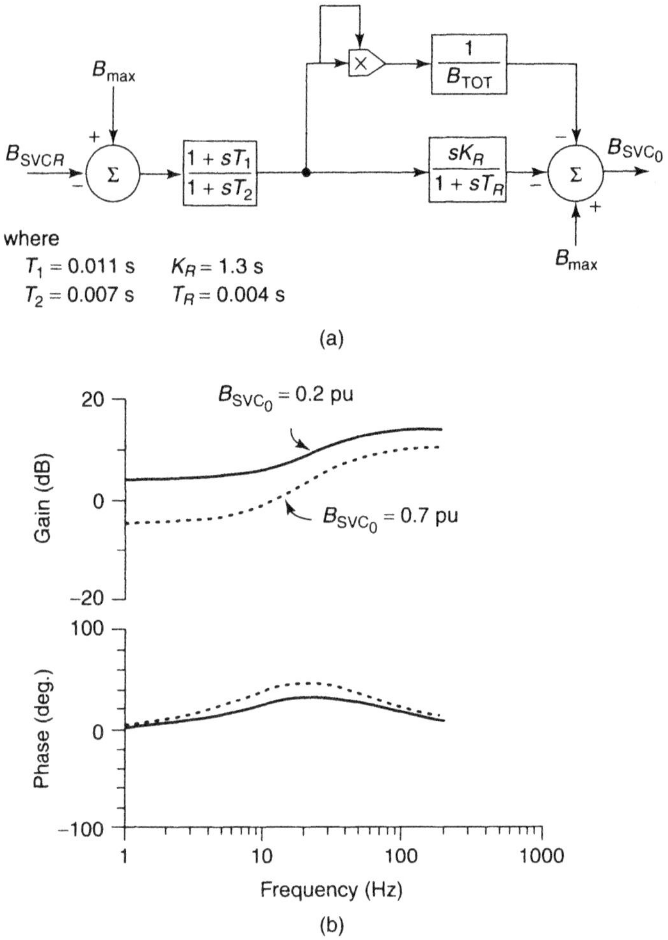

Figure 5.29 (a) A block diagram and (b) transfer function ($\Delta B_{SVC_0}/\Delta B_{SVCR}$) of a nonlinear series-dynamic compensator for the SVC voltage regulator.

between the SVC voltage regulator and the low-order network resonances is the installation of a notch filter on the ac side of the voltage-measurement system. This filter is tuned to the critical-resonant frequency in the 80–100 Hz range and is designed to provide minimum phase lag at 60 Hz to not significantly slow the response of SVC. The notch filter blocks the network-resonant frequency that otherwise translates to a lower frequency from the demodulation effect of the voltage transducer, and after circulating through the voltage-regulation loop, the frequency reenters the network in such a way that it causes resonance and consequent instability. Thus the effect of the notch filter is to increase the stability margin of the voltage-regulation loop.

The system loads exercise a significant damping influence on the network

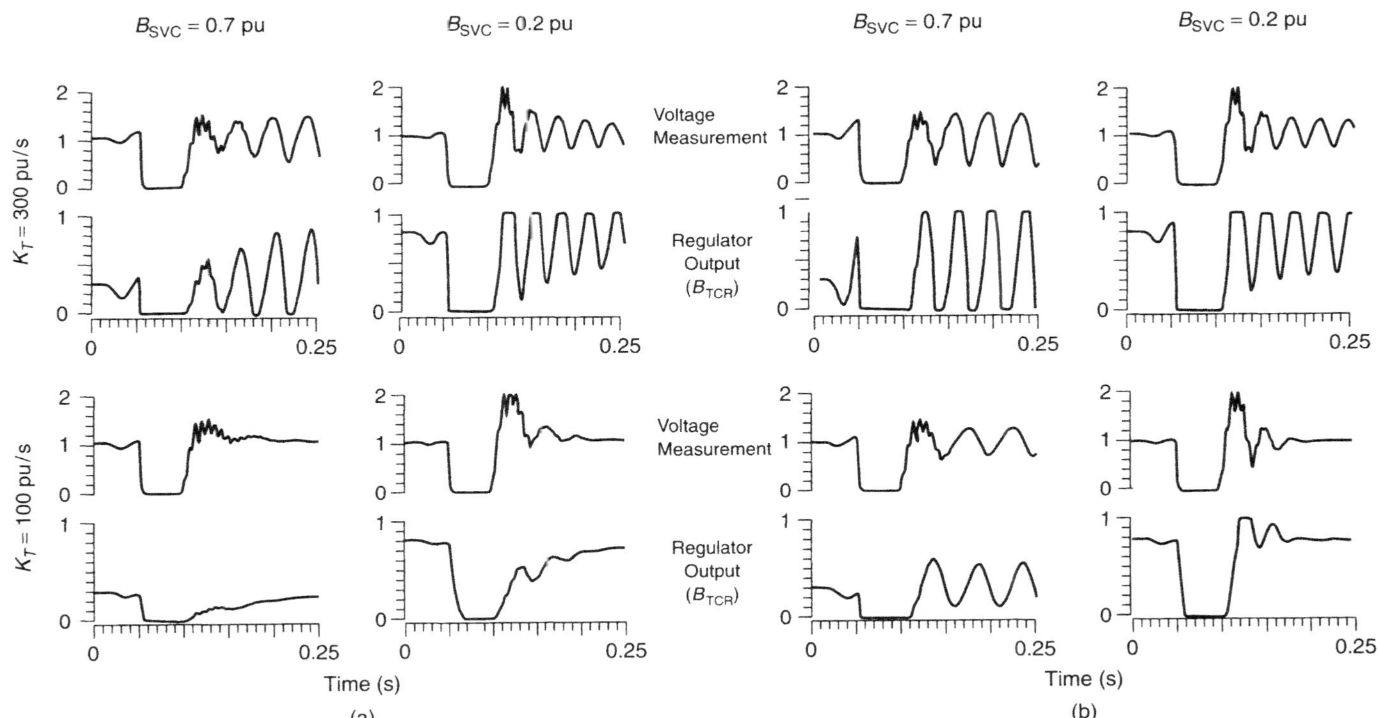

Figure 5.30 The effect of nonlinear series-dynamic compensation on the SVC transient response with a weak ac system (F_r = 80 Hz, ESCR = 0.7): (a) with nonlinear series-dynamic compensation and (b) with the simple voltage regulator.

resonance. For this reason, they should be accurately modeled in a TNA or other digital-simulation study to obtain a realistic estimate of the system behavior under the resonant conditions.

Figure 5.31 presents a comparison of the SVC responses both with and without the ac notch filter [2]. Both responses have the same regulator-gain setting, and both correspond to the clearing of a single line-to-ground fault. In the absence of the ac filter, the system comes close to instability. The ac system has a nearly 80-Hz resonant frequency, causing an oscillation in B_{ref} at a frequency of about 20 Hz (80 − 60 Hz). When the ac filter is installed, the SVC response becomes very well damped.

In designing these filters, the bandwidth must be carefully selected to allow for any deviations in the system frequency.

5.4 THE 2ND HARMONIC INTERACTION BETWEEN THE SVC AND AC NETWORK

The 2nd harmonic instability [2], [5], [12], [18]–[20] is a major problem experienced by SVCs during events that result in excessive production of 2nd harmonic currents, such as fault clearing, reactor/transformer switching, operation at significantly high voltage levels, and noise or imbalance in control signals or during severe geomagnetic disturbances. An elaborate analysis of this instability phenomenon is presented in ref. [5].

5.4.1 Influence of the 2nd Harmonic Voltage on the TCR

The TCR current is an integral of voltage across the TCR during the conduction interval and is thus very sensitive to any distortion in the TCR voltage. Figure 5.32 depicts the single-phase TCR current as affected by the superposition of a large 2nd harmonic voltage component on the fundamental voltage across the TCR while the issuing of firing pulses continues in an equidistant manner. Two cases are illustrated for a single-phase TCR; in each case, the dashed line represents the normal situation and the solid line represents the conditions during the 2nd harmonic voltage distortion. In case the 2nd harmonic voltage, V_2, is in phase with the fundamental-voltage wave, V_1, the area of the positive-side-current loop is less than the negative side, resulting in a negative direct current. If, however, the V_2 is phase-shifted by 90°, a positive dc current results, the magnitude of which is dependent on the magnitude of distortion and its phase relationship with the fundamental. In addition to the 2nd harmonic, other even-harmonics are produced, but their magnitudes are much lower than that of the 2nd harmonic.

In a 3-phase delta-connected TCR, two distinct kinds of voltage distortion result—positive sequence and negative sequence—depending on the phase sequence of the 2nd harmonic voltage V_2. The positive and negative sequences of voltage distortions are depicted in Figs. 5.33 and 5.34. Figure 5.33 illus-

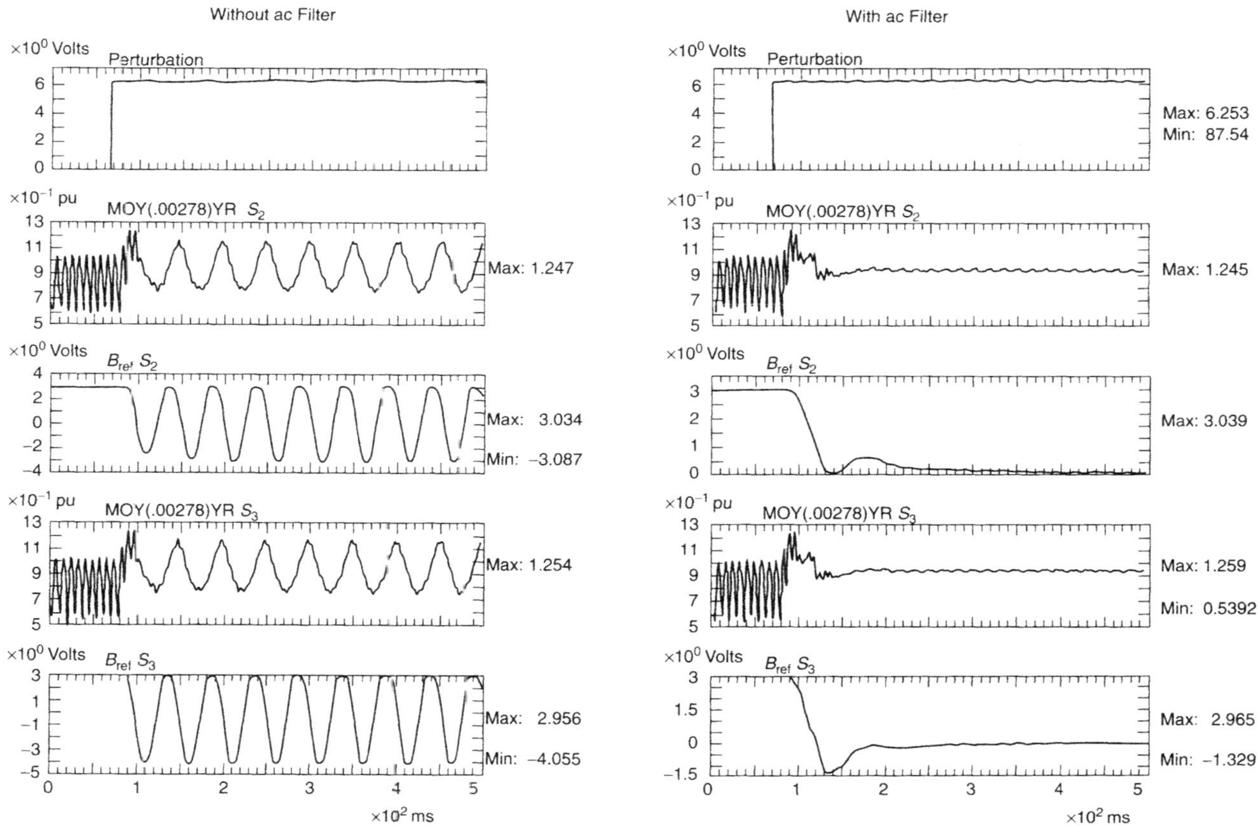

Notes: (1) Same regulator setting. (2) Phase-A-ground-fault clearing.
(3) For small perturbation, the compensators are marginally stable without the ac filter.

Figure 5.31 Effect of the ac filter on the compensator response for a network resonance near the fundamental frequency.

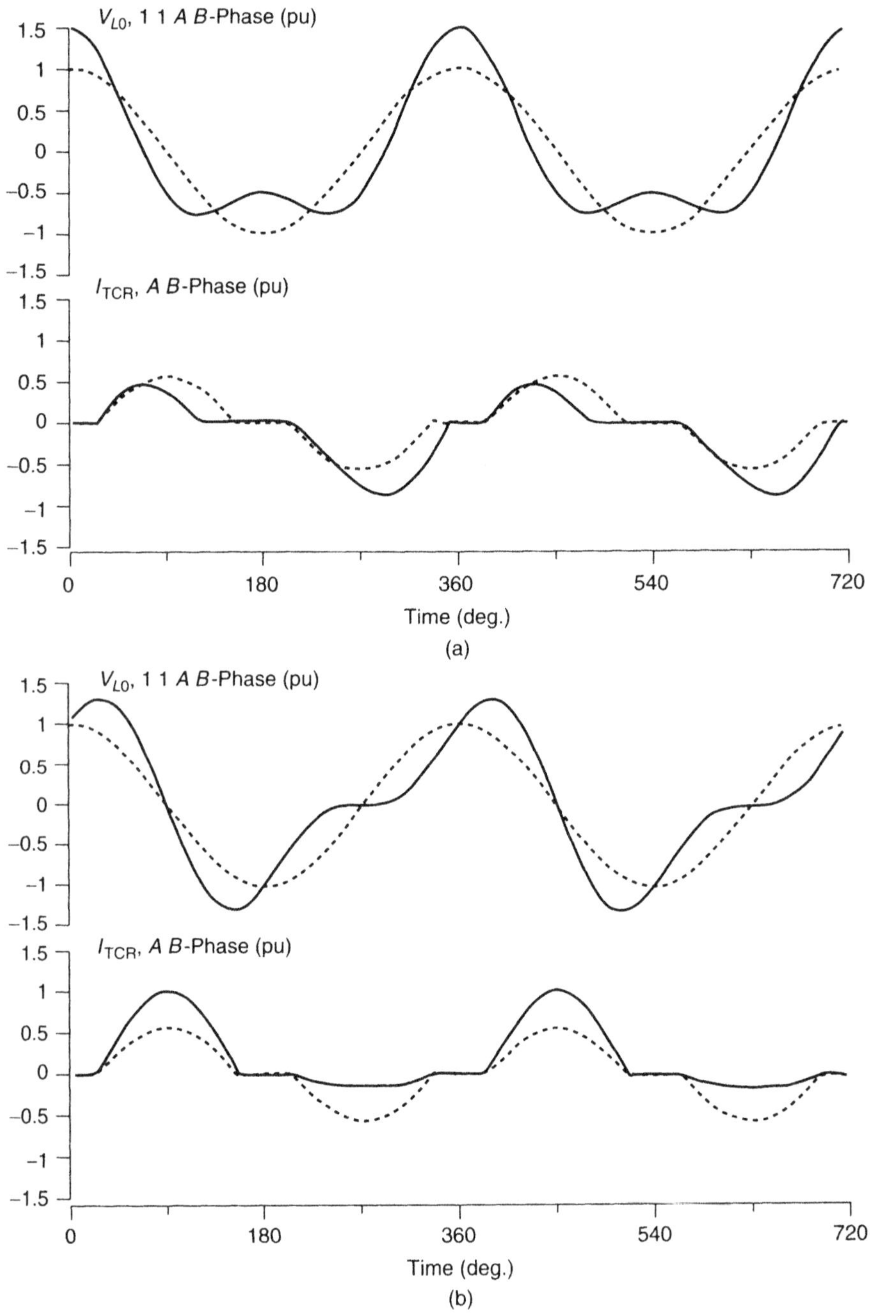

Figure 5.32 Natural response to the 2nd harmonic voltage: (a) V_2 in phase with V_1 and (b) V_2 lagging behind V_1 by 90°.

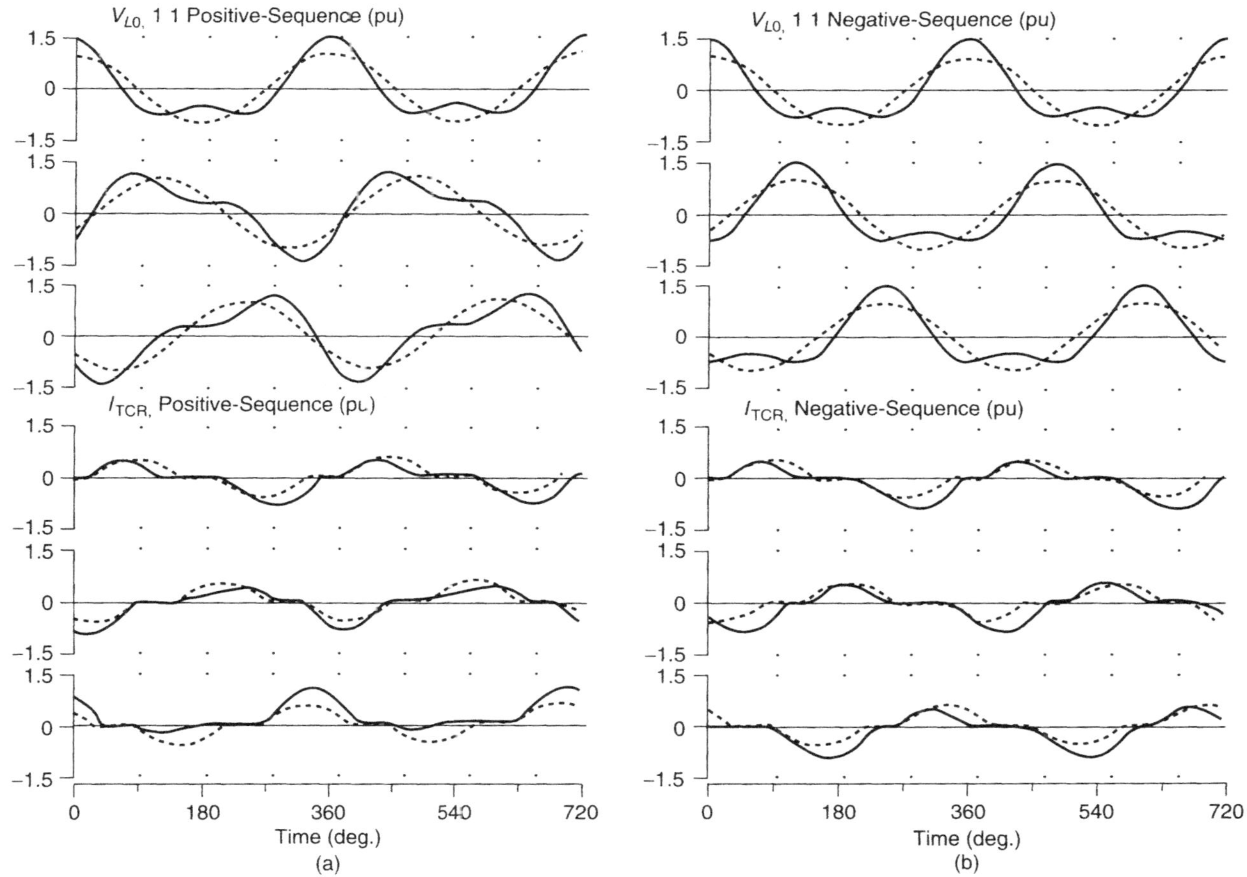

Figure 5.33 Natural response to the 2nd harmonic voltage for V_2 in phase with V_1 (the solid line corresponds to 50% V_2; the dashed line corresponds to $V_2 = 0$): (a) positive-sequence V_2 and (b) negative-sequence V_2.

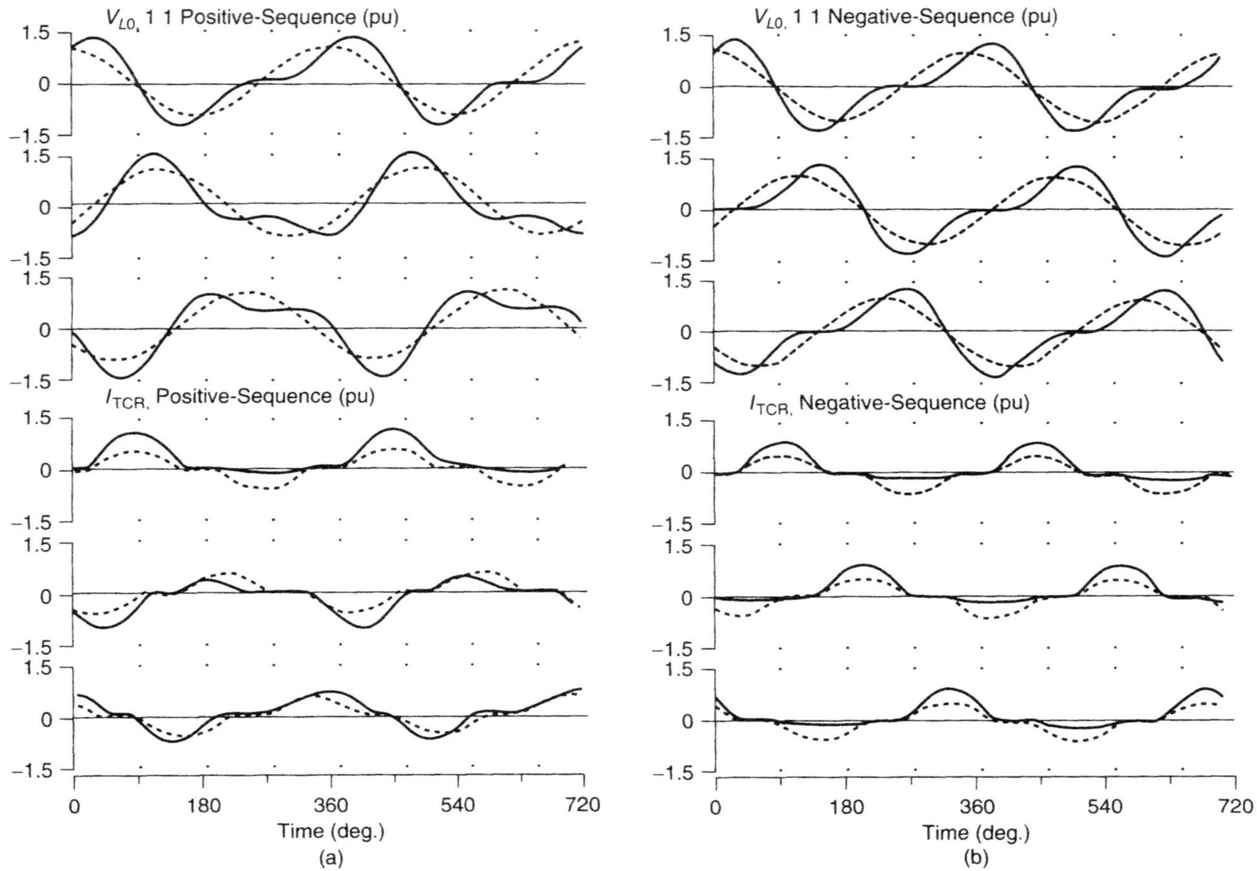

Figure 5.34 Natural response to the 2nd harmonic voltage for V_2 that is 90° out of phase with V_1 (the solid line corresponds to 50% V_2; the dashed line corresponds to $V_2 = 0$): (a) positive-sequence V_2 and (b) negative-sequence V_2.

trates when the V_2 is added in phase with the fundamental; Fig. 5.34, when V_2 is 90° phase–displaced from V_1. In positive-sequence distortion, the voltages and currents in all the three phases are different. However, in negative-sequence distortion, the voltage and current waveforms in all the three phases are same except that they are phase-displaced by 120°.

5.4.2 Causes of 2nd Harmonic Distortion

The 2nd harmonic interactions between the TCR and the ac system occur either during or after major system disturbances. This problem may also be sometimes experienced from spurious 60-Hz oscillations that may be induced on the dc side of the voltage-measurement system. Another possibility of exacerbation of the 2nd harmonic distortion is when the thyristor firing is unsymmetric.

5.4.2.1 Fault Clearing When faults take place in the ac system and are subsequently cleared, the transformers may develop a considerable amount of dc-offset flux. This flux results in a magnetizing current that is not symmetrical but is beset with a substantial dc component, leading to a voltage distortion with 2nd harmonics.

A 120-Hz component in the ac voltage causes a 60-Hz oscillation on the dc side of the voltage transducer from the demodulating effect of the measurement system. This 60-Hz frequency component is reflected in the susceptance reference signal B_{ref}, which causes an unequal firing of the thyristor. An asymmetrical TCR current results, with a high-dc component. The dc current flows into the SVC coupling transformer, saturating it unevenly and producing a magnetizing current with a large 2nd harmonic current. This magnetizing current, as it is drawn from the ac system, generates 2nd harmonic voltage components, the magnitude of which depends on the 2nd harmonic equivalent impedance of the network. If the ac system is weak and exhibits high impedance at the 2nd harmonic frequency, the overall 2nd harmonic voltage distortion is sustained or even amplified, resulting in 2nd harmonic instability.

An example of 2nd harmonic instability is shown in Fig. 5.35. (Also, see ref. [5].) It is interesting to note that the TCR currents are offset in a direction opposite that of the magnetizing current of the SVC coupling transformer in each phase. Both these currents take about one second to stabilize with a high dc offset, which can either persist indefinitely or decay with a very large time constant from the small, resistive component of the TCR and the coupling transformer. The severity of this voltage distortion is less with the TSC–TCR type of SVC compared to the FC–TCR type, as a smaller reactor size in the TSC–TCR generates a much lower magnitude of dc offset for the same asymmetry in firing angles.

It should be noted that as long as the dc offset persists in the transformer flux, harmonics in the magnetizing current continue to be injected in the network, which may interact adversely with network-resonant impedances.

The 2nd harmonic instability is somewhat alleviated by connecting 60-Hz

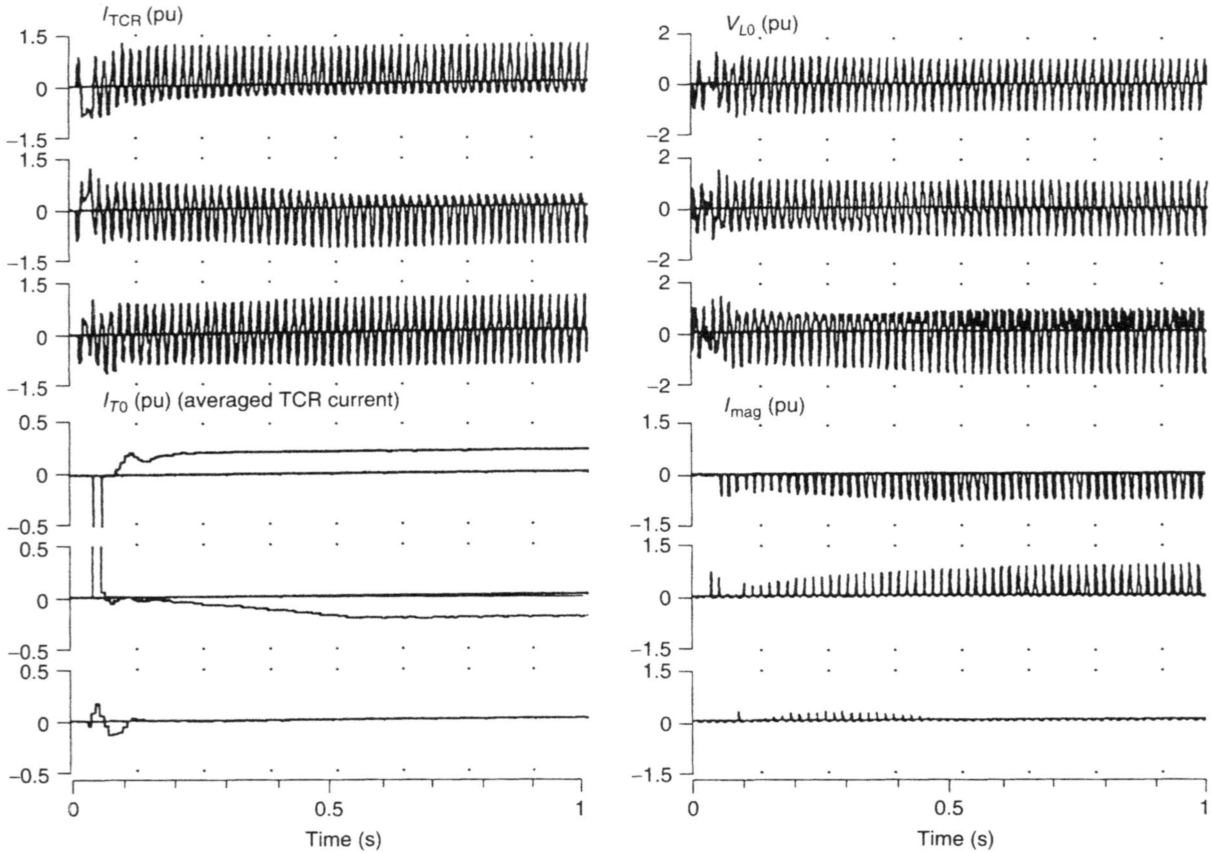

Figure 5.35 A 3-phase fault for a weak ac system (no TCR balance control).

notch filters on the dc side of the measurement system to prevent any 60-Hz oscillations from entering into the voltage-regulator control loop. At the same time, it must be ensured that no feedback of the 2nd harmonics can take place through the synchronization system. Doing so is generally achieved by employing phase-locked loop (PLL)–based synchronization systems, which have now become almost standard in SVC synchronization systems.

The aforementioned countermeasures are still not adequate, as 2nd harmonic interactions can occur through the TCR. Even through the firing is symmetrical, the TCR current can still be asymmetrical and have a finite dc component. The reason is that the TCR current is obtained as the integral of the voltage that may have been already distorted with 2nd harmonics.

Balancing the conduction period in both the positive and negative half-cycles also fails to offer an adequate solution. The only remedial measure that can be implemented through the control system is a TCR balance circuit, which balances the integral of the TCR current in each polarity and thus annuls the production of any dc current. (The operation of this circuit is discussed in Section 5.4.3.)

Another solution to this 2nd harmonic instability problem is by means of a 2nd harmonic power filter [18], [20], which prevents the passage of 2nd harmonic currents into the network. This is, however, an expensive solution.

5.4.2.2 Reactor/Transformer Switching Near an SVC When a large reactor or a transformer is energized near an SVC, the SVC coupling transformer can undergo saturation [20]. This activity may cause a considerable magnitude of dc current to flow between the SVC coupling transformer and the switched transformer or reactor. This phenomenon is slightly different than that obtained as a result of ac system faults, in which case the dc current flows only between the TCR and the coupling transformer.

The generation of dc current is dependent on the point on the voltage wave at which the reactor is switched. If this switching occurs at the voltage zero, maximal dc offset will be obtained. A case of reactor switching at voltage zero in the vicinity of an open secondary transformer (which, in our case, is the SVC coupling transformer) is shown in Fig. 5.36.

The reactor-switching–produced dc current flows through the source impedance, setting up a dc voltage component in the bus voltage V_T. As this dc voltage appears at the terminals of the transformer, a dc component develops in both the transformer flux and the resulting magnetizing current. A dc-offset current eventually starts circulating between the reactor and the transformer, and the resulting 2nd harmonic distortion of the bus voltage is sustained for an extended period of time (typically, tens of seconds), for the decay-time constant of the dc current is very large because of the low resistance in the dc path.

In situations when the transformer happens to be the coupling transformer of an SVC, a sustained 2nd harmonic voltage distortion at its terminals can cause a 2nd harmonic instability, as discussed earlier.

To accelerate the decay of the dc-offset voltages, preinsertion resistors are

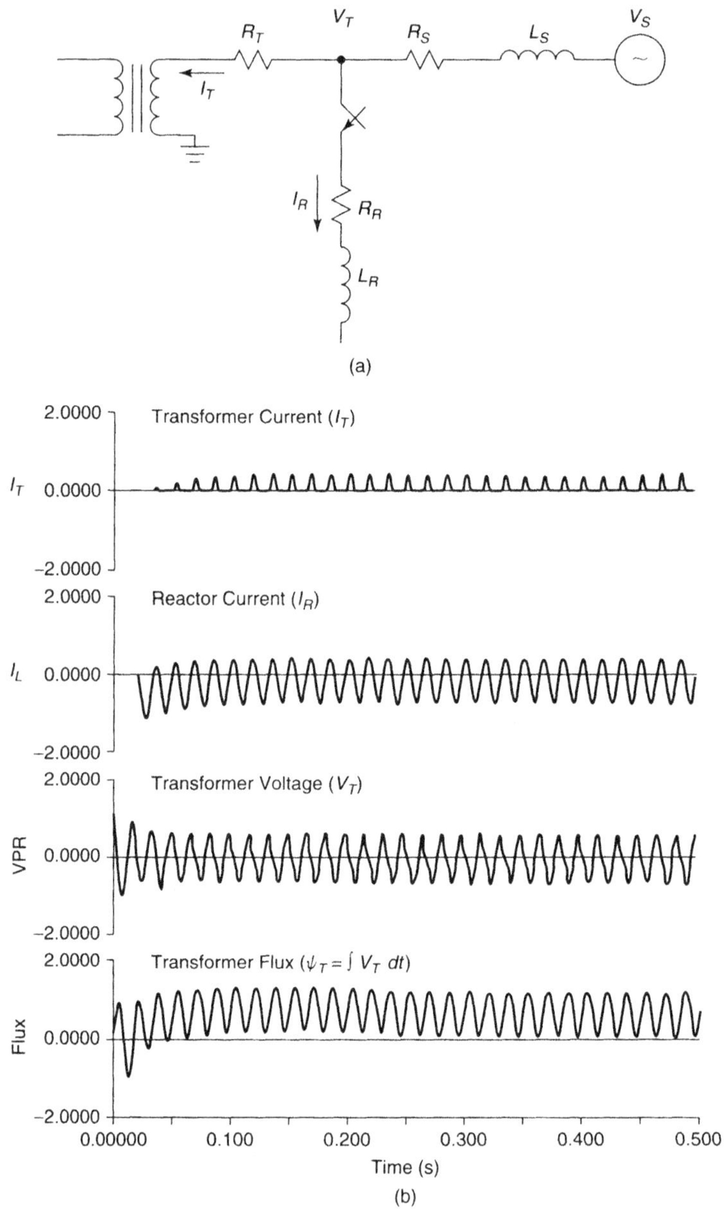

Figure 5.36 An example of reactor switching near the transformer: (a) a circuit diagram and (b) a transient response.

used while they energize large reactors on ac systems. The production of dc offset can be avoided by employing breakers, the switching times of which can be controlled accurately.

5.4.2.3 Geomagnetically Induced Currents The influence of solar magnetic disturbances on power systems and, specifically, on SVCs is well documented in [19], [21], and [22]. The earth's magnetic field is highly sensitive to solar eruptions. These eruptions create a quasi-direct potential difference along the surface of the earth. Consequently, slowly varying geomagnetically induced direct currents flow in the power systems through the solidly grounded neutrals of the transformers, causing dc saturation of transformers in the network that may lead to significant 2nd harmonic distortion.

A distortion in the TCR voltage causes a dc offset to develop in the TCR current that is equal in all the three phases and flows in the closed-delta configuration of the TCR. This dc offset, incidentally, is in a direction opposite that of the transformer flux.

A distinctive feature of the geomagnetically induced voltage distortion is that any kind of TCR control does not have a diminishing effect on this distortion, as it is an externally induced phenomenon that continues to persist. On the other hand, the 2nd harmonic distortion caused by such system disturbances as faults or reactor switching can be influenced by TCR control, as these disturbances are essentially finite-duration events.

5.4.2.4 Noise or Imbalance in the Control Systems Sometimes, spurious 60-Hz oscillations, or noise, is induced on the dc side of the voltage transducer, causing a dc component to develop in the TCR currents that may not be the same in all the three delta-connected phases. Therefore, a finite dc current flows in the SVC coupling transformer, eventually leading to 2nd harmonic distortion of the bus voltage and associated undesirable effects.

A similar voltage distortion also results in certain situations when an unequal spacing between firing angles produces unequal TCR conduction in the positive and negative half-cycles.

5.4.3 TCR Balance Control

One cause of 2nd harmonic voltage distortion is the dc component in TCR current, for which reason the SVC control systems in many applications are provided with a TCR balance control (TCRBC) [5]. The principle of the TCRBC is to modulate the firing angle to allow a different amount of conduction in the positive and negative polarity. Eventually, this activity reduces the dc component to zero. The TCRBC control is designed to have a large time constant; hence its operation does not conflict with the fast control dynamics of the SVC voltage regulator.

A block diagram of the TCRBC is shown in Fig. 5.37. The dc component of the TCR current is first measured through an averaging process. The instantaneous TCR current is integrated over each half-cycle, and the integral quantity obtained is added to those corresponding to the preceding two half-cycles. The average current, I_{T0}, is then computed and updated every half-cycle. The averaged TCR current is integrated to generate the balance regulator output signal,

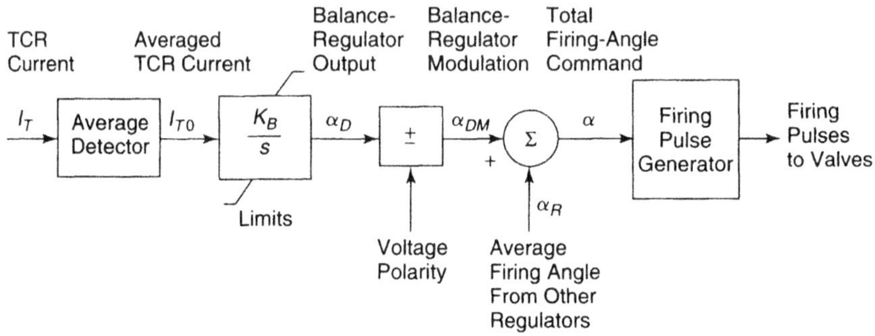

Figure 5.37 A block diagram of a basic TCR balance control (TCRBC).

α_D, that determines the magnitude and polarity of the firing-angle modulation. The firing angle is modulated by a signal that alternates at each half-cycle.

The operation of a TCRBC [5] is demonstrated in Fig. 5.38. A large 2nd harmonic voltage distortion is created on the low-voltage side of the SVC coupling transformer at 0.05 s by adding a 50% 2nd harmonic component to the fundamental voltage. Doing so causes a substantial dc offset to appear in the TCR current, with equidistant firing control. The TCRBC operation is initiated at 0.1 s, which gradually reduces the dc component in the TCR current to zero.

The effectiveness of a TCRBC in the fault situation of Fig. 5.35 is illustrated in Fig. 5.39, in which the 3-phase outputs of each major signal are shown. The high speed of the TCRBC response is achieved mainly because of the large gain of the balance regulator, which may not be possible to implement in all situations owing to stability considerations.

One way of accelerating the decay of dc offset in the TCR current is by incorporating an additional feedback of transformer flux. Because dc offset in the TCR current is caused by distortion in the transformer voltage, which, in turn, is caused by changes in the transformer flux, it is logical to expect that a speedup in the TCRBC response can be achieved by modulating the firing angle in response to the transformer flux change in addition to the average value of the TCR current.

The flux change is, in fact, derived by passing the transformer flux through a washout circuit. However, it can be equivalently obtained as the output when the TCR branch voltage is passed through a low-pass filter. This equivalence is illustrated in Fig. 5.40(a). The flux-change signal is filtered to remove any fundamental-frequency component using a notch filter. Such filtering improves the signal-to-noise ratio. The filtered-flux change signal is multiplied by a gain and added with the averaged TCR current, as depicted in Fig. 5.40(b). The enhancement in performance of the TCRBC with auxiliary flux-change input is illustrated in Fig. 5.41. The offset in the TCR current is mitigated in a much shorter time.

The considerations for design of the TCRBC to ensure stability under all realistically possible system configurations are discussed in ref. [5].

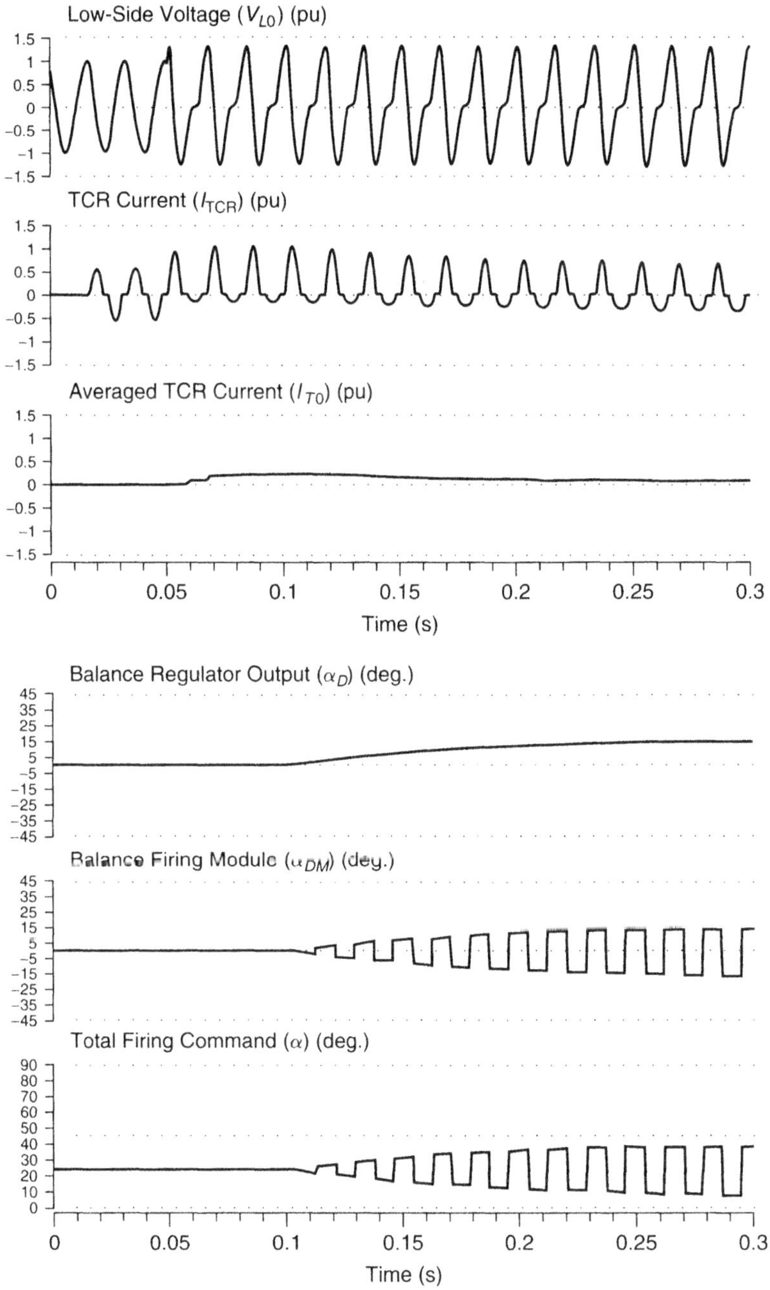

Figure 5.38 Natural response to the 2nd harmonic voltage for 50% V_2 that is 90° out of phase with V_1 (TCRBC is released at 0.1 s).

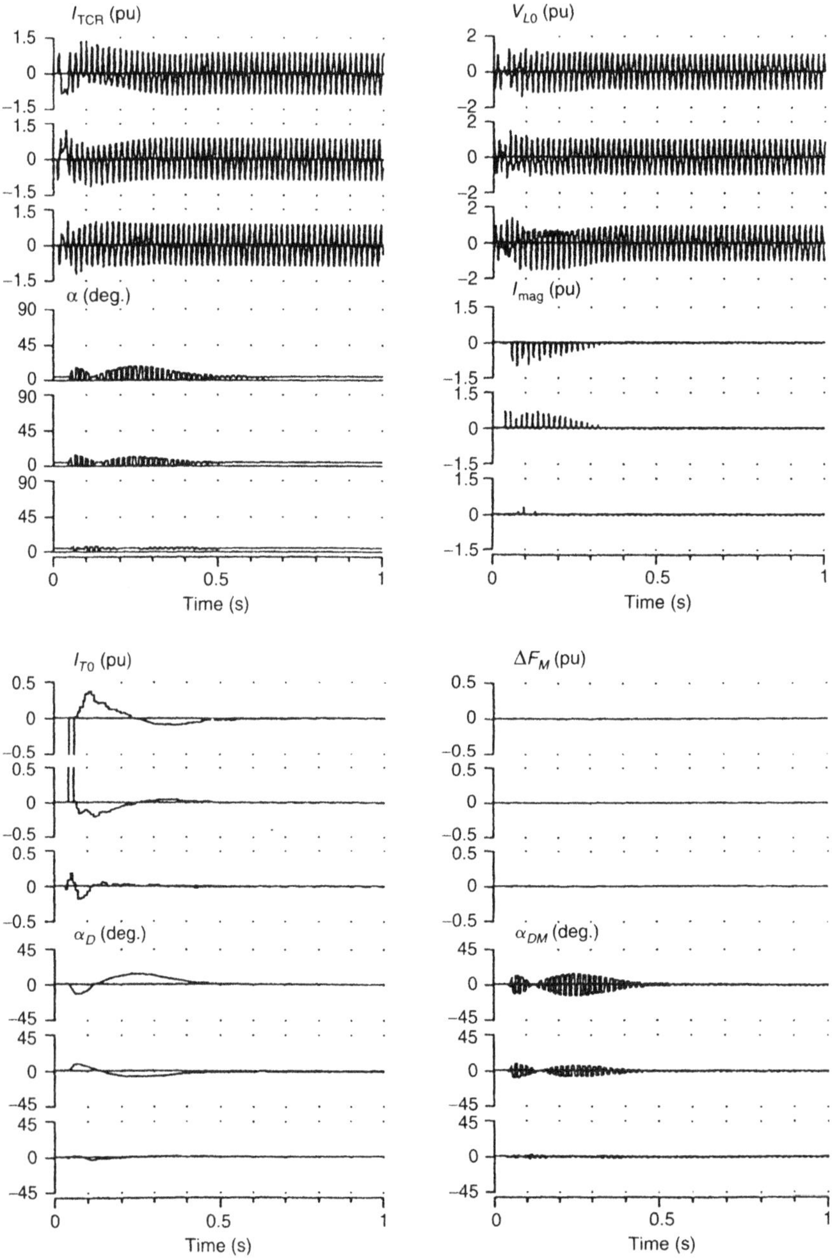

Figure 5.39 A 3-phase fault for a weak ac system (TCR balance control with I_{T0} input only).

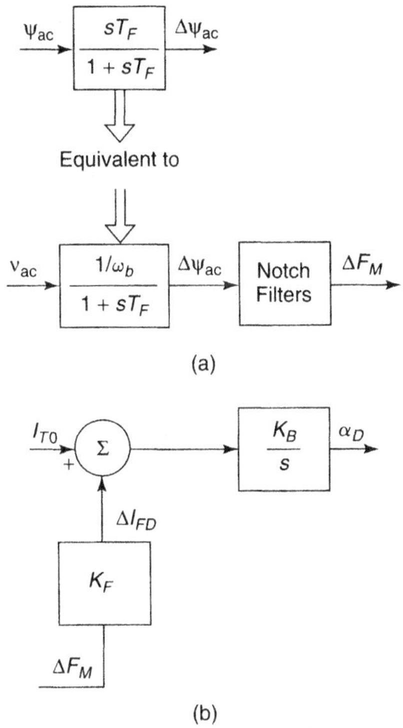

Figure 5.40 (a) Practical measurement of flux change and (b) addition of the flux-change signal to TCRBC.

5.5 APPLICATION OF THE SVC TO SERIES-COMPENSATED AC SYSTEMS

The ac transmission networks are increasingly compensated by series capacitors to enhance power-transmission capability and system stability. Hence, in many applications, SVCs are installed in series-compensated ac systems. Special problems are encountered from the interaction between SVC controls and series-compensated networks [23]–[25], both of which are discussed here.

5.5.1 ac System–Resonant Modes

Three kinds of resonances, as follows, are found in series-compensated ac systems.

5.5.1.1 Shunt-Capacitance Resonance This resonance is caused by the interaction between the series inductance, contributed by the lines, transformers, and generators, and the combined shunt capacitance, contributed by line charging, power-factor-correction capacitors, and SVCs. This resonant mode is

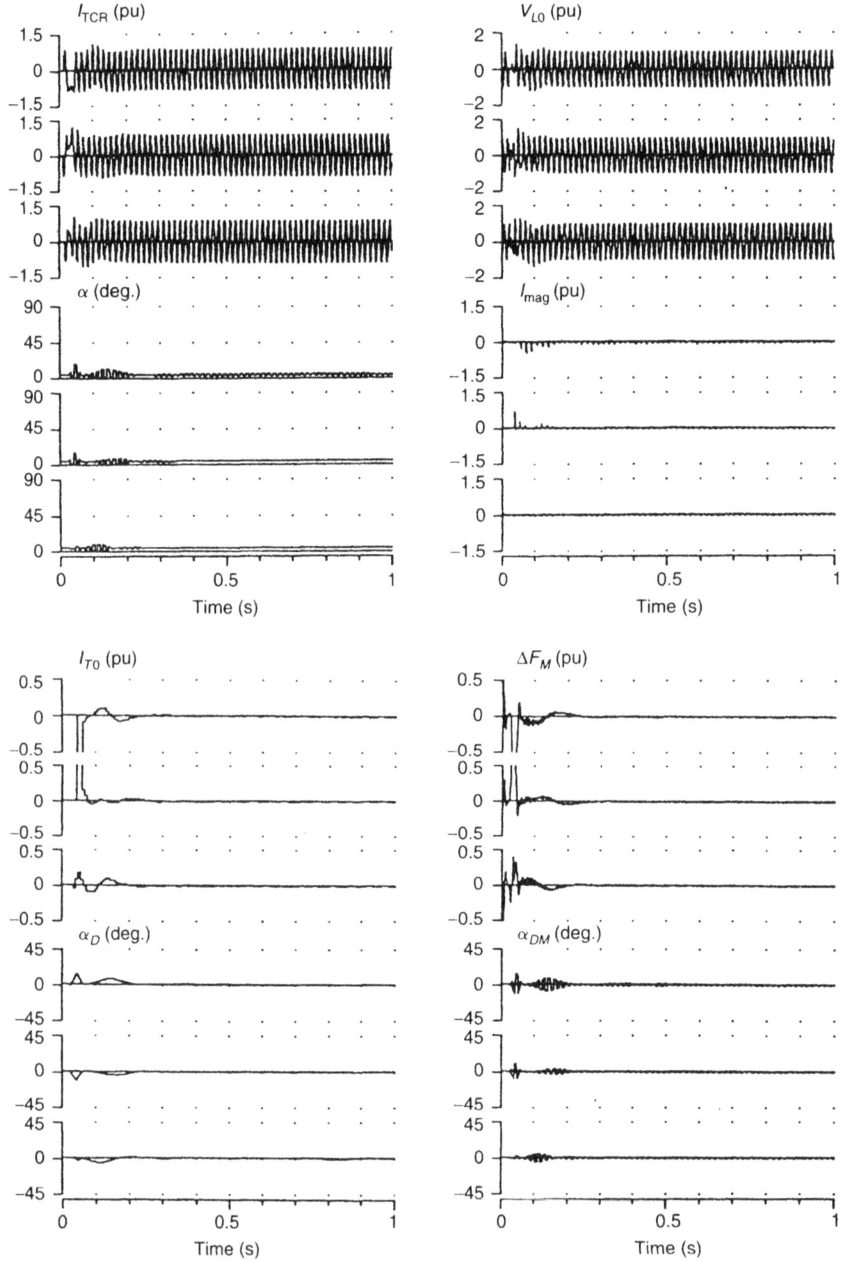

Figure 5.41 A 3-phase fault for a weak ac system (TCR balance control with both I_{T0} and ΔF_M inputs).

a feature of all transmission networks; it is not specific to series-compensated lines. The typical frequency range of this mode is 80–100 Hz; the exact value for a system is dependent on the network contingency. For instance, during intense power swings the SVCs tend to become more capacitive and the resonant frequency decreases. A very high impedance is offered by the network at this shunt-resonant-mode (pole) frequency.

5.5.1.2 Series-Line Resonance This resonance results from the interaction or energy interchange between the total line inductance and the series capacitance installed in the line. The frequency of this series-resonant mode increases with the level of series compensation, varying from zero with no series compensation to typically 25 Hz. In the Hydro-Quebec James Bay system, for instance, this resonant mode lies between 18 and 28 Hz. A characteristic of this series-resonant mode (zero) is that minimum impedance is offered by the network at the resonant frequency.

5.5.1.3 Shunt-Reactor Resonance The interaction between series capacitors and the total of shunt inductors located at different intermediate buses causes the shunt-reactor mode (pole). The frequency of this mode varies from several Hz to about 20 Hz, depending on the amount of shunt inductance. The mode frequency increases during light load conditions, when the level of shunt-inductive compensation employed is quite high. This mode is usually associated with a very low damping, as the only resistance that can introduce damping is provided by the line and shunt reactor, both of which have small losses (or high Q-factors). Also, loads do not generate enough damping, for the low series inductance of the line effectively appears in parallel to them.

A graphical illustration of these modes in a simplified representation of the Hydro-Quebec network under weak load conditions [23] is presented in Fig. 5.42. Note that the shunt-capacitance and shunt-reactor modes cause peak impedances, whereas the series line-resonant mode produces minimum impedance.

The influence of these modes on SVC performance is explained with reference to the representative system [24] depicted in Fig. 5.43. This test system models two parallel 735-kV lines, which are about 1000 km long. The equivalent systems at both the sending end and the receiving end are assumed to have a short-circuit capacity of 8000 MVA. The impedance angles of the loads at the sending and receiving ends, corresponding to 90 Hz, are 86° and 80°, respectively.

The impedance-versus-frequency characteristics of the study system, as observed from the receiving-end bus and the SVC bus, are depicted in Figs. 5.44 and 5.45. Figure 5.44 illustrates the effect of introducing series compensation in lines in the absence of shunt reactors, whereas Fig. 5.45 illustrates the influence of adding shunt reactors, R, in the series-compensated network. It is important to note that from the impedance-versus-frequency plots, different impedance characteristics are observed at different locations in the network.

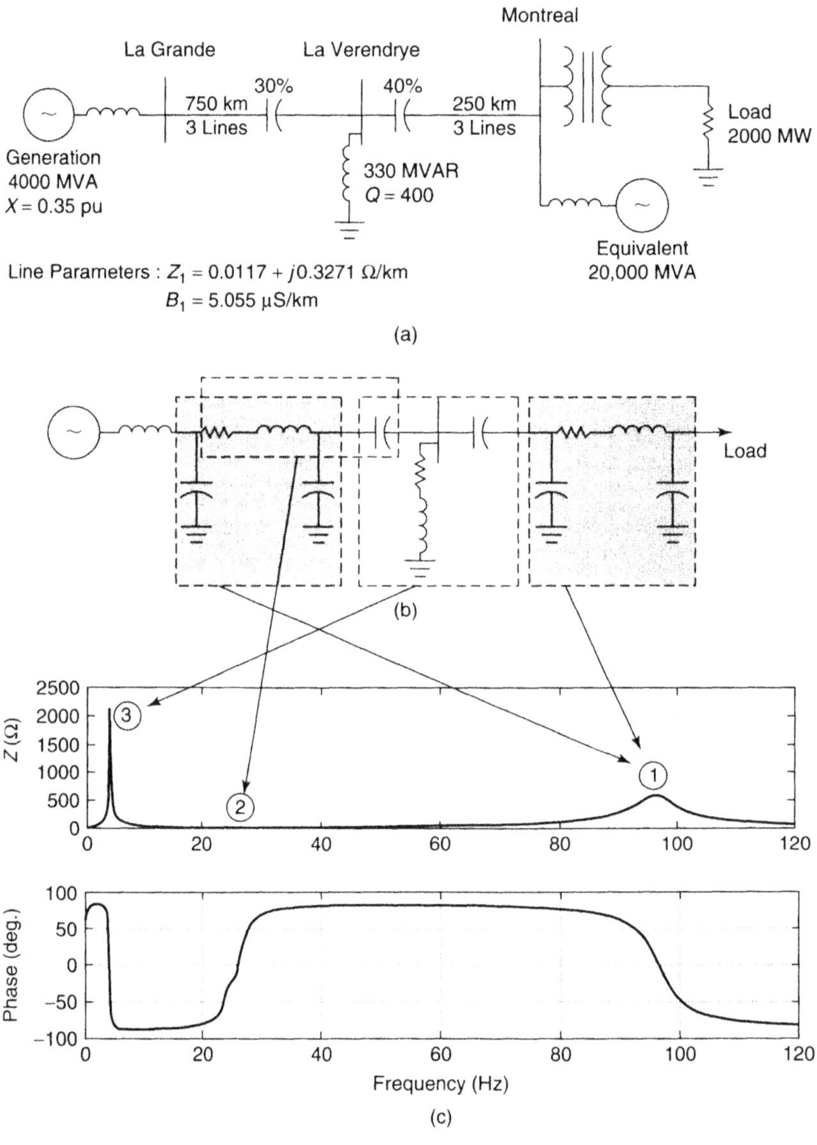

Figure 5.42 (a) A simplified Hydro-Quebec network during degraded conditions; (b) an equivalent representation using π models of transmission lines; and (c) the magnitude and phase of positive-sequence impedance at the La Verendrye substation.

Also, the influence of the various resonant modes is a function of the location where the concerned equipment is installed. Many resonant modes may exist in complex interconnected networks, each exercising a different influence at different locations.

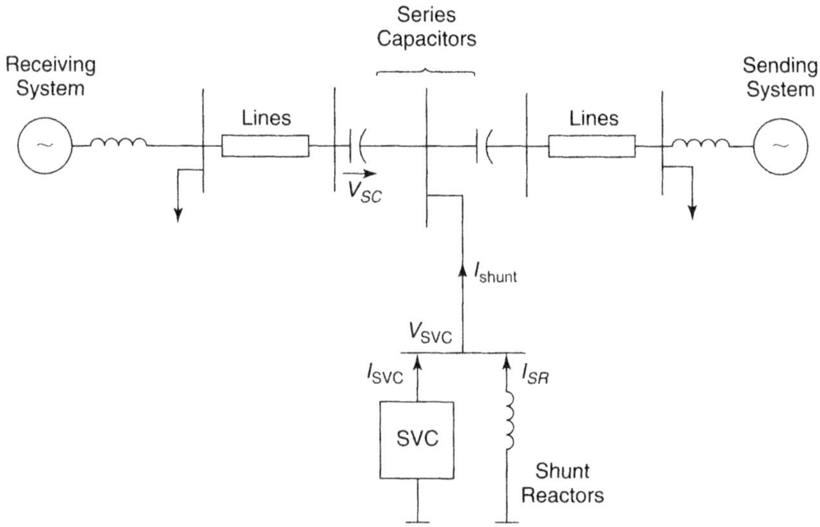

Figure 5.43 The representative ac system.

It is seen from Fig. 5.44 that the series-line, L, mode has a frequency of 24 Hz, whereas the shunt-capacitance mode has a frequency 80 Hz. The effect of the series-line mode is more profound at the line midpoint than at the receiving end. The series-compensated network offers a large impedance at low frequencies when viewed from the midpoint. No such phenomenon, however, is seen to occur at the receiving end.

Similarly, Fig. 5.45 shows that the shunt-reactor mode is much more critical when viewed from the line midpoint than at the receiving bus. These resonances play an important role in determining control-system stability and the choice of voltage-regulator gains.

5.5.2 SVC Transient Response With Series-Compensated ac Transmission Lines

The interaction between the SVC and the coupling-transformer saturation in series-compensated ac transmission networks is discussed in this section. The transformer saturation is dependent on the nature of the disturbance—small or large. For instance, a reactor switching close to the SVC is considered a small disturbance, whereas a 3-phase fault adjacent to the SVC constitutes a large disturbance. Saturation of the transformer consequential to the disturbances causes the generation of the full range of harmonics, creating overvoltages and distortion of the ac voltage that may last for many seconds. This phenomenon is witnessed in networks irrespective of whether they are series compensated. The only distinctive features of series-compensated networks are that the harmonic currents generated by transformers and other magnetic devices are much

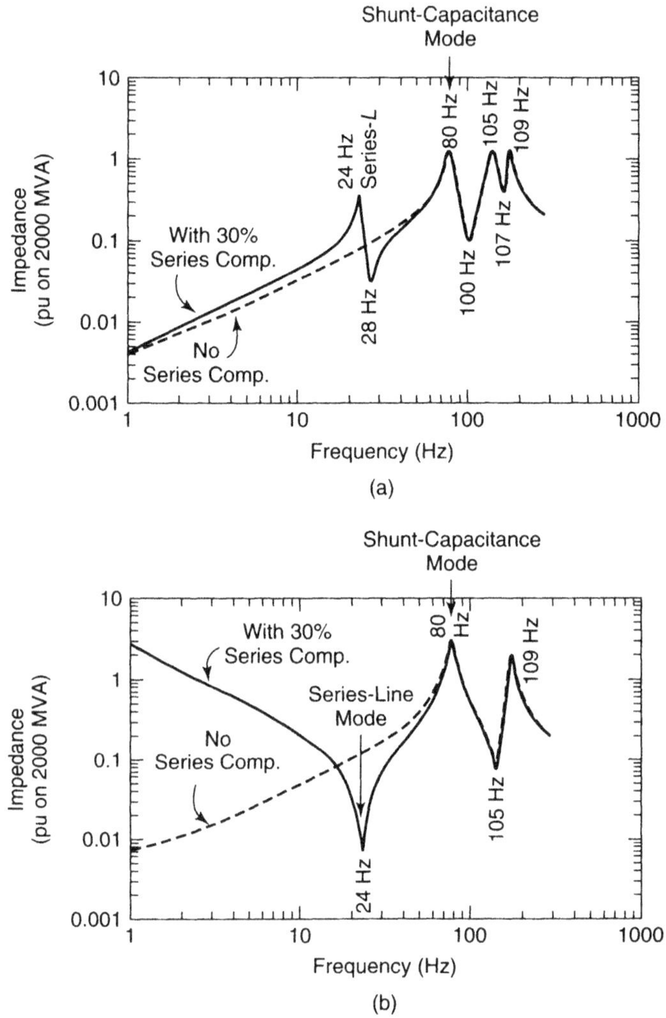

Figure 5.44 The effect of adding series compensation to lines without shunt reactors (a) at the receiving bus and (b) at the midpoint of 735-kV lines.

larger and exhibit a pulsating or modulated nature. As with non–series-compensated lines, the control-system stability is largely sensitive to the TCR operating point.

5.5.2.1 Reactor Switching A comparison of the effects of reactor switching in series-compensated and non–series-compensated systems is shown in Fig. 5.46. In systems without series compensation, the transformer-magnetizing current may not be large enough to cause saturation; hence the line current (the sum of the reactor current and the SVC current) decays monotonically.

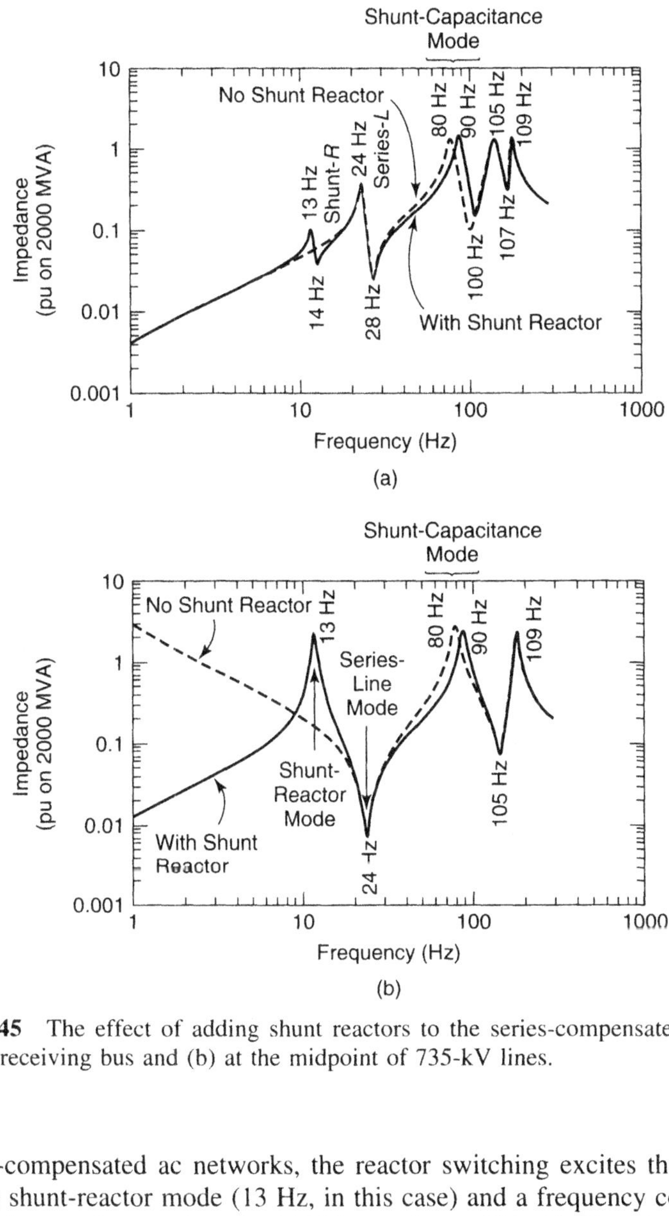

Figure 5.45 The effect of adding shunt reactors to the series-compensated network (a) at the receiving bus and (b) at the midpoint of 735-kV lines.

In series-compensated ac networks, the reactor switching excites the subsynchronous shunt-reactor mode (13 Hz, in this case) and a frequency component of 13 Hz is superimposed on the SVC bus voltage. This 13-Hz voltage generates a 13-Hz flux component in the transformer that, when added to the initial transformer flux, leads to saturation. In case of small disturbances, the total flux is such that it causes only asymmetrical saturation, and the transformer starts drawing a magnetizing current that is modulated at 13 Hz and comprises pulse trains at a frequency of 13 Hz. This magnetization current, in turn, produces a modulation of SVC bus voltage at 13 Hz. The SVC measurement system,

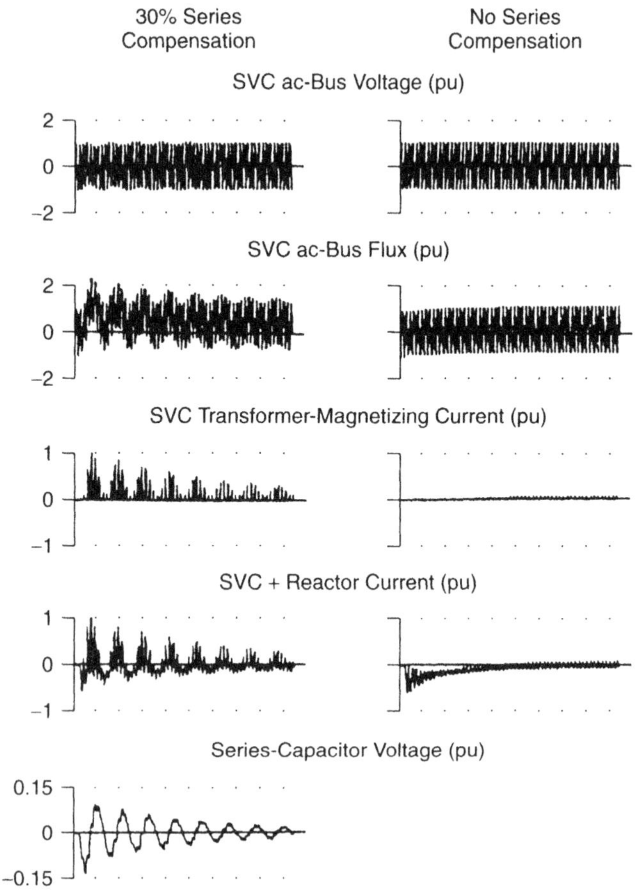

Figure 5.46 Reactor switching with and without series compensation.

because of its demodulating effect, generates on the dc-side frequency components of 47 Hz (60 − 13 Hz) and 73 Hz (60 + 13 Hz) from the 13-Hz addition phenomenon. Simultaneously, because of the 13-Hz modulation of SVC bus voltage, the measurement system produces a 13-Hz frequency component on the dc side of the transducer. The addition/modulation phenomenon [23] is illustrated in Fig. 5.47.

Because of the modulated magnetizing currents of saturated SVC transformers, a substantial difference (of the order of 200 MVAR, in some cases) is noticed between the actual primary susceptance of SVCs and the susceptance-order signal produced at the voltage-regulator output [23].

In addition to employing preinsertion resistors or breakers with precise closing times, the installation of reactors on the line side of series capacitors helps alleviate the large, modulated harmonic currents.

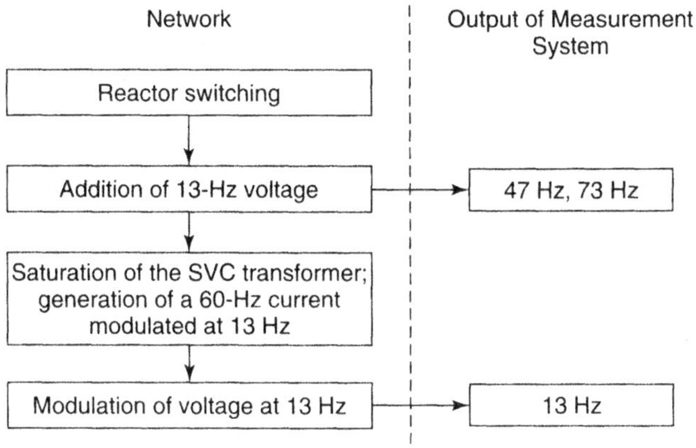

Figure 5.47 Explanation of the addition/modulation phenomenon.

5.5.2.2 Fault Application and Clearing When severe system faults occur near the SVC and are subsequently cleared, the transformer saturation is of a different nature than with reactor switching. The system response following the clearance of a 3-phase fault close to an SVC bus is depicted in Fig. 5.48. The magnitudes of the transformer-magnetizing current in both series-compensated and non–series-compensated systems are quite similar; however, the nature of the currents in both cases are very different.

The fault clearing results in the excitation of the shunt-reactor mode (13 Hz) in the SVC bus voltage. In some cases, however, the frequency of oscillation in the recovery voltage may increase slightly to, say, f_1 (15 Hz) from the alteration of circuit reactance by the reactance of the saturated transformer. This added oscillation at 15 Hz in the SVC bus voltage generates a 15-Hz flux in the transformer. An important feature with large disturbances is that the total transformer flux (15-Hz flux and initial flux) is sufficiently large to saturate the SVC transformer symmetrically. The magnetizing current then exhibits pulse trains separated by half the time period of 15 Hz. In other words, the magnetizing current is modulated at $2 \cdot 15$ Hz $= 30$ Hz, which also causes the modulation of SVC bus voltage at 30 Hz.

In a network without series compensation, only odd- and even-harmonics are generated in the magnetizing current. However, in series-compensated networks, the 30-Hz modulation of the 60-Hz magnetization current is reflected in the frequency spectrum as a 60-Hz line, with two sidebands at $60 - 30 = 30$ Hz and $60 + 30 = 90$ Hz. Together with the fundamental, the odd harmonics are also modulated at 30 Hz. Although even-harmonics are not produced because of the symmetric saturation, sidebands do appear on either side of these frequencies. For instance, the frequency spectrum of the magnetizing current exhibits frequencies of $120 - 15 = 115$ Hz and $120 + 15 = 135$ Hz. The range of frequencies appearing in the transformer-magnetizing current is given in Fig. 5.49.

The demodulation effect of the SVC measurement system [23] on the differ-

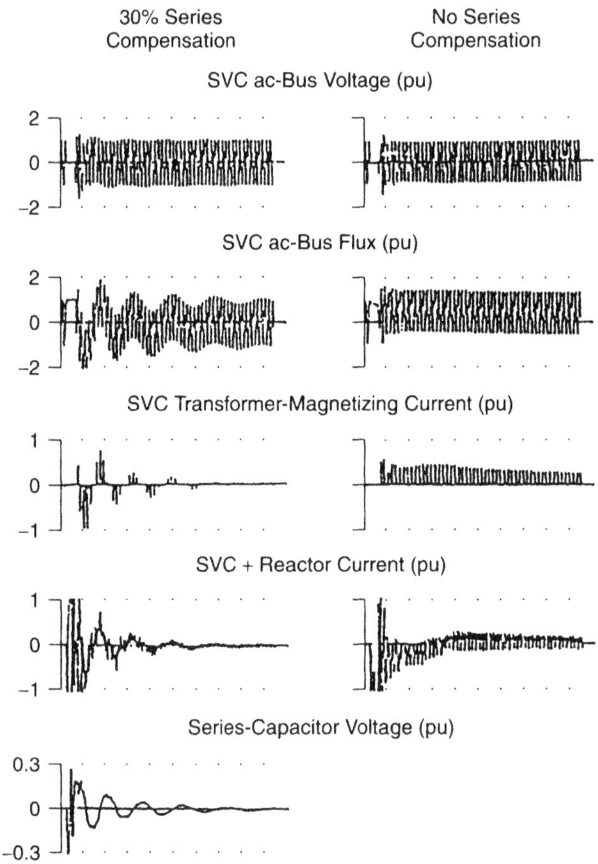

Figure 5.48 An ac fault with and without series compensation.

ent frequencies is explained in Fig. 5.50. The output of the voltage-measurement system and, consequently, the susceptance-order signal B_{ref} includes a 30-Hz component from the modulation of the 60-Hz SVC bus voltage (indicated in the figure by arrow 1) together with 45 Hz and 75 Hz from the addition of a 15-Hz component to the SVC bus voltage (indicated in the figure by arrow 2). Because the frequencies of 115 Hz and 135 Hz get added to the magnetization current, they get translated to 45 Hz and 75 Hz on the dc side of the voltage transducer (shown in the figure by arrow 3). If the ac network has a pole near the 2nd harmonic, then the aforementioned frequencies can play a critical role in determining the system stability.

Although from Fig. 5.48 it is seen that the total system current (SVC current + reactor current) decays very fast, this rapidity is caused only by the specific SVC control parameters chosen in this study. In a general situation, the current pulsations can be sustained and even amplified.

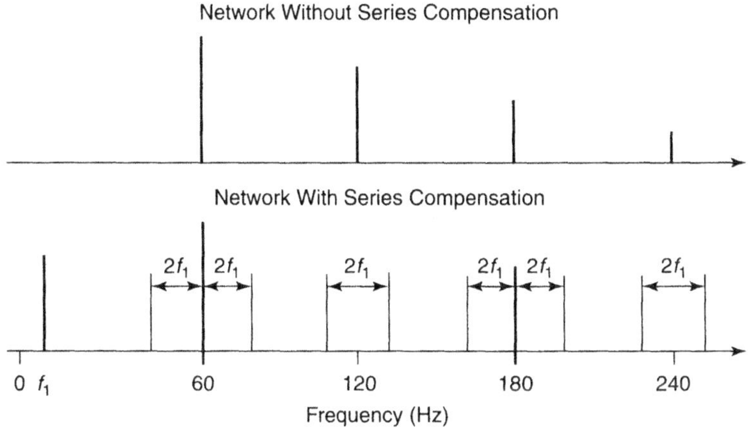

Figure 5.49 Spectrum of transformer-magnetizing current at fault clearing without and with series compensation (subsynchronous resonance at f_1).

5.5.3 Effect of the Shunt-Reactor Mode on the SVC Voltage Controller

The shunt-reactor mode can adversely influence the stability of the SVC voltage regulator. The nature of this influence and the measures for eliminating the

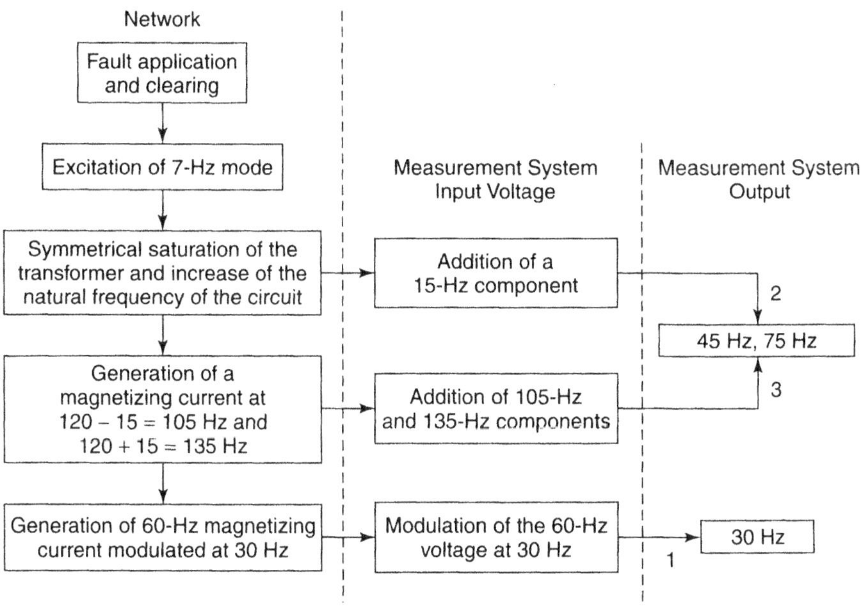

Figure 5.50 A schematic explanation of the addition and transformer-saturation phenomena at fault application and clearing.

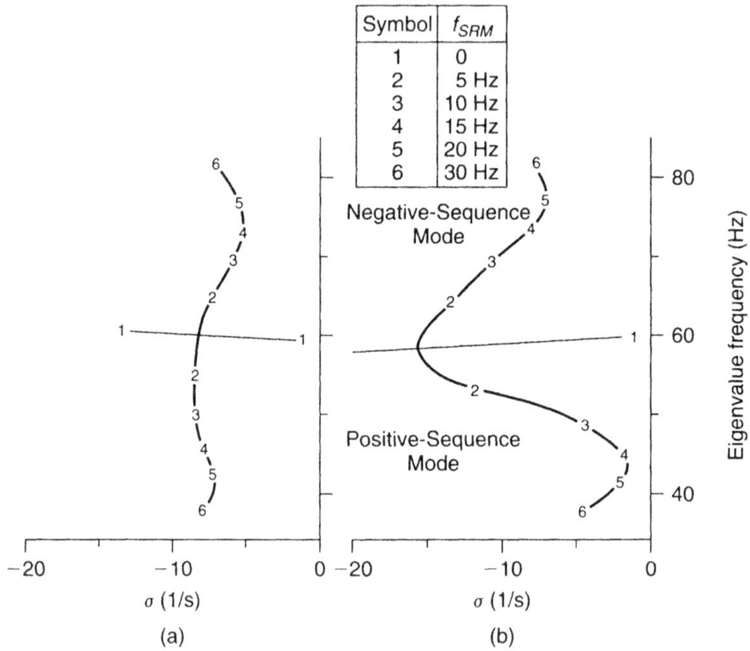

Figure 5.51 Root locus with varying shunt-reactor size for 30% series compensation (the SVC is at zero net output with one TSC and a near–full-conduction TCR): (a) no-SVC voltage regulator and (b) fast-SVC voltage regulator.

undesirable interaction are discussed here. The effect of the TCR conduction on this interaction is also discussed.

As noted previously, the demodulation effect of the voltage regulator produces frequency components $60 + f_1$ and $60 - f_1$ at the output in response to the shunt-reactor-mode frequency, f_1. The $60 + f_1$ Hz signal corresponds to the negative-sequence component of the shunt-reactor mode; the $60 - f_1$ Hz components correspond to the positive-sequence component of the shunt-reactor mode.

The effect of the voltage regulator on the stability of these modes, as obtained through eigenvalue analysis, is illustrated in Fig. 5.51. The variation in eigenvalues of these modes with changing inductances of the shunt reactor (or, alternatively, the frequency of the shunt-reactor mode, f_{SRM}) are depicted in Fig. 5.51. In the absence of the SVC voltage regulator, the generation of positive- and negative-sequence modes with increasing shunt inductance is demonstrated in Fig. 5.51(a). When a high-gain voltage regulator is considered, it is seen that the damping of the positive-sequence mode is reduced as the frequency of the shunt-reactor mode exceeds 7 Hz. However, the voltage regulator improves the damping of both modes if the $f_{SRM} < 7$ Hz. It is emphasized that this behavior may vary with one's choice of system parameters.

5.5.3.1 Effect of the TCR Operating Point The effect of the TCR operating point the on shunt-reactor mode is explained with reference to the study system considered in ref. [23]. The parameters of the study system correspond to the Hydro-Quebec system with one equivalent TCR of 440 MVA at 735 kV and three equal TSCs having a total rating of 1200 MVAR at 735 kV. Hence, to traverse over the full reactive-power range of the SVC, the TCR must go four times from minimum to maximum conduction. The SVC currents are expressed in per units on 1200 MVAR.

The sensitivity of the shunt-reactor mode to the TCR operating point is depicted in the root loci of Fig. 5.52. As soon as the TCR conduction goes below 50%, the positive-sequence mode (45 Hz) becomes unstable. The negative-sequence mode is, however, reasonably damped.

In the time domain, an example of this instability initiated by low TCR conduction is depicted in Fig. 5.53. This example corresponds to a step change in the angle across the ac network, causing the SVC to traverse from zero output (one TSC on and a fully conducting TCR) to about half the rated capacity (two TSCs on and TCR near-minimum conduction). The unstable positive-sequence mode having a 45-Hz frequency is reflected as increasing 45-Hz oscillations in the TCR susceptance signal. These oscillations temporarily stop growing when the limit-cycle operation is reached because of system nonlinearities.

5.5.3.2 Filtering of the Shunt-Resonant Mode As voltage-controller instability is experienced from the presence of the shunt-reactor-mode frequency at the input of the voltage-measurement system, an obvious choice for mitigating this effect is the installation of a high-pass filter between the potential-transformer outputs and the voltage transducer, as shown in Fig. 5.54. The 80-Hz and 96-Hz notch filters are already in service in the Hydro-Quebec system for counteracting the adverse interaction with shunt-capacitance-resonant modes [23], [24]. The frequency response of these filters shows that the inserted high-pass filter reduces the gain at low frequencies, but also that it increases the phase.

The effectiveness of the high-pass filter is demonstrated by both frequency-domain and time-domain analyses. Figure 5.55 shows the improvement in damping of the positive-sequence mode (45 Hz) that leads to its stabilization, although at the expense of negative-sequence-mode damping. The system response with the 15-Hz high-pass filter is shown in Fig. 5.56. The system has been stabilized, but the damping of both positive- and negative-sequence modes is low.

Increasing the break frequency of the high-pass filter reduces the effect of the shunt-reactor-resonant mode, but it may have an adverse impact on the damping of the shunt-capacitance mode. Therefore, a trade-off is needed to select the break frequency, for which studies have shown that 15–25 Hz constitutes a good choice for the break frequency.

The interaction between the shunt-reactor-resonant mode and the TCR balance control also needs to be investigated thoroughly in each SVC application,

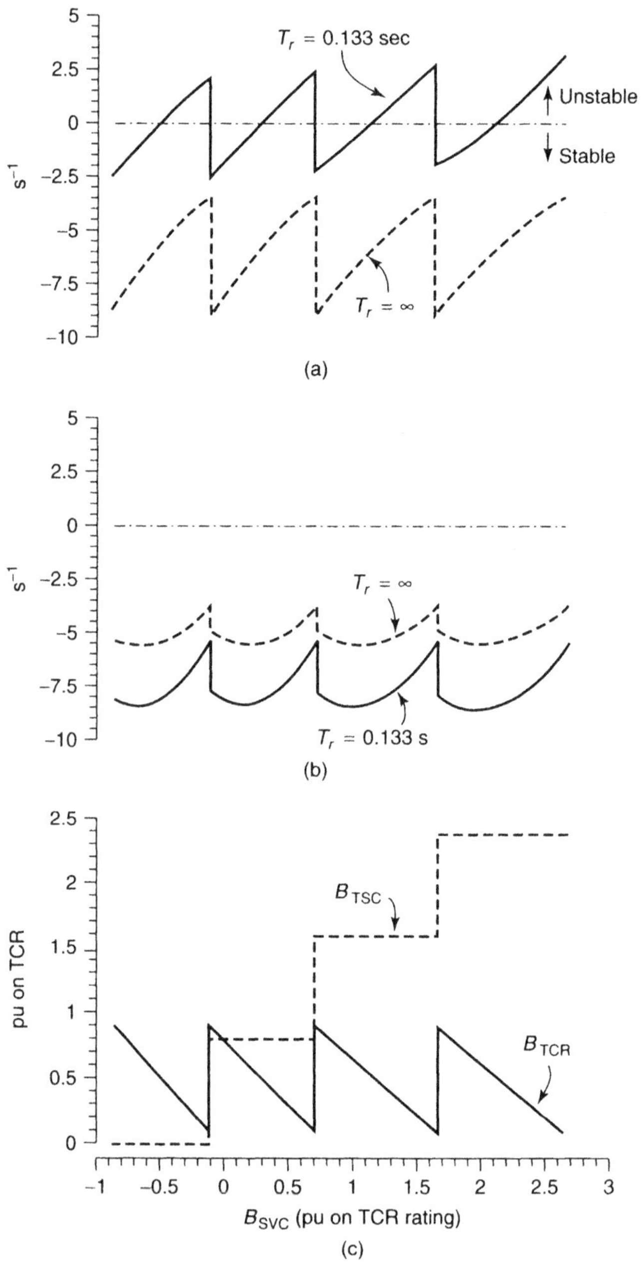

Figure 5.52 Damping of the shunt-reactor modes versus the SVC operating point (30% series compensation, 15-Hz shunt-reactor mode): (a) σ of the positive-sequence mode (45 Hz); (b) σ of the negative-sequence mode (75 Hz); and (c) the SVC operating point.

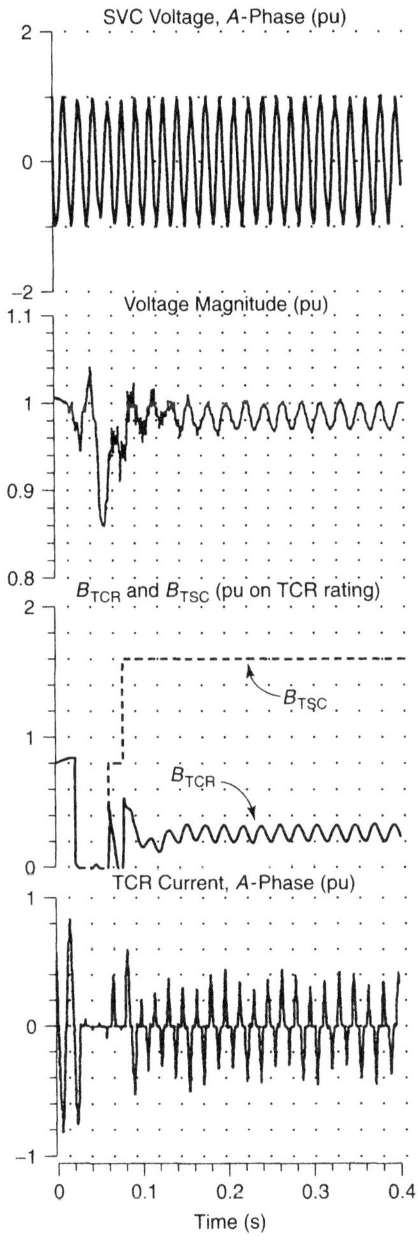

Figure 5.53 An example of unstable interaction with the shunt-reactor mode.

for situations may arise when the bandwidth of the TCR balance control is close to the shunt-reactor-resonant frequency and the system stability may be endangered.

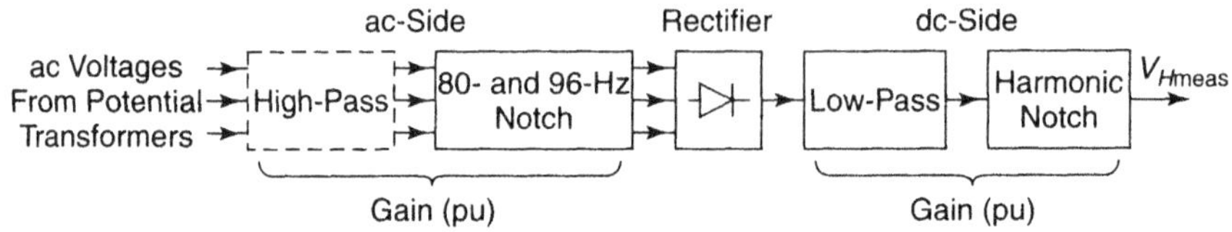

Figure 5.54 The filtering of measurement voltages (solid lines correspond to existing filters; dashed lines correspond to additional filters needed for series-compensated ac lines).

5.6 3RD HARMONIC DISTORTION

Field tests have demonstrated that in the steady state, a small amount of unbalance in voltage at the SVC bus can lead to a substantially high degree of unbalance in the TCR currents, especially in systems exhibiting a high impedance near the 3rd harmonic. In a specific situation, a 1% unbalance in the fundamental voltage translates into a 4% unbalance in TCR current.

Because the size of the TCR inductors is large, a small amount of unbalance in the SVC bus voltage results in the generation of a high amount of positive- and negative-sequence 3rd harmonic components in the TCR current. If the effective system impedance, as viewed from the TCR terminals, is quite high at the 3rd harmonic frequency, positive- and negative-sequence 3rd harmonic voltages appear on the low-voltage SVC bus.

A 3rd harmonic distortion in the SVC bus voltage causes an increased unbalance in TCR currents more than what would be caused by a simple unbalance in supply voltages with a linear-system impedance. A sustained distortion in the SVC bus voltage may lead to the maloperation of the TCR valves, which may prevent them from conducting. (Even if they do conduct, they may not stay in conduction for some cycles.) The whole process may aggravate the voltage distortion even further, and the harmonics may take an inordinately long time to decay.

The 3rd harmonic resonance can be mitigated by providing a 3rd harmonic filter on the low-voltage SVC bus. This could be achieved by converting a part of the fixed-capacitor banks into a damped 3rd harmonic filter. This technique has been found to be highly successful in eliminating SVC tripping from harmonic overload and minimizing the unbalance in the 60-Hz TCR current.

Another means for alleviating the 3rd harmonic distortion is by providing current-balancing circuits in the SVC control system. These circuits change the firing angle in each phase of the TCR to maintain an equal fundamental component of reactive current in all phases. These circuits are slow-acting to avoid interfering with the fast dynamics of the voltage regulator. However, it must be noted that this method of individual-phase control may itself produce noncharacteristic harmonics, including the 3rd, that must be taken care of appropriately.

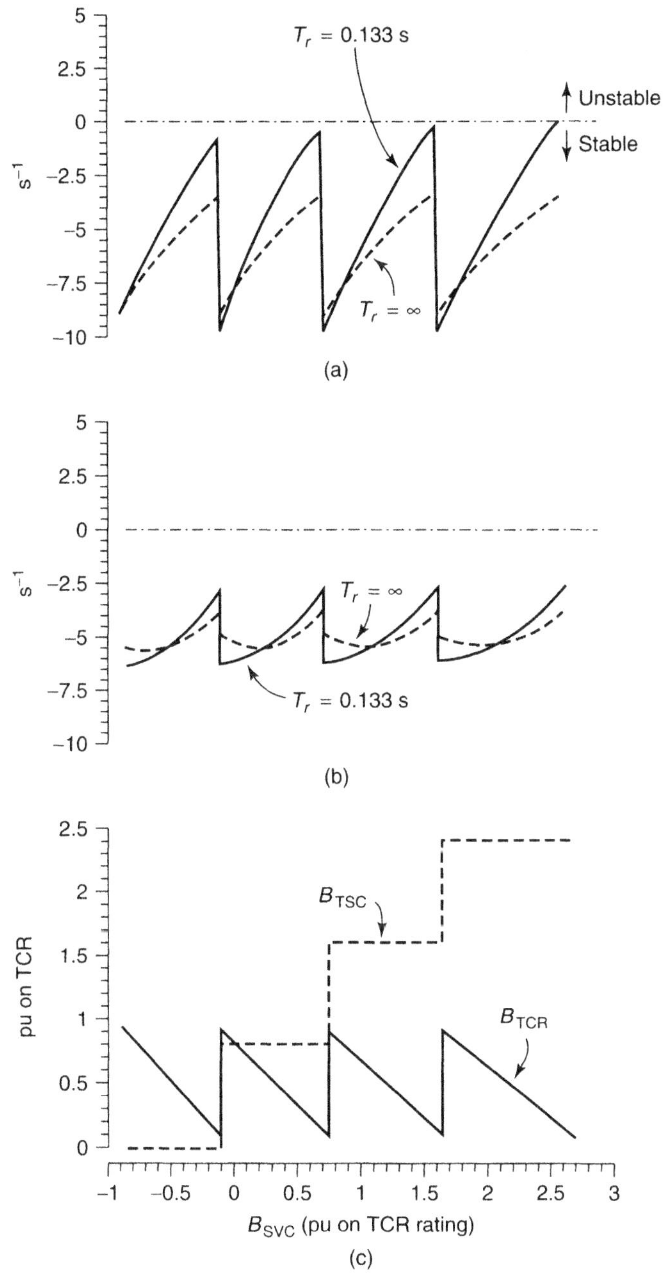

Figure 5.55 Damping of the shunt-reactor modes versus the SVC operating point with a 15-Hz high-pass filter (30% series compensation, 15-Hz shunt-reactor mode).

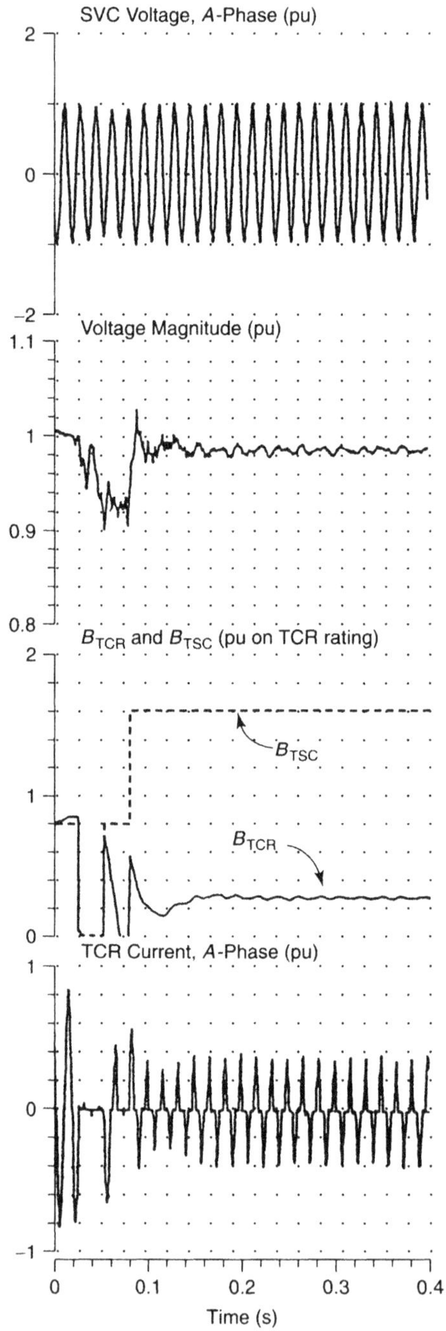

Figure 5.56 An example of unstable interaction with the shunt-reactor mode with a 15-Hz high-pass filter.

5.7 VOLTAGE-CONTROLLER DESIGN STUDIES

The proper design of the SVC voltage controller is critically dependent on the proper modeling of the involved devices and the network-resonant phenomena that might adversely affect the frequency range corresponding to the bandwidth of the voltage regulator [23]–[25]. Additional caution needs to be exercised while investigating SVC control action in series-compensated systems.

5.7.1 Modeling Aspects

Actual equipment models must be used for voltage-measurement systems, TCR firing controls, synchronization systems, TCR balance control, and the voltage regulator. Appropriate modeling of the network and loads is important to get a realistic estimate of the damping provided by these components. In series-compensated networks, the shunt-reactor mode damping is affected mainly by the low-frequency (typically 20-Hz) resistances of series-compensated transmission lines, shunt reactors, generators, motors, transformers, and feeders. For passive elements, the dc resistance must be used instead of the 60-Hz values. Furthermore, in this frequency range of 20 Hz, the generators and motors exhibit apparent negative resistance, thereby reducing the overall positive resistance of the circuit and, consequently, the damping of the shunt-reactor mode.

5.7.2 Special Performance-Evaluation Studies

Because the performance of SVC control is strongly dependent on TCR conduction, for each system condition (corresponding to a contingency), the controller action should be analyzed over the full range of SVC, from the zero-output to the maximum-output condition. The most stringent cases under which SVC stability must be examined are during zero-TCR conduction or when the SVC is fully capacitive.

In series-compensated systems, the effect of TCR conduction on shunt-reactor-mode damping should be investigated over the entire range of shunt-reactor and series-capacitor compensation levels. Doing so ensures control stability over the full frequency range of the shunt-reactor mode. It is recommended that the control solution to shunt-reactor mode instability be obtained first, as this solution may significantly influence the SVC performance in other contingencies [24].

5.7.3 Study Methodologies for Controller Design

The design and optimization of SVC voltage controller are usually performed employing the following frequency-domain and time-domain techniques:

5.7.3.1 Impedance-Versus-Frequency Computations

These studies are conducted at different buses for different contingencies, such as faults and the loss of generation, loads, or critical lines. These investigations are then performed for different system configurations, such as summertime networks and wintertime networks in which the generation and load patterns differ significantly. The results of these studies help identify the potential resonance problems that the SVC voltage controller can experience and for which adequate measures must be taken.

5.7.3.2 Eigenvalue Analyses These studies use linearized models of the SVC and the interconnected system about operating points. For series-compensated systems, a network model valid over the frequency range 0–300 Hz should be used. Eigenvalue analyses are simple to perform; as such, they are used widely for performing preliminary controller optimization.

5.7.3.3 Simulation Studies Once optimal controller parameters have been determined based on a small-signal eigenvalue analysis, the SVC performance is investigated under transient situations caused by faults, for example, when all system nonlinearities and transformer saturations are modeled. Problems such as 2nd harmonic interaction can only be studied through nonlinear simulation studies, not through eigenvalue analyses.

In the past, physical simulation studies were performed on the TNA using scaled-down models of system components to correlate and validate the linearized analysis results. However, TNA studies are quickly giving way to real-time digital simulation for SVC study and design because of its reduced cost, setup time, and model losses and damping. The digital simulation studies are usually performed with EMTP/EMTDC programs, giving a greater level of confidence in the controller design.

Finally, the designed SVC controller is field-tested [25] before it is commissioned and put into service.

5.8 SUMMARY

This chapter presented different control issues related to the voltage-control function of the SVC. The procedures for design of voltage regulator were described, and the influences of network resonances and harmonic resonances on the performance of SVC voltage control were discussed. (Modern power systems are increasingly becoming series-compensated to enhance power transfer.) This chapter also illustrated the impact of series compensation on the SVC voltage control and its design.

REFERENCES

[1] T. J. E. Miller, Ed., *Reactive Power Control in Electric Power Systems*, John Wiley and Sons, New York, 1982.

[2] R. M. Mathur, Ed., *Static Compensators for Reactive Power Control*, Canadian Electrical Association, Cantext Publications, Winnipeg, Manitoba, 1984.

[3] I. A. Erinmez, Ed., "Static Var Compensators," CIGRE Working Group 38-01, Task Force No. 2 on SVC, Paris, 1986.

[4] IEEE Power System Engineering Committee, "Application of SVC for System Dynamic Performance," IEEE Special Publication 87TH0187-5-PWR, 1987.

[5] The Electric Power Research Institute (EPRI) Report TR-100696, "Improved Static Var Compensator Control," Final Report of Project 2707-01, Prepared by General Electric (GE) Company, Schenectady, NY, June 1992.

[6] IEEE Power Engineering Society/CIGRE, *FACTS Overview*, Publication 95TP108, IEEE Press, New York, 1995.

[7] N. G. Hingorani and L. Gyugyi, *Understanding FACTS*, IEEE Press, 1999.

[8] Y. H. Song and A. T. Johns, Eds., *Flexible AC Transmission Systems (FACTS)*, IEE Press, London, U.K., 1999.

[9] D. Dickmander, B. Thorvaldsson, G. Stromberg, and D. Osborn, "Control System Design and Performance Verification for the Chester, Maine, Static Var Compensator," *IEEE Transactions on Power Delivery*, Vol. 7, No. 3, July 1992, pp. 1492–1503.

[10] IEEE Special Stability Controls Working Group, "Static Var Compensator Models for Power Flow and Dynamic Performance Simulation," *IEEE Transactions on Power Systems*, Vol. 9, No. 1, February 1994, pp. 229–239.

[11] H. E. Schweickardt, G. Romegialli, and K. Reichert, "Closed Loop Control of Static Var Sources (SVS) on EHV Transmission Lines," Paper A 78 135-6, Presented at IEEE/PES 1978 Winter Meeting, New York, January 29–February 3, 1978.

[12] K. R. Padiyar and R. K. Varma, "Damping Torque Analysis of Static Var System Controllers," *IEEE Transactions on Power Systems*, Vol. 6, No. 2, May 1991, pp. 458–465.

[13] K. R. Padiyar and R. K. Varma, "Concepts of Static Var System Control for Enhancing Power Transfer in Long Transmission Lines," *Electric Machines and Power Systems*, Vol. 18, No. 4–5, July–October 1990, pp. 337–358.

[14] L. J. Bohmann and R. H. Lasseter, "Equivalent Circuit for Frequency Response of a Static Var Compensator," *IEEE Transactions on Power Systems*, Vol. PWRS-1, No. 4, November 1986, pp. 68–74.

[15] J. Belanger, G. Scott, T. Anderson, and S. Torseng, "Gain Supervisor for Thyristor-Controlled Shunt Compensators," CIGRE Paper 38-01, Paris, 1984.

[16] CIGRE Working Group 14-29, "Coordination of Controls of Multiple FACTS/HVDC Links in the Same System," CIGRE Technical Brochure No. 149, Paris, December 1999.

[17] G. Sybille, P. Giroux, S. Dellwo, R. Mazur, and G. Sweezy, "Simulator and Field Testing of Forbes SVS," *IEEE Transactions on Power Delivery*, Vol. 11, No. 3, July 1996.

[18] P. V. Goosen, R. D. MacFarlane, H. Pesch, P. Sieber, and M. Schubert, "FC/TCR Type Static Compensators in ESKOM's 132 kV Network," CIGRE Paper 38-09, Paris, 1984.

[19] E. V. Larsen and J. M. Cutler, "Effect of GICs on SVCs," *Proceedings of EPRI GIC Conference*, San Francisco, November 8, 1989.

[20] H. Puente, M. L. Burgess, E. V. Larsen, and H. Elahi, "Switching of Large Shunt Reactors Near Static Var Compensators and HVDC Converters," *IEEE Transactions on Power Delivery*, Vol. 6, No. 1, January 1989, pp. 629–636.

[21] V. D. Albertson, "General Overview: Geomagnetic Storms and Electric Power System Effects," *Proceedings of EPRI GIC Conference*, San Francisco, November 8, 1989.

[22] H. Bilodeau, S. R. Chano, and J. P. Chayer, "Effect of Geomagnetically Induced Currents on Static Var Compensator Protection Systems," *CEA Transactions of Engineering and Operating Division*, Paper 92-SP-178, 1992.

[23] E. V. Larsen, D. H. Baker, A. F. Imece, L. Gerin-Lajoie, and G. Scott, "Basic Aspects of Applying SVCs to Series-Compensated ac Transmission Lines," *IEEE Transactions on Power Delivery*, Vol. 5, No. 3, July 1990, pp. 1466–1473.

[24] L. Gerin-Lajoie, G. Scott, S. Breault, E. V. Larsen, D. H. Baker, and A. F. Imece, "Hydro-Quebec Multiple SVC Application Control Stability Study," *IEEE Transactions on Power Delivery*, Vol. 5, No. 3, July 1990, pp. 1543–1551.

[25] IEEE Substations Committee, "IEEE Guide for Static Var Compensator Field Tests," IEEE Standard 1303–1994.

CHAPTER 6

SVC Applications

6.1 INTRODUCTION

Static var compensators (SVCs) constitute a mature technology that is finding widespread usage in modern power systems for load compensation as well as transmission-line applications. In high-power networks, SVCs are used for voltage control and for attaining several other objectives such as damping and stability control. Although the voltage-control issue was discussed extensively in Chapter 5, the concepts of SVC control in such applications as stability enhancement, damping subsynchronous oscillations, and improvement of HVDC link performance are presented in this chapter. The principles of SVC control in such applications are enunciated together with example cases. Basic issues relating to the design of SVC controllers in different applications are explained, and the essentials of design methodologies are presented. Also, factors involved in the determination of SVC ratings are described.

6.2 INCREASE IN STEADY-STATE POWER-TRANSFER CAPACITY

An SVC can be used to enhance the power-transfer capacity of a transmission line, which is also characterized as the steady-state power limit. Consider a single-machine infinite-bus (SMIB) system with an interconnecting lossless tie-line having reactance X shown in Fig. 6.1. Let the voltages of the synchronous generator and infinite bus be $V_1\angle\delta$ and $V_2\angle 0^\circ$, respectively. The power transferred from the synchronous machine to the infinite bus is expressed as

$$P = \frac{V_1 V_2}{X} \sin \delta \tag{6.1}$$

For simplicity, if $V_1 = V_2 = V$, then

$$P = \frac{V^2}{X} \sin \delta \tag{6.2}$$

The power thus varies as a sinusoidal function of the angular difference of the

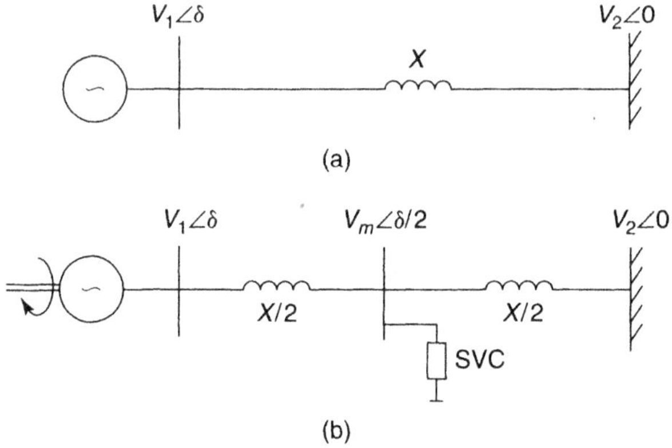

Figure 6.1 The single-machine infinite-bus (SMIB) system: (a) an uncompensated system and (b) an SVC-compensated system.

voltages at the synchronous machine and infinite bus, as depicted in Fig. 6.2. The maximum steady-state power that can be transferred across the uncompensated line without SVC corresponds to $\delta = 90°$; it is given by

$$P_{\max} = \frac{V^2}{X} \tag{6.3}$$

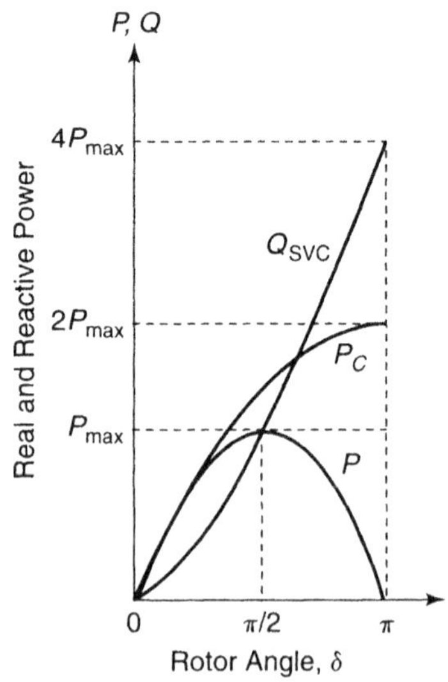

Figure 6.2 The variation of line real-power flow and SVC reactive-power flow in a SMIB system.

Let the transmission line be compensated at its midpoint by an ideal SVC. The term *ideal* corresponds to an SVC with an unlimited reactive-power rating that can maintain the magnitude of the midpoint voltage constant for all real-power flows across the transmission line. The SVC bus voltage is then given by $V_m \angle \delta/2$.

The electrical power flow across the half-line section connecting the generator and the SVC is expressed as

$$P_C = \frac{V_1 V_2}{X/2} \sin \frac{\delta}{2} \tag{6.4}$$

The power transfer in the other half-line section interconnecting the SVC, and the infinite bus is also described by a similar equation. Assuming further that $V_m = V_1 = V_2 = V$, Eq. (6.4) can be rewritten as

$$P_C = \frac{2V^2}{X} \sin \frac{\delta}{2} \tag{6.5}$$

which is depicted graphically in Fig. 6.2. The maximum transmittable power across the line is then given by

$$P_{C\max} = \frac{2V^2}{X} \tag{6.6}$$

which is twice the maximum power transmitted in the uncompensated case and occurs at $\delta/2 = 90°$. In other words, the midpoint-located ideal SVC doubles the steady-state power limit and increases the stable angular difference between the synchronous machine and the infinite bus from 90° to 180°.

If the transmission line is divided into n equal sections, with an ideal SVC at each junction of these sections maintaining a constant-voltage magnitude (V), then the power transfer (P'_c) of this line can be expressed theoretically by

$$P'_c = \frac{V^2}{X/n} \sin \frac{\delta}{n} \tag{6.7}$$

The maximum power, $P'_{c\max}$, that can be transmitted along this line is nV^2/X. In other words, with n sections the power transfer can be increased n times that of the uncompensated line. It may be understood that this is only a theoretical limit, as the actual maximum power flow is restricted by the thermal limit of the transmission line.

It can be shown that the reactive-power requirement, Q_{SVC}, of the midpoint SVC for the voltage stabilization is given by Eq. (6.8) [1].

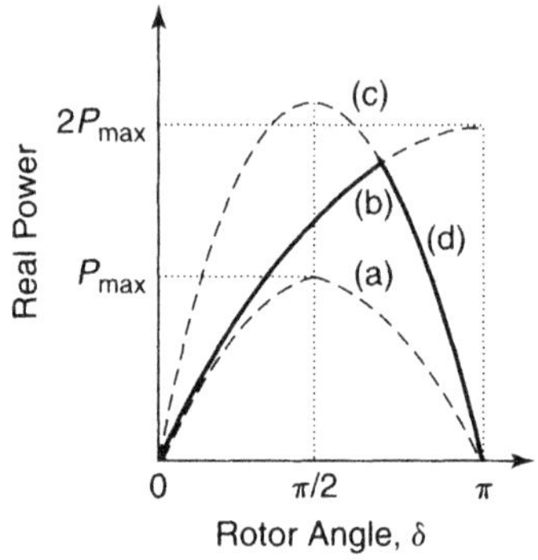

Figure 6.3 Power-angle curves of a SMIB system: curve (a) for an uncompensated case; curve (b) for an ideal midpoint SVC with unlimited rating ($Q_{SVC} > 4P_{max}$); curve (c) for a fixed capacitor connected at its midpoint; and curve (d) for a midpoint SVC with limited rating ($Q_{SVC} \approx 2P_{max}$).

$$Q_{SVC} = \frac{4V^2}{X}\left(1 - \cos\frac{\delta}{2}\right) \tag{6.8}$$

Figure 6.2 also depicts the variation of Q_{SVC} with δ. It is seen that to double the power transfer to $2P_{max}$, the required reactive-power rating of the SVC is four times the maximum power transfer in an uncompensated case, that is, $4P_{max}$. Such high-rated SVCs may not be economically feasible.

The power-transfer increase achievable with realistic SVCs of limited ratings [1] is depicted in Fig. 6.3. Curve (a) shows the power-angle relationship for the uncompensated case. Curve (b) shows the same relationship for an ideal SVC of a large reactive-power rating Q_{SVC} in excess of $4P_{max}$. Curve (c) represents the power-angle curve for a midline-located fixed capacitor. This curve is based on the corresponding equivalent reactance between the synchronous generator and the infinite bus. If an SVC incorporating a limited-rating capacitor as in the preceding text ($Q_{SVC} \cong 2P_{max}$) is connected at the line midpoint, it ensures voltage regulation until its capacitive output reaches its limit. In case the system voltage declines further, the SVC cannot provide any voltage support, and behaves as a fixed capacitor. Curve (d) of Fig. 6.3 represents the power-angle curve that shows this fixed-capacitor behavior and demonstrates that the realistic maximum power transfer will be much lower than the theoretical limit of $2P_{max}$ if the SVC has a limited reactive-power rating.

6.3 ENHANCEMENT OF TRANSIENT STABILITY

An SVC significantly enhances the ability to maintain synchronism of a power system, even when the system is subjected to large, sudden disturbances.

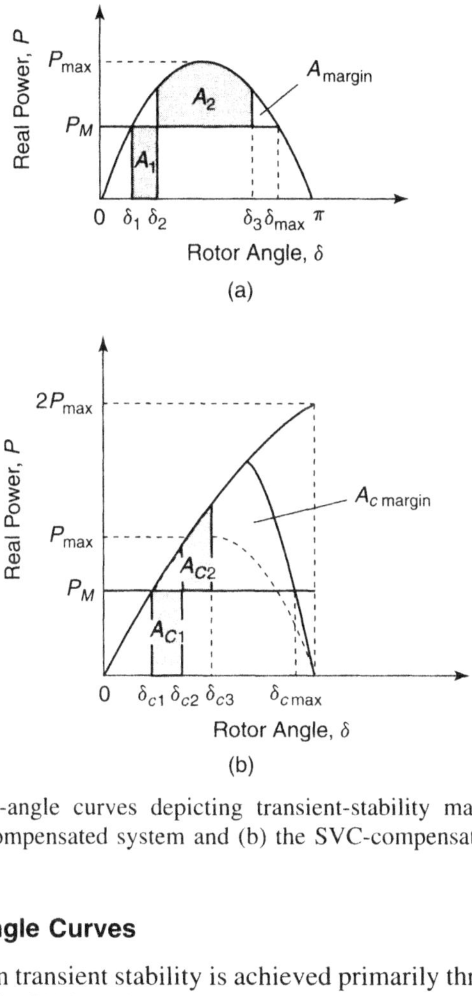

Figure 6.4 Power-angle curves depicting transient-stability margins in the SMIB system: (a) the uncompensated system and (b) the SVC-compensated system.

6.3.1 Power-Angle Curves

An enhancement in transient stability is achieved primarily through voltage control exercised by the SVC at the interconnected bus. A simple understanding of this aspect can be obtained from the power-angle curves [2], [3] of the uncompensated and midpoint SVC–compensated SMIB system.

Consider both the uncompensated and SVC-compensated power system depicted in Fig. 6.1. Assume that both systems are transmitting the same level of power and are subject to an identical fault at the generator terminals for an equal length of time. The power-angle curves for both systems are depicted in Fig. 6.4 [1]. The initial operating point in the uncompensated and compensated systems are indicated by rotor angles δ_1 and δ_{c1}. These points correspond to the intersection between the respective power-angle curves with the mechanical-input line P_M, which is same for both the cases.

In the event of a 3-phase-to-ground fault at the generator terminals, even though the short-circuit current increases enormously, the active-power output from the generator reduces to zero. Because the mechanical input remains unchanged, the generator accelerates until fault clearing, by which time the rotor angle has reached values δ_2 and δ_{C2} and the accelerating energy, A_1 and A_{C1}, has been accumulated in the uncompensated and compensated system, respectively. When the fault is isolated, the electrical power exceeds the mechanical-input power, and the generator starts decelerating. The rotor angle, however, continues to increase until δ_3 and δ_{c3} from the stored kinetic energy in the rotor. The decline in the rotor angle commences only when the decelerating energies represented by A_2 and A_{C2} in the two cases, respectively, become equal to the accelerating energies A_1 and A_{C1}.

The power system in each case returns to stable operation if the post-fault angular swing, denoted by δ_3 and δ_{C3}, does not exceed the maximum limit of $\delta_{\max}$ and $\delta_{c\max}$, respectively. Should these limits be exceeded, the rotor will not decelerate. The farther the angular overswing from its maximum limit, the more transient stability in the system. An index of the transient stability is the available decelerating energy, termed the *transient-stability margin*, and is denoted by areas A_{margin} and $A_{c\,\text{margin}}$ in the two cases, respectively. Clearly, as $A_{c\,\text{margin}}$ significantly exceeds A_{margin}, the system-transient stability is greatly enhanced by the installation of an SVC.

The increase in transient stability is thus obtained by the enhancement of the steady-state power-transfer limit provided by the voltage-control operation of the midline SVC.

6.3.2 Synchronizing Torque

A mathematical insight into the increase in transient stability can be obtained through the analysis presented in the text that follows. (See also ref. [4].)

The synchronous generator is assumed to be driven with a mechanical-power input, P_M. The transmission line is further assumed to be lossless; hence the electrical power output of the generator, P_E, and the power received by the infinite bus are same. The swing equation of the system can be written as

$$M\,\frac{d^2\delta}{dt^2} = P_M - P_E \tag{6.9}$$

where M = the angular momentum of the synchronous generator
δ = the generator-rotor angle

For small-signal analysis, the Eq. (6.9) is linearized as

$$M\,\frac{d^2\Delta\delta}{dt^2} = \Delta P_M - \Delta P_E \tag{6.10}$$

The mechanical-input power is assumed to be constant during the time of analysis; hence $\Delta P_M = 0$. The linearized-swing equation then becomes

$$M \frac{d^2\Delta\delta}{dt^2} = -\Delta P_E \tag{6.11}$$

or

$$\frac{d^2\Delta\delta}{dt^2} = -\frac{1}{M}\left(\frac{\partial P_E}{\partial\delta}\right)\Delta\delta = -\frac{K_S}{M}\Delta\delta \tag{6.12}$$

where K_S = the synchronizing power coefficient
= the slope of the power-angle curve
= $\partial P_E/\partial\delta$

or

$$\frac{d^2\Delta\delta}{dt^2} + \frac{K_s}{M}\Delta\delta = 0 \tag{6.13}$$

The characteristic equation of the differential equation (6.13) provides two roots:

$$\lambda_1, \lambda_2 = \pm\sqrt{K_s/M} \tag{6.14}$$

If the synchronizing torque K_S is positive, the resulting system is oscillatory with imaginary roots:

$$\lambda_1, \lambda_2 = \pm j\omega_s \tag{6.15}$$

where

$$\omega_s = \sqrt{K_S/M} \tag{6.16}$$

On the other hand, if the synchronizing torque K_S is negative, the roots are real. A positive real root characterizes instability. The synchronizing-torque coefficient is now determined for both the uncompensated and SVC-compensated systems. The details of the analysis are presented in Appendix B of this book.

6.3.2.1 Uncompensated System The electrical power, P, transferred from the generator across the lossless uncompensated tie-line is given by Eq. (6.1). The corresponding synchronizing torque is expressed by

$$K_{SU} = \frac{\partial P}{\partial\delta} = \frac{V_1 V_2}{X}\cos\delta \tag{6.17}$$

6.3.2.2 SVC-Compensated System The power transfer, P_C, from the generator across the lossless uncompensated tie-line is given by Eq. (6.4). It can be alternatively expressed in terms of an equivalent transfer reactance, X_T, between the generator bus and the infinite bus. (See Appendix B.)

$$P_C = \frac{V_1 V_2}{X} \sin \delta \tag{6.18}$$

where

$$X_T = X - \frac{X^2}{4} B_S \tag{6.19}$$

The net SVC susceptance, B_S, is given by

$$B_S = \frac{\alpha_c}{X_c} - \frac{\alpha_i}{X_i} \tag{6.20}$$

where $X_i = V_{\text{nom}}^2/Q_{ir}$ = the total inductive reactance of the SVC
$X_c = V_{\text{nom}}^2/Q_{cr}$ = the total capacitive reactance of the SVC
V_{nom} = the nominal voltage
Q_{ir}, Q_{cr} = the inductive- and capacitive-reactive-power rating of the SVC
α_i = the conducting fraction of the TCR
α_c = the conducting fraction of the TSC (= 1 for the fixed capacitor)

The SVC adjusts α_i and α_c to maintain a constant voltage, V_m, at the connecting bus.

The synchronizing-torque coefficient of the uncompensated system (see Appendix B) is expressed as

$$K_S = \frac{\partial P_E}{\partial \delta} = \frac{V_1 V_2 \cos \delta}{X_T} + \left(\frac{V_1 V_2 \sin \delta}{V_m X_T} \right)^2 \frac{X^2}{4X_T} \tag{6.21}$$

Thus the pure-voltage control operation of the SVC increases the synchronizing-torque coefficient by the following amount:

$$\Delta K_S = K_S - K_{SU} \tag{6.22}$$

Substituting from Eq. (6.17) in Eq. (6.21) gives

$$\Delta K_S = \frac{V_1 V_2 \cos \delta}{X X_T} (X - X_T) + \left(\frac{P}{V_m} \right)^2 \frac{X^2}{4X_T} \tag{6.23}$$

The frequency of oscillation also increases by a factor

$$\sqrt{1 + \frac{\Delta K_S}{K_S}} \tag{6.24}$$

An enhanced synchronizing torque implies an increase in the transient-stability margin of the power system. An SVC thus augments the transient stability of the power system.

6.3.3 Modulation of the SVC Bus Voltage

We have seen from the previous section that an SVC can improve the transient stability of a power system by maintaining the midpoint voltage constant. Now, however, it is shown that if an appropriate modulation of the SVC bus voltage is permitted, the transient stability can be substantially augmented as compared to the constant-voltage control strategy of SVC [4].

This concept is illustrated through the set of power-angle curves depicted in Fig. 6.5. Curve (a) illustrates the power-angle curve of the system without SVC, whereas curve (b) illustrates the same for the system compensated by an ideal SVC of an unlimited reactive-power rating. When the real-power output

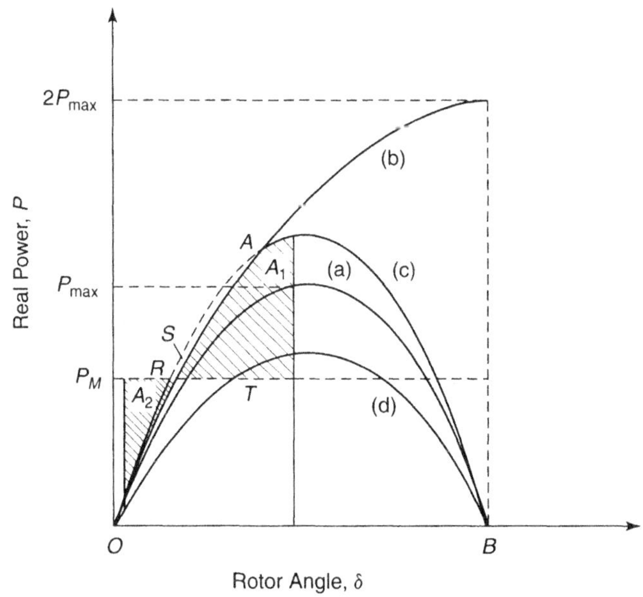

Figure 6.5 Power-angle curves of a SMIB system: curve (a) for an uncompensated case; curve (b) with an ideal midpoint-connected SVC; curve (c) with a midpoint-connected fixed capacitor; and curve (d) with a midpoint-connected fixed inductor.

of the synchronous generator gradually exceeds the surge-impedance loading (SIL), the SVC tends to become increasingly capacitive. As long as the SVC remains within its capacitive-controllable range, the power-angle curve remains the same as curve (b) until point A is reached, when the capacitance limit of SVC is attained. Beyond point A, the power-angle curve switches to curve (c), or curve $ORAB$, which corresponds to the power-angle curve of a fixed capacitor having the full rating of the SVC capacitor. This curve relates to an effective transfer reactance X_T, with positive B_S, that is less than the transmission-line reactance X [see Eq. (6.19)].

However, when the power transfer is less than the SIL, the SVC is inductive, with continuously varying levels of inductive-reactive power. If the SVC reactance is fixed at some inductive value, the power-angle curve changes to curve (d), which is below curve (a). In this case, the transfer reactance X_T becomes more than X because of the negative B_S.

To begin, let us examine the case of first-swing stability, in which the rotor angle increases following a fault and goes through an overswing. The decelerating energy, which also represents the synchronizing coefficient, is indicated by the hatched area A_1. This behavior relates to a constant-voltage control strategy of the SVC. If a higher voltage is established momentarily by making the SVC more capacitive, additional decelerating energy, shown by area ARS, would be made available. The full capacitor-swing curve is chosen only to illustrate this concept. Increasing the voltage temporarily thus restricts overswing and allows higher critical fault-clearing time.

Once the rotor angle reaches its maximum value, it tends to reverse, or backswing. It is important to minimize this backswing to ensure transient stability. For a constant-voltage control of the SVC, the developed accelerating power is indicated by A_2. However, if the SVC reactive power is rapidly changed to establish a slightly lower terminal voltage—momentarily—at the instant of maximum overswing, an additional accelerating torque, indicated by area OST, becomes available. This reduces the magnitude of backswing.

A control strategy of modulating the SVC bus voltage instead of keeping it strictly constant thus aids in substantially improving the overall transient stability of the study system. An example of the advantage achieved by adopting the voltage-modulation control strategy, in comparison to constant-voltage regulation, is presented in ref. [4]. Figure 6.6 depicts the performance of the SVC following a severe fault in a power system. The time variations in the generator-rotor angle, real-power transfer, bus voltage, and SVC susceptance are depicted for the two SVC control strategies, and the behavior of the uncompensated system is also presented. In the absence of an SVC, the fault clearing results in severe voltage depression followed by system instability. A voltage-modulation control strategy rapidly stabilizes the oscillations in the rotor angle, power transfer, and terminal voltage, as compared to the constant-voltage control of SVC. Thus, higher power transfer becomes feasible with enhanced transient stability.

Voltage-modulation control strategies are implemented through auxiliary

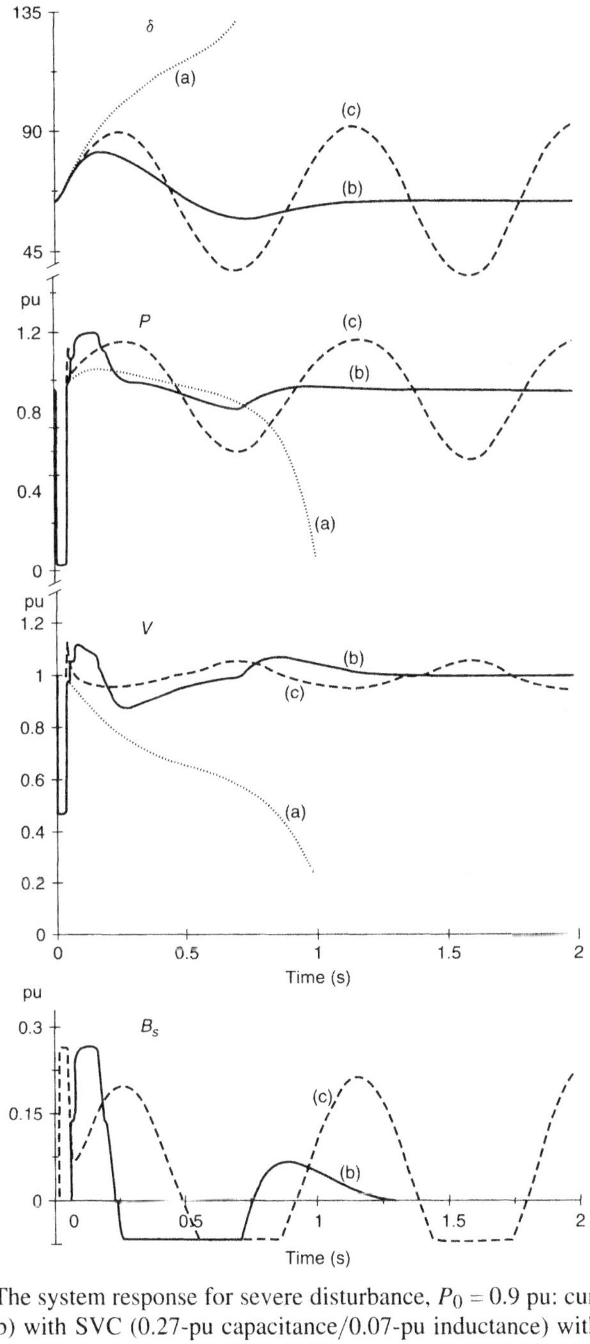

Figure 6.6 The system response for severe disturbance, $P_0 = 0.9$ pu: curve (a) without SVC; curve (b) with SVC (0.27-pu capacitance/0.07-pu inductance) with the optimum control strategy; and curve (c) with SVC (0.27-pu capacitance/0.07-pu induction) with the constant-voltage control strategy.

controllers of the SVC, which are discussed in detail in the next section (although within the context of damping improvement).

6.4 AUGMENTATION OF POWER-SYSTEM DAMPING

The power-transfer capacity along a transmission corridor is limited by several factors; for example, the thermal limit, the steady-state stability limit, the transient-stability limit, and system damping. In certain situations, a power system may have inadequate—even negative—damping; therefore, a strong need arises to enhance the electrical damping of power systems to ensure stable, oscillation-free power transfer. A typical scenario of the magnitude of various limits, especially where damping plays a determining role [5], is depicted graphically in Fig. 6.7.

Oscillations in power systems are caused by various disturbances. If the system is not series-compensated, the typical range of oscillation frequencies extends from several tenths of 1 Hz to nearly 2 Hz. Several modes of oscillation may exist in a complex, interconnected power system.

The behavior of generator oscillations is determined by the two torque components: the *synchronizing torque* and *damping torque*. The synchronizing torque ensures that the rotor angles of different generators do not drift away following a large disturbance. (In other words, the synchronizing torque binds the different generators into synchronism, assuring transient stability.) In addition, the magnitude of the synchronizing torque determines the frequency of oscillation. Meanwhile, damping torque influences the decay time of oscillations. Even if a power system is stable, the oscillations may be sustained for a long period without adequate damping torque.

Examined in this chapter is the concept of the SVC's ability to not only impart but also enhance the damping torque in a power system through SVC controllers.

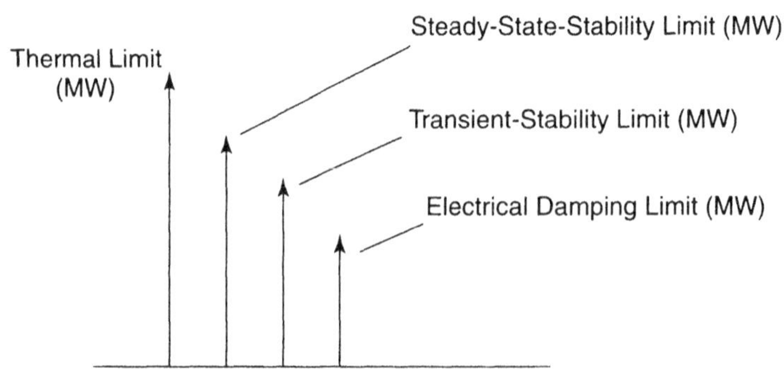

Figure 6.7 Comparison of different limits on the power flow.

6.4.1 Principle of the SVC Auxiliary Control

The SVCs are employed primarily for voltage control; they do not contribute to system damping. In the following analysis, it is shown that by incorporating auxiliary control, an SVC can significantly improve the electrical damping of power systems [1].

Consider the SMIB system depicted in Fig. 6.1. The SVC regulates the midpoint voltage to V_m. Let

$$
\begin{aligned}
V_1 &= |V_1| \sin(\omega t + \delta) \\
V_m &= |V_m| \sin\left(\omega t + \frac{\delta}{2}\right) \\
V_2 &= |V_2| \sin(\omega t)
\end{aligned}
\tag{6.25}
$$

where, for the sake of simplicity, it is assumed that $|V_1| = |V_2| = V$.

The linearized-swing equation for this system with constant mechanical-input power is given by Eq. (6.11). The electrical power P_E (= P_C) transmitted across the line, although given by Eq. (6.4), is rewritten as follows for a sending-end voltage magnitude, V.

$$
P_E = \frac{V V_m}{X/2} \sin \frac{\delta}{2}
\tag{6.26}
$$

An incremental change in electrical power is obtained by linearizing Eq. (6.26) as

$$
\Delta P_E = \frac{\partial P_E}{\partial V} \Delta V + \frac{\partial P_E}{\partial V_m} \Delta V_m + \frac{\partial P_E}{\partial \delta} \Delta \delta
\tag{6.27}
$$

It is further assumed that the sending-end voltage is constant; thus, $\Delta V = 0$. Substituting ΔP_E from Eq. (6.27) in Eq. (6.11) results in

$$
M \frac{d^2(\Delta\delta)}{dt^2} + \frac{\partial P_E}{\partial V_m} \Delta V_m + \frac{\partial P_E}{\partial \delta} \Delta\delta = 0
\tag{6.28}
$$

Equation (6.28) describes the small-signal dynamic behavior of the system, in which the effect of the SVC is represented by the middle term

$$
\left[\left(\frac{\partial P_E}{\partial V_m}\right) \Delta V_m\right]
\tag{6.29}
$$

If the SVC is operated to maintain the midpoint voltage V_m strictly constant, as is done for voltage control, ΔV_m becomes zero, in which case Eq. (6.28) reduces to

$$M \frac{d^2(\Delta\delta)}{dt^2} + \frac{\partial P_E}{\partial \delta} \Delta\delta = 0 \tag{6.30}$$

The characteristic equation corresponding to Eq. (6.30) is

$$s^2 + \frac{1}{M} \left. \frac{\partial P_E}{\partial \delta} \right|_0 = 0 \tag{6.31}$$

The roots of Eq. (6.31) lie on the imaginary axis that result in undamped oscillations in the rotor angle δ with a frequency of

$$\omega_n = \sqrt{\frac{1}{M} \left. \frac{\partial P_E}{\partial \delta} \right|_0} \tag{6.32}$$

where $\partial P_E / \partial \delta$ = the synchronizing power coefficient

It is clearly evident that an SVC operating on a pure-voltage control mode is unable to provide any system damping. However, the SVC can contribute to system damping if it is allowed to modulate the midpoint voltage instead of maintaining it strictly constant. Specifically, the midpoint voltage can be modulated as a function of $d(\Delta\delta)/dt$, that is,

$$\Delta V_m = K \frac{d(\Delta\delta)}{dt} \tag{6.33}$$

where K = the constant

Substituting Eq. (6.33) in Eq. (6.28) results in a modified incremental-swing equation:

$$M \frac{d^2(\Delta\delta)}{dt^2} + \left. \frac{\partial P_E}{\partial V_m} \right|_0 K \frac{d(\Delta\delta)}{dt} + \left. \frac{\partial P_E}{\partial \delta} \right|_0 \Delta\delta = 0 \tag{6.34}$$

The corresponding characteristic equation is

$$s^2 + 2\xi s + \omega_n^2 = 0 \tag{6.35}$$

where

$$2\xi = \frac{K}{M} \frac{\partial P_E}{\partial V_m} \tag{6.36}$$

With the introduction of voltage-modulating control, the SVC transforms the power system into a positively damped system. From Eq. (6.35), it is seen that the roots now lie on the left half of the s plane, and any oscillations in the rotor angle will decay with time. Such an additional control feature is termed *auxiliary control*, *supplementary control*, or *power-swing damping control* (PSDC). One effective signal for improving system damping, as seen in this analysis, is the variation in bus frequency f, for it relates to the rotor-angle oscillations:

$$\frac{d(\Delta\delta)}{dt} \cong \Delta f \tag{6.37}$$

An intuitive insight into the mechanism of PSDC is obtained as follows: At any instant, if $d(\Delta\delta)/dt$ or Δf is positive, it is implied that the rotor is tending to accelerate from the buildup of kinetic energy within it. This buildup can be alleviated by adjusting the SVC control so that the generator electrical power output is increased during that particular time interval, which requires an increase in the SVC terminal voltage as explained from Eq. (6.4). On the other hand, if $d(\Delta\delta)/dt$ is negative, indicating rotor deceleration, the electrical power output of the generator must be decreased by reducing the SVC terminal voltage through damping-controller action.

In some situations, to achieve rapid damping, the output of the auxiliary damping controller is fed directly to the firing controller. The voltage regulator is completely bypassed, and the delays associated with the voltage controller are eliminated.

The auxiliary controllers of the SVC greatly enhance the flexibility and performance capability of the SVC without requiring any increase in its rating.

6.4.2 Torque Contributions of SVC Controllers

The influence of dynamic devices on the behavior of different electromechanical modes can be explained through the associated synchronizing and damping torques. A modal analysis approach [6], [7] is now explained, wherein the system variables are separated into their modal components and each electromechanical mode is treated independently. The responses of each mode are finally combined to result in the overall system behavior.

6.4.2.1 Effect of the Power System A simple representation of an electromechanical mode in a power system excluding SVC is depicted in Fig. 6.8. In this figure, M_i denotes the effective inertia of all machines impacting the ith mode; D_i and K_i denote the modal damping and synchronizing coefficients, respectively; ω_b denotes the system-base frequency; and δ_i and ω_i denote the angle and frequency of the ith electromechanical mode.

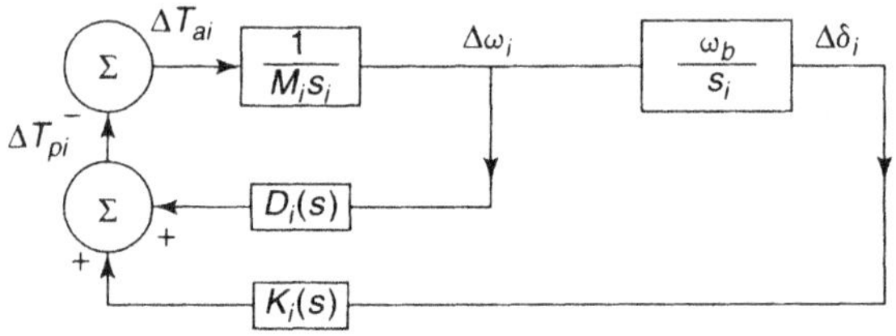

Figure 6.8 A block diagram depicting the modal damping- and synchronizing-torque contributions of a power system.

$$\omega_i = \sqrt{\omega_b \frac{K_i}{M_i}} \tag{6.38}$$

Also,

$$s = j2\pi f_i \tag{6.39}$$

where f_i = the modal frequency

It is emphasized that the synchronizing and damping torques are the combined contribution of controllers of all dynamic devices present in the system, including generator-excitation systems. The contribution of the SVC is derived separately.

6.4.2.2 Effect of the SVC The influence of SVC is integrated in the response of the *i*th electromechanical mode [6], [7], as shown in the modified block diagram Fig. 6.9. This diagram comprehensively demonstrates the individual effects of the SVC voltage regulator as well as the SVC auxiliary PSDC. It has already been shown that the SVC can provide damping to the power system only if the auxiliary damping controllers are incorporated in the SVC control, which modulate the bus voltage in response to a control signal sensitive to power oscillations.

Although both synchronizing- and damping-torque coefficients are influenced by the generator-excitation systems, only the damping torque is affected by the system loads and turbine governors.

SVC Voltage Regulator This generates a susceptance-reference signal, ΔB_{SVCi}, that primarily causes the bus voltage to change by ΔV_{Bi} through a function $K_{VBi}(s)$ representing the network response. The same susceptance output, ΔB_{SVCi}, also generates a synchronizing-torque contribution by acting through a function $K_{TBi}(s)$. The SVC voltage regulator is represented by the transfer function $V_{regi}(s)$:

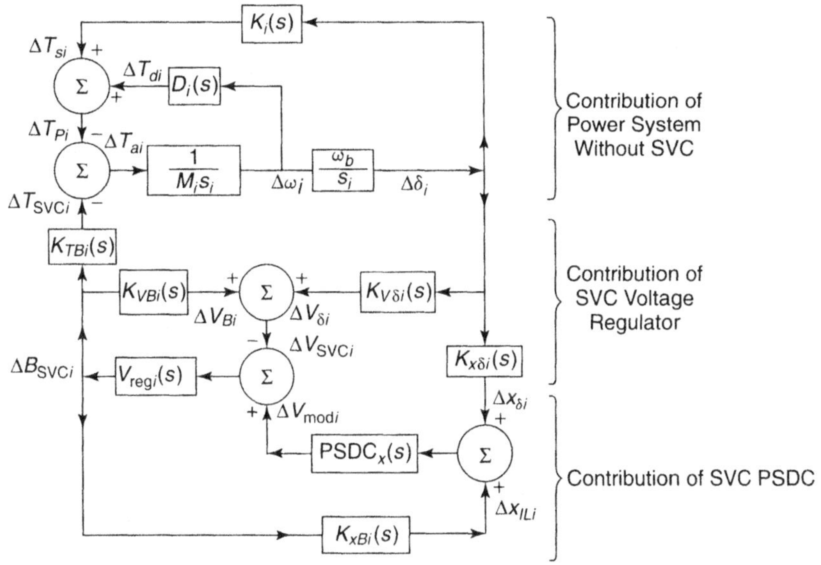

Figure 6.9 A block diagram showing the modal damping- and synchronizing-torque contributions of the power system, the SVC voltage regulator, and the SVC PSDC.

$$V_{\text{reg}i}(s) = \frac{K_R}{1 + s_i T_R} \tag{6.40}$$

The modal voltage at the SVC bus is influenced by the modal speed ω_i and, consequently, by the modal angle δ_i, both of which impart their contribution through the frequency-dependent function $K_{V\delta i}(s)$,

SVC PSDC An auxiliary control signal, $\Delta x_{\delta i}$, is provided as input to the SVC PSDC. As explained previously, this signal must be a function of modal speed or modal angle $\Delta\delta_i$, to which its relationship must be expressed through the transfer function $K_{x\delta i}(s)$. The SVC PSDC is modeled by the transfer function $\text{PSDC}_x(s)$ that contributes an additional modulating input, $\Delta V_{\text{mod}i}$, to the voltage regulator. The SVC susceptance, $\Delta B_{\text{SVC}i}$, generates an inner-loop response, ΔX_{ILi}, which influences the auxiliary signal through the transfer function $K_{xBi}(s)$.

The power system of Fig. 6.9 with SVC voltage and auxiliary controllers is simplified to the system depicted in Fig. 6.10 to clearly illustrate the damping and synchronizing effect of each system component.

The contribution of the power system without SVC to the modal-accelerating torque is obtained through the transfer function $K_{Pi}(s)$:

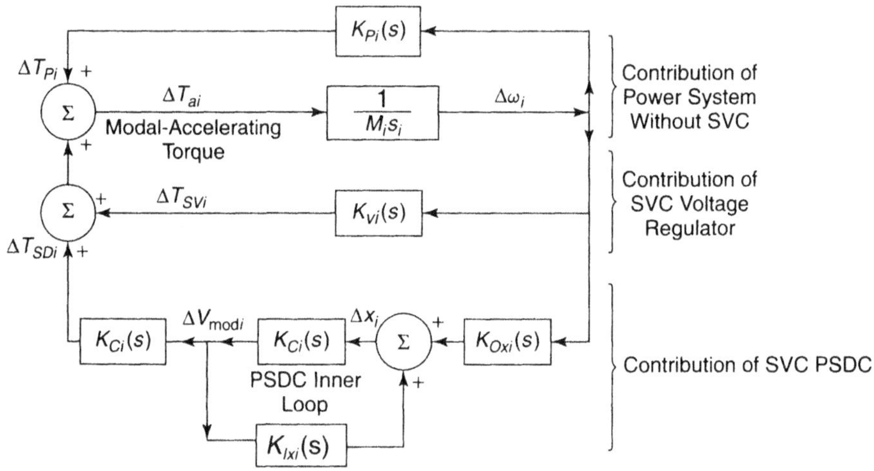

Figure 6.10 A simplified block diagram showing the damping- and synchronizing-torque contributions of the different controllers on the *i*th power-swing mode.

$$K_{Pi}(s) = D_i(s) + K_i(s)\,\frac{\omega_b}{s_i} \tag{6.41}$$

SVC Voltage Control The closed-loop transfer function of the voltage controller is given by

$$V_{\text{loop}i} = K_{VBi}\,\frac{V_{\text{reg}i}}{1 + V_{\text{reg}i}K_{VBi}} \tag{6.42}$$

Because the bandwith of the voltage-regulator loop is much greater than the frequency of the electromechanical-swing mode, $V_{\text{loop}i}$ will generally have a magnitude close to unity and a phase lag of 5°–20°.

The net modal-synchronizing-torque component, ΔT_{svi}, contributed by the SVC voltage regulator in response to modal speed is obtained as follows through the transfer function:

$$K_{Vi} = K_{TBi}K_{V\delta i}\,\frac{V_{\text{loop}i}}{K_{VBi}} \tag{6.43}$$

where K_{Vi} has a phase angle close to −90°.

SVC PSDC The effectiveness of the SVC auxiliary PSDC is dependent mainly on the following three factors:

1. *Controllability*, in which the ability of the auxiliary input to control the modal-accelerating torque is represented through the block $K_{OXi}(s)$:

$$K_{Ci} = K_{TBi} \frac{V_{\text{loop}i}}{K_{VBi}} \tag{6.44}$$

2. *Observability*, in which the sensitivity of the selected auxiliary signal to the modal-speed variations is modeled through the transfer function $K_{OXi}(s)$:

$$K_{OXi} = K_{X\delta i} + K_{V\delta i} K_{XBi} \frac{V_{\text{loop}i}}{K_{VBi}} \tag{6.45}$$

3. *Inner-loop sensitivity*, in which the sensitivity of the selected auxiliary signal to the output of the PSDC is manifested through the block $K_{IXi}(s)$:

$$K_{IXi} = K_{XBi} \frac{V_{\text{loop}i}}{K_{VBi}} \tag{6.46}$$

In general, all the transfer functions depicted in Figs. 6.8–6.10 are frequency-dependent for a general power system with several controllable devices. However, if the SVC is located far away from the other controllers, these transfer functions reduce to real functions for the relevant range of frequencies.

6.4.3 Design of an SVC PSDC

The design of an SVC PSDC is explained through a case-study power system, as presented in refs. [6] and [7]. This power system comprises two similarly large areas interconnected by a tie-line and compensated at the midpoint by an SVC, as shown in Fig. 6.11. Each area is modeled by an equivalent generator with an excitation system, a power-system stabilizer, and a turbine governor. The equivalent generators in each area generate about 1 pu (10 GW) of power that is used largely to supply the local load. A maximum of 0.1 pu power can be transmitted along the tie-line corresponding to 90° angular separation between the areas. This tie-line has a reactance of 10 pu on the area base.

The SVC is installed to enhance the power-transfer capability of the tie-line. The system exhibits an inter-area mode of 0.5 Hz, and correspondingly, the SVC voltage-regulator parameters are set at $K_R = 20$ pu/pu (slope = 5%) and

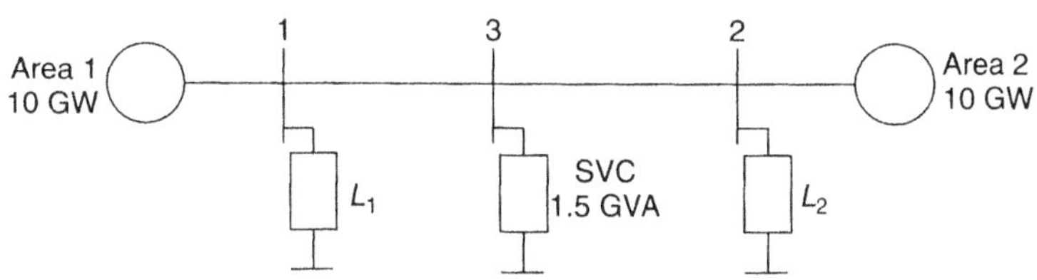

Figure 6.11 The two-area case-study power system.

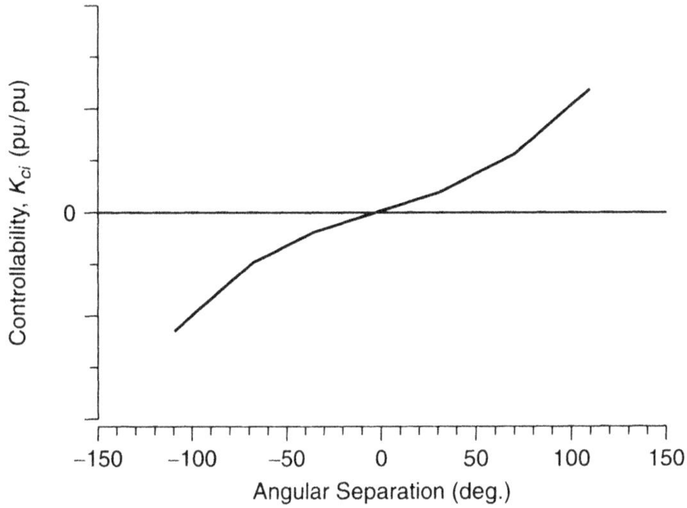

Figure 6.12 Controllability K_{ci} as a function of inter-tie power transfer.

$T_R = 0.2$ s. The TCR susceptance is rated at 0.15 pu on the area base; the SVC capacitance is chosen to ensure 1 pu voltage for all levels of power flow.

6.4.3.1 Controllability The effectiveness of an SVC in controlling power swings is examined through the plot of controllability, K_{ci}, as a function of inter-tie power transfer [6] illustrated in Fig. 6.12. The power transfer is represented by the angular separation between areas 1 and 2 of the two-area case-study power system. If power flows from area 1 to area 2, the angle of separation is positive. The controllability is seen to increase with transmitted power, indicating that the SVC becomes more effective in damping-power oscillations, which is a very desirable feature. However, a concern is the change in the sign of controllability when the direction of power flow reverses. This change must be considered when damping controllers are designed.

6.4.3.2 Influence of SVC Sites and the Nature of Loads The SVCs are normally located at the following sites:

1. Buses experiencing maximum voltage excursions.
2. Buses where an SVC of a given rating can impart maximum electrical damping to the power system.
3. Buses where the critical voltage-sensitive loads are connected.
4. The electrical midpoint of transmission lines connecting two areas, where maximum voltage swings are likely in the absence of an SVC.

The natural availability of a step-down transformer with the appropriate sec-

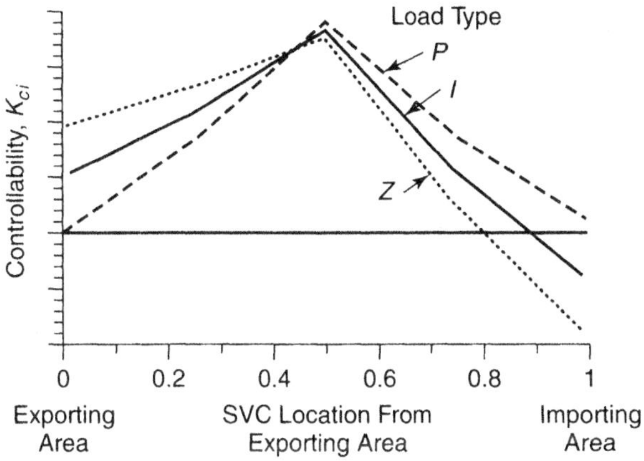

Figure 6.13 Controllability K_{ci} as a function of SVC location and load types.

ondary or tertiary voltage levels also serves as an economic criterion for SVC placement at a bus.

Although the guidelines in the preceding list are generally followed, they are intuitive; thus they need mathematical validation. In the case-study system of Fig. 6.12, a controllability plot is obtained as a function of an SVC's distance from the exporting area—with the load types as the parameter [6]—as shown in Fig. 6.13. It is evident from the controllability variation that the midpoint-located SVC is most effective in damping the power swings. Furthermore, if the loads are constant impedance or constant current, the siting of an SVC close to the receiving end or importing area may result in negative damping, thus jeopardizing the system.

It may not be advisable to modulate the SVC bus voltage if the SVC is located in a load area [6], [7]. Varying the voltage changes both the load power and the generator-output power that flows in the line. The two effects may conflict with each other as the operating point changes, for which reason it may not be possible to damp the local swing mode in a robust manner. A beneficial effect can still be imparted by stiffening the system with the pure voltage control of SVC.

In this example, controllability has been obtained in respect of the only one existing power-swing mode. In multimachine systems, controllability variation needs to be examined for the different electromechanical or swing nodes that occur in the system. It may be noted that the ideal SVC location for damping one node may be different from that corresponding to another mode. In such a situation, a careful choice fo location must be made that is based on system requirements and objectives.

Sometimes, multiple SVCs on the distribution side are more effective than a large SVC at the transmission level [8]. One reason is that multiple SVCs improve the power quality; another is that they offer direct reactive-power support to the loads during power swings, especially to the induction-motor loads. The SVCs should, however, be separated by substantial impedances as far as possible to avoid local controller interactions, as explained in Chapter 9.

6.4.3.3 Selection Criteria for PSDC Input Signals An appropriate choice of control signal for an auxiliary PSDC is critical in securing requisite levels of electrical damping in power systems. The control input can comprise only one signal or a combination of signals for a given application. In either case, the following factors must be considered for selecting an auxiliary signal:

1. The sensitivity of the control signal to the power-swing modes—the *observability*, as discussed in Section 6.4.2.2. This sensitivity must be great, particularly at higher power levels; it is only at these levels that the real need for damping arises.
2. The effectiveness, or *robustness*, of the control signal in making positive-damping contributions over widely varying system-operating conditions. Positive damping must be ensured, even if the direction of the power flow reverses.
3. The choice between locally measurable signals and remotely communicated signals. Usually preferred over remote signals, locally measurable signals are more easily and reliably measured without any communication delays. However, there is, another school of thought that is in favor of remote signals, often espousing the highly reliable and successful operation of HVDC links based on telecom signals, as well as the recent advances in implementing very fast communication systems.

   Signals such as accelerating power of critical generators could still prove to be the best controller signals. It is argued [9] that despite the cost and risk of transmitting such signals to remote controller locations, they may still provide better performance than the locally measured signals. In the worst case, these high-quality remote signals may still be used as reference for comparing the performance of local signals or even as a backup in case the local signals prove inadequate.
4. The vulnerability to local PSDC inner-loop instability. The signal should achieve a required level of damping through the auxiliary damping controller, without inducing any local instabilities in the SVC controller itself, by keeping the inner-loop gain low. This local instability can occur if the selected control signal is highly sensitive to the SVC output.
5. A minimum sensitivity of the control signal to other power-system swing modes. For instance, in a transmission-line application, the SVC must respond only to the power swing on the line; it must not respond to some other local oscillation modes in the areas connected to the line, for a substantial control range may be lost in damping those modes for which the SVC may not be able to control satisfactorily.
6. The possibility of adverse interaction with other controllable power-sys-

tem devices. The control signal must have as low a sensitivity to the response of other stabilizing controllers as possible for avoidance of local instabilities within the controller bandwidth.

7. A small sensitivity of the signal to random noises occurring in the power system. Some fraction of this noise may lie with the bandwidth of the power swings. Thus the controller must be ensured to respond only to true power-system swings, not to any noise component.

6.4.3.4 Input Filtering The control inputs must be filtered appropriately to avoid interacting with the electrical phenomena outside the required control bandwidth of the SVC [8].

A high-pass filter is usually provided on the low-frequency side to prevent a large response during power damping, or long-term settling after a generator outage, or even a load rejection. Low-pass filtering is incorporated to eliminate interaction with the high-frequency torsional-oscillation modes. Adverse interaction with specific resonant modes, such as network resonances, is precluded by means of the application of notch filters.

A high-frequency cutoff around 0.05 Hz (3-s washout time constant) and a double low-pass cutoff of about 4 Hz (40-ms time constant) has been found from experience to be a good starting point for filter design. The filters can be fine-tuned after the complete controller design has been accomplished.

6.4.3.5 General Characteristics of PSDC Input Signals Some of the locally measurable individual signals that can be potential inputs of the SVC PSDC are as follows:

SVC Bus Frequency This relates directly to the changes in the rotor speed if they are measured close to the generating bus. However, its sensitivity to actual speed variation diminishes with increasing distance from the generator, especially so at the line midpoint where the SVC is installed.

SVC Bus Voltage This is the main SVC control signal, as the device is primarily used in power systems for voltage regulation. The voltage magnitude is a function of the angular difference across the tie-line connecting the two areas. The drawback with using bus voltage as a damping-control signal is that it is highly sensitive to the SVC output susceptance and results in an undesirably high inner-loop sensitivity.

Active-Power Flow The integral of incremental active power is dependent on the rate of angle change. This fact is evident from the linearized-swing equation of a generator given in Eq. (6.10), assuming that the mechanical-input power remains constant over the duration of controller action.

Active Component of the Line Current Line-current flows are also influenced by the angular difference across the tie-lines. However, the specific line from which this signal should be extracted needs to be chosen judiciously.

Line-Current Magnitude The utility of the line-current magnitude signal lies in its inclusion of the effects of all other current components that are caused by system disturbances and which may be superimposed on the main fundamental-current wave.

6.4.3.6 Performance of PSDC Input Signals The traits of the aforementioned signals are examined in the context of the two-area case-study system of Fig. 6.11. The active power and the active line current are considered positive when they flow from area 1 to area 2, and the power and current signals are measured on the line connected to area 1.

The observability K_{OXi} and magnitude of inner-loop sensitivity for the different signals are plotted in Fig. 6.14 as a function of power transfer, with the line impedance as parameter [7]. Positive angular separation indicates the power transfer from area 1 to area 2; negative angular separation, the power transfers from area 2 to area 1.

The observability K_{OXi} is a complex quantity, of which only the magnitude is plotted in Fig. 6.14(a). In fact, the phase angle of observability does not change greatly with an operating condition other than the sign reversal. However, this is not true with the phase angle of inner-loop sensitivity K_{IXi}, which undergoes marked variations. Figure 6.14(b) shows the magnitude plot of inner-loop sensitivity K_{IXi}. The following observations are made:

Observability The line-current magnitude, voltage, and frequency all display increasing observability with power transfer, which is a desirable feature. There is, in fact, a change of sign when the direction of power flow reverses. The magnitude of observability is highest when the line-current magnitude signal is followed closely by the voltage; however, this magnitude for the bus frequency is two orders lower than that of the line-current magnitude signal. The active-power flow and the line-current active component show diminished observability for large levels of power transfer—without, however, any sign reversal when the power-flow direction changes.

Inner-Loop Sensitivity Magnitude The factor K_{IXi} is highly sensitive to the operating conditions. The bus voltage, frequency, power flow, and active current all display increasing K_{IXi} with the power flow, whereas the line-current magnitude shows an opposite trend. The bus-voltage signal exhibits the maximum level of unacceptable inner-loop sensitivity. The highest value of inner-loop sensitivity for the bus voltage is obtained with maximum shunt compensation, provided to ensure the maximum power transfer corresponding to the weakest network. On the other hand, a similar scenario appears for the line-current magnitude signal with minimum shunt compensation corresponding to the strongest network configuration.

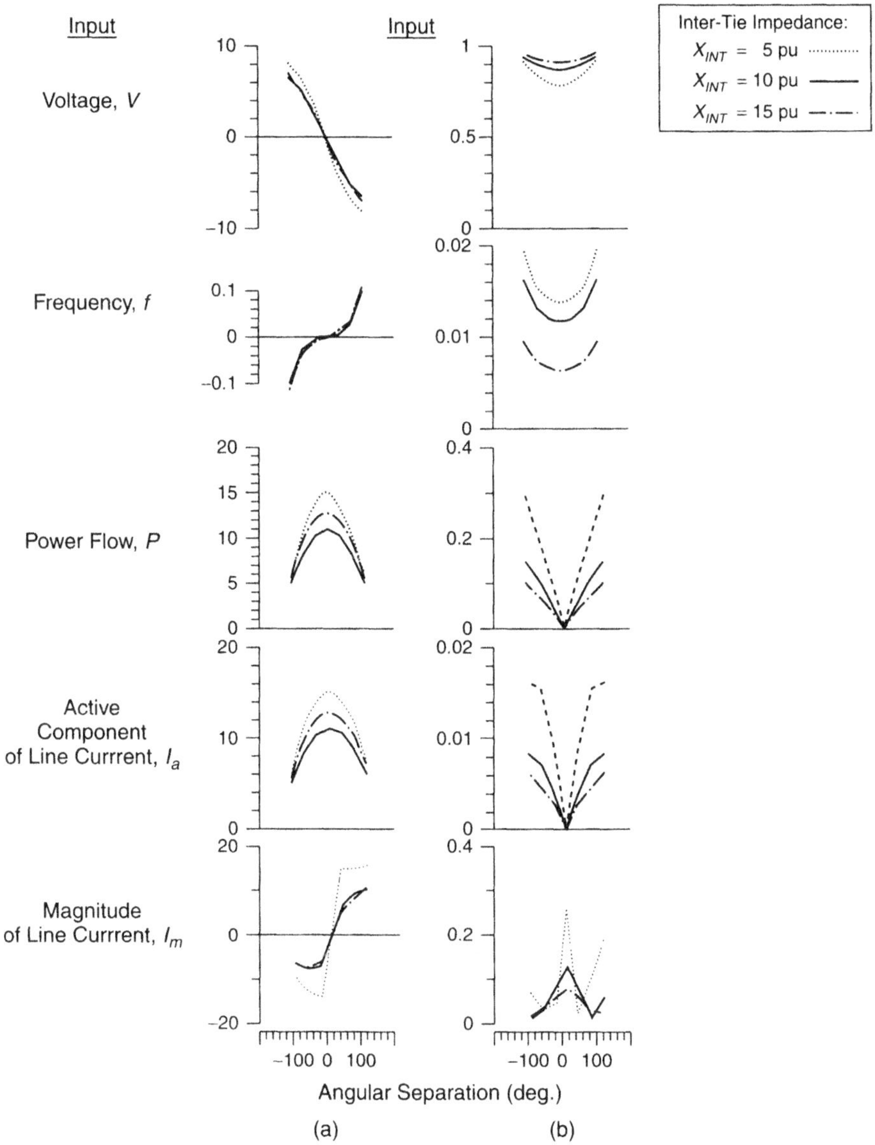

Figure 6.14 Observability and inner-loop sensitivity of various input signals versus the inter-tie power transfer and impedance: (a) observability K_{OXi} (pu/pu) and (b) inner-loop sensitivity $|K_{IXi}|$ (pu/pu).

6.4.3.7 SVC PSDC Requirements The transfer function of the PSDC for a control signal, x, relating modal torque to modal speed, as obtained from Fig. 6.10, is given by

$$G_{DCi}(s) = K_{OXi}(s) \frac{\mathrm{PSDC}_x(s)}{1 - \mathrm{PSDC}_x(s) \cdot K_{IXi}(s)} K_{Ci}(s) \frac{\omega_b}{s} \tag{6.47}$$

Two features are obvious from Eq. (6.47):

1. The PSDC can impact the system damping significantly if the inner-loop gain is low—even less than unity—for all power-system–operating conditions.

$$|\mathrm{PSDC}_x(s)|\ |K_{IXi\max}| \ll 1 \tag{6.48}$$

where $|K_{IXi\max}|$ is the maximum value of inner-loop gain magnitude under all system conditions. In such a situation, the damping produced by the auxiliary controller is directly proportional to the PSDC transfer function. Since both the gain and phase of the inner-loop sensitivity undergo significant variation with the operating condition, it is a desirable that the PSDC damping contribution be made impervious to the inner-loop behavior. This is ensured by the fulfillment of the condition in Eq. (6.48).

2. The damping contribution is rendered independent of the PSDC if the inner-loop gain is high, that is, if

$$|\mathrm{PSDC}_x(s)|\ |K_{IXi\min}| \gg 1 \tag{6.49}$$

where $K_{IXi\min}$ is the minimum value of the inner-loop gain magnitude under all system-operating conditions. The damping torque then becomes a function of the system characteristics as reflected in observability K_{OXi} and controllability K_{Ci}.

For the auxiliary controller to generate positive damping, the transfer function, $G_{DCi}(s)$, should be essentially real and positive. If a margin value of the inner-loop gain can now be specified, the maximum gain of the PSDC can be computed for a reference-case power-system–operating condition.

$$\mathrm{PSDC}_{xi(D\max)} = \frac{e^{j\theta_{RXi}}}{GM_{IL}|K_{IXi\max}|} \tag{6.50}$$

$$\theta_{RXi} = -\mathrm{angle}\{K_{CiR}K_{OXiR}\} \tag{6.51}$$

where $\mathrm{PSDC}_{xi(D\max)}$ = the frequency response of transfer function $\mathrm{PSDC}_x(s)$ at complex frequency, s_i, for imparting maximum damping to the ith mode employing the auxiliary control signal x

$s_i = j2\pi f_i$

f_i = the frequency of ith power-swing mode

θ_{RXi} = the angle of PSDC_{xi} for the reference case that ensures the validity of $G_{DCi}(s)$, thereby contributing positive-damping torque (the reference-case power-operating condition is usually chosen to correspond to a heavy-power transfer scenario)

GM_{IL} = the prespecified gain margin of the inner loop, typically greater than 10 dB

K_{CiR}, K_{OXiR} = the controllability and observability constants for the reference case

Performance of Different Signals Every control signal provides a different level of maximum damping to the system. The various control signals, x, employable as inputs to the PSDC are compared in terms of the maximum damping contribution, $D\max_{xi}$, which they can make corresponding to ith swing mode for varying operating conditions.

$$D\max_{xi} = \text{Real}\{K_{ci}\text{PSDC}_{xi(D\max)}\} \tag{6.52}$$

where $D\max_{xi}$ = the maximum damping achievable with control signal x corresponding to swing mode i.

At any operating condition, $D\max_{xi}$ is computed using $\text{PSDC}_{xi(D\max)}$ corresponding to an angular separation of 70° as a reference case and an inner-loop gain margin of 10 dB. Plots are obtained of $D\max_{xi}$ as a function of power transfer between the two areas for each signal [7]; these are depicted in Fig. 6.15.

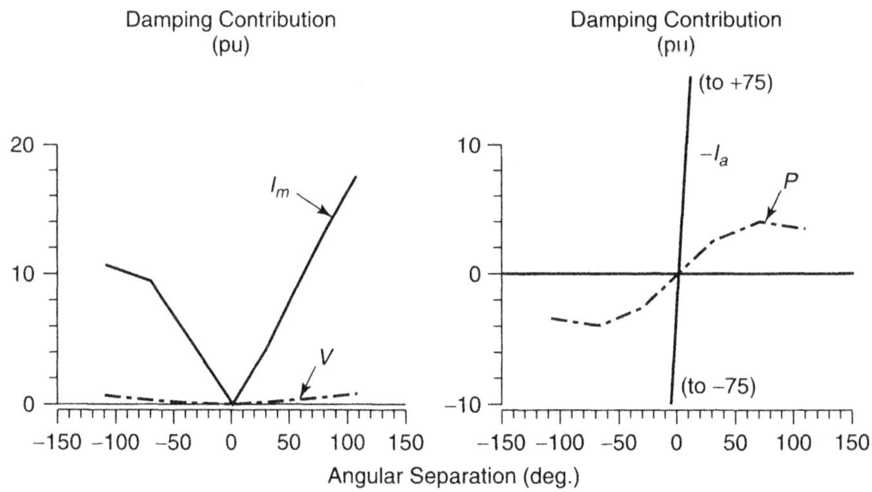

Figure 6.15 The maximum PSDC damping contributions with a 10-dB inner-loop gain margin.

The line active current offers the maximum damping of all signals because of a very small inner-loop sensitivity. However, both line active current and power change the sign of their damping contributions if the direction of power flow reverses. The bus-voltage signal provides positive damping for all levels of power transfer in either direction, just like the line-current magnitude, but the level of generated damping is seriously restricted by a very high inner-loop sensitivity. The line-current magnitude provides substantial positive damping irrespective of the power-flow direction. It is relatively insensitive to SVC control action and thus becomes the preferred signal among those considered.

The general performance of a controller-transfer function can be simply explained in terms of its gain-and-phase variations with frequency [8]. The controller generates a pure-damping influence if the controller has a purely proportional gain, that is, the phase shift between the output and control signal is zero. As the phase shift increases, the synchronizing influence improves with a corresponding increment in the swing-mode frequency. The damping effect, on the other hand, declines with phase shift and becomes negative when the phase shift exceeds 90°. The phase-lead provides a desynchronizing effect, and it correspondingly reduces the synchronizing torque, which is highly undesirable.

An effective controller-transfer function should generally provide a phase shift of 0 to 45° over the frequency range of the swing mode to be damped [10]. The gain should be selected so it can use the full rating of the SVC during the most critical exigency.

6.4.3.8 Design Procedure for a PSDC The following procedure is suggested in refs. [6] and [7] for the design of the auxiliary PSDC.

1. The controller is designed primarily for the dominant swing-mode frequency.
2. The desired phase angle of the controller-transfer function is obtained from Eq. (6.51) corresponding to the pure-damping condition. This phase angle is a function of the controllability and observability constants.
3. An operating point signifying a heavy-power transfer scenario is chosen, and a specific magnitude of system damping is selected for this scenario. The desired controller gain is that which ensures the specific magnitude of damping for the chosen operating point subject to the following conditions:
    a. an inner-loop gain margin of at least 10 dB is satisfied for the most constraining network configuration;
    b. a maximum level of interaction with subsynchronous modes is ensured; and
    c. a noise amplification beyond an acceptably small limit is not permitted.
4. The efficacy of the PSDC controller must be established for both forward and reverse power flow in the tie-line.

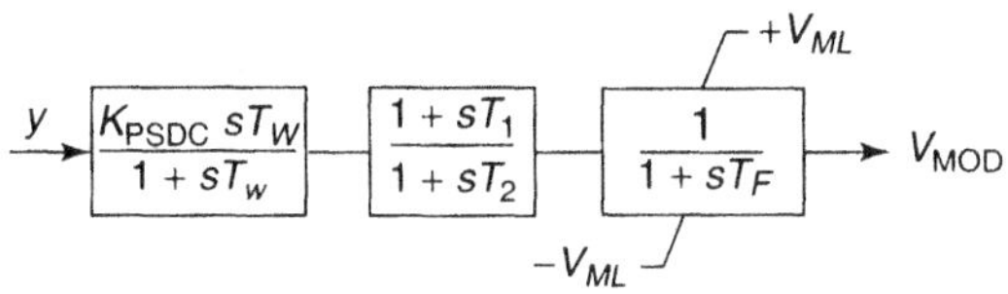

Figure 6.16 The structure of a simple PSDC.

A tentative value of controller gain can be obtained by performing stability simulations for the worst system configuration in the absence of PSDC and noting the maximum variation in the auxiliary signal magnitude. The gain may be chosen as that which can cause the SVC reactive power to traverse its entire controllable range for this peak variation in the auxiliary control signal.

A typical PSDC controller comprises a lead-lag stage, a washout stage, and a high-frequency–filtering stage, together with a gain [7] as shown in Fig. 6.16. The filter is designed to pass the swing-mode-frequency signal while allowing for any variation in this frequency from system conditions. It rejects frequencies associated with non–power-swing modes, such as subsynchronous torsional oscillations and modes relating to noise signals that override the auxiliary control signals. In some cases, this noise may be within the bandwidth of the power-swing frequencies. The control system, therefore, needs to be designed carefully; too high a gain must be avoided.

The technique for PSDC controller design, although described for a two-area system, is also valid for a three-area system [6], [7] or a multi-area system. In this cases, however, more than one power swing mode is likely to exist. The effectiveness of the same SVC is dependent on the location of the loads as well as its own placement. The controllability of a mode may improve if the SVC is located close to the midpoint of that mode shape. In the event that the midpoints of different modes are at different locations, the damping benefit, which the SVC can provide for one mode, will not be the same as that for the other modes.

6.4.3.9 Case Study The PSDC design for a midpoint-located SVC in the two-area system depicted in Fig. 6.11 is presented in the following case study [6], [7]. The performance of different auxiliary control signals is also described.

A large power-transfer scenario is chosen in which a 70° angular difference exists across the tie line. The desired damping level is selected as that which the system inherently has with zero power transfer across the tie-line.

The gain-and-phase data for various auxiliary control signals are computed according to the technique described previously and are compiled in Table 6.1. It may be noted that the stipulated damping level is not achieved by bus voltage and frequency signals because of the inner-loop gain-margin constraint.

After resorting to careful engineering judgment, the controllers for each auxiliary signal are designed to satisfy the gain-and-phase requirements. The

TABLE 6.1 Damping-Controller Characteristics (0.5 Hz)

Input Signal	Pure-Damping Phase Shift	Gain (pu/pu) for Desired Damping	Gain (pu/pu) for Inner-Loop Constraint	Inner-Loop Worst-Case Gain Margin
Voltage	−121°	2.5[a]	0.35	10 dB
Frequency	+162°	345[a]	20	10 dB
Power	+93°	1.4	1.4	15 dB
I active	+91°	1.5	1.5	39 dB
I magnitude	+91°	1.5	1.5	22 dB

[a] This gain would achieve the desired damping if the inner-loop gain margin were >10 dB; however, the actual inner-loop gain would prevent the attainment of the desired damping, even with a much higher damping- control gain.

magnitude and phase characteristics for the controller based on the line-current-magnitude signal, as an example, are depicted in Fig. 6.17, and the controller parameters are indicated as well. The PSDC gain is usually fine-tuned on the basis of root-loci studies of the critical system modes.

The performance of two control signals—active power and line-current magnitude—in response to large disturbances is obtained through time-domain simulation and is illustrated in Fig. 6.18. The power-reversal scenario is depicted as well.

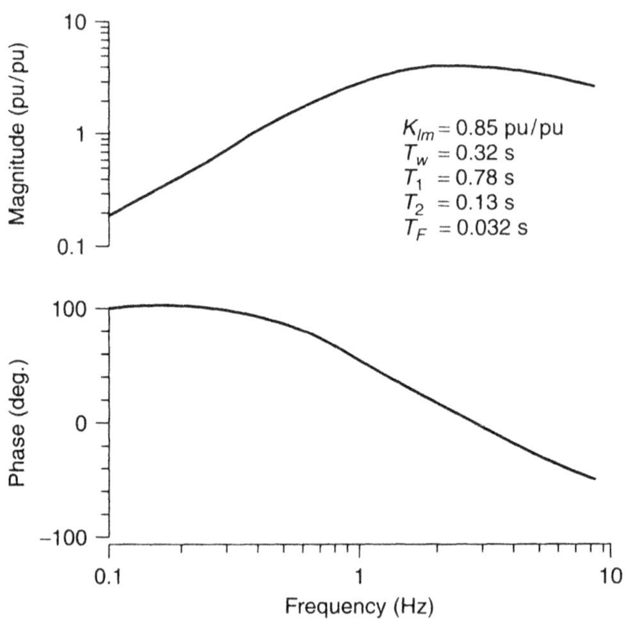

Figure 6.17 The sample transfer function of damping controller with I_m input.

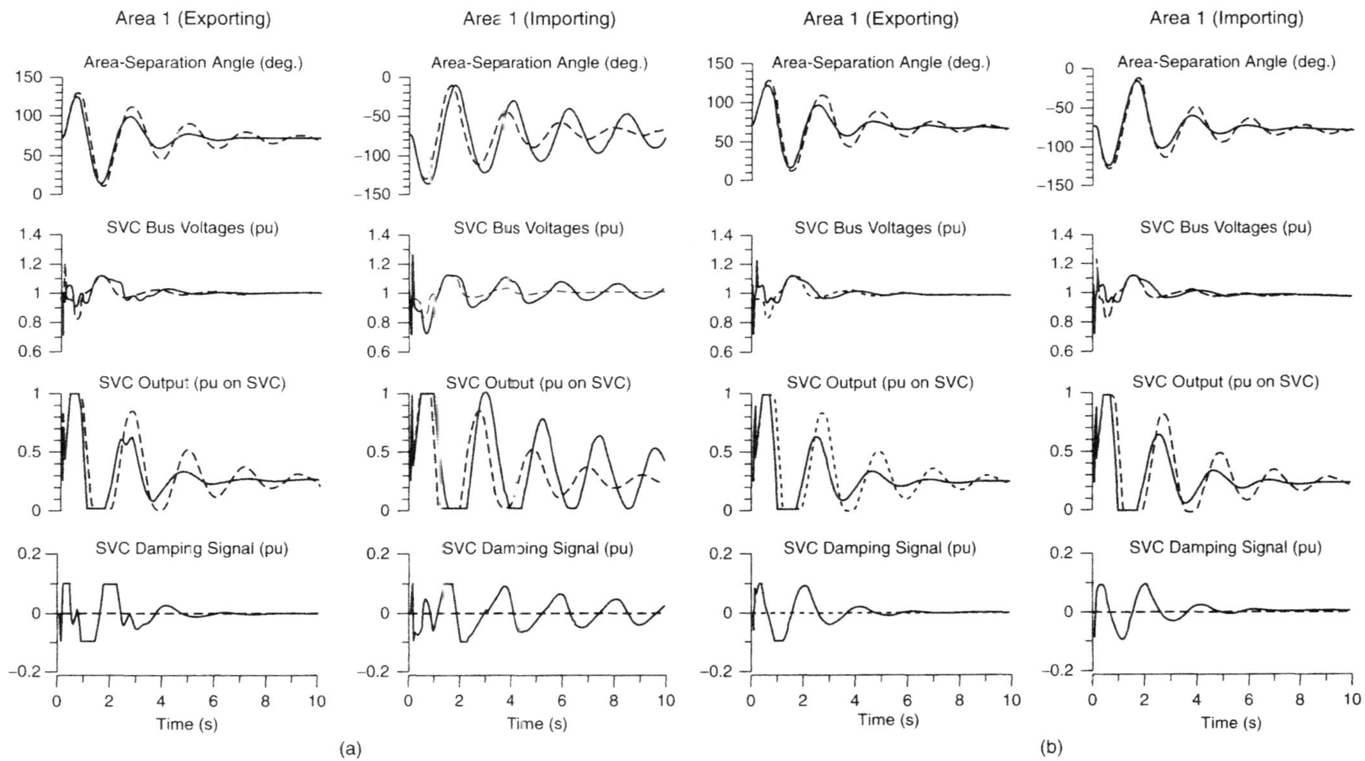

Figure 6.18 Influence of the power-transfer direction on PSDC damping performance. (Note: Solid curve, with PSDC; dashed curve, without PSDC.) (a) the power input and (b) the current input.

The following observations are made:

1. With the damping controller, the first-swing oscillations in both the area-separation angle and the midpoint bus-voltage signals are decreased. This decrease shows that the PSDC controller can also be effective in improving first-swing stability, especially if the disturbance causes the system to oscillate close to its stability limit.
2. Some additional oscillations are visible in the active-power control signal, have a frequency higher than the power-swing frequency, and are enhanced by inner-loop dynamics. It may be noted that the gain margin with the line-current magnitude controller is higher than with the active-power controller; therefore, the oscillations generated by inner-loop dynamics have a greater tendency to appear in the active-power damping signal.
3. During power reversal, the damping provided by the active-power controller becomes negative while the line-current magnitude controller continues to impart positive damping to the system. The imparted damping is a product of the controllability and observability indices from Eq. (6.47). For this system, it is noted that from Fig. 6.12, the system controllability changes sign with power reversal. The observability of the line-current magnitude also correspondingly changes sign, although not that of the active power; hence this activity leads to a positive damping for the line-current magnitude signal but not for active power. Therefore, an adaptive controller may be required for the active-power signal that changes the sign of the PSDC controller output with the direction of the power flow.

6.4.4 Composite Signals for Damping Control

A detailed analysis of SVC control with several single individual-control signals, together with the associated controller design, was presented in the previous section. In some situations, not one but a combination of two or more signals may prove advantageous. The auxiliary controller output would then be a composite sum of the outputs of two different controllers, each having a different input-control signal.

In an alternative situation, two or more locally available signals may be algebraically manipulated to compute a new composite signal that can be fed to a single auxiliary PSDC controller. A computed signal, which has been shown to be of significant value, is described in the following text.

6.4.4.1 Frequency of Remotely Synthesized Voltage This involves the frequency computation of a remote-area ac voltage or a remote-generator internal voltage using locally measurable SVC bus voltage and transmission-line-current signals, together with the knowledge of the intervening impedance between the SVC bus and the remote bus [7], [11], [12].

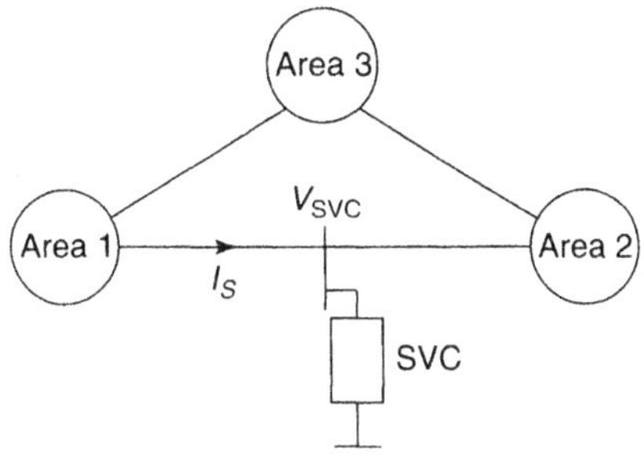

Figure 6.19 An SVC installed at the midpoint of a critical line in a multi-area system.

Consider the example system shown in Fig. 6.19. The synthesized voltage, $V_{1\,\text{syn}}$, of the remote bus (area 1) is expressed as

$$\begin{aligned}\overline{V}_{1\,\text{syn}} &= \overline{V}_{\text{SVC}} + \overline{Z}_{\text{syn}}\overline{I}_S \\ &= \overline{V}_{\text{SVC}} + (R_{\text{syn}} + jX_{\text{syn}})\overline{I}_S \end{aligned} \tag{6.53}$$

where R_{syn} and X_{syn} represent the resistance and reactance, respectively, of the synthesizing impedance.

The relation given in Eq. (6.53) can be rewritten in terms of phase quantities as

$$\begin{aligned} V_{1\,\text{syn}a} &= V_{\text{SVC}a} + R_{\text{syn}}I_{Sa} + X_{\text{syn}}(i_{Sc} - i_{Sb}) \\ V_{1\,\text{syn}b} &= V_{\text{SVC}b} + R_{\text{syn}}I_{Sb} + X_{\text{syn}}(i_{Sa} - i_{Sc}) \\ V_{1\,\text{syn}c} &= V_{\text{SVC}c} + R_{\text{syn}}I_{Sc} + X_{\text{syn}}(i_{Sb} - i_{Sa}) \end{aligned} \tag{6.54}$$

A frequency transducer is then employed to determine the frequency of $V_{1\,\text{syn}a}$. This PSDC signal responds primarily to the swings of the concerned remote area, but it also responds to some extent to the swings of all other areas connected to the bus.

The synthesizing impedance is chosen to be the actual value or a representative measure of the transmission-line impedance between the SVC bus and the inertial center of the area. As this impedance may vary over a period of time because of the network reconfiguration, an average value is generally chosen. The magnitude of this impedance may also be made a function of system status—namely, the breaker position.

The synthesized-frequency signal is highly effective for two reasons: (1) the signal virtually resembles modal speed, resulting in a high-observability $K_{OXi}(s)$, and (2) the inner-loop gain $K_{IXi}(s)$ is minimal with this signal.

An implementation of a PSDC with a synthesized-frequency signal is depicted in Fig. 6.20. Appropriate low-pass filtering is built in to avoid interaction with other oscillatory modes, such as those of torsional vibrations. A high-pass filter is also built in to preclude any steady-state offset. The bandpass filter is designed to eliminate any system noise overriding the control signal.

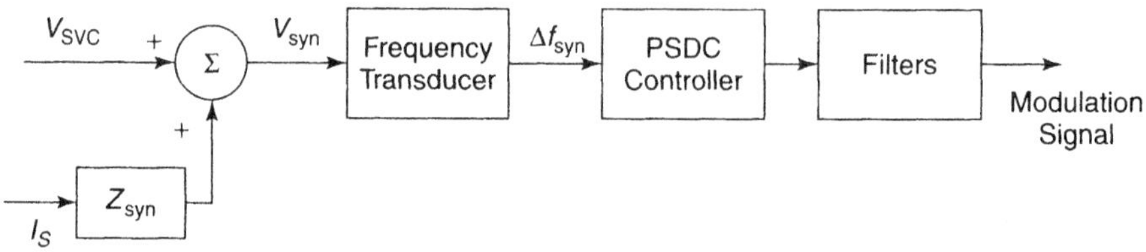

Figure 6.20 Implementation of a PSDC with a synthesized-frequency signal.

6.4.4.2 Case Study The application of a multiple-input, synthesized, modulating signal in improving steady-state stability is presented in this case study [11], [12] that relates to the system of Example 5.2 (Chapter 5). The damping torque technique is employed to examine the efficacy of different control signals for reactive-power modulation of a midline-located SVC in a SMIB system. The electrical damping torque results are correlated with those obtained from eigenvalue analysis.

The synthesized signal is termed *computed internal frequency* (CIF) [11], [12] and involves the computation of the internal-voltage frequency of the remotely stationed generator, using locally measured bus voltage, transmission-line current at the SVC bus, and the equivalent reactance between the generator internal voltage and SVC bus.

The efficacy of the CIF in improving power-transfer capability is compared with two other signals—line-reactive power flowing into the SVC bus and the SVC bus frequency, f_s. The PSDC auxiliary controller $G(s)$ is assumed to be a first-order lead-lag transfer function of a design based on root-loci studies.

$$G(s) = K_B\left(\frac{1 + sT_1}{1 + sT_2}\right) \tag{6.55}$$

Root-Loci Analysis As the objective of SVC control is to enhance power-transfer capacity, the auxiliary controller parameters are determined at an operating-generator power level of (P_g) 800 MW that is close to the network limit of 980 MW. For different auxiliary controllers, the loci of critical eigenvalues are obtained with varying controller parameters, each taken one at a time. Based on these root loci, a range is selected for different parameters (K_B, T_1, and T_2) in which a high degree of stabilization is provided to the critical-system nodes. Power-transfer limits are then evaluated for different combinations of parameters in these ranges. The controller parameters, which result in the maximum power transfer, are chosen as optimal for the particular auxiliary controller. A comparison of different auxiliary controllers is presented in Table 6.2.

The CIF auxiliary controller results in the highest power transfer of 980 MW, implying the full use of transmission-network capacity and representing a 60% enhancement over the power-transfer limit of 610 MW attained by pure voltage control of the SVC. It is further found [11], [12] that even with an error of ±10%

TABLE 6.2 Comparative Study of Different SVC Auxiliary Controllers

Auxiliary Control Signal	None (Voltage Controller)	Bus Frequency	Line Reactive Power	CIF
Optimal parameters				
K_B	—	−0.065	−0.035	−0.4
T_1	—	0	0.044	0
T_2	—	0.03	0.02	0.03
Eigenvalues with optimal parameters for P_g = 800 MW				
Generator	$0.186 \pm j4.34$	$-0.132 \pm j4.31$	$-0.046 \pm j5$	$-4.131 \pm j7.52$
	$-32.0 \pm j0.874$	−36.9, −2.89	−32.3, −28.6	−30.9, −1.29
	−2.89	$-27.7 - j10$	−3.02	−3.3
	$-0.893 \pm j0.97$	$-0.9 + j0.97$	$-0.62 + j0.85$	$-0.584 + j1.25$
AVR	$-0.893 - j0.97$	$-0.9 - j0.97$	$-0.62 - j0.85$	$-0.584 - j1.25$
	$-26.0 \pm j24.1$	$-26.3 \pm j22.72$	$-26.1 \pm j24.3$	$-25.9 \pm j22.9$
Network	$-3.2 \pm j3496$	$-3.2 \pm j3496$	$-3.2 \pm j3496$	$-3.2 \pm j3496$
	$-13.2 \pm j2470$	$-13.2 \pm j2470$	$-13.2 \pm j2470$	$-13.2 \pm j2469$
	$-3.2 \pm j2868$	$-3.2 \pm j2868$	$-3.2 \pm j2868$	$-3.2 \pm j2868$
	$-14.9 \pm j1841$	$-14.9 \pm j1841$	$-14.9 \pm j1842$	$-14.0 \pm j1842$
	$-12.9 \pm j1000$	$-12.91 \pm j1003$	$-13.0 \pm j1000$	$-13.3 \pm j1002$
	$-23.8 \pm j369$	$-1.16 \pm j362$	$-26.8 \pm j379$	$-23.1 \pm j355$
	$-23.5 \pm j313.5$	$-23.3 \pm j314.1$	$-24.9 \pm j313$	$-25.0 \pm j316$
SVC	$-544.3 \pm j65.8$	$-540.3 \pm j87.8$	$-544.0 \pm j73.3$	$-539.0 \pm j80.1$
	$-5.63 \pm j313.2$	$-5.66 \pm j307$	$-0.32 \pm j323$	$-7.7 \pm j293$
	$-49.5 \pm j99.8$	$-78.0 \pm j95$	$-8.0 \pm j55.9$	$-26.7 \pm j159$
		$-27.7 \pm j10$	−138.1	−108.3
Power limit (MW)	610.2	911	868.5	980

in the estimation of equivalent inductance L_E, the efficacy of the signal did not get degraded.

Damping-Torque Analysis As the stability of the study system is predominantly dependent on the low-frequency (1–2 Hz) rotor oscillation mode, the electrical torque components are evaluated in the frequency range 0–15 rad/s. The variation of electrical damping torque with frequency for the pure-voltage controller (without auxiliary control) and with different auxiliary SVC controllers is depicted in Fig. 6.21. The corresponding variation of synchronizing torque is illustrated in Fig. 6.21. The rotor-mode eigenvalues for various controllers obtained from Table 6.2, together with the damping and synchronizing torques corresponding to these rotor-mode frequencies, are displayed in Table 6.3.

It is observed that damping torque is positive for line-reactive power, bus frequency, and CIF auxiliary controllers, all of which exhibit stable rotor modes. Damping torque is, however, negative for voltage controller, which has an unstable rotor mode, and the magnitude of damping torque is higher if the rotor-mode eigenvalue for any controller has a more negative real part. Therefore, a correlation is seen between the damping torque and eigenvalue analysis with respect to the stability of the rotor mode. The high-synchronizing torque associated with the CIF signal results in a higher rotor-mode frequency.

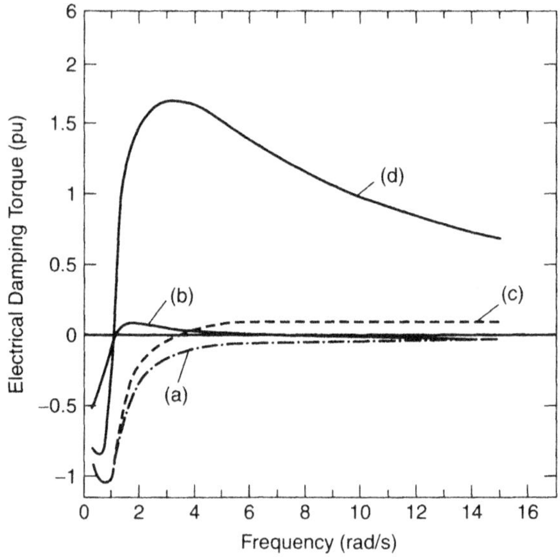

Figure 6.21 The electrical damping torque for an SVC voltage controller and different auxiliary controllers: (a) the voltage controller; (b) the line-reactive power; (c) the bus frequency; and (d) the CIF.

This study demonstrates that in a SMIB system, auxiliary signals based on frequency, such as bus frequency or CIF, lead to higher power transfers relative to other signals. The CIF provides maximum damping to the rotor mode at the considered operating point and results in the highest power transfer of all the signals investigated, clearly indicating the importance of high observability of the mode to be damped in the auxiliary control signal. The synthesized signal CIF contains the effect of the swing mode in the highest measure and is thus most effective.

6.4.5 Alternative Techniques for the Design of SVC Auxiliary Controllers

Once an appropriate location of SVC has been determined, the auxiliary controller design can be accomplished by any of the following techniques:

- phase-and-gain margin technique (described in this chapter)
- pole-placement technique (see refs. [3] and [13])
- linear quadratic Gaussian (LQG) control (see ref. [14])
- optimal robust control and H_∞ optimization (see refs. [15]–[17])
- fuzzy control (see refs. [18] and [19])
- artificial neural network (ANN) control (see ref. [20])
- genetic algorithms (see ref. [21])

TABLE 6.3 Electrical Damping and Synchronizing Torques for SVC Voltage Controllers and Different Auxiliary Controllers

SVC No.	Auxiliary Signal	Rotor-Mode Eigenvalue	Electrical Damping Torque (pu)	Synchronizing Torque (pu)
1	None (voltage controller)	$0.19 \pm j4.34$	0.084	4.28
2	Line reactive power	$0.044 \pm j5.00$	0.02	5.7
3	SVC bus frequency	$0.133 \pm j4.31$	0.06	4.2
4	CIF	$4.13 \pm j7.53$	1.2	9.4

6.5 SVC MITIGATION OF SUBSYNCHRONOUS RESONANCE (SSR)

Subsynchronous resonance (SSR) is a well-understood phenomenon [22]–[27]. It is defined by the IEEE as an

> electric power condition where the electric network exchanges energy with a turbine/generator at one or more of the natural frequencies of the combined system below the synchronous frequency of the system.

Subsynchronous oscillations in the torsional system of the steam turbine–generator may be caused by series compensation of transmission lines or by adverse interaction with HVDC controls, speed governors, and generator-excitation controls. If uncontrolled, these oscillations may lead to a permanent damage in the turbine-shaft system.

The SSR manifests itself through three forms [26]:

1. Induction generator effect (involving the electrical system alone).
2. Torsional interaction (involving both the electrical and mechanical systems).
3. Transient torque (involving both the electrical and mechanical systems but initiated by severe disturbances, such as faults).

The SSR alleviation strategies were developed and tested on the IEEE's First SSR Benchmark System [28] that presented severe SSR problems with four torsional modes spread over a wide range of frequencies. The decrement factor (i.e., the real part of the eigenvalues of the critical torsional modes with varying line-series compensation levels) is depicted in Fig. 6.22. The complete system eigenvalues for the four critical series-compensation levels are given in Table 6.4.

6.5.1 Principles of SVC Control

One of the means for obliterating torsional interaction between the torsional system of the generator and the series-compensated network is to install an SVC

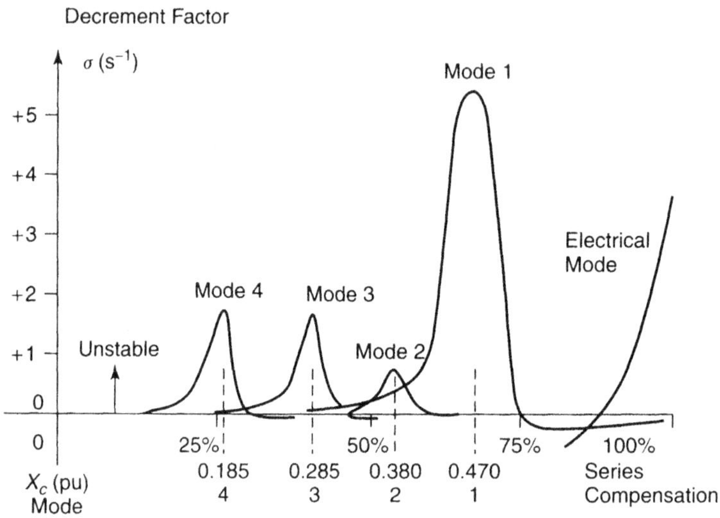

Figure 6.22 The real part of unstable eigenvalues of torsional and electrical modes for the IEEE's First Benchmark System (P_g = 1 pu, pf = 0.9 lag).

at the generator terminals. The SVC is essentially a shunt-connected inductance, through which the current is modulated in response to the rotor oscillations or generator speed.

The generator speed is chosen as the control signal because it includes components of all prime-mover modes that need to be damped. Such an SVC has been referred to in the literature [29]–[32] as *dynamic stabilizer* and *modulated-inductance stabilizer* (MIS). This SVC is equipped with only a speed-deviation controller, not the voltage regulator that constitutes an integral component of normal SVCs. An interesting feature is that even though the SVC reactive current acts through the transmission system, the SVC control action is not dependent on the generally varying transmission-system reactances [29]. In the absence of generator torsional oscillations, the SVC acts as a steady and continuous reactive load.

The SVC for SSR stabilization provides current amplification at resonance, for which reason it is placed at the generator terminals for maximum efficacy. A positive damping is ensured if the SVC reactive current is modulated 180° out of phase with the generator-rotor velocity deviation. In other words: If the rotor velocity increases, the SVC reactive current is made to decrease, and vice versa [29].

It may be recalled that this control philosophy is similar to that employed for enhancing generator damping to improve the power transfer. When the rotor velocity increases, the reduction in the SVC inductive current enhances the electrical output of the generator. For a constant mechanical input, this enhancement leads to a reduction in the kinetic energy of the rotor, resulting in an eventual

TABLE 6.4 System Eigenvalues of IEEE's First SSR Benchmark System Without SVC (P_g = 11 pu, pf = 0.9 lag)

Mode Identification	$X_{C(4)}$ = 0.185 pu	$X_{C(3)}$ = 0.285 pu	$X_{C(2)}$ = 0.380 pu	$X_{C(1)}$ = 0.470 pu
Mode 5	$-0.0 \pm j298.16$	$-0.0 \pm j298.16$	$-0.0 \pm j298.18$	$-0.0 \pm j298.18$
Mode 4	$+1.705 \pm j202.34$	$-0.077 \pm j202.62$	$-0.057 \pm j202.74$	$-0.047 \pm j202.78$
Mode 3	$+0.054 \pm j160.71$	$+1.633 \pm j160.18$	$-0.055 \pm j160.30$	$-0.040 \pm j160.41$
Mode 2	$+0.007 \pm j127.06$	$+0.024 \pm j127.12$	$+0.753 \pm j126.74$	$-0.016 \pm j126.90$
Mode 1	$+0.044 \pm j99.20$	$+0.112 \pm j99.47$	$+0.369 \pm j100.26$	$+5.406 \pm j98.62$
Mode 0	$-0.366 \pm j8.27$	$-0.425 \pm j9.32$	$-0.533 \pm j10.54$	$-0.707 \pm j12.01$
Electrical mode	$-5.497 \pm j202.96$	$-4.981 \pm j160.81$	$-3.781 \pm j126.6$	$-7.421 \pm j99.76$
Supersynchronous mode	$-4.614 \pm j551.22$	$-4.651 \pm j593.26$	$-4.655 \pm j626.72$	$-4.675 \pm j654.73$
Generator-rotor circuits and controllers	−0.157	−0.234	−0.307	−0.381
	−3.348	−3.589	−3.869	−4.220
	−21.32	−21.381	−21.475	−21.573
	−32.99	−33.216	−33.512	−33.905

decrease in the rotor velocity. The difference in the two schemes, however, lies in the behavior of the SVC controllers. In case of power-transfer augmentation, the damping is introduced in the lower inertial frequency range 0.1–2 Hz, whereas in the case of SSR suppression, the damping is initiated in the higher torsional-frequency range 10–40 Hz.

The effectiveness of SVC control in damping the SSR is critically dependent on the phase shift between the SVC control input (generator-speed deviation) and the SVC-implemented susceptance change at the interconnected bus. These phase shifts are caused mainly by the inherent thyristor-firing delay (see Chapter 4) and the filter in the speed-deviation measurement system.

From ref. [32], it is shown that, ideally, a zero phase shift in the control system will result in positive damping for all the torsional modes. In practical circuits, the phase shifts are minimized to less than 10° by introducing the appropriate phase lead-lag circuits in the control system. This narrow range of acceptable phase shift can be substantially enhanced by using an SVC with a primary-voltage control loop and an auxiliary controller with generator-speed deviation as the control signal [33].

An SVC having only the speed-auxiliary controller damps all the torsional modes, thereby ensuring the mitigation of torsional interaction. However, it has a tendency to reduce the damping associated with the electrical mode, which may result in an SSR from the induction-generator action. To preclude such an eventuality, poleface-damper windings are provided simultaneously in the synchronous generators. A successful implementation of the speed-signal filtering has been reported in the field [34].

The deterioration of the electrical mode damping with a speed-auxiliary signal is compensated by the addition of the voltage regulator to the generator speed-auxiliary control. The bus-voltage signal contributes the electrical state to the SVC control. This SVC configuration can therefore counteract successfully the torsional oscillations at all critical levels of series compensation, as well as

TABLE 6.5 System Eigenvalues With SVC Control (P_g = 1 pu, pf = 0.9 lag)

Mode Identification	$X_{C(4)} = 0.185$ pu	$X_{C(3)} = 0.285$ pu	$X_{C(2)} = 0.380$ pu	$X_{C(1)} = 0.470$ pu
Mode 5	$0.0 \pm j298.16$	$-0.0 \pm j298.16$	$-0.0 \pm j298.16$	$-0.0 \pm j298.16$
Mode 4	$-2.020 \pm j213.00$	$-0.307 \pm j207.04$	$-0.143 \pm j206.53$	$-0.077 \pm j206.35$
Mode 3	$-0.049 \pm j160.99$	$-1.845 \pm j157.27$	$-0.403 \pm j162.29$	$-0.233 \pm j161.99$
Mode 2	$-0.021 \pm j127.11$	$-0.004 \pm j127.03$	$-0.405 \pm j126.43$	$-0.157 \pm j127.38$
Mode 1	$-0.258 \pm j99.94$	$-0.127 \pm j99.75$	$-0.008 \pm j99.16$	$-0.491 \pm j94.97$
Mode 0	$-0.550 \pm j8.40$	$-0.623 \pm j9.38$	$-0.710 \pm j10.40$	$-0.837 \pm j11.69$
Electrical mode	$-3.187 \pm j200.55$	$-3.090 \pm j165.74$	$-4.452 \pm j131.37$	$-4.105 \pm j106.97$
Supersynchronous mode	$-3.716 \pm j551.86$	$-3.694 \pm j594.31$	$-3.698 \pm j625.07$	$-3.714 \pm j652.93$
Electrical system	$-5.653 \pm j356.53$	$-5.611 \pm j356.62$	$-5.548 \pm j356.65$	$-5.430 \pm j356.78$
Generator-rotor	-0.640	-0.633	-0.622	-0.602
circuits and	$-3.825 \pm j2.51$	$-3.841 \pm j2.31$	$-3.862 \pm j2.09$	$-3.890 \pm j1.74$
controllers	-20.540	-20.577	-20.620	-20.679
	$-29.958 \pm j0.981$	-29.286	-28.530	-27.883
	-154.517	-30.715	-31.571	-32.350
		-154.434	-154.644	-155.153
	-655.01	-655.13	-655.38	-655.593

the electrical self-excitation from the induction-generator effect that occurs at high levels of series compensation.

The damped eigenvalues of the IEEE's First SSR Benchmark Model, with complete SVC controls for the four critical levels of series compensation, are given in Table 6.5. The time-domain simulation of SVC performance in damping mode 3, as an example, is illustrated in Fig. 6.23.

6.5.2 Configuration and Design of the SVC Controller

The speed-auxiliary controller is a proportional controller when it is used both alone [29]–[32], [34] and in conjunction with a voltage regulator [33]. A reactive-power controller proposed in ref. [35] uses a torsional monitor to measure medium-pressure–turbine speed, two low-pressure–turbine speeds, and generator-shaft speed in the IEEE's First Benchmark System to obtain the modal speeds for damping the torsional oscillations. A reliable speed measurement is done through a multitoothed wheel that generates a carrier signal in which the frequency is many orders of magnitude different than that of the modulating-signal frequency [33]. In case speeds of different shaft reactions need to be obtained, several optical disc pickups are installed at these shaft locations.

An SVC with a single-input–signal-output (SISO) proportional–integral derivative (PID) auxiliary-speed controller, in conjunction with a voltage regulator, is proposed in ref. [36] for damping torsional oscillations. The auxiliary PID controller is designed by shifting two SSR eigenvalues to prespecified locations. Because pole placement is possible for only two SSR eigenvalues with an SISO controller, a single-input–multiple-output (SIMO) controller is suggested for allowing the assignment of more SSR modes [37]. This SVC auxiliary con-

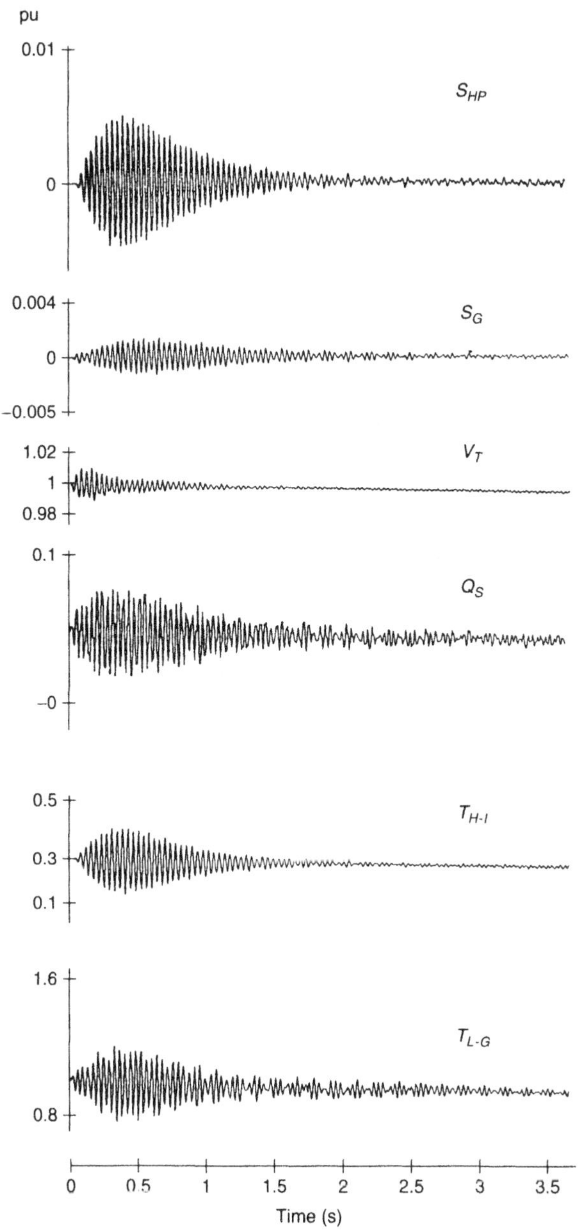

Figure 6.23 Damping of torsional mode 3 with an SVC, where S_{HP} = the speed deviation of the HP turbine rotor mass; S_G = the speed deviation of the generator rotor mass; V_T = the generator terminal voltage; Q_s = the SVC reactive power; T_{H-I} = the mechanical shaft torque in HP-IP section; and $T_{L\text{-}G}$ = the mechanical shaft torque in LP-GEN section.

troller is implemented by a multiple proportional–integral (MPI) controller with one medium-pressure–turbine speed, two low-pressure–turbine speeds, and one generator-shaft speed as the feedback signals. The supplementary control signal resulting from this control scheme contains the components of four torsional-mode frequencies. With appropriately designed controllers, satisfactory damping is provided for all the torsional modes under different operating conditions and a wide range of line-series compensation levels. In fact, any SVC controller designed for SSR mitigation must damp all the torsional modes for all critical levels of series compensation, as well as control the induction generator effect–based SSR that occurs at high levels of series compensation.

The SVCs described in the preceding text for damping subsynchronous oscillations are connected at the generator terminals for the sole purpose of SSR mitigation. An alternative concept is described in refs. [38] and [39], in which a midline-located SVC in a series-compensated SMIB system is used for both power-transfer improvement and SSR stabilization. The SVC employs a voltage regulator and a combination of two auxiliary controllers. One auxiliary signal is the magnitude of the line current, whereas the other is the CIF of the generator, and the root-loci technique is used for designing both controllers. This control scheme requires adaptive controllers, which need careful implementation.

The SVC used for SSR mitigation offers the following advantages: As a shunt device, it need not carry the full generator current, and it provides damping of all torsional oscillations, initiated by any means, within the bandwidth of the controller. In addition, it is insensitive to system frequency variations.

Because this device is critical in damping torsional oscillations, care must be exercised to minimize faults in the zone of its installation. Due consideration is also necessary for circumventing the harmonics generated by the SVC. A multiwinding coupling transformer is usually employed [31] to reduce the injected 5th and 7th harmonics.

6.5.3 Rating of an SVC

An SVC sizing relationship is proposed in ref. [29] according to the average shunt-reactor inductance, L_s, as given by

$$L_s = \frac{mL_g}{\Delta} \tag{6.56}$$

where Δ = the deviation in the shaft-oscillation angle

m = the SVC modulation index ($m = 1$ for full SVC modulation)

L_g = the generator inductance

A pessimistic estimate of the maximum rotor-angle deviation that the SVC is expected to control is taken to be that which can cause permanent deformation of the generator shaft. The SVC is assumed to be modulated completely, that is, the full rating of the SVC is used for SSR damping.

If a generator-rotor velocity deviation, $\omega_n\Delta$, of 0.01 pu is permitted for a lower torsional mode having frequency $\omega_n = 0.412$ pu, the average SVC inductance is obtained as $(0.412)(0.19)/0.01 = 7.828$ pu for a generator inductance of 0.19 pu. This inductance necessitates an SVC of a steady-state MVA rating $(1.0)^2/7.828$ pu $= 0.128$ pu of the generator rating, corresponding to 1-pu terminal voltage. As the rating relates to average conditions, an SVC of twice the rating, that is, 0.256 pu, may be used. It should be noted that because the rating was computed for one torsional mode, the rating would be different for the other modes; hence an appropriate size may be selected that can damp all the torsional modes.

6.6 PREVENTION OF VOLTAGE INSTABILITY

Voltage instability, the subject of intense research since 1980 [3], is caused by the inadequacy of the power system to supply the reactive-power demand of certain loads, such as induction motors. A drop in the load voltage leads to an increased demand for reactive power that, if not met by the power system, leads to a further decline in the bus voltage. This decline eventually leads to a progressive yet rapid decline of voltage at that location, which may have a cascading effect on neighboring regions that causes a system voltage collapse.

6.6.1 Principles of SVC Control

The voltage at a load bus supplied by a transmission line is dependent on the magnitude of the load, the load-power factor, and the impedance of the transmission line. Consider an SVC connected to a load bus, as shown in Fig. 6.24(a). The load has a varying power factor and is fed by a lossless radial transmission line. The voltage profile at the load bus, which is situated at the receiver end of the transmission line, is depicted in Fig. 6.24(b).

For a given load-power factor, as the transmitted power is gradually increased, a maximum power limit is reached beyond which the voltage collapse takes place. In this typical system, if the combined power factor of the load and SVC is appropriately controlled through the reactive-power support from the SVC, a constant voltage of the receiving-end bus can be maintained with increasing magnitude of transmitted power, and voltage instability can be avoided.

6.6.1.1 A Case Study An SVC can be used successfully to prevent voltage instability. The case study presented here demonstrates the application of SVC to mitigate voltage instability in a radial system loaded by a large composite load of induction motors and static loads, all under steady-state and transient conditions [40].

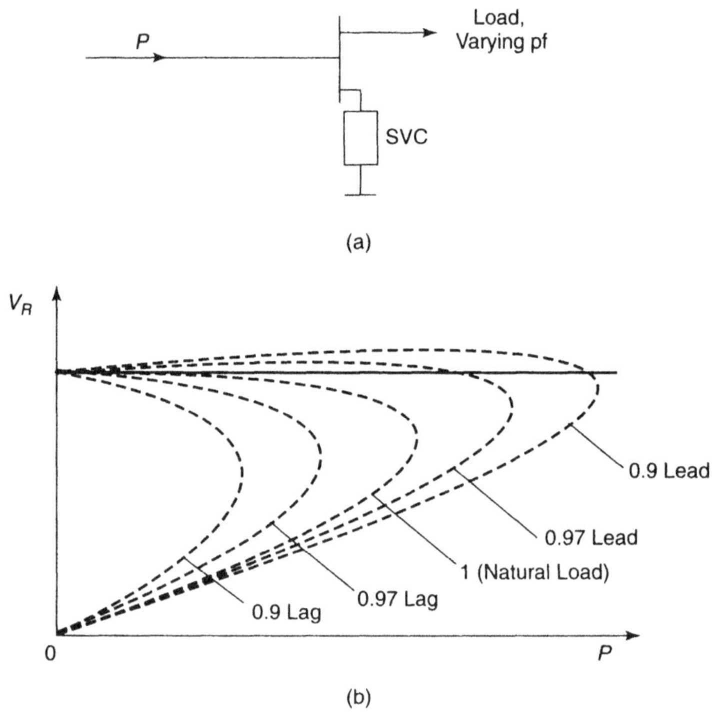

Figure 6.24 (a) An SVC connected at the load bus by a radial transmission line supplying a load and (b) the voltage profile at the receiving end of a loaded line with a varying power factor load.

The 400-kV radial case-study system [40] shown in Fig. 6.25 involves power supply over a double-circuit transmission line to a load center that comprises a 50% large induction motors (IM) and 50% static loads. An FC–TCR SVC is connected to the tertiary of a 3-winding load transformer, and the SVC voltage controller is of the PI type.

The instability is caused by tripping one of the transmission lines and is detected from eigenvalue analysis. The postdisturbance response for 1 s period is shown in Fig. 6.26. In the absence of the SVC, the load voltage falls to a level of 0.8 pu in 80 ms after the initial transients and falls further to a magnitude of 0.57 pu in less than 1 s. The onset of induction-motor instability occurs at a voltage of 0.8 pu. With falling terminal voltage, the induction-motor load-reactive power starts increasing rapidly, leading to eventual voltage collapse. If the induction motor loads are completely replaced by static loads of same value, voltage instability does not occur.

When an appropriately designed SVC is connected at the load bus, the post-disturbance system performance alters to that depicted in Fig. 6.27. After the damping of fast initial transients, the load voltage stabilizes in about 50 ms.

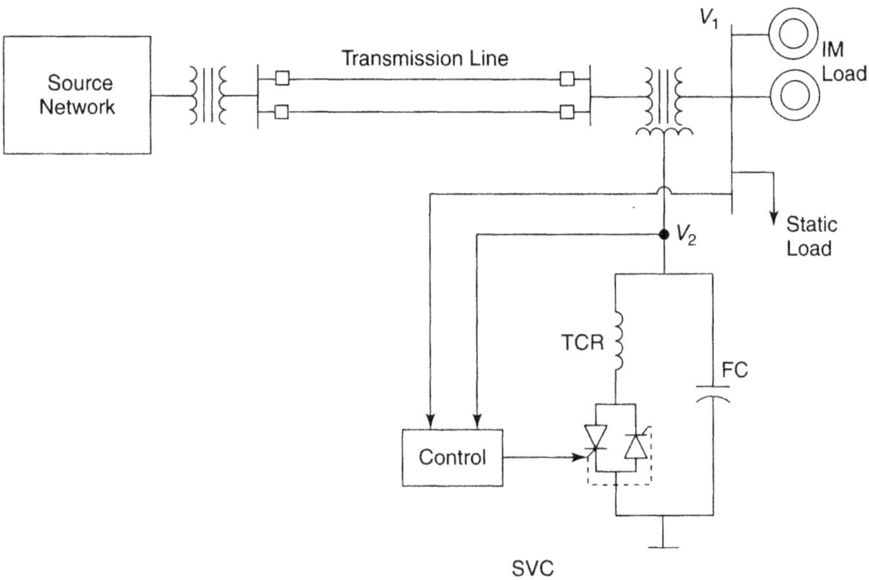

Figure 6.25 A case-study system.

The final value of the stabilized load voltage is a function of the capacitive-reactive-power rating of the SVC, which can be improved further by additional steady-state voltage-regulating devices. It is evident that this voltage stabilization is achieved only from the rapid response offered by the SVC. A breaker-switched shunt capacitor of equivalent rating as the SVC is unable to prevent voltage collapse.

Voltage instability is also illustrated through voltage-susceptance diagrams, where it is shown that if the slope of the voltage-susceptance characteristic is negative, voltage collapse may result. A detailed analysis presented in ref. [41] demonstrates that an SVC can successfully extend stable operation into that zone.

6.6.2 Configuration and Design of the SVC Controller

As the primary purpose of an SVC is voltage control, a PI-type voltage regulator is generally sufficient. The controller parameters are optimized using eigenvalue analysis to give fast, stable responses over the full range of expected network impedances and also without any adverse interactions with the power-system oscillation modes.

In some situations, voltage dips may also be accompanied by system oscillations, as in the case of critical synchronous motor loads supplied by a distribution feeder [42]. An auxiliary damping control may then need to be installed along with the voltage regulator. A small-signal Hopf bifurcation analysis [43] has shown that an SVC with auxiliary control increases system loadability. The

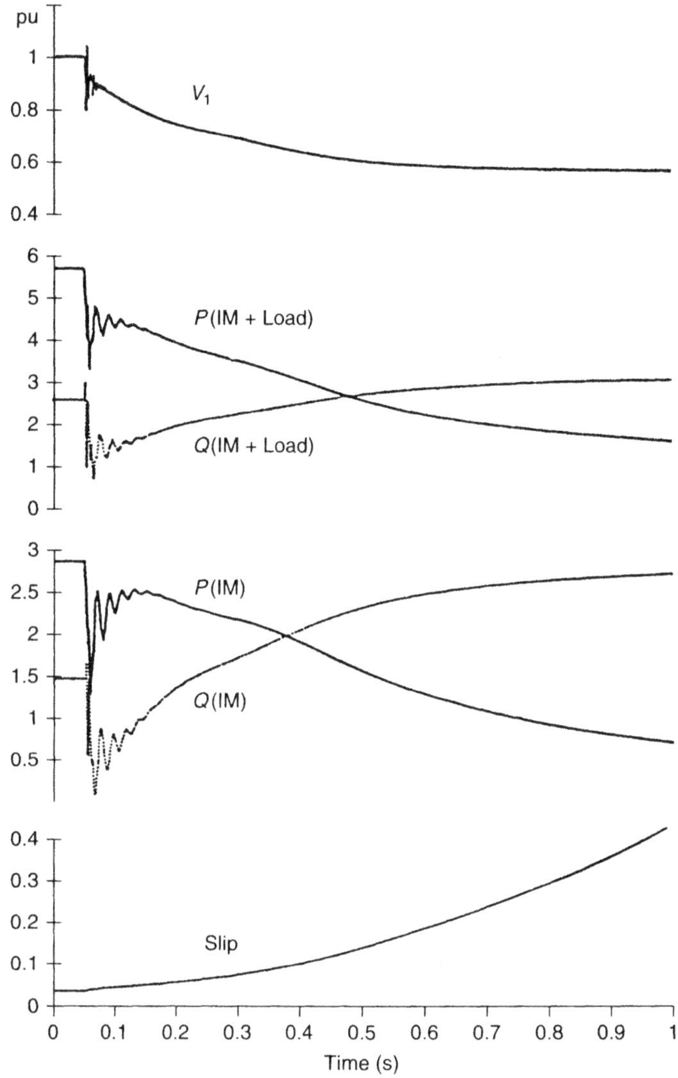

Figure 6.26 The system transient response for opening one circuit in Fig. 6.25 (50% IM load without a SVC).

principles of voltage-controller design were presented in Chapter 5; the concepts of auxiliary power-swing damping control were discussed in Section 6.4.

6.6.3 Rating of an SVC

Steady-state considerations determine the capacitive rating of an SVC. During a critical outage, the capacitive-reactive power needed to regulate the load voltage

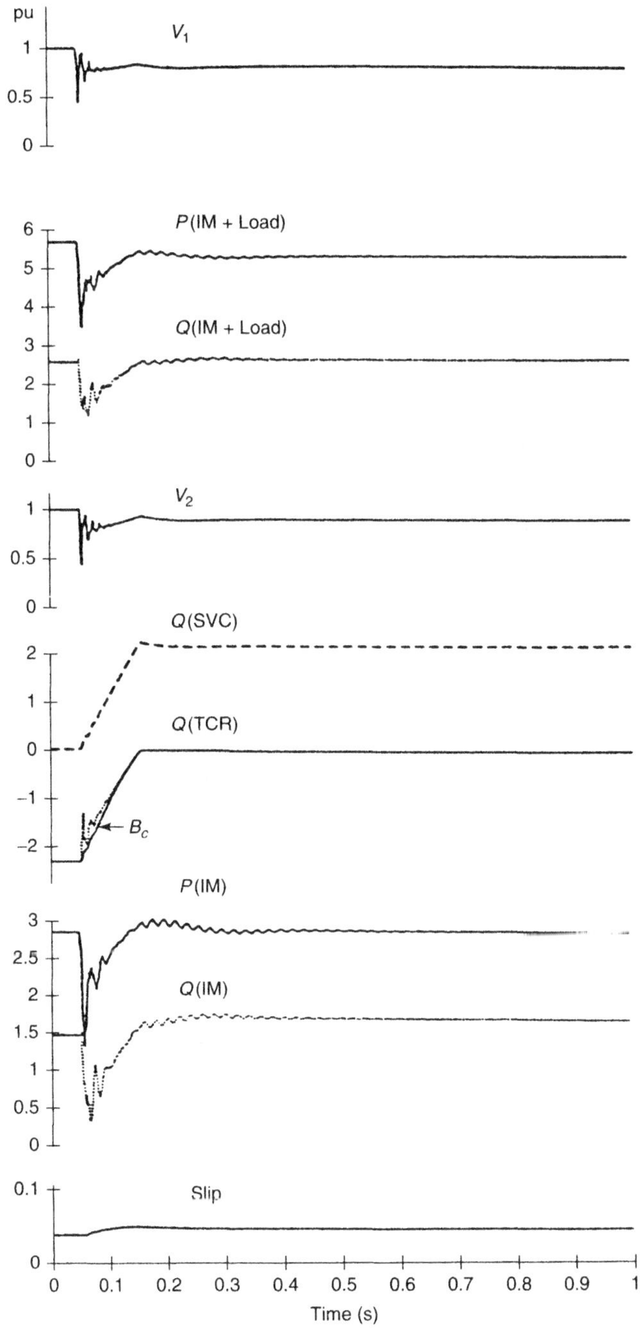

Figure 6.27 The system transient response for opening one circuit in Fig. 6.25 [50% IM load with a SVC (TCR–FC)].

to a marginally stable level is selected as the capacitive range of SVC. Alternatively, once the critical bus that needs reactive-power support is identified, the SVC rating is chosen based on the capacitive-reactive power required to maintain the bus voltage at the minimum estimated SVC voltage-control range for the specified maximum loading condition or the voltage-collapse point [44]. The collapse is indicated by the system Jacobian's increasing singularity at that loading point and is obtained through load-flow studies. The inductance rating is chosen to be that which can restrict the dynamic overvoltages at the SVC bus to 10%. This is determined from transient studies for critical-load rejections.

It is shown in ref. [44] that the system loading cannot be increased beyond a maixmum value, irrespective of the size of the SVC connected at the critical bus. One means of obtaining the optimal SVC rating is maximization of a performance index, f_p, where

$$f_p = \frac{\lambda_0(\text{MW})}{Q_{\text{SVC}}(\text{MVAR})} \tag{6.57}$$

where λ_0 = the maximum system loading

Q_{SVC} = the SVC MVAR rating

The point of maximum f_p corresponds to the maximum load increase at the minimum MVAR compensation level. This reactive-power level is chosen to be the optimal SVC rating.

6.7 IMPROVEMENT OF HVDC LINK PERFORMANCE

The HVDC converters consume significant amounts of reactive power—as high as 60% of the rated active power [45]. Although a part of this reactive component is supplied by the ac harmonic filters, the remainder must be pooled in by the ac system. If the ac system is strong, that is, its short-circuit level is typically greater than four times the dc link power, the ac system has a reasonably adequate capability to supply the balance of reactive power to the HVDC converter.

However, an HVDC transmission system connected to weak ac systems presents unique problems of stability, temporary overvoltages, and recovery after faults. Dynamic reactive-power support at the HVDC terminals can alleviate some of these problems. Traditionally, synchronous compensators have been employed for this purpose, as they provide continuously controllable reactive power as well as improve the system short-circuit level. Equipping them with advanced excitation systems enhances their speed [46].

An SVC offers the advantage of rapid response, lower maintenance, higher reliability, easier installation, and a lack of contribution to the fault current as compared to synchronous compensators. For these reasons, the SVC has emerged as an attractive technoeconomic alternative for providing voltage control at HVDC terminals.

6.7.1 Principle and Applications of SVC Control

The major benefits of SVC application in HVDC transmission systems are

1. voltage regulation,
2. support during recovery from large disturbances, and
3. suppression of temporary overvoltages.

6.7.1.1 Voltage Regulation Reactive-power support is needed to maintain nominal voltage at the HVDC terminals, especially during peak power transfers. An optimal voltage is also required at the HVDC bus to minimize losses in the ac line feeding the HVDC link. Mechanically switched capacitors, of which more than one is needed for supplying reactive power, do not constitute an effective solution because of their discrete nature of operation and their finite switching capability. An SVC is an efficient means of providing voltage regulation as well as minimizing line loss because of the SVC's smooth, rapid control with no limitation on the number of switchings [46].

6.7.1.2 Suppression of Temporary Overvoltages An SVC can efficiently curb the temporary overvoltages resulting from load rejections. It is reported [46] that in the event of simultaneous blocking of all the HVDC poles in a 1000-MW link, a ±200-MVAR SVC can limit the overvoltage to about 1.1 pu; in the absence of the SVC, the overvoltage will reach levels of 1.3–1.4 pu. The installation of the SVC allows a greater flexibility in HVDC link operation during weak network situations; otherwise, several operating restrictions need to be imposed for limiting overvoltages.

6.7.1.3 Support During Recovery from Large Disturbances Contingencies in the power system, such as faults, have a debilitating effect on the power transfer in the HVDC link, leading to either temporary or total load rejections. The major contribution of the SVCs is their stabilizing effect on the voltage that aids greatly in the fast recovery of the HVDC link after fault clearing [47].

A performance comparison of fixed and dynamic compensation at the HVDC inverter terminal of a very weak ac system is presented in the following case studies [48]. A fixed-compensation choice relates to fixed capacitors, whereas the dynamic-compensation choice comprises three schemes: a synchronous compensator (SC), a static var compensator (SVC), and a combined SC–SVC. The capacitive ranges of all schemes are identical. One case study corresponds to a permanent dc block; the other, to a 3-phase-to-ground inverter fault.

In the first case study, the inverter ac root mean square (rms) voltage during

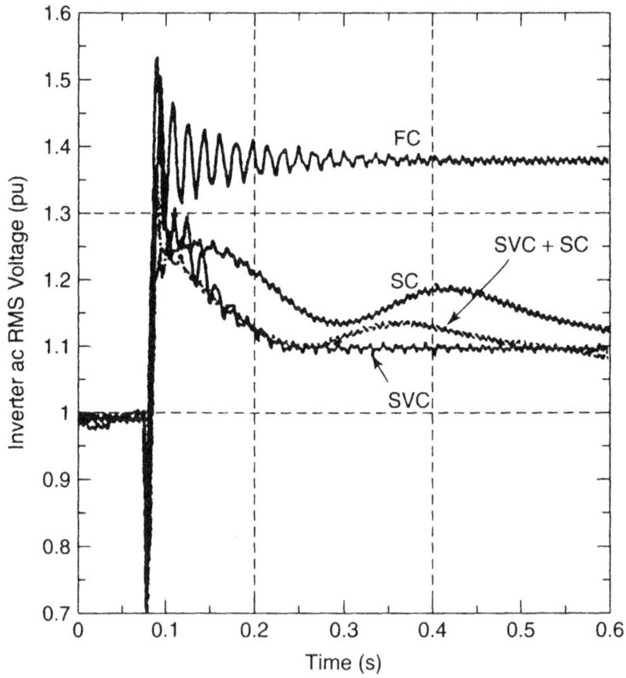

Figure 6.28 The inverter ac bus voltage during a permanent inverter block.

a permanent inverter block is as depicted in Fig. 6.28. An initial overvoltage peak of 1.7 pu results. Fixed capacitors do not provide any voltage control action, but the voltage reduces to 1.37 pu from the transformer saturation. The SVC provides the fastest overall control of overvoltage, although it is unable to influence the first peak. The SC successfully reduces the first overvoltage peak to 1.3 pu, but it provides a sluggish response later (similar observations have been reported in refs. [46] and [49]). This first overvoltage peak reduction suffers because the SVCs in closed-loop control cannot respond faster than one-and-a-half cycles, whereas an SC can provide a near-instantaneous response by increasing the short-circuit level of the system. The SC also introduces low-frequency (2-Hz) oscillations from its rotor dynamics. The SVC–SC combination not only reduces the overvoltage peak to an intermediate value, but it also provides a response time similar to that of the SVC case. The rotor-mode oscillations are also reduced substantially.

In the second case study, the inverter dc power during a 5-cycle, 3-phase-to-ground fault at the inverter bus for the four schemes is as illustrated in Fig. 6.29. Although the SVC is the fastest acting device by itself, it provides the slowest response. This behavior brings forth a special characteristic of the SVC—that in very weak ac systems, the TSC component of the SVC tends to decrease the effective short-circuit ratio (ESCR), which can cause system instability by

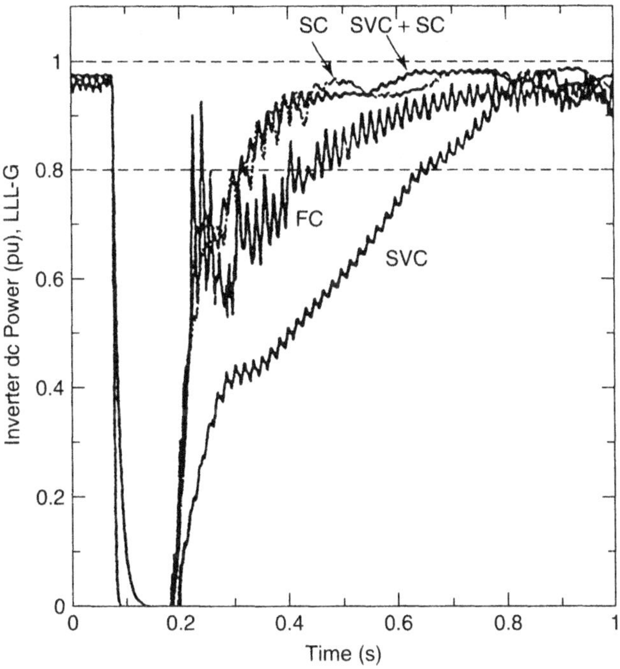

Figure 6.29 The inverter dc power during a 5-cycle, 3-phase-to-ground fault at the inverter bus.

repeated commutation failures as the TCR firing angle is changed. To obviate this instability, the firing angle of the inverters is reduced and then gradually ramped back. Doing so significantly slows the recovery process with the SVC. The power recovery with the fixed capacitor is quicker, but the fastest power recovery occurs in the case of SC–SVC combination.

Where the ac system is not as weak as considered in the second case study, the SVC installation allows a greater flexibility in the HVDC link operation [46].

6.7.2 Configuration and Design of the SVC Controller

The SVC control system normally includes a voltage regulator. The gain is determined from root-loci investigations, and the choice of the gain–time constant is influenced by ac system strength, as discussed in Chapter 5. It may be recalled that a large regulator gain tends to make the control system unstable as the system short-circuit impedance increases [50].

The ESCR of the ac system changes with the reactive-power output of the SVC [48]. To provide voltage support during low voltages, the SVC increases its capacitive-reactive-power output by lowering its capacitive reactance. This decrease leads to a reduction in the ESCR, which may be crucial in case of weak ac-receiving systems. In one study [48], the ESCR was shown to vary from 1.37 to 0.94 as the SVC traversed from a fully inductive to fully capacitive state.

6.7.2.1 Interactions Between the SVC and the HVDC The SVC voltage regulator has been shown to favorably impact the HVDC current controller at the lower frequencies (<10 Hz), where the overall stability improves in the inverter mode operation [46], [50]. In some cases, saturation of the SVC transformer during energization leads to an accompanied saturation of the HVDC converter transformer, which is caused by the bus-voltage distortion.

6.7.3 Rating of the SVC

The inductive rating of the SVC is chosen to be that which can reduce the likely highest temporary overvoltage following a disturbance (such as permanent dc block) to within acceptable limits. The fixed capacitors and filters are assumed to be connected at the inverter bus in this evaluation.

The capacitive rating of the SVC must be such that it can completely fulfill the reactive-power demand of the inverter terminal at the rated power and voltage, thereby ensuring a unity power factor at the ac bus. Additional capacitance-reactive power may also be provided for improved handling of undervoltage situations.

6.8 SUMMARY

Static var compensators (SVCs) have been in the power industry for nearly the past three decades. It is therefore a mature technology that now has wide applications [51]–[54], [55]–[59]. This chapter presented the various applications of SVCs, such as enhancement of power transfer, improvement of transient stability, increase in system damping, suppression of SSR, alleviation of voltage instability, and dynamic compensation of HVDC converters. The principles of control and methodology of SVC controller design were presented, with examples for each application.

REFERENCES

[1] L. Gyugyi, "Fundamentals of Thyristor-Controlled Static Var Compensators in Electric Power System Applications," IEEE Special Publication 87TH0187-5-PWR, *Application of Static Var Systems for System Dynamic Performance*, 1987, pp. 8–27.

[2] W. D. Stevenson, *Elements of Power System Analysis*, McGraw-Hill, New York 1982.

[3] P. Kundur, *Power System Stability and Control*, McGraw-Hill, New York, 1993.

[4] A. E. Hammad, "Applications of Static Var Compensators in Utility Power Sys-

tems," IEEE Special Publication 87TH0187-5-PWR, *Application of Static Var Systems for System Dynamic Performance*, 1987, pp. 2–35.

[5] A. Olwegard, K. Walve, G. Waglund, H. Frank, and S. Torseng, "Improvement of Transmission Capacity by Thyristor Controlled Reactive Power," *IEEE Transactions on Power Apparatus and Systems*, Vol. 100, August 1981, pp. 3930–3939.

[6] E. V. Larsen and J. H. Chow, "SVC Control Design Concepts for System Dynamic Performance," IEEE Special Publication 87TH0187-5-PWR, *Application of Static Var Systems for System Dynamic Performance*, 1987, pp. 36–53.

[7] The Electric Power Research Institute (EPRI) Report TR-100696, "Improved Static VAR Compensator Control," Final Report of Project 2707-01, Prepared by General Electric (GE) Company, Schenectady, NY, June 1992.

[8] E. V. Larsen, "Control Aspects of FACTS Applications," *Proceedings of EPRI Conference on FACTS*, Cincinnati, OH, 1990.

[9] J. F. Hauer, "Operational Aspects of Large Scale FACTS Controllers," *Proceedings of EPRI Conference on FACTS*, Cincinnati, OH, 1990.

[10] E. V. Larsen and D. A. Swann, "Applying Power System Stabilizers, Part II: Performance, Objectives, and Tuning Concepts," *IEEE Transactions on Power Apparatus and Systems*, Vol. PAS–100, No. 6, June 1981, pp. 3017–3046.

[11] K. R. Padiyar and R. K. Varma, "Damping Torque Analysis of Static Var System Controllers," *IEEE Transactions on Power Systems*, Vol. 6, No. 2, May 1991, pp. 458–465.

[12] K. R. Padiyar and R. K. Varma, "Static Var System Auxiliary Controllers for Improvement of Dynamic Stability," *International Journal of Electrical Power and Energy Systems*, Vol. 12, No. 4, October 1990, pp. 287–297.

[13] S. Lee and C. C. Liu, "An Output Feedback Static Var Controller for the Damping of Generator Oscillations," *Electric Power Systems Research*, Vol. 29, 1994, pp. 9–16.

[14] J. R. Smith, D. A. Pierre, D. A. Rudberg, I. Sadighi, and A. P. Johnson, "An Enhanced LQ Adaptive Var Unit Controller for Power System Damping," *IEEE Transactions on Power Systems*, Vol. 4, No. 2, May 1989, pp. 443–451.

[15] H. Kwakernaak, "Robust Control and H_∞ Optimization," Tutorial paper, *Automatica*, Vol. 29, No. 2, 1993, pp. 255–273.

[16] M. Parniani and M. R. Iravani, "Optimal Robust Control Design of Static Var Compensator," *IEEE Transactions on Power Delivery*.

[17] Q. Zhao and J. Jiang, "Robust SVC Controller Design for Improving Power System Damping," *IEEE Transactions on Power Systems*, Vol. 10, No. 4, November 1995.

[18] P. K. Dash, S. Mishra, and A. C. Liew, "Fuzzy Logic Based Var Stabilizer for Power System Control," *IEE Proceedings on Generation, Transmission, and Distribution*, Vol. 142, No. 6, November 1995.

[19] H. J. Haubrich, T. Seitz, and D. Povh, "Static Var Compensators with Fuzzy Control for Improvement of Power System Performance," *CIGRE 1995 Symposium on Power Electronics in Electric Power Systems*, Tokyo, May 1995.

[20] K. A. Ellithy and S. M. Al-Alwai, "Tuning a Static Var Compensator Controller Over a Wide Range of Load Models Using an Artificial Neural Network," *Electric Power Systems Research*, Vol. 38, No. 2, August 1996, pp. 97–104.

[21] P. Ju, E. Handschin, and F. Reyer, "Genetic Algorithm–Aided Controller Design With Application to SVC," *IEE Proceedings on Generation, Transmission, and Distribution*, Vol. 143, No. 3, 1996, pp. 258–262.

[22] IEEE Committee Report, "A Bibliography for the Study of Subsynchronous Resonance Between Rotating Machines and Power Systems," *IEEE Transactions on Power Apparatus and Systems*, Vol. 95, No. 1, January–February 1976, pp. 216–218.

[23] IEEE Committee Report, "First Supplement to a Bibliography for the Study of Subsynchronous Resonance Between Rotating Machines and Power Systems," *IEEE Transactions on Power Systems*, Vol. 98, No. 6, November–December 1979, pp. 1872–1875.

[24] IEEE Committee Report, "Second Supplement to a Bibliography for the Study of Subsynchronous Resonance Between Rotating Machines and Power Systems," *IEEE Transactions on Power Systems*, Vol. 104, No. 2, February 1985, pp. 321–327.

[25] IEEE Committee Report, "Third Supplement to a Bibliography for the Study of Subsynchronous Resonance Between Rotating Machines and Power Systems," *IEEE Transactions on Power Systems*, Vol. 6, No. 2, May 1991, pp. 830–834.

[26] IEEE Committee Report, "Reader's Guide to Subsynchronous Resonance," *IEEE Transactions on Power Systems*, Vol. 7, No. 1, February 1992.

[27] IEEE Torsional Issues Working Group, "Fourth Supplement to a Bibliography for the Study of Subsynchronous Resonance Between Rotating Machines and Power Systems," *IEEE Transactions on Power Systems*, Vol. 12, No. 3, August 1997, pp. 1276–1282.

[28] IEEE SSR Task Force, "First Benchmark Model for Computer Simulation of Subsynchronous Resonance," *IEEE Transactions on Power Apparatus and Systems*, Vol. 96, September/October 1977, pp. 1565–1572.

[29] T. H. Putman and D. G. Ramey, "Theory of the Modulated Reactance Solution for Subsynchronous Resonance," *IEEE Transactions on Power Apparatus and Systems*, Vol. 101, No. 6, June 1982, pp. 1527–1535.

[30] IEEE SSR Working Group, "Countermeasures to Subsynchronous Resonance Problems," Presented at IEEE/PES Summer Meeting, Vancouver, British Columbia, 1979.

[31] D. G. Ramey, D. S. Kimmel, J. W. Dorney, and F. H. Kroening, "Dynamic Stabilizer Verification Tests at the San Juan Station," *Proceedings of the Symposium on Countermeasures for Subsynchronous Resonance*, Presented at IEEE/PES 1981 Summer Meeting, 1981.

[32] N. C. Abi Samra, R. F. Smith, T. E. McDermott, and M. B. Chidester, "Analysis of Thyristor-Controlled Shunt SSR Countermeasures," *IEEE Transactions on Power Apparatus and Systems*, Vol. 104, No. 3, March 1985, pp. 584–597.

[33] A. E. Hammad and M. El-Sadek, "Application of a Thyristor-Controlled Var Compensator for Damping Subsynchronous Oscillations in Power Systems," *IEEE Transactions on Power Apparatus and Systems*, Vol. 103, No. 1, January 1984, pp. 198–211.

[34] D. G. Ramey, D. S. Kimmel, J. W. Dorney, and F. H. Kroening, "Dynamic Stabilizer Verification Tests at the San Juan Station," *IEEE Transactions on Power Apparatus and Systems*, Vol. 100, December 1981, pp. 5011–5019.

[35] O. Wasynczuk, "Damping Subsynchronous Resonance Using Reactive Power Control," *IEEE Transactions on Power Apparatus and Systems*, Vol. 100, No. 3, March 1981, pp. 1096–1104.

[36] Y. Y. Hsu and C. J. Wu, "Design of PID Static Var Controller for the Damping of Subsynchronous Oscillations," *IEEE Transactions on Energy Conversion*, Vol. 3, No. 2, June 1988, pp. 210–216.

[37] S. Lee and C. C. Liu, "Damping Subsynchronous Resonance Using a SIMO Shunt Reactor Controller," *IEEE Transactions on Power Systems*, Vol. 9, No. 3, August 1994, pp. 1253–1262.

[38] K. R. Padiyar and R. K. Varma, "Static Var System Auxiliary Controller for Damping Torsional Oscillations," *International Journal of Electrical Power and Energy Systems*, Vol. 12, No. 4, October 1990, pp. 271–286.

[39] K. R. Padiyar, R. K. Varma, S. Gupta, and M. Kumar, "A Novel Static Var System Control for Damping Subsynchronous Oscillations," *Proceedings of VII National Power Systems Conference*, New Delhi, India, December 1994, pp. 331–335.

[40] A. E. Hammad and M. Z. El-Sadek, "Prevention of Transient Voltage Instabilities Due To Induction Motor Loads by Static Var Compensators," *IEEE Transactions on Power Systems*, Vol. 4, No. 3, August 1989, pp. 1182–1190.

[41] I. A. Hiskens and C. B. McLean, "SVC Behaviour Under Voltage Collapse Conditions," *IEEE Transactions on Power Systems*, Vol. 7, No. 3, August 1992, pp. 1078–1087.

[42] O. T. Tan and R. Thottapillil, "Static Var Compensators for Critical Synchronous Motor Loads During Voltage Dips," *IEEE Transactions on Power Systems*, Vol. 9, No. 3, August 1994, pp. 1517–1523.

[43] M. J. Laufenberg, M. A. Pai, and K. R. Padiyar, "Hopf Bifurcation Control in Power Systems with Static Var Compensators," *International Journal of Electrical Power and Energy Systems*, Vol. 19, No. 5, 1997, pp. 339–347.

[44] C. A. Canizares and Z. T. Faur, "Analysis of SVC and TCSC Controllers in Voltage Collapse," *IEEE Transactions on Power Systems*, Vol. 14, No. 1, February 1999, pp. 158–165.

[45] J. D. Ainsworth, A. Gavrilovic, and H. L. Thanawala, "Static and Synchronous Compensators for HVDC Transmission Converters Connected to Weak AC Systems," CIGRE Paper 31-01, 1980.

[46] B. Thorvaldsson, B. Arnlov, E. Saethre, and T. Ohnstad, "Joint Operation HVDC/SVC," *Proceedings of 6th International Conference on AC and DC Transmission*, London, 1996, Conference Publication No. 423, pp. 281–284.

[47] I. A. Erinmez, Ed., "Static Var Compensators," CIGRE Working Group 38-01, Task Force No. 2 on SVC, Paris, 1986.

[48] O. B. Nayak, A. M. Gole, D. G. Chapman, and J. B. Davis, "Dynamic Performance of Static and Synchronous Compensators at an HVDC Inverter Bus in a Very Weak AC System," Paper 93 SM 447-3 PWRD, Presented at IEEE/PES 1993 Summer Meeting, Vancouver, British Columbia, July 1993.

[49] S. Nyati, S. R. Atmuri, V. Koschik, R. M. Mathur, and D. Flueckiger, "Effectiveness of Var Controllers Connected at the Converter Bus of an HVDC Link Feeding into a Weak AC System," *Proceedings of 4th International Conference*

on AC and DC Power Transmission, IEE Conference Publication 255, London, September 1985.

[50] S. A. Miske Jr., R. J. Piwko, J. B. Tice, and J. A. Whitacre, "Control Modelling and Simulation Studies for a Static Var Compensator Applied to Weak AC System Incorporating an HVDC Installation," CIGRE Paper 31-01, 1980.

[51] W. H. Litzenberger, Ed., "An Annotated Bibliography of High Voltage Direct Current Transmission 1989–91," Published by the Bonneville Power Administration (BPA), Portland, OR, and the Western Area Power Administration, 1992.

[52] W. H. Litzenberg, Ed., "An Annotated Bibliography of High Voltage Direct Current Transmission and Flexible AC Transmission System (FACTS) Devices 1991–93," Published by the Bonneville Power Administration (BPA), Portland, OR, and the Western Area Power Administration, 1994.

[53] W. H. Litzenberger and R. K. Varma, Eds., "An Annotated Bibliography of High Voltage Direct Current Transmission and FACTS Devices 1994–95," Published by the Bonneville Power Administration (BPA), Portland, OR, Western Area Power Administration and the U.S. Department of Energy, 1996.

[54] W. H. Litzenberger, R. K. Varma, and J. D. Flanagan, Eds., "An Annotated Bibliography of High Voltage Direct Current Transmission and FACTS Devices 1996–97," Published by The Electric Power Research Institute (EPRI) and the Bonneville Power Administration (BPA), Portland, OR, 1998.

[55] CIGRE Task Force 38.05.04, "Analysis and Optimization of SVC Use in Transmission System," CIGRE Technical Brochure No. 77, 1993.

[56] N. G. Hingorani and L. Gyugyi, *Understanding FACTS*, IEEE Press, New York, 1999.

[57] Y. H. Song and A. T. Johns, Eds., *Flexible AC Transmission Systems (FACTS)*, IEE Press, London, 1999.

[58] IEEE Power Engineering Society, *FACTS Applications*, Publication 96TP116-0, IEEE Press, New York, 1996.

[59] CIGRE Task Force 38.02.16, "Impact of Interactions Among Power System Controls," CIGRE Technical Brochure No. 166, Paris, August 2000.

CHAPTER 7

The Thyristor-Controlled Series Capacitor (TCSC)

7.1 SERIES COMPENSATION

7.1.1 Fixed-Series Compensation

Series capacitors offer certain major advantages over their shunt counterparts. With series capacitors, the reactive power increases as the square of line current, whereas with shunt capacitors, the reactive power is generated proportional to the square of bus voltage. For achieving the same system benefits as those of series capacitors, shunt capacitors that are three to six times more reactive-power–rated than series capacitors need to be employed [1]–[3]. Furthermore, shunt capacitors typically must be connected at the line midpoint, whereas no such requirement exists for series capacitors.

Let Q_{se} and Q_{sh} be the ratings of a series and shunt capacitor, respectively, to achieve the same level of power transfer through a line that has a maximum angular difference of δ_{max} across its two ends. Then

$$\frac{Q_{se}}{Q_{sh}} = \tan^2\left(\frac{\delta_{max}}{2}\right) \tag{7.1}$$

Specifically, for δ_{max} of 35°, Q_{se} will be approximately 10% of Q_{sh}. Even though series capacitors are almost twice as costly as shunt capacitors (per-unit var) because of their higher operating voltages, the overall cost of series compensation is lower than shunt compensation.

7.1.2 The Need for Variable-Series Compensation

Compensation of transmission lines by series capacitors is likely to result in the following [4]:

1. enhanced base-power flow and loadability of the series-compensated line;
2. additional losses in the compensated line from the enhanced power flow; and

3. increased responsiveness of power flow in the series-compensated line from the outage of other lines in the system.

Studies [4] have revealed that with increasing level of fixed-series compensation, even though the losses in remaining transmission lines decrease, the overall system losses are exacerbated from the enhanced losses in the series-compensated line. Also, the increased sensitivity or responsiveness of the compensated line to other network outages may cause a line loading that exceeds the enhanced loadability level of the line itself. These undesirable effects can be avoided by employing variable levels of series compensation instead of fixed compensation. Series compensation can be varied, depending on the enhancement of power transfer desired at that time, without affecting other system-performance criteria.

7.1.3 Advantages of the TCSC

Use of thyristor control in series capacitors potentially offers the following little-mentioned advantages:

1. Rapid, continuous control of the transmission-line series-compensation level.
2. Dynamic control of power flow in selected transmission lines within the network to enable optimal power-flow conditions and prevent the loop flow of power.
3. Damping of the power swings from local and inter-area oscillations.
4. Suppression of subsynchronous oscillations. At subsynchronous frequencies, the TCSC presents an inherently resistive–inductive reactance. The subsynchronous oscillations cannot be sustained in this situation and consequently get damped.
5. Decreasing dc-offset voltages. The dc-offset voltages, invariably resulting from the insertion of series capacitors, can be made to decay very quickly (within a few cycles) from the firing control of the TCSC thyristors.
6. Enhanced level of protection for series capacitors. A fast bypass of the series capacitors can be achieved through thyristor control when large overvoltages develop across capacitors following faults. Likewise, the capacitors can be quickly reinserted by thyristor action after fault clearing to aid in system stabilization.
7. Voltage support. The TCSC, in conjunction with series capacitors, can generate reactive power that increases with line loading, thereby aiding the regulation of local network voltages and, in addition, the alleviation of any voltage instability.
8. Reduction of the short-circuit current. During events of high short-circuit current, the TCSC can switch from the controllable-capacitance to

the controllable-inductance mode, thereby restricting the short-circuit currents.

7.2 THE TCSC CONTROLLER

The basic conceptual TCSC module comprises a series capacitor, C, in parallel with a thyristor-controlled reactor, L_S, as shown in Fig. 7.1(a). However, a practical TCSC module also includes protective equipment normally installed with series capacitors, as shown in Fig. 7.1(b).

A metal-oxide varistor (MOV), essentially a nonlinear resistor, is connected across the series capacitor to prevent the occurrence of high-capacitor over-

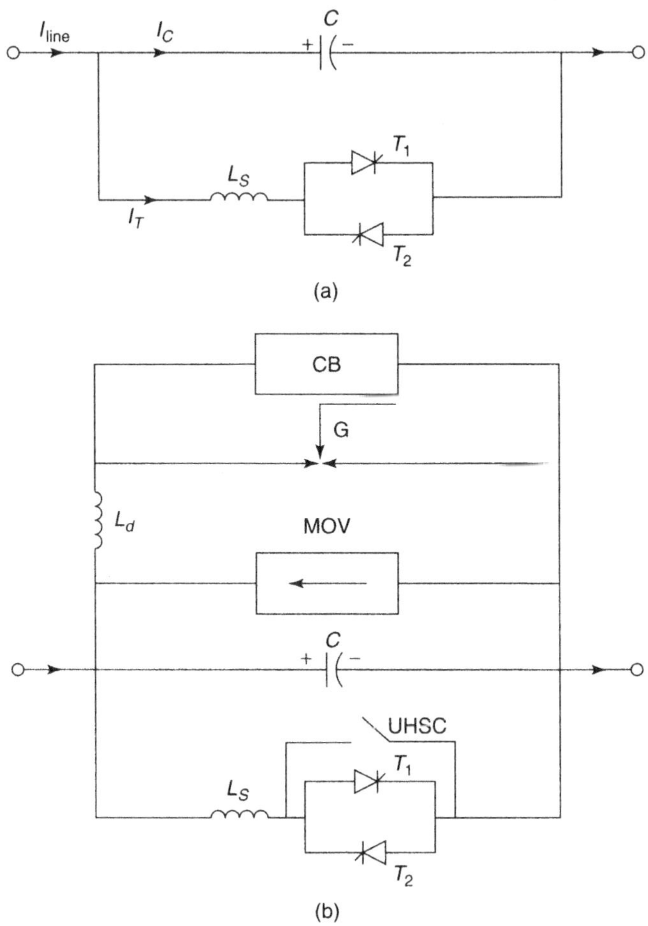

Figure 7.1 A TCSC module: (a) a basic module and (b) a practical module.

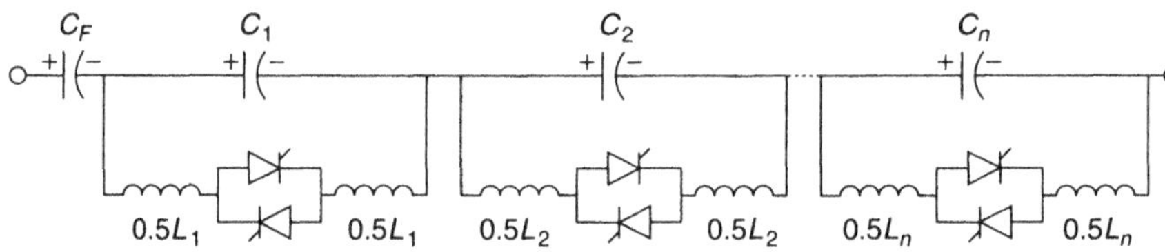

Figure 7.2 A typical TCSC system.

voltages. Not only does the MOV limit the voltage across the capacitor, but it allows the capacitor to remain in circuit even during fault conditions and helps improve the transient stability.

Also installed across the capacitor is a circuit breaker, CB, for controlling its insertion in the line. In addition, the CB bypasses the capacitor if severe fault or equipment-malfunction events occur. A current-limiting inductor, L_d, is incorporated in the circuit to restrict both the magnitude and the frequency of the capacitor current during the capacitor-bypass operation.

If the TCSC valves are required to operate in the fully "on" mode for prolonged durations, the conduction losses are minimized by installing an ultra–high-speed contact (UHSC) across the valve. This metallic contact offers a virtually lossless feature similar to that of circuit breakers and is capable of handling many switching operations. The metallic contact is closed shortly after the thyristor valve is turned on, and it is opened shortly before the valve is turned off. During a sudden overload of the valve, and also during fault conditions, the metallic contact is closed to alleviate the stress on the valve.

An actual TCSC system usually comprises a cascaded combination of many such TCSC modules, together with a fixed-series capacitor, C_F. This fixed-series capacitor is provided primarily to minimize costs. A conceptual TCSC system with basic TCSC modules is shown in Fig. 7.2. The capacitors—C_1, C_2, . . . , C_n—in the different TCSC modules may have different values to provide a wider range of reactance control. The inductor in series with the antiparallel thyristors is split into two halves to protect the thyristor valves in case of inductor short circuits.

7.3 OPERATION OF THE TCSC

7.3.1 Basic Principle

A TCSC is a series-controlled capacitive reactance that can provide continuous control of power on the ac line over a wide range. From the system viewpoint, the principle of variable-series compensation is simply to increase the fundamental-frequency voltage across an fixed capacitor (FC) in a series-compensated line through appropriate variation of the firing angle, α [5]. This enhanced voltage changes the effective value of the series-capacitive reactance.

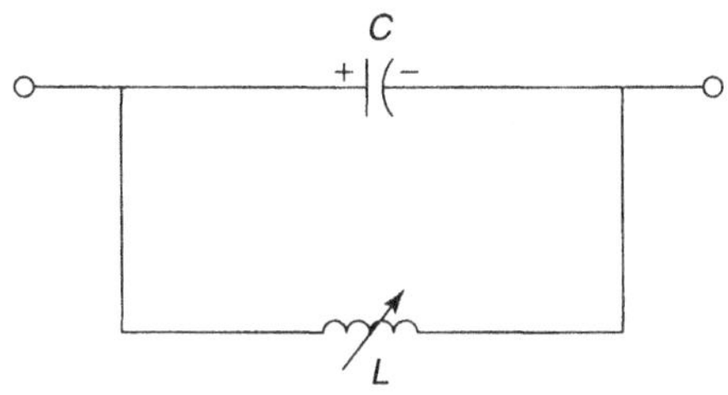

Figure 7.3 A variable inductor connected in shunt with an FC.

A simple understanding of TCSC functioning can be obtained by analyzing the behavior of a variable inductor connected in parallel with an FC, as shown in Fig. 7.3. The equivalent impedance, Z_{eq}, of this LC combination is expressed as

$$Z_{eq} = \left(j\,\frac{1}{\omega C}\right) \| \,(j\omega L) = -j\,\frac{1}{\omega C - \dfrac{1}{\omega L}} \tag{7.2}$$

The impedance of the FC alone, however, is given by $-j(1/\omega C)$.

If $\omega C - (1/\omega L) > 0$ or, in other words, $\omega L > (1/\omega C)$, the reactance of the FC is less than that of the parallel-connected variable reactor and that this combination provides a variable-capacitive reactance are both implied. Moreover, this inductor increases the equivalent-capacitive reactance of the LC combination above that of the FC.

If $\omega C - (1/\omega L) = 0$, a resonance develops that results in an infinite-capacitive impedance—an obviously unacceptable condition. If, however, $\omega C - (1/\omega L) < 0$, the LC combination provides inductance above the value of the fixed inductor. This situation corresponds to the inductive-vernier mode of the TCSC operation.

In the variable-capacitance mode of the TCSC, as the inductive reactance of the variable inductor is increased, the equivalent-capacitive reactance is gradually decreased. The minimum equivalent-capacitive reactance is obtained for extremely large inductive reactance or when the variable inductor is open-circuited, in which the value is equal to the reactance of the FC itself.

The behavior of the TCSC is similar to that of the parallel LC combination. The difference is that the LC-combination analysis is based on the presence of pure sinusoidal voltage and current in the circuit, whereas in the TCSC, because of the voltage and current in the FC and thyristor-controlled reactor (TCR) are not sinusoidal because of thyristor switchings. The exact analysis of this condition is presented in Section 7.5.

7.3.2 Modes of TCSC Operation

There are essentially three modes of TCSC operation [3], [6]–[8]; these are illustrated in Fig. 7.4 and described in the following text.

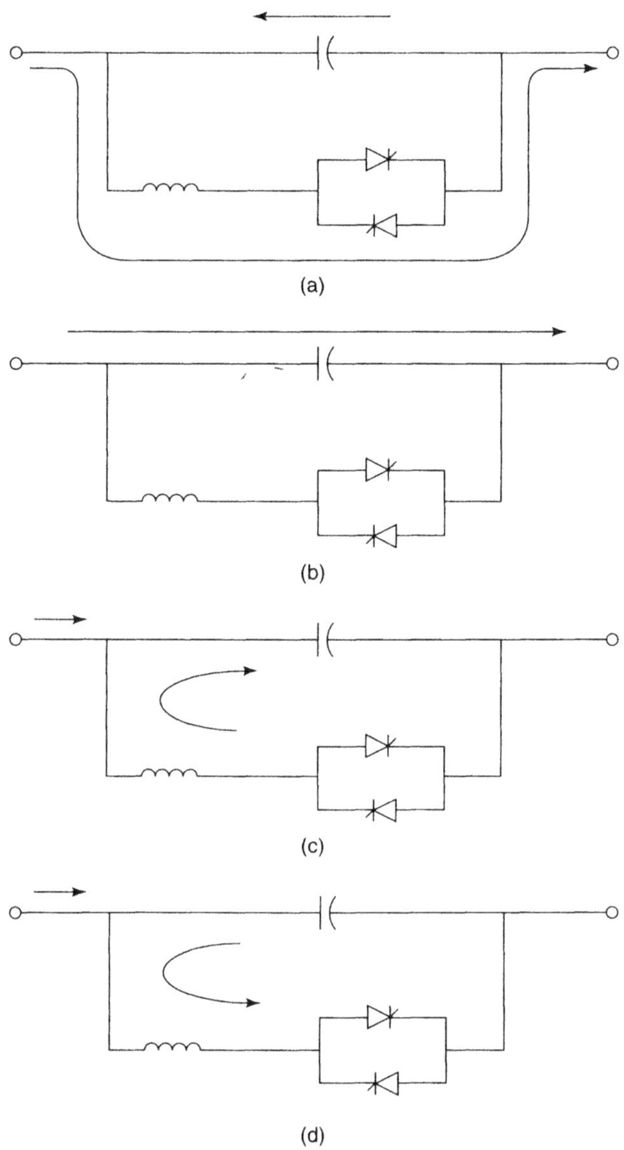

Figure 7.4 Different operating modes of a TCSC: (a) the bypassed-thyristor mode; (b) the blocked-thyristor mode; (c) the partially conducting thyristor (capacitive-vernier) mode; and (d) the partially conducting thyristor (inductive-vernier) mode.

7.3.2.1 Bypassed-Thyristor Mode In this bypassed mode, the thyristors are made to fully conduct with a conduction angle of 180°. Gate pulses are applied as soon as the voltage across the thyristors reaches zero and becomes positive, resulting in a continuous sinusoidal of flow current through the thyristor valves. The TCSC module behaves like a parallel capacitor–inductor combination. However, the net current through the module is inductive, for the susceptance of the reactor is chosen to be greater than that of the capacitor.

Also known as the *thyristor-switched-reactor* (TSR) mode, the bypassed-thyristor mode is distinct from the *bypassed-breaker* mode, in which the circuit breaker provided across the series capacitor is closed to remove the capacitor or the TCSC module in the event of TCSC faults or transient overvoltages across the TCSC.

This mode is employed for control purposes and also for initiating certain protective functions. Whenever a TCSC module is bypassed from the violation of the current limit, a finite-time delay, T_{delay}, must elapse before the module can be reinserted after the line current falls below the specified limit.

7.3.2.2 Blocked-Thyristor Mode In this mode, also known as the *waiting* mode, the firing pulses to the thyristor valves are blocked. If the thyristors are conducting and a blocking command is given, the thyristors turn off as soon as the current through them reaches a zero crossing. The TCSC module is thus reduced to a fixed-series capacitor, and the net TCSC reactance is capacitive. In this mode, the dc-offset voltages of the capacitors are monitored and quickly discharged using a dc-offset control [9] without causing any harm to the transmission-system transformers.

7.3.2.3 Partially Conducting Thyristor, or Vernier, Mode This mode allows the TCSC to behave either as a continuously controllable capacitive reactance or as a continuously controllable inductive reactance. It is achieved by varying the thyristor-pair firing angle in an appropriate range. However, a smooth transition from the capacitive to inductive mode is not permitted because of the resonant region between the two modes.

A variant of this mode is the *capacitive-vernier-control* mode, in which the thyristors are fired when the capacitor voltage and capacitor current have opposite polarity. (Refer to Fig. 7.8, to be discussed later.) This condition causes a TCR current that has a direction opposite that of the capacitor current, thereby resulting in a loop-current flow in the TCSC controller. The loop current increases the voltage across the FC, effectively enhancing the equivalent-capacitive reactance and the series-compensation level for the same value of line current. To preclude resonance, the firing angle α of the forward-facing thyristor, as measured from the positive reaching a zero crossing of the capacitor voltage, is constrained in the range $\alpha_{\min} \leq \alpha \leq 180^\circ$. This constraint provides a continuous vernier control of the TCSC module reactance. The loop current increases as α is decreased from 180° to $\alpha_{\min}$. The maximum TCSC reactance permissible with $\alpha = \alpha_{\min}$ is typically two-and-a-half to three times the capacitor reactance at fundamental frequency.

Another variant is the *inductive-vernier mode*, in which the TCSC can be operated by having a high level of thyristor conduction. In this mode, the direction of the circulating current is reversed and the controller presents a net inductive impedance.

Based on the three modes of thyristor-valve operation, two variants of the TCSC emerge:

1. *Thyristor-switched series capacitor* (TSSC), which permits a discrete control of the capacitive reactance.
2. *Thyristor-controlled series capacitor* (TCSC), which offers a continuous control of capacitive or inductive reactance. (The TSSC, however, is more commonly employed.)

7.4 THE TSSC

A TSSC scheme consists of a series connection of multiple TCSC modules together with a fixed-series capacitor, as shown in Fig. 7.2. The thyristor pairs operate either in the blocked mode or the bypassed mode, thus acting as switches that are off or on, respectively. The inductors $L_1, L_2, \ldots, L_n$ in Fig. 7.2 are replaced by small current-limiting inductors for suppressing any transient current flow through the thyristor valves during switching. With each valve switching, the corresponding series capacitor is either inserted or removed from the transmission-line circuit.

Many capacitive-reactance steps are made possible by installing capacitors of different reactances. One such scheme has $n - 1$ capacitors, each having a reactance of $(0.5X_C/n)$ and one capacitor of reactance $0.5X_C$. Appropriate switchings can result in the following combinations of effective reactances:

$$X_{\text{eff}} = \frac{0.5X_C}{n}\,p, \qquad p = 0, 1, 2, \ldots, 2n \tag{7.3}$$

The TSSC offers the following benefits compared to mechanically switched series capacitors:

1. The thyristor switches allow an unlimited number of operations without any wear. This capability is used to alter the degree of line compensation more frequently and to achieve a greater control over the power flow.
2. Exact switching instants (point-of-voltage waveforms) can be selected with thyristors, which significantly minimizes the switching transients. In contrast, the switching of mechanical breakers is unsynchronized.
3. A very rapid speed of response, in which the time between the initiation of a control signal and a capacitor insertion, or bypass, is typically less than a half-cycle (8 ms for 60 Hz). Thus, in case a major tie-line suffers an outage, the power-transfer capability of an alterative line can be increased rapidly through the TSSC.
4. No generation of harmonics.

5. A possibility that minimizing thyristor-valve losses in a TSSC can be accomplished by using UHSCs.

The TSSC scheme is quite satisfactory when only stepwise control of transmission-line reactance is considered adequate. However, when continuous control is desired, the other variant—the TCSC—is employed.

7.5 ANALYSIS OF THE TCSC

The analysis of TCSC operation in the vernier-control mode [3] is performed based on the simplified TCSC circuit shown in Fig. 7.5. Transmission-line current is assumed to be the independent-input variable and is modeled as an external current source, $i_S(t)$. It is further assumed that the line current is sinusoidal, as derived from actual measurements demonstrating that very few harmonics exist in the line current. However, the analysis presented in the following text may be erroneous to the extent that the line current deviates from a purely sinusoidal nature. Operating conditions resulting in this phenomenon are rare, and the expressions derived in the following text are used widely.

The current through the fixed-series capacitor, C, is expressed as

$$C \frac{dv_C}{dt} = i_S(t) - i_T(t) \cdot u \tag{7.4}$$

The switching variable $u = 1$ when the thyristor valves are conducting, that is, when the switch S is closed. On the other hand, $u = 0$ when the thyristors are blocked, that is, when switch S is open. The thyristor-valve current, $i_T(t)$, is then described by

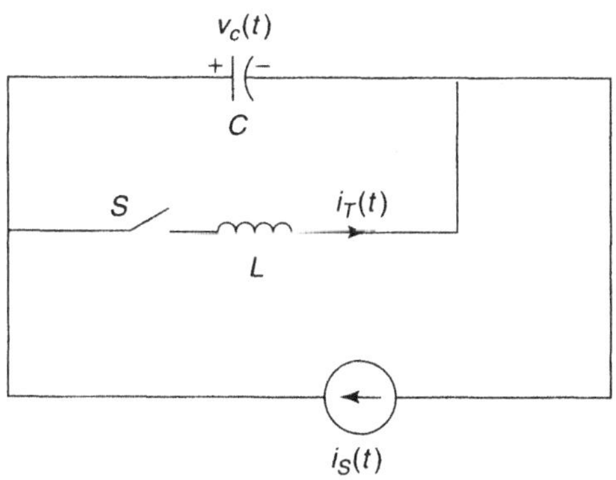

Figure 7.5 A simplified TCSC circuit.

$$\frac{Ldi_T}{dt} = v_C \cdot u \tag{7.5}$$

Let the line current, $i_S(t)$, be represented by

$$i_S(t) = I_m \cos \omega t \tag{7.6}$$

Equations (7.4) and (7.5) can be solved with the knowledge of the instants of switching. In equidistant firing-pulse control, for balanced TCSC operation, the thyristors are switched on twice in each cycle of line current at instants t_1 and t_3, given by

$$t_1 = -\frac{\beta}{\omega} \tag{7.7}$$

$$t_3 = \frac{\pi - \beta}{\omega} \tag{7.8}$$

where β is the angle of advance (before the forward voltage becomes zero). Or,

$$\beta = \pi - \alpha; \qquad 0 < \beta < \beta_{\max} \tag{7.9}$$

The firing angle α is generated using a reference signal that can be in phase with the capacitor voltage. The thyristor switch S turns off at the instants t_2 and t_4, defined as

$$t_2 = t_1 + \frac{\sigma}{\omega} \tag{7.10}$$

$$t_4 = t_3 + \frac{\sigma}{\omega} \tag{7.11}$$

where α is the conduction angle, which is assumed to be the same in both the positive and the negative cycle of conduction. Also,

$$\sigma = 2\beta \tag{7.12}$$

Solving the TCSC equations (7.4)–(7.6) results in the steady-state thyristor current, i_T, as

$$i_T(t) = \frac{k^2}{k^2 - 2} \operatorname{Im}\left[\cos \omega t - \frac{\cos \beta}{\cos k\beta} \cos \omega_r t\right]; \qquad -\beta \le \omega t \le \beta \tag{7.13}$$

where

$$\omega_r = \frac{1}{\sqrt{LC}} \tag{7.14}$$

$$k = \frac{\omega_r}{\omega} = \sqrt{\frac{1}{\omega L} \cdot \frac{1}{\omega C}} = \sqrt{\frac{X_C}{X_L}} \tag{7.15}$$

and X_C is the nominal reactance of the FC only. The steady-state capacitor voltage at the instant $\omega t = -\beta$ is expressed by

$$v_{C1} = \frac{\text{Im } X_C}{k^2 - 1} (\sin \beta - k \cos \beta \tan k\beta) \tag{7.16}$$

At $\omega t = \beta$, $i_T = 0$, and the capacitor voltage is given by

$$v_C(\omega t = \beta) = v_{C2} = -v_{C1} \tag{7.17}$$

The capacitor voltage is finally obtained as

$$v_C(t) = \frac{\text{Im } X_C}{k^2 - 1} \left(-\sin \omega t + k \frac{\cos \beta}{\cos k\beta} \sin \omega_r t \right); \qquad -\beta \le \omega t \le \beta \tag{7.18}$$

$$v_C(t) = v_{C2} + \text{Im } X_C(\sin \omega t - \sin \beta); \qquad \beta < \omega t < \pi - \beta \tag{7.19}$$

Because the nonsinusoidal capacitor voltage, v_C, has odd symmetry about the axis $\omega t = 0$, the fundamental component, V_{CF}, is obtained as

$$V_{CF} = \frac{4}{\pi} \int_0^{\pi/2} v_C(t) \sin \omega t \; d(\omega t) \tag{7.20}$$

The equivalent TCSC reactance is computed as the ratio of V_{CF} to I_m:

$$X_{\text{TCSC}} = \frac{V_{CF}}{I_m} = X_C - \frac{X_C^2}{(X_C - X_L)} \frac{2\beta + \sin 2\beta}{\pi} + \frac{4X_C^2}{(X_C - X_L)} \frac{\cos^2 \beta}{(k^2 - 1)} \frac{(k \tan k\beta - \tan \beta)}{\pi} \tag{7.21}$$

Alternatively, the net reactance of the TCSC in per units of X_C, denoted by $X_{\text{net}}(= X_{\text{TCSC}}/X_C)$, can be expressed as

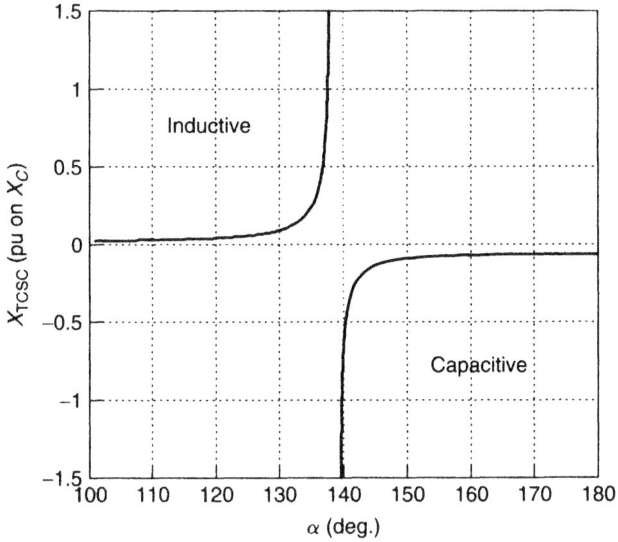

Figure 7.6 Variation of the TCSC reactance with firing angle α.

$$X_{\text{net}} = 1 - \frac{X_C}{(X_C - X_L)} \frac{\sigma + \sin\sigma}{\pi} + \frac{4X_C}{(X_C - X_L)} \frac{\cos^2(\sigma/2)}{(k^2 - 1)} \cdot \frac{[k\tan(k\sigma/2) - \tan(\sigma/2)]}{\pi} \tag{7.22}$$

The variation of per-unit TCSC reactance, (X_{TCSC}/X_C), as a function of firing angle α is depicted in Fig. 7.6. It is noted from Eq. (7.21) that a parallel resonance is created between X_L and X_C at the fundamental frequency, corresponding to the values of firing angle α_{res}, given by

$$\alpha_{\text{res}} = \pi - (2m - 1)\frac{\pi\omega}{2\omega_r}; \qquad m = 1, 2 \tag{7.23}$$

or alternatively,

$$\beta_{\text{res}} = (2m - 1)\frac{\pi\omega}{2\omega_r}; \qquad m = 1, 2 \tag{7.24}$$

The different resonances can be reduced to one by a proper choice of $k = (\omega_r/\omega)$ in the range $90^\circ < \alpha < 180^\circ$ or $0 < \beta < 90^\circ$. For instance, in the Kayenta TCSC [10], [11], the choice of inductance as 0.0068 H (henries) across the 15-Ω (ohms) series capacitor ($C = 177\ \mu$F) results in only one resonance at $\alpha = 143^\circ$. If, however, the inductance is 0.0034 H, two resonances at $\alpha = 160^\circ$ and at $\alpha = 101^\circ$ will occur.

At the resonant point, the TCSC exhibits a very large impedance and results in a significant voltage drop. This resonant region is avoided by installing limits on the firing angle. The typical value of the firing angle resulting in resonance is 145°. Filters are provided in the synchronizing and timing circuits, ensuring that any transients or distortions in the ac system voltage do not affect the TCSC control-system performance. Nevertheless, it is clear that the vernier operation of the TCSC can only enhance the apparent reactance in both capacitive and inductive domains. It is not possible for the TCSC to reduce the reactances.

Because the TCSC is used mainly as a capacitive device, the convention is to define positive reactance as capacitive and negative reactance as inductive—just the opposite of the convention used in circuit analysis and load-flow studies. The implication of different values of X_{net} can be given in terms of the root mean square (rms) value of line current I_{line} [6]. For instance,

$X_{net} = +1$ pu implies that the thyristors are not conducting.

$X_{net} = +2$ pu implies that the thyristors are fired so that the resulting circulating current in the FC–TCR loop causes a 60-Hz voltage of $2.0 \cdot X_C I_{line}$ pu to appear across the FC, which lags behind the line current by 90° (i.e., *capacitive*).

$X_{net} = -1$ pu implies that the thyristors are fired so that the resulting circulating current in the FC–TCR loop causes a 60-Hz voltage of $1.0 \cdot X_C I_{line}$ to appear across the FC, which leads the line current by 90° (i.e., *inductive*).

The rms-capacitor voltage V_C, the rms-capacitor current, I_C, and rms-thyristor current, I_T, are plotted in Fig. 7.7(a) as functions of the reactor in per-unit conduction [6]. In the capacitive mode, $I_C > I_T$; in the inductive mode, $I_C < I_T$. The resulting circulating current in the capacitive mode causes the capacitor current, capacitor voltage, and (consequently) the apparent capacitive reactance to increase with the thyristor conduction. The corresponding variation of the TCSC net reactance X_{net} with per-unit reactor conduction is also superimposed for comparison in Fig. 7.7(b) [6]. The permissible range of the firing angle is therefore determined by the FC voltage and current ratings.

The TCSC voltage, line current, capacitor current, TCR current, and valve voltage for the capacitive-vernier and inductive-vernier modes are depicted in Figs. 7.8(a) and (b), respectively; the actual TCSC voltage and thyristor current for different operating points in the capacitive and inductive regions are plotted in Fig. 7.9(a) and (b), respectively [6]. In Fig. 7.9(a), the solid line indicates the TCSC operating in the blocked-thyristor mode, with an X_{net} of 1 pu; the dashed lines indicate the thyristor current values, giving $X_{net} = 1.5$ pu and 2 pu. In Fig. 7.9(b), the solid line indicates the bypassed-thyristor mode, with an $X_{net} = -0.15$ pu; the dashed lines indicate the thyristor current levels, giving $X_{net} = -0.5$ pu and -1 pu.

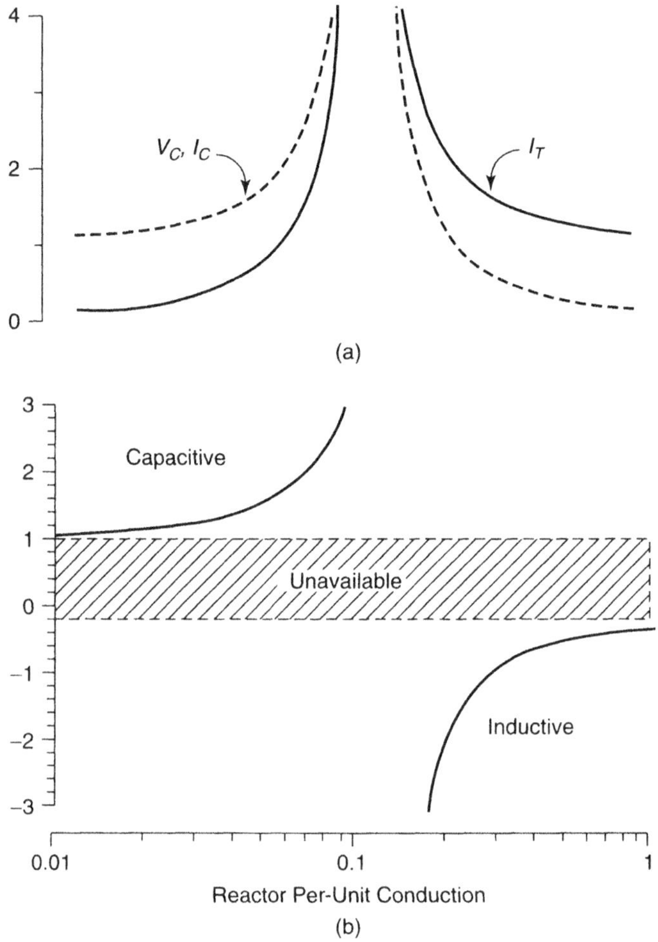

Figure 7.7 Variation of the TCSC voltage, currents, and reactance with per-unit TCR conduction: (a) currents (pu on I_{line}), voltage (pu on $X_C \cdot I_{\text{line}}$); (b) X_{net} (pu on X_C).

7.6 CAPABILITY CHARACTERISTICS

Although TCSC design is based on the application requirements, the operational limits are determined by the characteristics of different TCSC components. The important limits are described in the following list:

1. *Voltage limits*, of which the maximum amount across any operating equipment (including series capacitors) is determined by the equipment's insulation level. The constraint on voltage may vary with the duration of voltage application. For short durations (typically less than 2 s), the overvoltage limit of the MOV is more critical than that of the capacitor.

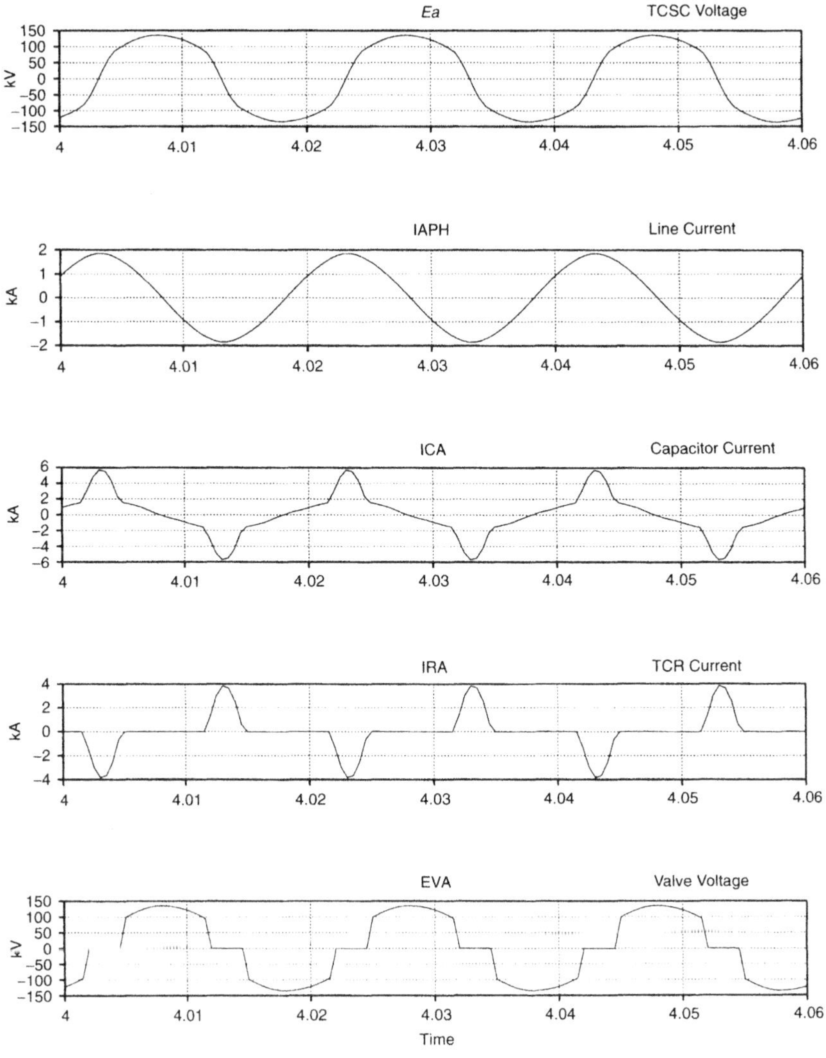

Figure 7.8(a) TCSC waveforms in the capacitive mode of operation ($\alpha = 150^\circ$).

2. *Current limits*, which may need to be imposed on the currents in the thyristor valve, FC, and surge inductor to prevent overheating. Harmonics also cause heating and therefore have a constraining effect on the TCSC operation.

3. *Firing-angle limits* of the thyristors, which must be carefully restricted so that the TCSC does not venture into the resonant region (even temporarily).

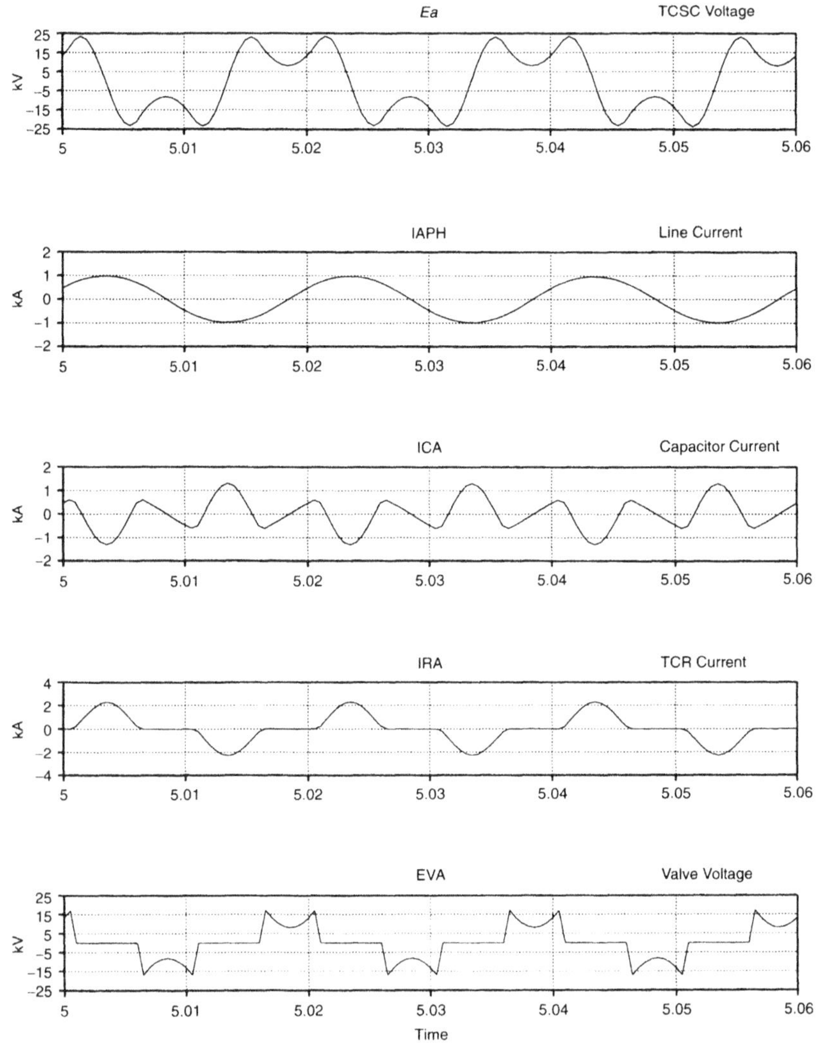

Figure 7.8(b) TCSC waveforms in the inductive mode of operation ($\alpha = 130°$).

7.6.1 The Single-Module TCSC

The capability characteristics of a single-module TCSC in the voltage line–current plane is depicted in Fig. 7.10 [6]. These characteristics are illustrated for continuous-time applications, short-duration implementations (30 min), and 1–10 s. In both capacitive and inductive zones, the operation is generally constrained between the minimum and maximum reactance limits. Certain other voltage-, current-, and harmonic-related limits also exist that constrain the operating range, as described in the following text.

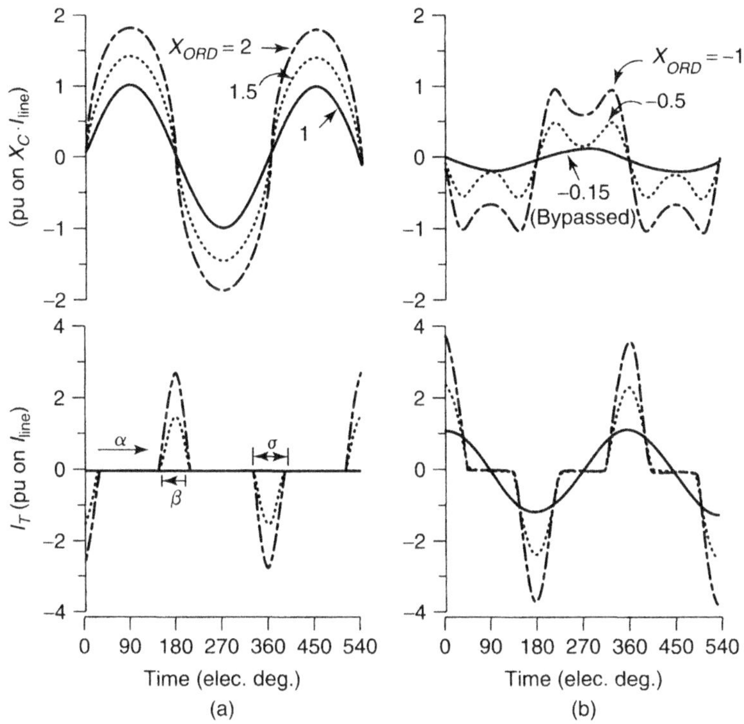

Figure 7.9 Variation of the TCSC voltage and thyristor current with an operating point: (a) capacitive and (b) inductive.

In the capacitive region, the maximum apparent capacitive reactance is chosen based on the TCSC design so that the TCSC does not venture close to or into the inherently unstable resonant point. The maximum X_{TCSC} is typically 2–3 pu, as expressed in per units of X_C. This restriction is imposed in the form of a maximum advance angle, β, limit. The minimum TCSC capacitive reactance is obtained when the thyristors are blocked, corresponding to $\alpha = 180°$ and the absence of thyristor-current flows and relating to $X_{TCSC} = 1$ pu. As the line current increases, the TCSC voltage increases until the maximum voltage limit of the TCSC is reached, as indicated in Fig. 7.10. This limit is dependent on the duraction of the voltage application.

In the inductive-reactance zone, the maximum reactance limit is also selected to prevent the TCSC from operating in the resonant region. This limit is attained at low line currents and is expressed in terms of the maximum firing-delay angle. A maximum inductive reactance of 2 pu is typical; the minimum inductive reactance limit is reached when the thyristors are fully conducting, corresponding to $\alpha = 90°$. With increasing line current, harmonics are generated that cause not only the heating of the reactor and thyristors but also peak voltages close to the voltage limits of the capacitors and MOV, which is indicated

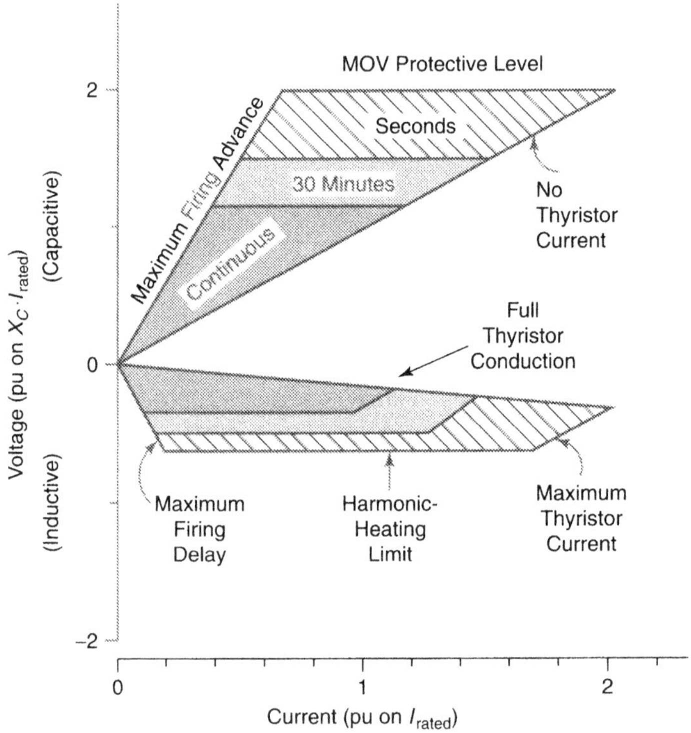

Figure 7.10 The *V-I* capability characteristics for a single-module TCSC.

by a time duration–dependent constant-voltage limit in the capability curve. At high line currents, the maximum thyristor current in the inductive-vernier mode constrains the TCSC operation.

Alternatively, the TCSC capability can be expressed in a reactance–line-current plane, as shown in Fig. 7.11. The dynamic range of the TCSC reactance is reduced with increasing line current. As discussed previously, a smooth transition from the inductive to capacitive region is not possible. The TCSC usually operates in the first quadrant of both the *V-I* and *X-I* characteristics.

7.6.2 The Multimodule TCSC

In several power-system applications, such as power-flow control and damping enhancement, a smooth variation in the line reactance is desirable that can be achieved by splitting a single TCSC into multiple modules and operating them independently in the inductive and capacitive modes.

Splitting a single TCSC into two modules, each with a half-MVA rating, results in a *V-I* capability curve with both similar and dissimilar operation, as noted in Fig. 7.12 [6]. Similar operation of both modules in either the capaci-

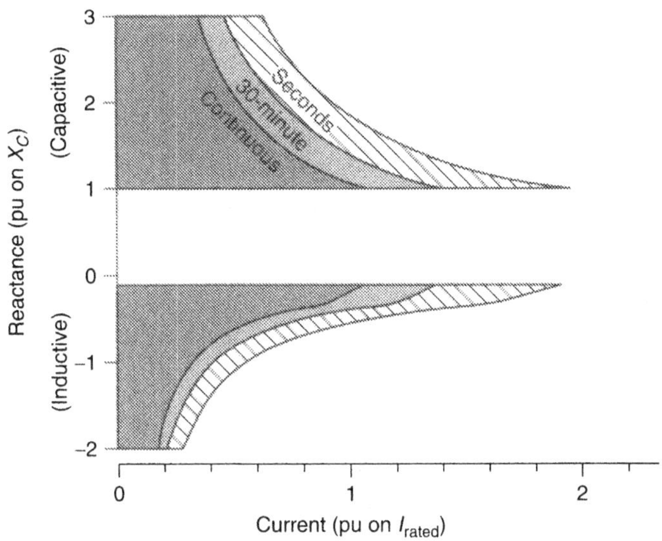

Figure 7.11 The *X-I* capability characteristic for a single-module TCSC.

tive or inductive mode produces a capability curve identical to that of a single-module TCSC, as expected, whereas dissimilar operation results in an intermediate characteristic, as noted in Fig. 7.12. In the figure, both the minimum capacitive- and minimum inductive-reactance limits vanish, and a continuous transition from the capacitive to the inductive domain becomes feasible.

The *V-I* and *X-I* capability curves for a multimodule TCSC are depicted in Fig. 7.13 and Fig. 7.14, respectively [6]. (The details of *X-I* characteristics are presented in Section 7.10.1.) It is evident that the greater the number of TCSC modules, the more expansive the controllable range of the TCSC for the same total MVA rating of the TCSC, (although achieved at a greater cost).

In the multimodule case, the modules are only switched in or out if the desired TCSC reactance is lower than the nominal capacitive reactance of the bank. The vernier operation is resorted to only when the reactance order is greater than the rated capacitive reactance of the bank.

7.7 HARMONIC PERFORMANCE

The harmonic currents generated in the TCR branch during vernier operation tend to flow in the parallel combination of the FC and the equivalent harmonic impedance of the ac power system. Components of the harmonic-current flow in both the FC and the ac system determine the harmonic voltages introduced therein. However, as the impedance of the FC is very low compared to the equivalent ac system, the harmonics circulate predominantly in the TCR–FC

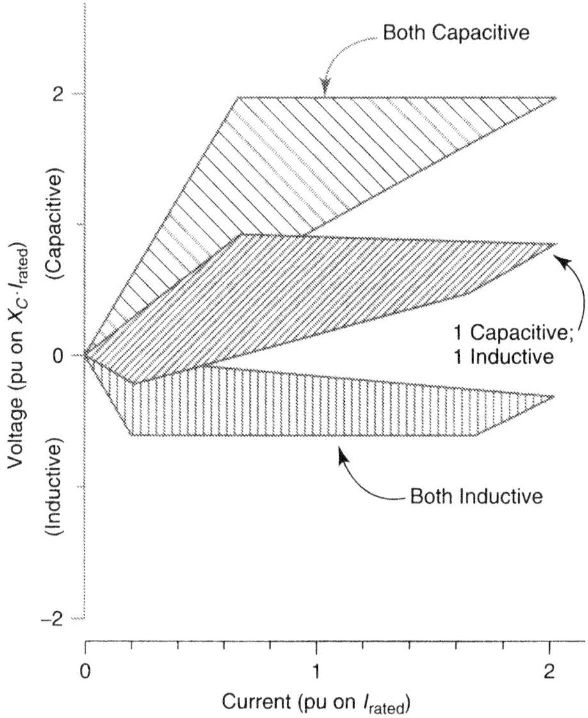

Figure 7.12 The *V-I* capability characteristics for a two-module TCSC.

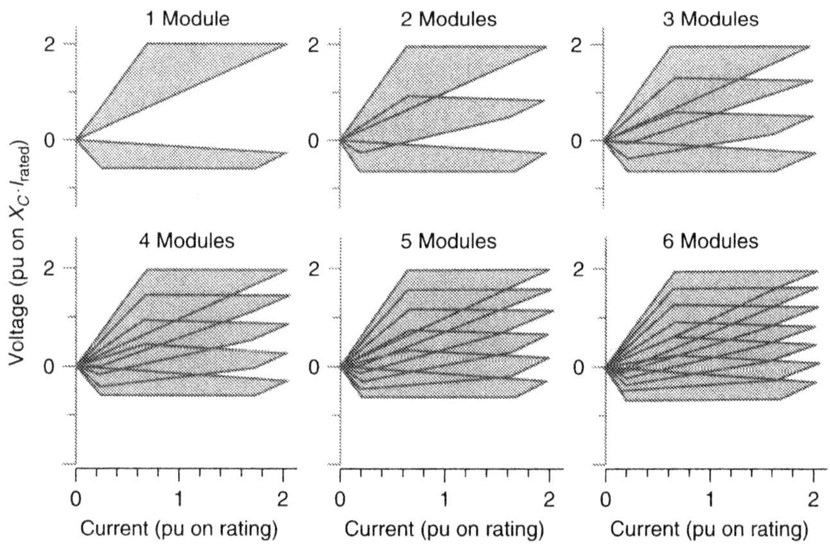

Figure 7.13 The *V-I* capability characteristic for a multimodule TCSC.

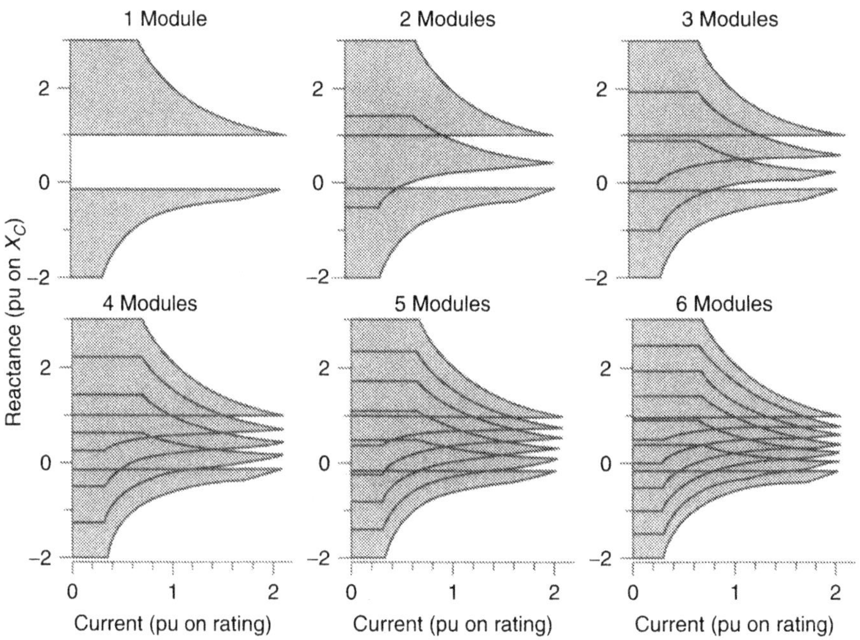

Figure 7.14 The *X-I* capability characteristic for a multimodule TCSC.

loop and do not leak into the ac system. Hence the line current in the TCSC system continues to remain sinusoidal, which is also substantiated by actual field measurements; for example, the total voltage-harmonic distortion in the ac system from a TCSC installation [12], corresponding to a capacitive operation at $\alpha = 147^\circ$, has been reported to be less than 1.5%.

Nonetheless, the capacitor voltage does contain odd harmonics of the fundamental frequency of the order

$$h = 2m - 1; \qquad m = 1, 2 \tag{7.25}$$

Magnitudes of different harmonic voltages, $V_C(h)$, expressed as per units of the fundamental component, V_{CF}, are given by

$$\frac{V_C(h)}{V_{CF}} = \frac{0.77}{h}\left[\frac{\sin\beta(1-h)}{(1-h)} + \frac{\sin\beta(1+h)}{(1+h)}\right] - \frac{0.77}{h}\frac{\cos\beta}{\cos k\beta}\left[\frac{\sin\{\beta(k-h)\}}{(k-h)} + \frac{\sin\{\beta(k+h)\}}{(k+h)}\right] \tag{7.26}$$

The harmonic components of the capacitor voltage are plotted in Fig. 7.15 as a function of the apparent TCSC reactance over its complete range, and

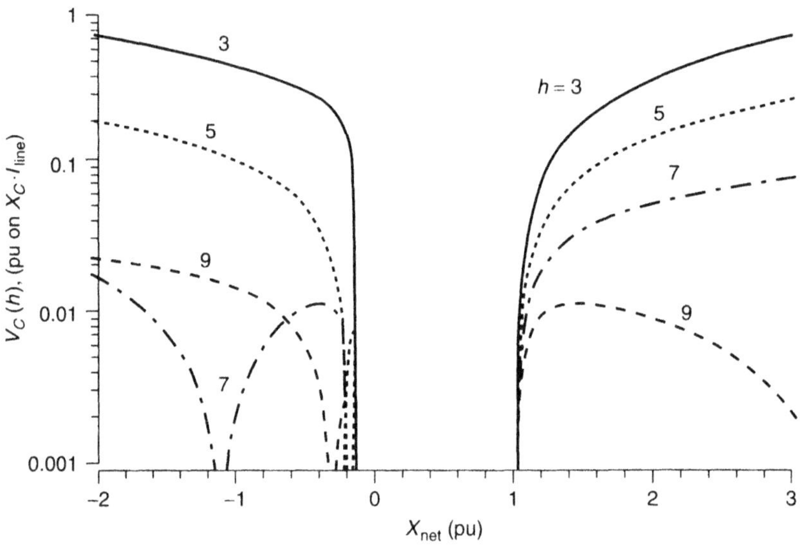

Figure 7.15 Harmonics in the capacitor voltage.

they are expressed in per units of the capacitor voltage occurring without any thyristor current. The capacitor offers a very low impedance path for higher harmonics, thereby reducing their appearance in any significant measure in the capacitor voltage. Only the lower-order harmonics—namely, the 3rd, 5th, and 9th—predominate. The variation of different harmonic components as a function of line current is illustrated in Fig. 7.16. In both capacitive and inductive regions, the maximum harmonic distortion of the capacitor voltage generally occurs at a value of line current that is lower than the rated current.

Practical TCSC projects do not employ a power-line filter, as the harmonic content introduced by the TCSC in the line is very minimal. In the Slatt installation [6], the worst-case harmonic current is comparable to the exciting current of a large transformer of similar rating.

7.8 LOSSES

The TCSC losses [6]–[8] comprise the following:

1. the series-capacitor losses;
2. the reactor-conduction loss;
3. the thyristor-conduction loss; and
4. the switching losses from heating the snubber resistor when voltage step occurs across the snubber capacitor, as well as from heating the thyristor during the recovery process.

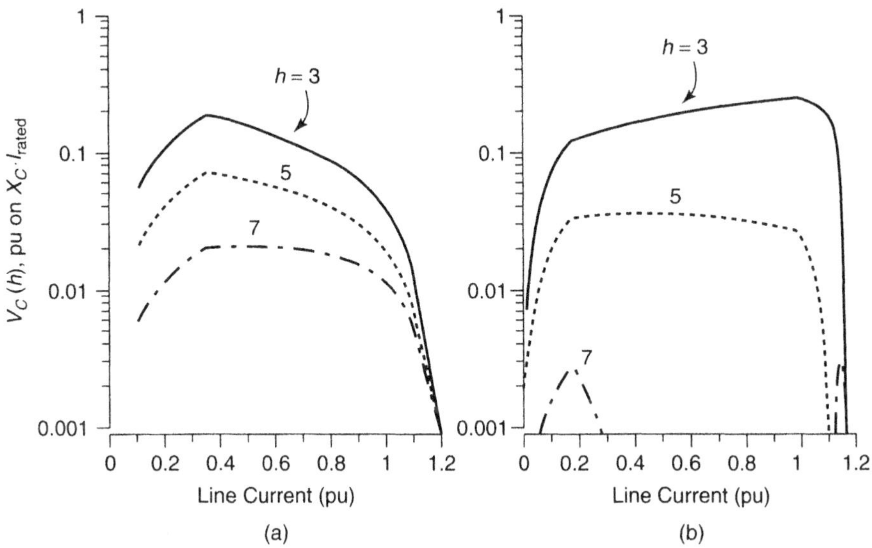

Figure 7.16 The TCSC harmonic voltages as a function of line current: (a) the maximum capacitive vernier and (b) the maximum inductive vernier.

The various component losses and the resulting total loss are illustrated in Fig. 7.17 [6] for the different operating modes of TCSC. Figure 7.17(a) shows the FC losses during the blocked-thyristor operation mode; Fig. 7.17(b) shows the losses in the capacitive-vernier operation mode corresponding to the maximum capacitive-reactance limit; Fig. 7.17(c) shows the different loss components in the bypassed thyristor operation mode for the minimum inductive-reactance operation; and Fig. 7.17(d) shows the losses corresponding to the maximum inductive-reactance limit operation in the inductive-vernier mode. The inductive operation results in increased losses relative to the capacitive operation. These losses are expressed as a percent of the continuous TCSC MVA rating—$\mathrm{MVAR_{nom}}$—defined as

$$\mathrm{MVAR_{nom}} = 3I_{\mathrm{rated}} \cdot V_{\mathrm{rated}} \tag{7.27}$$

When multiple modules are used, the minimum loss that occurs corresponds to the blocked-thyristor operation mode of all thyristor modules, that is, with $X_{\mathrm{net}} = 1$ pu. At other operating points, the losses depend on the reactance order. The multimodule TCSC loss characteristic with six modules is shown in Fig. 7.18 for two magnitudes of line current. Figure 7.18(a) depicts the losses when the line current is half the rated value, whereas Fig. 7.18(b) depicts the losses when the line current equals the rated magnitude.

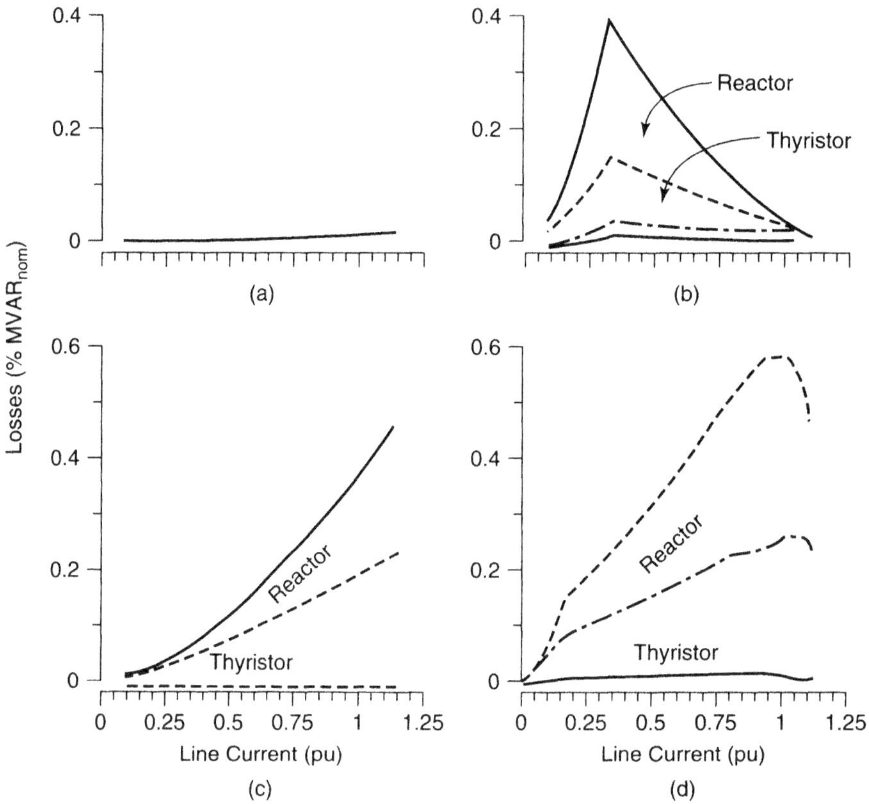

Figure 7.17 The losses in a single-module TCSC: (a) blocked or $X_{\text{net}} = 1$ pu; (b) the maximum capacitive vernier; (c) the bypassed; and (d) the maximum inductive vernier.

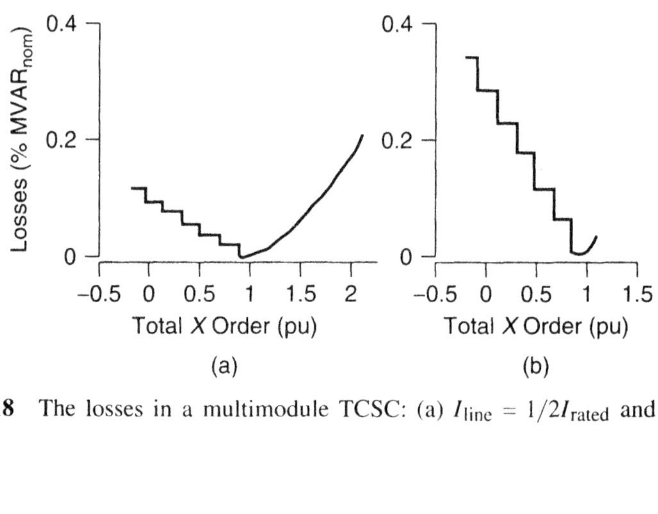

Figure 7.18 The losses in a multimodule TCSC: (a) $I_{\text{line}} = 1/2 I_{\text{rated}}$ and (b) $I_{\text{line}} = I_{\text{rated}}$.

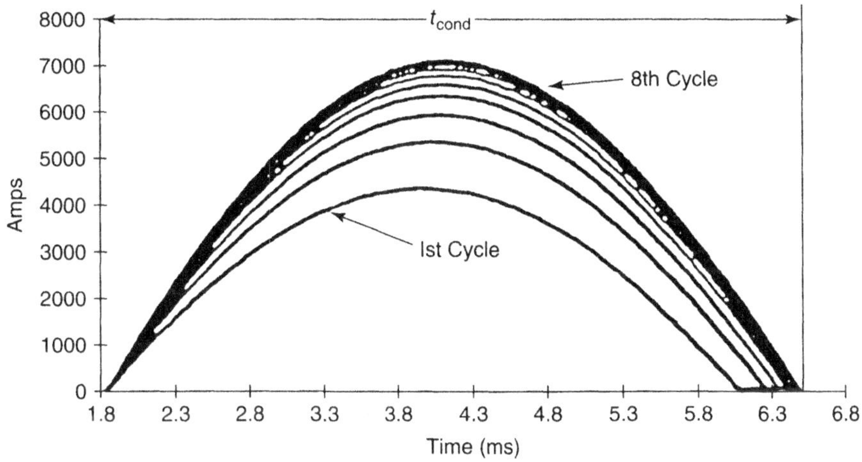

Figure 7.19 Cycle-by-cycle buildup of TCR inductor current.

7.9 RESPONSE OF THE TCSC

The steady state of the TCSC after an input change is usually achieved after several cycles. The steady state is reached when the half-cycle pulse of the inductor current becomes symmetrical about the zero crossing of the capacitor voltage. Also at this time, the capacitor voltage at the instant of thyristor firing becomes equal in magnitude—albeit with opposite polarity—to the capacitor voltage at the instant of thyristor turn-off. An 8-cycle buildup of inductor current [13] for a firing angle of 130° is shown in Fig. 7.19, in which only the positive half-cycle is displayed. The factors affecting the response time of the TCSC [10], [12], [14] are as follows:

1. *Initial operating point.* Illustrated in Fig. 7.20 [14] is the line-current response for the Kayenta TCSC system, as obtained through EMTP simulation for equal step change in the TCSC conduction angle at different initial operating points. The effects of synchronization with the TCSC voltage and the line current are exemplified by parts (a) and (b), respectively, of the figure. Note that this open-loop TCSC response resembles that of a fourth-order transfer function, $Z(s)$ [12].

$$Z(s) = \frac{\omega^2}{(1 + s\tau)^2(s^2 + 2\omega\xi s + \omega^2)} \tag{7.28}$$

2. *Magnitude of step change in firing angle.* A step change in the firing angle at a high level of TCR conduction causes a sluggish response, whereas the same firing-angle step change at a low level of TCR conduction results in a fast, oscillating response. This substantial varia-

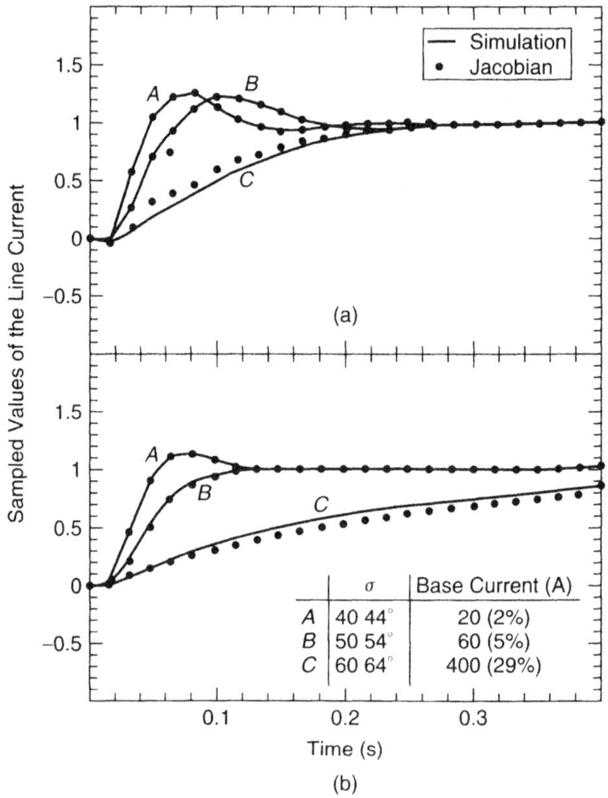

Figure 7.20 The line-current response to equal step change in the conduction angle at different operating points for a Kayenta TCSC: (a) synchronization with capacitor voltage and (b) synchronization with line current.

tion in TCSC response time is caused by the TCR interaction with the TCSC capacitance–transmission-line reactance combination. The TCSC response differs slightly when the firing-angle synchronization is done with respect to the zero crossings of the line current instead of the zeroes of the TCSC voltage. The difference is more visible when step change is made at higher initial-conduction angles. In practical systems, synchronization with respect to zeroes of line current is preferred, as the line-current signal is quite stable and free from harmonics, unlike the capacitor-voltage signal.

3. *Resulting level of series compensation.* The TCSC response slows down if the requested change in capacitive reactance results in very high levels of series compensation [10].

4. *System impedance.* Shown in Fig. 7.21 is the TCSC response in a closed-loop-current control mode as a function of system strength. The response

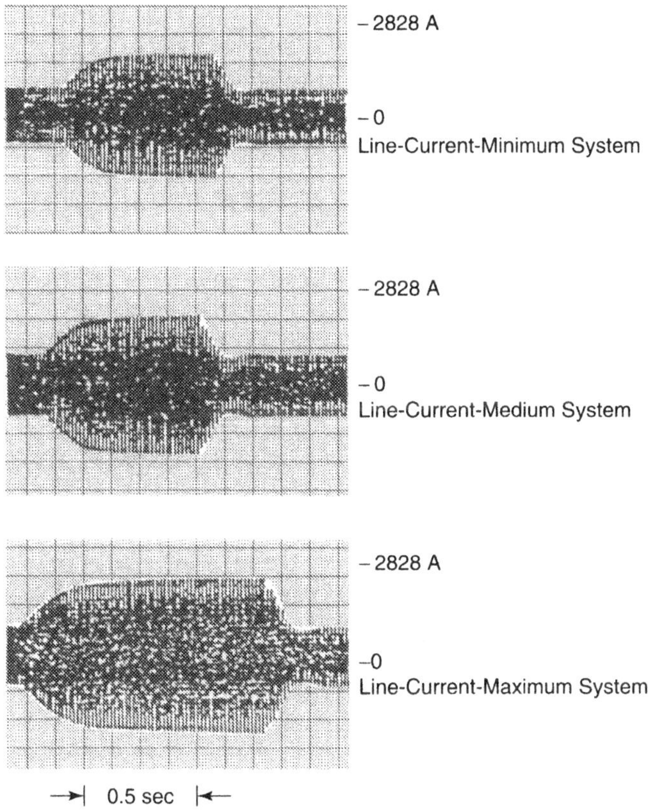

Figure 7.21 The TCSC current-controller response for different system strengths.

time is seen to increase with the ac system strength [12]—in other words, the TCSC responds faster in weak ac systems.

a. *Unbalanced operation.* The TCSC can operate successfully with symmetrical control in all the three phases, even when a significantly large unbalance exists in the line loadings of the three phases.

b. *Choice of reactor size.* The maximum capacitance impedance can be realized with different reactor sizes, but the component ratings and the firing-angle control range will differ in each case. For some TCSC capacitors, a large reactor results in a lower circulating current and, consequently, diminished current ratings for the reactor and the thyristor valves. In addition, the range of the firing angle is also enhanced [13]. The eventual choice is based on an economic analysis of each individual case.

The reactor size is also chosen so that the resonance between the FC and the inductor is below the 3rd harmonic. The ratio between the reactance of inductor and that of the FC, as adopted in practical TCSC installations, is typically around 0.133 [6]. The natural frequency of the FC and the reactor typically ranges between 120 and 150 Hz. The Kayenta TCSC installation uses 145 Hz; the Slatt TCSC substation, 127 Hz [15], [16].

7.10 MODELING OF THE TCSC

A TCSC involves continuous-time dynamics, relating to voltages and currents in the capacitor and reactor, and nonlinear, discrete switching behavior of thyristors. Deriving an appropriate model for such a controller is an intricate task.

7.10.1 Variable-Reactance Model

A TCSC model for transient- and oscillatory-stability studies, used widely for its simplicity, is the variable-reactance model [17], [18] depicted in Fig. 7.22. In this quasi-static approximation model, the TCSC dynamics during power-swing frequencies are modeled by a variable reactance at fundamental frequency. The other dynamics of the TCSC model—the variation of the TCSC response with different firing angles, for example—are neglected [10], [12], [14].

It is assumed that the transmission system operates in a sinusoidal steady state, with the only dynamics associated with generators and PSS. This assumption is valid, because the line dynamics are much faster than the generator dynamics in the frequency range of 0.1–2 Hz that are associated with angular-stability studies.

As described previously, the reactance-capability curve of a single-module TCSC, as depicted in Fig. 7.11, exhibits a discontinuity between the inductive and capacitive regions. However, this gap is lessened by using a multimode TCSC. The variable-reactance TCSC model assumes the availability of a continuous-reactance range and is therefore applicable for multimodule TCSC configurations. This model is generally used for inter-area mode analysis, and it provides high accuracy when the reactance-boost factor (= X_{TCSC}/X_C) is less than 1.5.

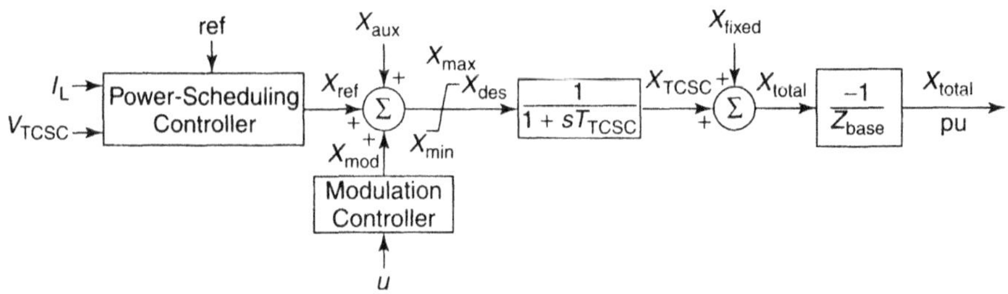

Figure 7.22 A block diagram of the variable-reactance model of the TCSC.

7.10.1.1 Transient-Stability Model In the variable-reactance model for stability studies, a reference value of TCSC reactance, X_{ref}, is generated from a power-scheduling controller based on the power-flow specification in the transmission line. The reference X_{ref} value may also be set directly by manual control in response to an order from an energy-control center, and it essentially represents the initial operating point of the TCSC; it does not include the reactance of FCs (if any). The reference value is modified by an additional input, X_{mod}, from a modulation controller for such purposes as damping enhancement. Another input signal, this applied at the summing junction, is the open-loop auxiliary signal, X_{aux}, which can be obtained from an external power-flow controller.

A desired magnitude of TCSC reactance, X_{des}, is obtained that is implemented after a finite delay caused by the firing controls and the natural response of the TCSC. This delay is modeled by a lag circuit having a time constant, T_{TCSC}, of typically 15–20 ms [19]. The output of the lag block is subject to variable limits based on the TCSC reactance-capability curve shown in Fig. 7.14. The resulting X_{TCSC} is added to the X_{fixed}, which is the reactance of the TCSC installation's FC component.

To obtain per-unit values, the TCSC reactance is divided by the TCSC base reactance, Z_{base}, given as

$$Z_{base} = \frac{(\mathrm{kV}_{TCSC})^2}{\mathrm{MVA}_{sys}} \tag{7.29}$$

where kV_{TCSC} = the rms line–line voltage of the TCSC in kilovolts (kV)
MVA_{sys} = the 3-phase MVA base of the power system

The TCSC model assigns a positive value to the capacitive reactance, so X_{total} is multiplied by a negative sign to ensure consistency with the convention used in load-flow and stability studies. The TCSC initial operating point, X_{ref}, for the stability studies is chosen as

$$X_{ref} = X_{total} - X_{fixed} \tag{7.30}$$

The reactance capability curve of the multimodal TCSC shown in Fig. 7.14 can be simply approximated by the capability curve shown in Fig. 7.23. This figure can be conveniently used for the variable-reactance model of TCSC, and the capability curve that the figure depicts includes the effect of TCSC transient-overload levels.

It should be noted that the reactance limit for high currents is depicted in Fig. 7.23 as a group of discrete points for the different modules. During periods of overcurrent, only some TCSC modules move into the bypassed mode, for the bypassing of a module causes the line current to decrease and thus reduces the need for the remaining TCSC modules to go into the bypass mode. However, for the case of modeling, only one continuous-reactance limit—denoted by a

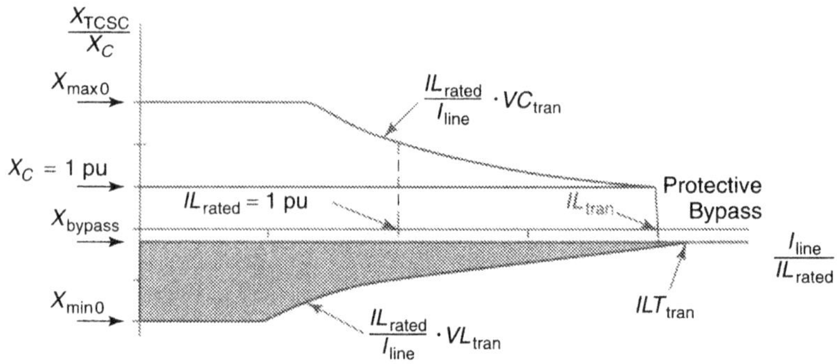

Figure 7.23 A simplified reactance-capability curve of a multimodule TCSC.

vertical line in Fig. 7.23—is considered for all TCSC modules. The typical TCSC data that can be used for stability studies are listed in Tables 7.1 and 7.2. All reactances are expressed in per units on X_C; all voltages, in per units on $IL_{rated} \cdot X_C$; and all currents, in amps.

In the capacitive region, the different TCSC reactance constraints are caused by the following:

1. The limit on the TCSC firing angle, represented by constant reactance limit $X_{max\,0}$.
2. The limit on the TCSC voltage VC_{tran}. The corresponding reactance constraint is give by

TABLE 7.1 Typical TCSC Data for Stability Studies (TCSC with Capacitive-Vernier Capability Only)

Input Data	Description	Units	Typical Value
kV_{rated}	Rated rms line–line voltage	kilovolts	—[a]
IL_{rated}	Rated rms line current	amperes	—[a]
X_C	Nominal TCSC reactance	ohms	—[a]
X_{fixed}	Fixed compensation	ohms	—[a]
T_{TCSC}	Time constant associated with the TCSC firing control	seconds	0.015
X_{bypass}	TCSC reactance for the bypass mode	pu of X_C	–0.15
$X_{max\,0}$	Upper limit of TCSC reactance	pu of X_C	3
VC_{tran}	Maximum voltage of the capacitive-vernier region for the transient overload	pu of $IL_{rated} \cdot X_C$	2
IL_{tran}	Maximum transient line current	pu of IL_{rated}	2
T_{delay}	Protective-bypass recovery delay	seconds	0.025

[a]— indicates that the value is dependent on a specific application.

TABLE 7.2 Typical TCSC Data for Stability Studies (TCSC with Inductive-Vernier Capability Only)

Input Data	Description	Units	Typical Value
$X_{\min 0}$	Lower limit of TCSC reactance	pu of X_C	−2
VL_{tran}	Maximum voltage of the inductive-vernier region for the transient overload	pu of $IL_{\text{rated}} \cdot X_C$	−0.6
IL_{tran}	Maximum line current for full-conduction operation for inductive-region transient overload	pu of $IL_{\text{rated}} \cdot X_C$	2

$$X_{\max\ VC} = (VC_{\text{tran}}) \frac{IL_{\text{rated}}}{I_{\text{line}}} \tag{7.31}$$

3. The limit on the line current (IL_{tran}) beyond which the TCSC transpires into the protective-bypass mode:

$$\begin{aligned} X_{\max\ I_{\text{line}}} &= \infty \qquad \text{for } I_{\text{line}} < IL_{\text{tran}} \cdot IL_{\text{rated}} \\ &= X_{\text{bypass}} \qquad \text{for } I_{\text{line}} > IL_{\text{tran}} \cdot IL_{\text{rated}} \end{aligned} \tag{7.32}$$

The effective capacitive-reactance limit is finally obtained as a minimum of the following limits:

$$X_{\max\ \text{limit}} = \min(X_{\max\ 0}, X_{\max\ VC}, X_{\max\ I_{\text{line}}}) \tag{7.33}$$

In the inductive region, the TCSC operation is restricted by the following limits:

1. The limit on the firing angle, represented by a constant-reactance limit $X_{\min\ 0}$.
2. The harmonics-imposed limit, represented by a constant-TCSC-voltage limit VL_{tran}. The equivalent-reactance constraint is given by

$$X_{\min\ VL} = (VL_{\text{tran}}) \frac{IL_{\text{rated}}}{I_{\text{line}}} \tag{7.34}$$

3. The limit on the fundamental component of current that is permitted to flow through the thyristors in the bypassed-thyristor mode during a transient. This current limit is also expressed as a minimum-reactance limit:

$$X_{\min\ ILT} = \left[1 - \frac{ILT_{\text{tran}} \cdot IL_{\text{rated}} \cdot (1 - X_{\text{bypass}})}{I_{\text{line}}} \right] \tag{7.35}$$

The final inductive-reactance limit in the inductive-vernier operation is obtained as a maximum of the foregoing constraints:

$$X_{\min\ \text{limit}} = \max(X_{\min\ 0}, X_{\min\ VL}, X_{\min\ ILT}) \tag{7.36}$$

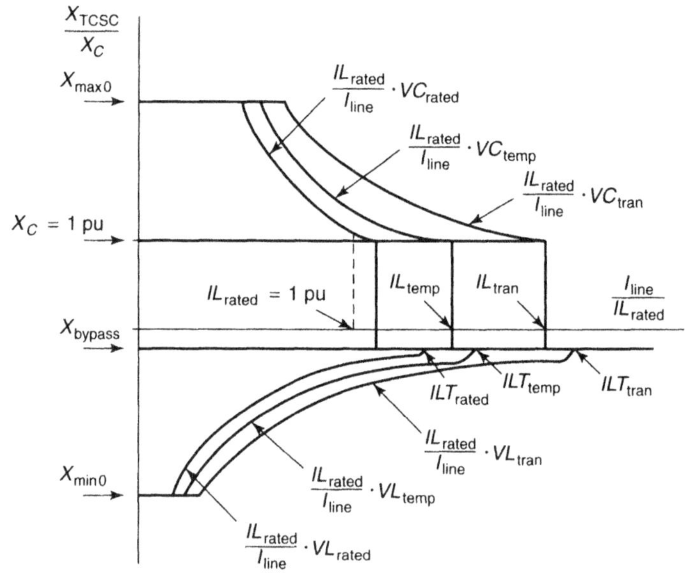

Figure 7.24 The *X-I* capability characteristic of a multimodule TCSC indicating time-dependent overload limits.

If a TCSC is not expected to operate in the inductive-vernier mode, the minimum-reactance limit is X_{bypass} irrespective of the line-current magnitude.

7.10.1.2 Long-Term-Stability Model The capability curves of the TCSC depend on the duration for which the voltage- and current-operating conditions persist on the TCSC. In general, two time-limited regions of TCSC operation exist: the *transient-overload region*, lasting 3–10 s, and the *temporary-overload region*, lasting 30 min; both are followed by the *continuous region*. The overall reactance-versus-line-current (*X-I*) capability curve of the TCSC is depicted in Fig. 7.24, with the relevant data presented in Table 7.3.

For long-term dynamic simulations, an overload-management function needs to be incorporated in the control system. This function keeps track of the TCSC variables and their duration of application, and it also determines the appropriate TCSC overload range, for which it modifies the $X_{\text{max limit}}$ and $X_{\text{min limit}}$. It then applies the same modifications to the controller.

The variable-reactance model does not account for the inherent dependence of TCSC response time on the operating conduction angle. Therefore, entirely incorrect results may be obtained for the high-conduction-angle operation of the TCSC or for whenever the power-swing frequency is high (>2 Hz) [15]. However, the model is used widely in commercial stability programs because of its simplicity, and it is also used for system-planning studies as well as for initial investigations of the effects of the TCSC in damping-power oscillations. A reason for the model's widespread use lies in the assumption that controls

TABLE 7.3 Typical TCSC *X-I* Curve Data for Long-Term-Stability Studies

Input Data	Description	Units	Typical Value
VC_{rated}	Maximum voltage of the capacitive-vernier region for the continuous operation	pu of $IL_{rated} \cdot X_C$	1.15
VC_{temp}	Maximum voltage of the capacitive-vernier region for the temporary overload	pu of $IL_{rated} \cdot X_C$	1.5
IL_{temp}	Maxium temporary line current	pu of IL_{rated}	1.35
VL_{rated}	Maximum voltage of the inductive-vernier region for continuous operation	pu of $IL_{rated} \cdot X_C$	−0.3
VL_{temp}	Maximum voltage of the inductive-vernier region for the temporary overload	pu of $IL_{rated} \cdot X_C$	−0.45
ILT_{rated}	Rated line current for full-conduction operation in the inductive region	pu of IL_{rated}	1.35
ILT_{temp}	Maximum line current for full-conduction operation for the inductive-region temporary overload	pu of IL_{rated}	1.35
T_{tran}	Maximum time for transient overload condition	seconds	3–10
T_{temp}	Maximum time for temporary overload condition	seconds	1800
$Reset_{tran}$	Reset for the transient-overload capability	multiple of T_{tran}	2
$Reset_{temp}$	Reset for the temporary-overload capability	multiple of T_{temp}	2

designed to compensate the TCSC response delay are always embedded in the control system by the manufacturer and are therefore ideal. Hence the response predicted by the model is a true replica of actual performance. In situations where this assumption is not satisfied, a more detailed stability model is required that accurately represents the inherent slow response of the TCSC.

7.10.2 An Advanced Transient-Stability Studies Model

An alternate TCSC model for transient-stability studies has been developed [15], [20] that effectively solves the differential equations pertaining to the TCSC capacitor and the TCR. The TCSC model is invoked at every half-cycle of the line current. A variable is used to store the instantaneous capacitor voltage at the line zero crossing—at the end of each half-cycle—to be used as the initial condition for the next sample process. The TCR is represented by a current source updated by the fundamental component of TCR current that the model calculates at each half-cycle. Also, the model incorporates the effects of both thyristor firing and synchronization. The triggering instant is a function of the signal that is used for synchronization, such as the TCSC voltage or line current.

The model is compatible with conventional transient-stability programs in that it updates the capacitor voltage at every half-cycle while the stability program updates the line current with the same frequency. It is also flexible enough to integrate not only controls for minimizing the TCSC-response delay but higher-order controls as well. Although slightly complex, the model correlated closely with EMTP results of TCSC performance.

7.10.2.1 TCSC Controller Optimization and TCSC Response-Time Compensation The fourth-order transfer function of Eq. (7.28), representing the open-loop response of the TCSC, is incorporated in the closed-loop model of the TCSC to optimize the controller parameters [12], [14]. An alternate method for TCSC modeling, one that is based on Poincare mapping [21], is proposed in ref. [14] in which the Jacobian of the Poincare map is used to calculate the open-loop dynamic response of the Kayenta TCSC. Figure 7.20 illustrates the response of the line current to identical step change in the conduction angle at different operating points. The Jacobian-based results correlate closely with those obtained from a detailed EMTP simulation demonstrating the efficacy of the alternate method.

A closed-loop, conduction-angle-control scheme is proposed in ref. [14] to minimize the sensitivity of TCSC response to the operating point, as shown in Fig. 7.25. The desired conduction angle is denoted by σ_{req}, whereas the measured conduction angle is denoted by σ_m. The difference between σ_{req} and σ_m is fed to a proportional–integral (PI) controller that attempts to reduce this difference to zero in the steady state. The measured σ_m is updated twice in a given time period, whereas the desired conduction signal can vary continuously in response to higher-level controllers.

A closed-loop response that is fairly independent of the operating point is obtained with the proportional-controller gain, K_p, and the integral-controller gain, K_i, as follows:

$$K_p = -10 \exp\left[-\sqrt{\frac{65 - \sigma_{\text{req}}}{2}}\right] \tag{7.37}$$

$$K_i = -23 - 24 \exp\lfloor(\sigma_{\text{req}} - 65)/2\rfloor \tag{7.38}$$

for $\sigma < 64°$, which typically results in $1 \text{ pu} < X_{\text{TCSC}} < 3 \text{ pu}$.

The performance of the closed-loop, conduction-angle controller for similar step changes in the conduction angle, as in the open-loop response shown in Fig. 7.20, is illustrated in Fig. 7.26 for synchronization with line current. Except for very high conduction angles, the closed-loop TCSC response is almost independent of the operating point in the capacitive region. Furthermore, the re-

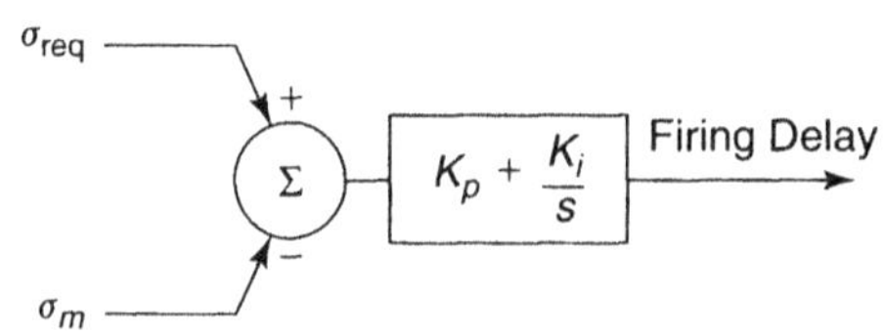

Figure 7.25 The TCSC closed-loop, conduction-angle controller.

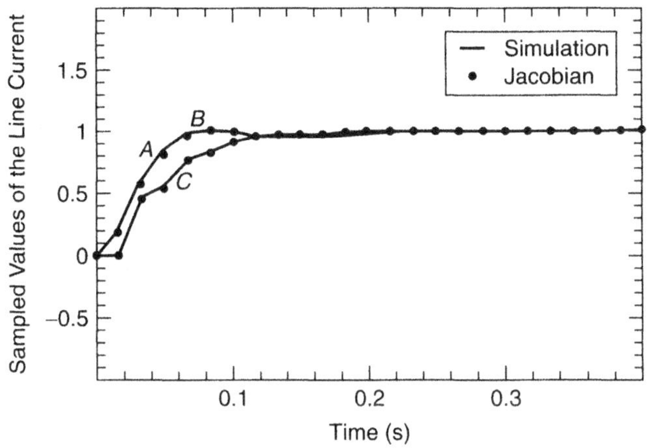

Figure 7.26 Response of the closed-loop, conduction-angle controller.

sponse is much less oscillatory; it settles in half the time taken in the open-loop case.

7.10.3 Discrete and Phasor Models

A linearized discrete-time model of the TCSC-compensated line is proposed in ref. [22] that uses samples at the peaks of capacitor voltage. Even though the discrete models are accurate, their derivative is somewhat complicated and may not easily interface with the phasor-based models of other power-system devices, such as generator and transmission lines. Simple continuous-time, large signal dynamic models for TCSC are proposed in ref. [23] that are based on the representation of voltages and currents as of the time-varying Fourier series type and use the dynamics of short term Fourier coefficients, with certain simplifications. These phasor models integrate well with the standard phasor models of power-system components, and they are almost of order of magnitude faster than that of any detailed time-domain simulation.

7.10.4 Modeling for Subsynchronous Resonance (SSR) Studies

Although EMTP models of the TCSC are highly accurate and used widely for SSR mitigation studies, these models cannot be used for small-signal-stability analysis and are too time-consuming for optimization of TCSC controller parameters.

For SSR studies, the variable-reactance model discussed previously does not provide accurate results unless the torsional oscillations occur at frequencies substantially lower than the synchronous frequency. The variable-reactance model is also inadequate because it does not represent the passive damping associated with the "open-loop" operation of the TCSC [24], [25], that is, when the

TCSC operates at a specified conduction angle irrespective of the generator rotor-angle oscillations. Passive damping, which varies with the steady-state conduction angle of the TCSC, has a substantial influence on the dynamics of the power system. Closed-loop operation with very low feedback gains is also assumed to resemble open-loop operation.

Poincare map linearization is used in ref. [24] to calculate the passive damping of the TCSC and its highly substantial effect on modal dampings of the subsynchronous torsional modes. A linearized TCSC model based on the same technique has been developed in ref. [26] for SSR studies. This model includes thyristor-triggering logic, the synchronization system, and such higher-level controls as power-oscillation damping loop, and it also effectively represents the passive damping of the TCSC controller as well as the active damping offered by the higher-order controls.

In the Poincare map–based approaches, the inputs are assumed to remain constant over one half-cycle period. By remaining constant, the inputs limit the TCSC model–effective bandwidth to 0–120 Hz, a range that is still adequate for SSR studies because the torsional modes lie within the 0–60 Hz range.

The overall modeling approach presented in ref. [26] is to develop a sampled data model of the TCSC and derive a continuous-time/linear-time–invariant model that correlates with the discrete model. This model is then used for the study of subsynchronous oscillations.

Independent modular modeling of the TCSC and the network is feasible about an operating point if the steady-state solution of the TCSC-compensated system can be approximated by nominal quantities in the d–q reference frame. This approximation is possible only if the TCSC constitutes a small percent of the total series compensation of the line [25], which in general is true for existing TCSC installations. However, modular modeling is not valid in systems where a strong harmonic interaction exists between the TCSC and the network [24]—a situation that usually happens when the entire series compensation is provided by the TCSC.

A modular small-signal TCSC model is derived in ref. [25] by linearizing the half-cycle-period Poincare map associated with sampling the TCSC capacitor voltage twice at every cycle. A modular, dynamic-phasor model of the TCSC has been developed for SSR studies in ref. [27]; it is used as an alternative to the sampled data–based models. The phasor TCSC model is efficient and can be integrated readily with phasor-based representations of other power-system components.

7.11 SUMMARY

This chapter described the basic operating principles of the (TCSCs). The capability characteristics of TCSCs we presented, as was a discussion on the harmonics and losses associated with the TCSCs. Different models of varying complexities were described for various power-system studies.

REFERENCES

[1] E. W. Kimbark, "Improvement of System Stability by Switched Series Capacitors," *IEEE Transactions on Power Apparatus and Systems*, Vol. 85, February 1966, pp. 180–188.

[2] J. J. Vithyathil, C. W. Taylor, M. Klinger, and W. A. Mittelstadt, "Case Studies of Conventional and Novel Methods of Reactive Power Control on AC Transmission System," CIGRE Paper 38-02, Paris, 1988.

[3] K. R. Padiyar, *Analysis of Subsynchronous Resonance in Power Systems*, Kluwer Academic Publishers, Boston, 1999.

[4] R. M. Maliszewski, B. M. Pasternack, H. N. Scherer, M. Chamia, H. Frank, and L. Paulsson, "Power Flow Control in a Highly Integrated Transmission Network," CIGRE Paper 37-303, Paris, 1990.

[5] N. Christl, P. Luelzberger, M. Pereira, K. Sadek, P. E. Krause, A. H. Montoya, D. R. Torgerson, and B. A. Vossler, "Advanced Series Compensation with Variable Impedance," *Proceedings of EPRI Conference on FACTS*, 1990.

[6] E. V. Larsen, K. Clark, S. A. Miske Jr., and J. Urbanek, "Characteristics and Rating Considerations of Thyristor-Controller Series Compensation," *IEEE Transactions on Power Delivery*, Vol. 9, No. 2, April 1994.

[7] N. G. Hingorani and L. Gyugyi, *Understanding FACTS: Concepts and Technology of Flexible AC Transmission Systems*, IEEE Press, New York, 1999.

[8] Y. H. Song and A. T. Johns, Eds., *Flexible AC Transmission Systems (FACTS)*, IEE Press, London, 1999.

[9] F. Unterlab, S. Weib, and K. Renz, "Control and Protection of Advanced Series Compensation," Paper 92-SP-179, *CEA Transactions of Engineering and Operating Division*, Vol. 31, Pt. 4, 1992.

[10] N. Christl, R. Hedin, R. Johnson, P. Krause, and A. Montoya, "Power System Studies and Modeling for the Kayenta 230 kV Substation Advanced Series Compensation," *IEE Fifth International Conference on AC and DC Power Transmission*, London, September 1991.

[11] S. G. Jalali and R. H. Lasseter, "Harmonic Instabilities in Advanced Series Copensators," *Proceedings of EPRI Conference on FACTS (FACTS 2)*, Boston, 1992.

[12] N. Christl, R. Hedin, K. Sadek, P. Lutzelberger, P. E. Krause, S. M. McKenna, A. H. Montoya, and D. Torgerson, "Advanced Series Compensation (ASC) With Thyristor-Controlled Impedance," Paper 14/37/38-05, CIGRE, Paris, 1992.

[13] S. G. Helbing and G. G. Karady, "Investigations of an Advanced Form of Series Compensation," *IEEE Transactions on Power Delivery*, Vol. 9, No. 2, April 1994, pp. 939–947.

[14] S. G. Jalali, R. H. Lasseter, and I. Dobson, "Dynamic Response of a Thyristor-Controlled Switched Capacitor," *IEEE Transactions on Power Delivery*, Vol. 9, No. 3, July 1994, pp. 1609–1615.

[15] S. G. Jalali, R. A. Hedin, M. Pereira, and K. Sadek, "A Stability Model for Advanced Series Compensator (ASC)," *IEEE Transactions on Power Delivery*, Vol. 11, No. 2, April 1996, pp. 1128–1133.

[16] J. Urbanek, R. J. Piwko, E. V. Larsen, B. L. Damsky, B. C. Furumasu, and W. Mittelstadt, "Thyristor-Controlled Series Compensation Prototype Installation at

the Slatt 500 kV Substation," *IEEE Transactions on Power Delivery*, Vol. 8, No. 3, July 1993, pp. 1460–1469.

[17] J. J. Paserba, N. W. Miller, E. V. Larsen, and R. J. Piwko, "A Thyristor-Controlled Series Compensation Model for Power System Stability Analysis," *IEEE Transactions on Power Delivery*, Vol. 10, No. 3, July 1995, pp. 1471–1478.

[18] CIGRE Working Group 14.29, "Coordination of Controls of Multiple FACTS/HVDC Links in the Same System," CIGRE Technical Brochure No. 149, Paris, December 1999.

[19] S. Nyati, C. A. Wegner, R. W. Delmerico, D. H. Baker, R. J. Piwko, and A. Edris, "Effectiveness of Thyristor-Controlled Series Capacitor in Enhancing Power System Dynamics: An Analog Simulator Study," *IEEE Transactions on Power Delivery*, Vol. 9, No. 2, April 1994, pp. 1018–1027.

[20] M. Pereira, K. Sadek, and S. G. Jalali, "Advanced Series Compensation (ASC) Model for Stability Programs," *CIGRE International Colloquium on HVDC and FACTS*, Montreal, September 18–19, 1995.

[21] J. M. T. Thompson and H. B. Steward, "Nonlinear Dynamics and Chaos: Geometrical Methods for Scientists and Engineers," John Wiley and Sons, London, 1987.

[22] A. Ghosh and G. Ledwich, "Modelling and Control of Thyristor-Controlled Series Compensators," *IEEE Proceedings on Generation, Transmission, and Distribution*, Vol. 142, No. 3, May 1995, pp. 297–304.

[23] P. Mattavelli, G. C. Verghese, and A. M. Stankovic, "Phasor Dynamics of Thyristor-Controlled Series Capacitor Systems," Paper 96 SM 532-2 PWRS, Presented at the IEEE/PES Summer Meeting, Denver, CO, July–August 1996.

[24] R. Rajaraman, I. Dobson, R. Lasseter, and Y. Shern, "Computing the Damping of Subsynchronous Oscillations Due To a Thyristor-Controlled Series Capacitor," *IEEE Transactions on Power Delivery*, Vol. 11, April 1996, pp. 1120–1127.

[25] B. K. Perkins and M. R. Iravani, "Dynamic Modelling of a TCSC with Application to SSR Analysis," *IEEE Transactions on Power Systems*, Vol. 12, No. 4, November 1997, pp. 1619–1625.

[26] H. A. Othman and L. Angquist, "Analysis Modelling of Thyristor-Controlled Series Capacitors for SSR Studies," *IEEE Transactions on Power Systems*, Vol. 11, No. 1, February 1996, pp. 119–127.

[27] P. Mattavelli, A. M. Stankovic, and G. C. Verghese, "SSR Analysis with Dynamic Phasor Model of Thyristor-Controlled Series Capacitor," *IEEE Transactions on Power Systems*, Vol. 14, No. 1, February 1999, pp. 200–208.

CHAPTER 8

TCSC Applications

8.1 INTRODUCTION

Thyristor-controlled series capacitors (TCSCs) can be used for several power-system performance enhancements, namely, the improvement in system stability, the damping of power oscillations, the alleviation of subsynchronous resonance (SSR), and the prevention of voltage collapse. The effectiveness of TCSC controllers is dependent largely on their proper placement within the carefully selected control signals for achieving different functions. Although TCSCs operate in highly nonlinear power-system environments, linear-control techniques are used extensively for the design of TCSC controllers.

In addition to discussing the foregoing TCSC uses, this chapter describes the performance of two recent TCSC installations: one in Sweden, the other in Brazil.

8.2 OPEN-LOOP CONTROL

The open-loop impedance control, is the most basic type of TCSC control, used primarily for power-flow control. The control system is depicted in Fig. 8.1. The desired steady-state level of series compensation or line power flow is expressed in the form of a reactance reference applied to the controller. The controller is modeled by a delay block that represents TCSC action; its magnitude is typically chosen as 15 ms [1]. The controller outputs a reactance-order signal that is linearized to obtain the necessary firing angle; then, the firing-angle signal is transmitted to the firing-pulse generator, which issues the gating pulses for the TCSC thyristors to implement the desired reactance.

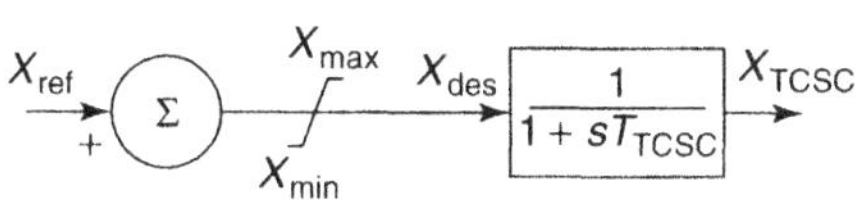

Figure 8.1 A TCSC constant-impedance controller.

8.3 CLOSED-LOOP CONTROL

8.3.1 Constant-Current (CC) Control

In the constant-current (CC) control, the desired line-current magnitude is fed as a reference signal to the TCSC controller, which strives to maintain the actual line current at this value. A typical TCSC CC-controller model [2] is depicted in Fig. 8.2. The 3-phase current is measured and rectified in the measurement unit. The rectified signal is passed through a filter block that comprises a 60-Hz and a 120-Hz notch filter as well as a high-pass filter. The emanating signal is then normalized to ensure per-unit consistency with the reference-current signal.

The controller is typically of the proportional–integral (PI) type that outputs the desired susceptance signal within the preset limits, as discussed in Chapter 4. A linearizer block converts the susceptance signal into a firing-angle signal.

An operation-mode selector unit is generally used for TCSC protection. During short-circuit conditions, at which time the current through the metal-oxide varistor (MOV) exceeds a threshold, the TCSC is made to switch to the bypassed-thyristor mode or the thyristor-switched reactor (TSR) mode. In this mode, the thyristors conduct fully ($\sigma = 180°$), reducing both the TCSC voltage and the current substantially and thereby reducing the stress on the MOV. During the clearance of faults, the "waiting mode" is implemented; when the capacitors are brought back into the circuit, a dc-voltage offset builds up that is discharged into this waiting mode.

The steady-state control characteristic of the CC control on the $V_{\text{TCSC}} - I_L$ (the TCSC voltage–line current) plane is depicted in Fig. 8.3(a). The convention used in the figure is to treat the capacitor voltage as positive (which is opposite of the convention used in load flow). The characteristic is divided into three segments: OA, AB, and BC. Segments OA and BC represent the maximum and

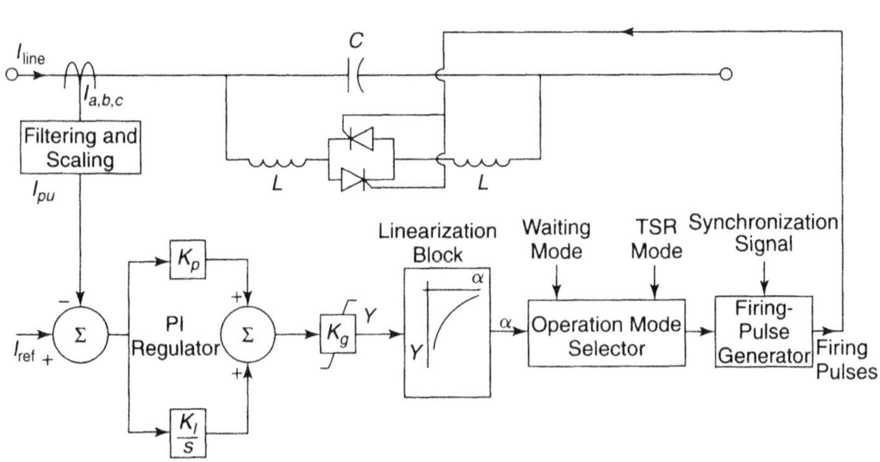

Figure 8.2 A TCSC constant-current (CC) controller model.

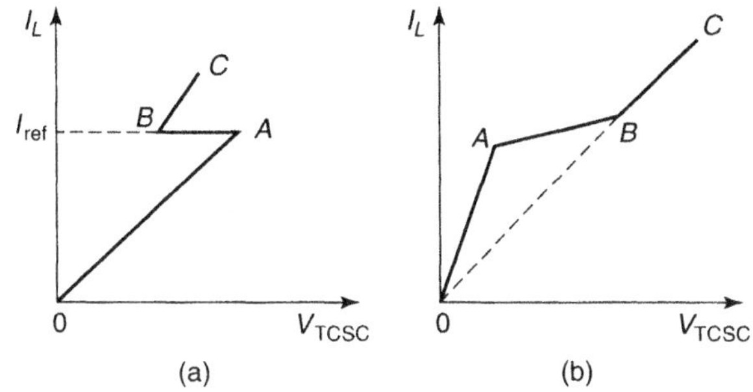

Figure 8.3 The TCSC control characteristics: (a) CC control and (b) CA control.

minimum TCSC reactance limits, respectively. Segment *AB* represents the control range in which the TCSC reactance is varied through the firing control to maintain a specified line current, I_{ref}.

8.3.2 Constant-Angle (CA) Control

This control is useful and relevant for situations in which transmission paths exist in parallel with the TCSC-compensated line. The control objective during transient or contingency situations is to keep the power flow unchanged in the parallel paths while allowing variations in the power transmitted across the TCSC-compensated line. To keep the power flow constant in shunt paths necessitates maintaining the angular difference constant across the lines, thus imparting the name *constant-angle control*, or CA control, to this strategy. If the voltage magnitudes at the two line ends are assumed to be regulated, then maintaining a constant angular difference implies maintaining a constant-voltage drop, V_L, across the line.

The control objective while neglecting the line resistance is expressed as [3]

$$V_L = I_L X_{LR} - V_{\text{TCSC}} = K = V_{L\,\text{ref}} \tag{8.1}$$

or

$$I_L = \frac{1}{X_{LR}}\,(K + V_{\text{TCSC}}) \tag{8.2}$$

or

$$I_{\text{ref}} - \left(I_L - \frac{V_{\text{TCSC}}}{X_{LR}}\right) = 0 \tag{8.3}$$

where K = the constant = $V_{L\,\text{ref}}$

I_L = the magnitude of current in the TCSC-compensated line

X_{LR} = the net line reactance, including the effect of the fixed capacitor $= X_L - X_{FC}$

V_{TCSC} = the voltage across the TCSC (positive for capacitive voltage; negative for inductive voltage)

I_{ref} = the reference current $= K/X_{LR} = V_{L\,\text{ref}}/X_{LR}$

X_L = the total-line inductive reactance

X_{FC} = the reactance of the fixed capacitor in the line (if any)

As mentioned previously, TCSCs are usually employed is conjunction with fixed capacitors for minimizing cost and improving control efficacy.

The control characteristic for CA control in the $V_{\text{TCSC}} - I_L$ plane is depicted in Fig. 8.3(b). Line segment AB represents the control range, having a slope X_{LR}. Segments OA and BC represent the minimum and maximum TCSC reactance limits, respectively. The CA control is highly effective in reducing power swings.

The TCSC control-system block diagram, incorporating features of both CC and CA control, is shown in Fig. 8.4. In this figure, T_m indicates the time constant associated with the measurement circuit, which is generally a first-order low-pass filter. Similar measurement circuits are assumed for both TCSC-voltage and line-current measurements. In CC control, the multiplier block S is set to zero, whereas in CA control, S is assigned the value $1/X_{LR}$. The regulator is primarily a PI controller that is occasionally in cascade with a phase-lead circuit, as shown in Fig. 8.5. If pure-integral action is required, K_P is set to zero.

For CC control, the integral gain K_I is considered positive. In this control scheme, a positive current-error signal implies that the TCSC capacitive reactance must be increased to enhance the line current and thereby reduce the error signal, and in the CA control, the gain K_I is treated negative. If the current-error signal is positive, it is noted from Eqs. (8.1)–(8.3) that the net voltage drop V_L in the line is less than the reference $V_{L\,\text{ref}}$, necessitating a decrease in

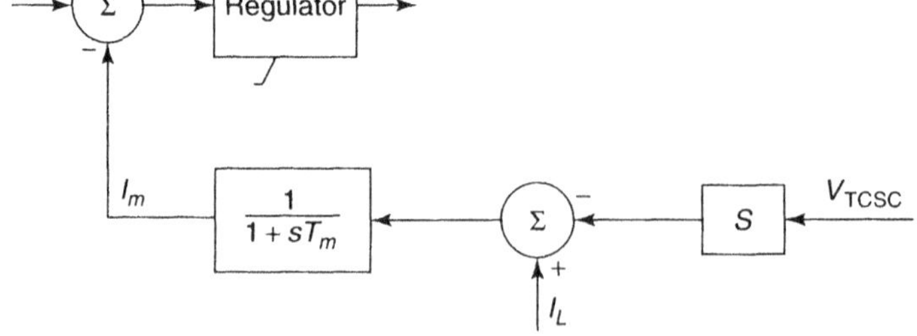

Figure 8.4 A block diagram of a CC or CA controller.

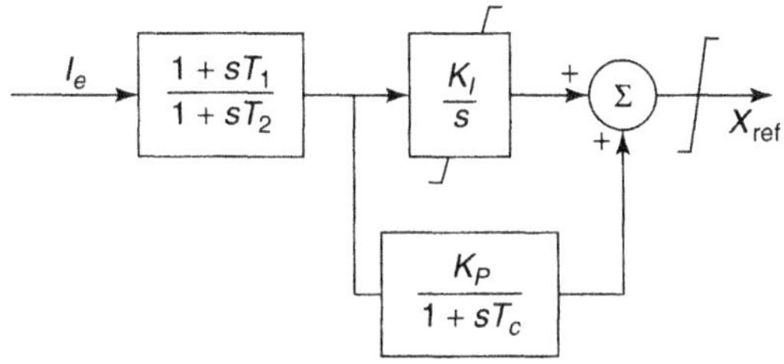

Figure 8.5 A block diagram of the regulator.

the TCSC voltage V_{TCSC} and consequently in the TCSC reactance X_{TCSC} (or X_{ref}). For this reason, K_I is assigned a negative sign in CA control.

Although the TCSC firing delays are modeled by a single time constant of 15 ms, they may be ignored in electromechanical-stability studies [1], [4], as their effect is insignificant.

An elaborate case study for a single-machine infinite-bus (SMIB) system that depicts the influence of constant line-power control and CA control of the TCSC is presented in ref. [5]. It is demonstrated that the CC and CA strategies are suitable only in SMIB systems having two or more parallel transmission paths. In case a critical contingency causes an outage of the parallel transmission paths, the line-power scheduling, controller must be disabled. If, however, the TCSC is equipped with additional damping controllers, the paths should be retained to ensure satisfactory damping levels during an outage.

8.3.3 Enhanced Current Control

To improve the damping of certain oscillatory modes, such as subsynchronous oscillations, an optimized, derivative line-current feedback is embedded in the TCSC controller, as depicted in Fig. 8.6 [2], [6]. In this control system, the voltage regulator is a simple PI controller slightly different from the one depicted in Fig. 8.2. The optimized current controller is shown to successfully damp subsynchronous oscillations for all levels of line-series compensation [2], unlike a conventional controller, which provides very low damping to an SSR mode.

8.3.4 Constant Power Control

The block diagram of a typical TCSC power controller is depicted in Fig. 8.7 [2], [6]. The line power flow is computed from the measured local voltage and current signals after the *abc*–$\alpha\beta 0$ transformation. The calculated power signal is converted into a per-unit quantity and filtered, then fed to the summing junction of the power controller. The reference signal, P_{ref}, denotes the desired level of real-power flow in the TCSC-compensated line, and the power controller has a PI structure. The remaining control-system components were described previously.

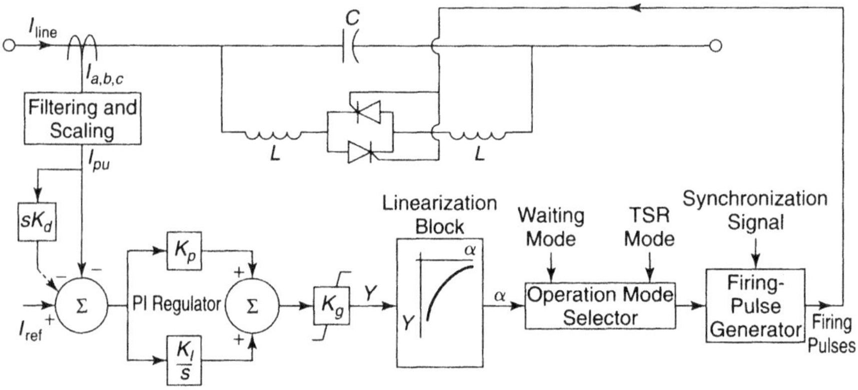

Figure 8.6 An optimized TCSC CC structure.

The TCSC power controllers are usually effective if used as slow controllers for damping power oscillations or subsynchronous oscillations. An attempt to increase the controllers' speed by reducing the power-controller time constant, T_p, renders the response oscillatory. Usually, T_p is set to 100 ms.

8.3.5 Enhanced Power Control

The need to keep TCSC power controllers slow is potentially detrimental to the power system, as it extends the post–fault system recovery period. A much-

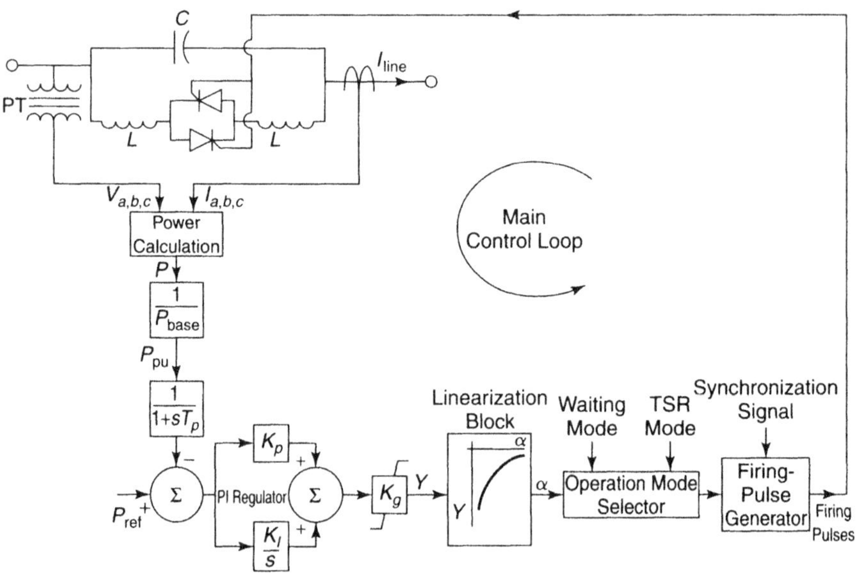

Figure 8.7 A conventional TCSC power-control structure.

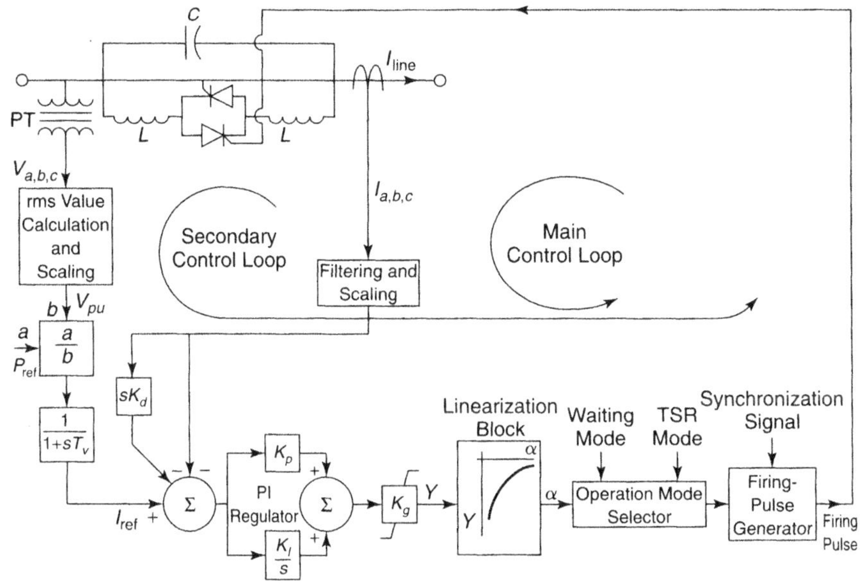

Figure 8.8 An enhanced TCSC power-control structure.

improved TCSC power controller that combines the beneficial influences of both power control and current control is depicted in Fig. 8.8 [2], [6]. It consists of two control loops—a fast, inner-current control loop and a slow, outer-power control loop. The power controller provides the current-reference signal for the current controller. Such a controller allows a fast TCSC response to system faults, yet it also allows a desired slow response to the electromechanical oscillations.

8.3.6 Firing Schemes and Synchronization

An equidistant firing scheme is most commonly employed in TCSC control. However, in some special situations in which the damping of the electrical self-excitation modes is needed, individual firing control is shown to be advantageous [2]. The TCSC firing controls are most effective when they are synchronized with line-current zero crossings primarily because the line current constitutes an almost sinusoidal signal. Synchronization with TCSC-voltage zero crossings may lead to erroneous results, as this voltage is beset with substantial harmonics and spurious noise signals.

8.4 IMPROVEMENT OF THE SYSTEM-STABILITY LIMIT

During the outage of a critical line in a meshed system, a large volume of power tends to flow in parallel transmission paths, which may become severely overloaded. Providing fixed-series compensation on the parallel path to augment the power-transfer capability appears to be a feasible solution, but it may increase the total system losses. Therefore, it is advantageous to install a TCSC in key transmission paths, which can adapt its series-compensation level to the instantaneous system requirements and provide a lower loss alternative to fixed-series compensation.

The series compensation provided by the TCSC can be adjusted rapidly to ensure specified magnitudes of power flow along designated transmission lines. This condition is evident from the TCSC's efficiency, that is, ability to change its power flow as a function of its capacitive-reactance setting:

$$P_{12} = \frac{V_1 V_2}{(X_L - X_C)} \sin \delta \tag{8.3}$$

where P_{12} = the power flow from bus 1 to bus 2

V_1, V_2 = the voltage magnitudes of buses 1 and 2, respectively

X_L = the line-inductive reactance

X_C = the controlled TCSC reactance combined with fixed-series-capacitor reactance

δ = the difference in the voltage angles of buses 1 and 2

This change in transmitted power is further accomplished with minimal influnece on the voltage of interconnecting buses, as it introduces voltage in quadrature. In contrast, the SVC improves power transfer by substantially modifying the interconnecting bus voltage, which may change the power into any connected passive loads. The freedom to locate a TCSC almost anywhere in a line is a significant advantage.

Power-flow control does not necessitate the high-speed operation of power-flow control devices. Hence discrete control through a TSSC may also be adequate in certain situations. However, the TCSC cannot reverse the power flow in a line, unlike HVDC controllers and phase shifters.

8.5 ENHANCEMENT OF SYSTEM DAMPING

The TCSC can be made to vary the series-compensation level dynamically in response to controller-input signals so that the resulting changes in the power flow enhance the system damping [7]–[13]. The power modulation results in a corresponding variation in the torques of the connected synchronous generators—particularly if the generators operate on constant torque and if passive bus loads are not installed [14].

The damping control of a TCSC or any other FACTS controller should generally do the following:

1. stabilize both postdisturbance oscillations and spontaneously growing oscillations during normal operation;
2. obviate the adverse interaction with high-frequency phenomena in power systems, such as network resonances; and
3. preclude local instabilities within the controller bandwidth.

In addition, the damping control should

1. be robust in that it imparts the desired damping over a wide range of system operating conditions, and
2. be reliable.

8.5.1 Principle of Damping

The concept of damping enhancement by line-power modulation can be illustrated with the two-machine system depicted in Fig. 8.9 [14]. The machine SM_1 supplies power to the other machine, SM_2, over a lossless transmission line. Let the speed and rotor angle of machine SM_1 be denoted by η_1 and φ_1, respectively; of machine SM_2, denoted by η_2 and φ_2, respectively.

During a power swing, the machines oscillate at a relative angle $\Delta\varphi$ ($= \varphi_2 - \varphi_1$). If the line power is modulated by the TCSC to create an additional machine torque that is opposite in sign to the derivative of the rotor-angle deviation, the oscillations will get damped. This control strategy translates into the

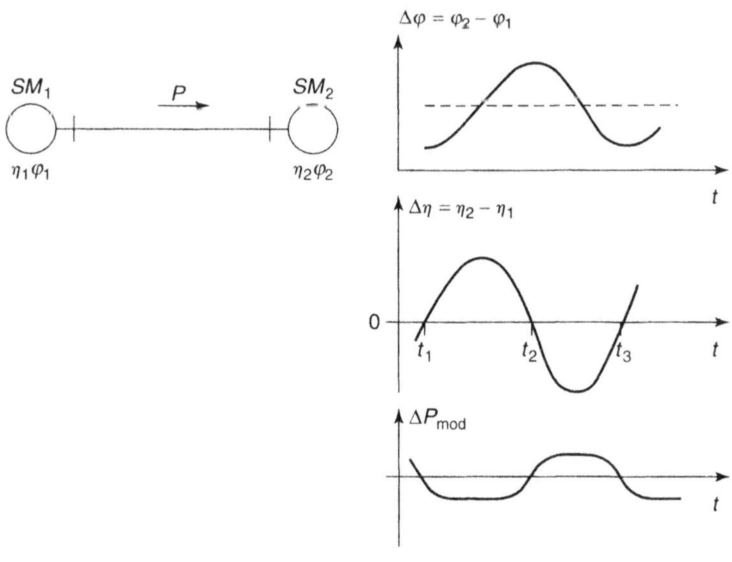

Figure 8.9 The TCSC line-power modulation for damping enhancement.

following actions: When the receiving end–machine speed is lower than the sending end–machine speed, that is, $\Delta\eta$ ($=\eta_2-\eta_1$) is negative, the TCSC should increase power flow in the line. In other words, while the sending-end machine accelerates, the TCSC control should attempt to draw more power from the machine, thereby reducing the kinetic energy responsible for its acceleration. On the other hand, when $\Delta\eta$ is positive, the TCSC must decrease the power transmission in the line. This damping control strategy is depicted in Fig. 8.9 through plots of the relative machine angle $\Delta\varphi$, the relative machine speed $\Delta\eta$, and the incremental power variation ΔP_{mod}. It may be recalled from Chapter 6 that the damping-control action of the SVC is also similar to that described in the preceding text.

The incremental variation of the line-power flow ΔP, given in megawatts (MW), with respect to ΔQ_{TCSC}, given in MVAR, is as follows [14]:

$$\frac{\Delta P}{\Delta Q_{\text{TCSC}}} = \frac{1}{2\tan\delta/2}\left(\frac{I}{I_N}\right)^2 \tag{8.5}$$

where δ = the angular difference between the line-terminal voltages
I = the operating-point steady-state current
I_N = the rated current of the TCSC

Thus the TCSC action is based on the variation of line-current magnitude and is irrespective of its location. Typically, the change in line-power transfer caused by the introduction of the full TCSC is in the range of 1–2, corresponding to an angular difference (δ) of 30°–40° across the line.

The influence of any bus load on the torque/power control of the synchronous generator is derived for the case of a resistive load and completely inductive generator impedance [14]. The ratio of change in generator power to the ratio of change in the power injected from the line into the generator bus is expressed as

$$\frac{\Delta P_m}{\Delta P} = \frac{\cos(\delta/2 \pm \alpha)}{\cos(\delta/2)} \tag{8.6}$$

where the + sign corresponds to the sending end; the − sign, the receiving end. Also,

where ΔP_m = the variation in generator power
ΔP = the variation in power injected from the transmission line into the machine bus
$\alpha = \tan^{-1}(X_{\text{source}}/R_{\text{load}})$ (it is assumed that $R_{\text{load}} \gg X_{\text{source}}$)

The effect of all practical passive loads is generally moderate, and the sign of generator power is not changed. In the absence of any bus load, $\Delta P_m = \Delta P$.

It is not necessary to make the entire series compensation in a line controllable; in fact, the effectiveness of a TCSC is shown to increase in presence of fixed series compensation. The required series compensation in a line is therefore usually split into a fixed-capacitor component and a controllable TCSC component. The controlled-to-fixed ratio of capacitive reactance in most applications is in the 0.05–0.2 range, the exact value determined by the requirements of the specific application.

8.5.2 Bang-Bang Control

Bang-bang control is a discrete control form in which the thyristors are either fully switched on ($\alpha = 90^\circ$) or fully switched off ($\alpha = 180^\circ$). Thus the TCSC alternates between a fixed inductor and a fixed capacitor, respectively, and it is advantageous that such control is used not only for minimizing first swings but for damping any subsequent swings as well. Bang-bang control is employed in face of large disturbances to improve the transient stability [3].

8.5.3 Auxiliary Signals for TCSC Modulation

The supplementary signals that could be employed for modulating TCSC impedance are listed in the text that follows and are discussed in refs. [14]–[16], [8]–[12].

8.5.3.1 Local Signals These signals constitute the following:

1. the line current,
2. the real-power flow,
3. the bus voltage, and
4. the local bus frequency.

8.5.3.2 Remote Signals These signals constitute the following:

1. the rotor-angle/speed deviation of a remote generator,
2. the rotor-angle/speed (frequency) difference across the system, and
3. the real-power flow on adjacent lines.

The angular difference between remote voltages can be synthesized by using local voltages at the two terminals of the TCSC and through the line current [15]. Alternatively, a recent approach [16] may be adopted wherein the phase angles of remote areas can be measured directly by using synchronized phasor-measurement units.

Adjacent-line real-power flow can be measured remotely and transmitted to the TCSC control system through telecommunication. Despite telecommunication delays, this signal can be used satisfactorily and economically for line-power scheduling, which itself is a slow control.

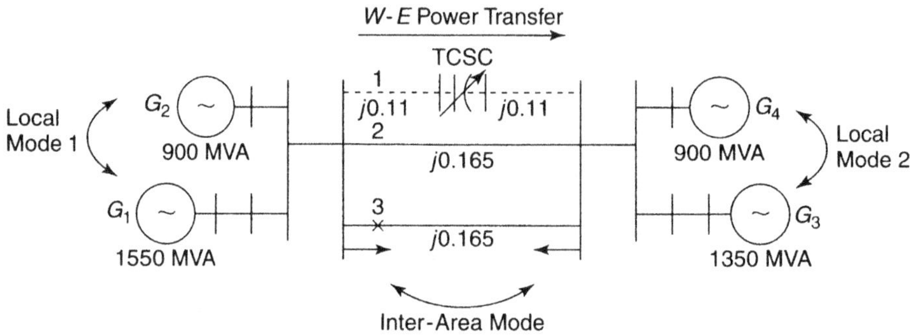

Figure 8.10 A two-area, four-machine system.

Selection of Input Signals It is a desirable feature that the TCSC controller-input signals can extend as far as possible without sensitivity to the TCSC output. This feature ensures that the control signals represent mainly the system conditions for which the TCSC is expected to improve. Local bus frequency is seen to be less responsive to system power swings as compared to the synthesized-voltage frequency [17], although both line current and bus voltage are also shown to be fairly effective.

8.5.4 Case Study for Multimodal Decomposition–Based PSDC Design

A two-area, four-machine system [18] is depicted in Fig. 8.10. Each area is represented by two machines of widely different MVA ratings. Although the basic system data is as it is noted in ref. [18], the MVA ratings and the tie-line impedances have been modified to the values shown in Fig. 8.10. A TCSC is placed in one of the tie-lines.

The system shows two local modes—one in each area—and an inter-area mode, as listed in Table 8.1. In the base system, 600 MW of power is transferred across the three tie-lines, whereas in the contingency situation, the same amount of power is transmitted through tie-lines 1 and 2, with tie-line 3 out of service.

8.5.4.1 Selection of the Measurement Signal The following criteria are adopted for selecting the measurement signal:

TABLE 8.1 Swing-Mode Frequency and Damping

Inter-Area Mode	Local Mode 1	Local Mode 2
	Frequency-and-Damping Ratio	
0.62 Hz, 0.17%	1.15 Hz, 11%	1.02 Hz, 18%
0.52 Hz, –0.13%	1.13 Hz, 10%	1.00 Hz, 18%

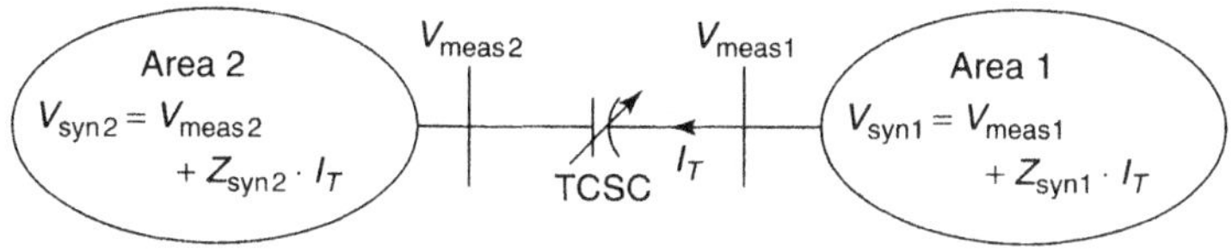

Figure 8.11 Synthesized voltages and the angular difference.

1. it should be derived from local measurements, and
2. it should be sensitive to an inter-area mode but insensitive to local modes.

Moreover, the power-swing damping controller (PSDC) that is based on this signal should not interact adversely with other controllers.

The angular difference of synthesized remote voltages on each side of the TCSC is chosen as the measurement signal. This signal has been shown from early studies [5], [17] to be highly efficient, and its concept can be explained from Fig. 8.11. The voltages on either side of the TCSC, namely, $V_{\text{meas}\,1}$ and $V_{\text{meas}\,2}$, are measured along with TCSC line current, I_T. Using these quantities and the appropriate magnitudes of synthesizing impedances $Z_{\text{syn}\,1}$ and $Z_{\text{syn}\,2}$, the remote voltages behind the synthesizing impedances are computed as follows:

$$V_{\text{syn}\,1} = V_{\text{meas}\,1} + Z_{\text{syn}\,1} I_T \tag{8.7}$$

$$V_{\text{syn}\,2} = V_{\text{meas}\,2} - Z_{\text{syn}\,2} I_T \tag{8.8}$$

The synthesized angular difference is subsequently obtained as

$$\theta_{12} = \angle V_{\text{syn}\,1} - \angle V_{\text{syn}\,2} \tag{8.9}$$

It is evident that θ_{12} represents the angular difference between the centers of the machines in the two areas. This signal comprises the inter-area mode and, to a very small extent, the local modes.

8.5.4.2 Selection of the Synthesizing Impedance In the case of a single-generator infinite-bus system considered in Example 5.2 of Chapter 5 (Section 5.3.4), the synthesized impedance was simply the transmission-line reactance added to the generator subtransient reactance. However, in a multimachine environment within an area, it is not straightforward to obtain the synthesizing impedances from the FACTS device terminal to the center of the equivalent machine in that area. An initial estimate is the driving-point impedance or the short-circuit impedance seen in both directions. The appropriate values of the synthesizing impedances [Eqs. (8.10) and (8.11)] are determined from a detailed analysis based on the damping-controller design indices: the maximum damping influence (MDI) and natural phase influence (NPI), as described in Appendix C.

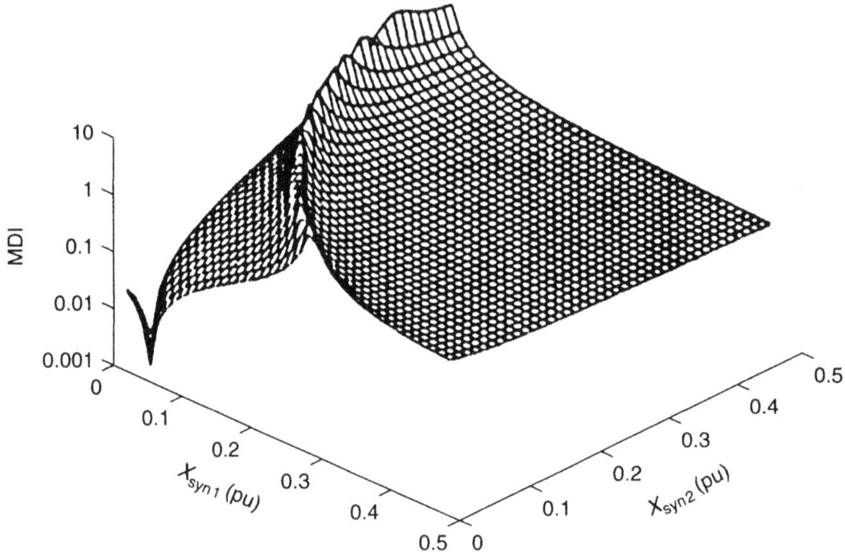

Figure 8.12 The MDI indices versus the synthesizing reactances for the inter-area mode of the base system.

$$Z_{\text{syn}\,1} = R_{\text{syn}\,1} + jX_{\text{syn}\,1} \tag{8.10}$$

$$Z_{\text{syn}\,2} = R_{\text{syn}\,2} + jX_{\text{syn}\,2} \tag{8.11}$$

The resistances are set at 10% of the respective reactances according to the typical X/R ratios in the transmission lines.

A three-dimensional plot of the MDI on the z-axis and $X_{\text{syn}\,1}$ and $X_{\text{syn}\,2}$ on the x- and y-axes, respectively, is depicted in Fig. 8.12. It is seen that a high MDI is obtained when the magnitudes of $X_{\text{syn}\,1}$ and $X_{\text{syn}\,2}$ are inversely related, that is, if one is small, the other should be large, and vice verse.

A similar three-dimensional plot of the NPI index on the z-axis over the $X_{\text{syn}\,1}$–$X_{\text{syn}\,2}$ plane is illustrated in Fig. 8.13. It has a ridge similar to that in the MDI plot. Desirable NPIs of -90° ($= 270^\circ$) result in a high synchronizing torque and are obtained from small values of the synthesizing reactance; undesirable NPIs of $+90^\circ$ result in a deterioration in the synchronizing torque and are obtained from large magnitudes of synthesizing reactance.

Because the designed PSDC controller must be robust with respect to the operating conditions, the $X_{\text{syn}\,1} - X_{\text{syn}\,2}$ regions of high MDI (MDI > 1) are plotted in Fig. 8.14 for both base and contingency systems. The high MDI region of the contingency system is wider from the removal of a parallel tie-line and also indicates that TCSC becomes more effective in damping power swings from a wider range of $X_{\text{syn}\,1}$ and $X_{\text{syn}\,2}$.

In Fig. 8.14, point A has an MDI of 0.3 and an NPI of -90°, which increases

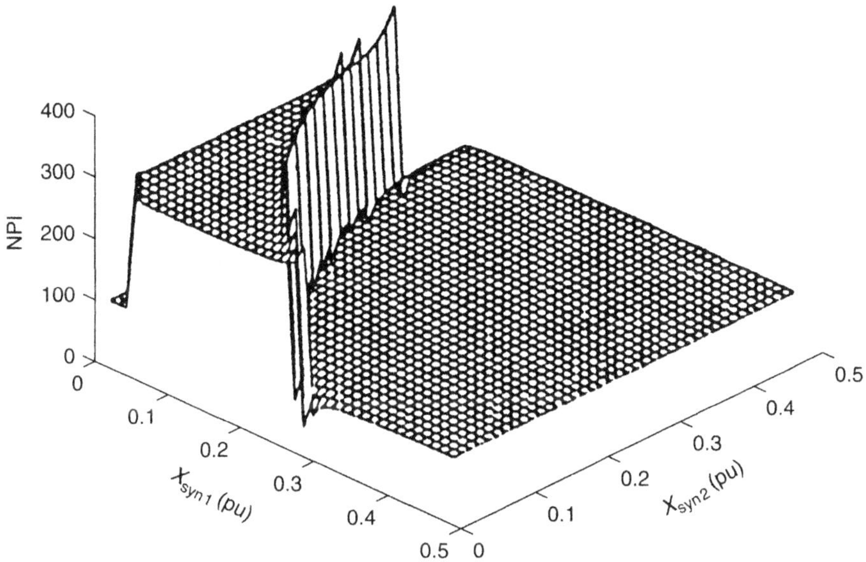

Figure 8.13 The NPI indices versus the synthesizing reactances for the inter-area mode of the base system.

the synchronizing torque, and point C has an MDI of 0.3 but an NPI of $+90^\circ$, (which decreases the synchronizing torque. Point B, on the other hand, gives the highest MDI. For the inter-area mode, the controller phase index (CPI) is obtained as -90°. The PSDC is desigend to have the following components:

1. A filtered derivative stage to convert the synthesized angular difference into a synthesized angular frequency-difference signal having a transfer function of

$$\left(\frac{s}{1+0.04s}\right)\left(\frac{1}{1+0.04s}\right) \tag{8.12}$$

2. A washout stage of $s/(1+s)$.
3. A gain K_s.

The overall PSDC is given by

$$K_{\text{PSDC}}(s) = -K_s \frac{1}{(1+0.04s)} \frac{s}{(1+s)} \frac{s}{(1+0.04s)} \tag{8.13}$$

The root loci of the inter-area mode [19], with varying PSDC gain K_s for the three choices of synthesizing reactances, is depicted in Fig. 8.15. For synthesis choices A and C, the root-loci curves approach their respective zeroes as the

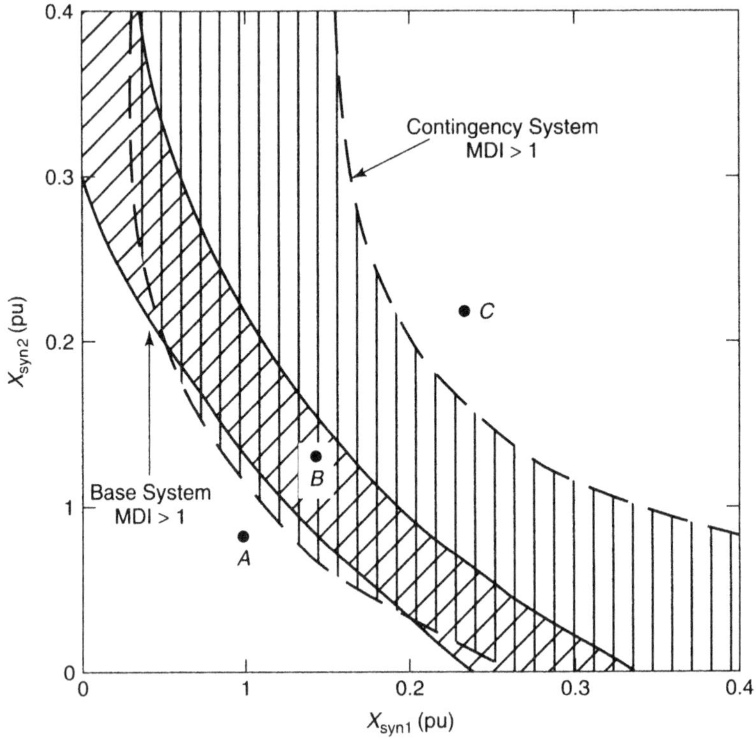

Figure 8.14 "Good" MDI regions for base and contingency cases.

controller gain is increased. Choice A leads to an increase in frequency, whereas choice C leads to a decrease in frequency (the damping improvement in both cases is limited to 0.3). These behaviors are to be expected from the NPI values; however, choice B generates substantial damping and is obviously the favored choice.

In a study given in ref. [20], a comparative evaluation of PSDC control signals shows that TCSC bus voltage performs better than the line power signal in the damping inter-area mode in a large power system. It may be understood that such results are system-dependent; they cannot be generalized. The synthesized angular difference was not considered in this study.

8.5.5 H_∞ Method–Based PSDC Design

The H_∞ method is an alternative technique for designing a robust controller that can operate satisfactorily over a wide range of operating conditions, even in the presence of uncertainties caused by system disturbances and errors in instrumentation. It is a frequency-domain method [21], [22], which necessarily requires an accurate, reduced dynamic model of the power system. The mini-

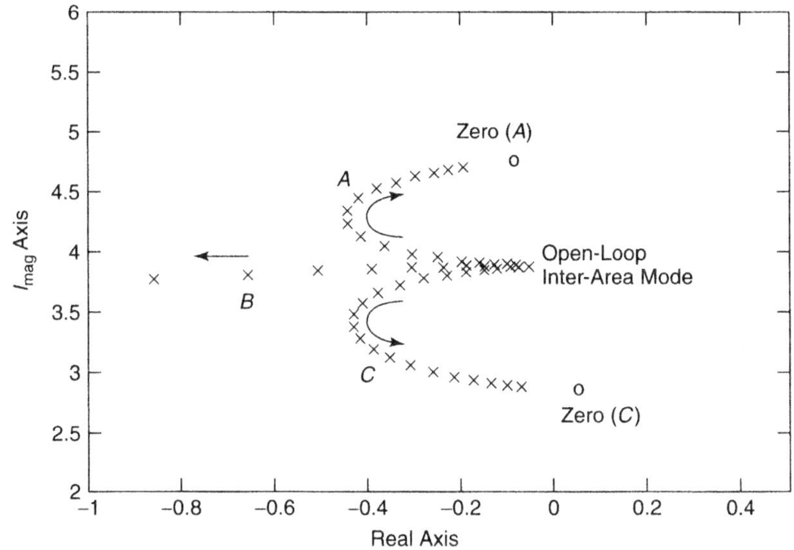

Figure 8.15 Root loci of the inter-area mode for damping controllers using measurement *A*, *B*, or *C*.

mized system model must exhibit the important dynamic characteristics of the system in the desired frequency range of the swing modes to be damped.

Modal analysis is used to first identify the various oscillation modes. Their frequency, damping ratios, and mode shapes are then computed for different operating conditions, including contingencies.

The TCSC modulating signal is chosen after examining the frequency responses of TCSC input with respect to various system quantities. Frequency response tests are highly useful in determining the effectiveness of different feedback signals for power-swing damping; indeed, the frequency response is an indicator of both observability and controllability. A potentially effective signal exhibits a significant peak (resonance) that encompasses the swing-mode frequency with reasonably far-removed troughs (zeroes) on either side. In one study [22], the TCSC bus voltage was chosen as a modulating signal for a damping 0.75-Hz mode that is based on the frequency response depicted in Fig. 8.16. The closeness of the responses between the full model and the reduced-order model indicates the acceptable accuracy of the model-reduction process up to 1 Hz. The details of H_∞ method–based TCSC controller design are provided in ref. [22]. This technique usually produces a fairly high order controller that can, however, be reduced to a much lower order controller.

With TCSC voltage as the modulating signal, a washout-filter stage is added to the controller-transfer function so that the PSDC does not respond to slow variations in the modulating signal. Although the PSDC significantly enhances the damping of the relevant swing mode in the post-fault situations, it may in

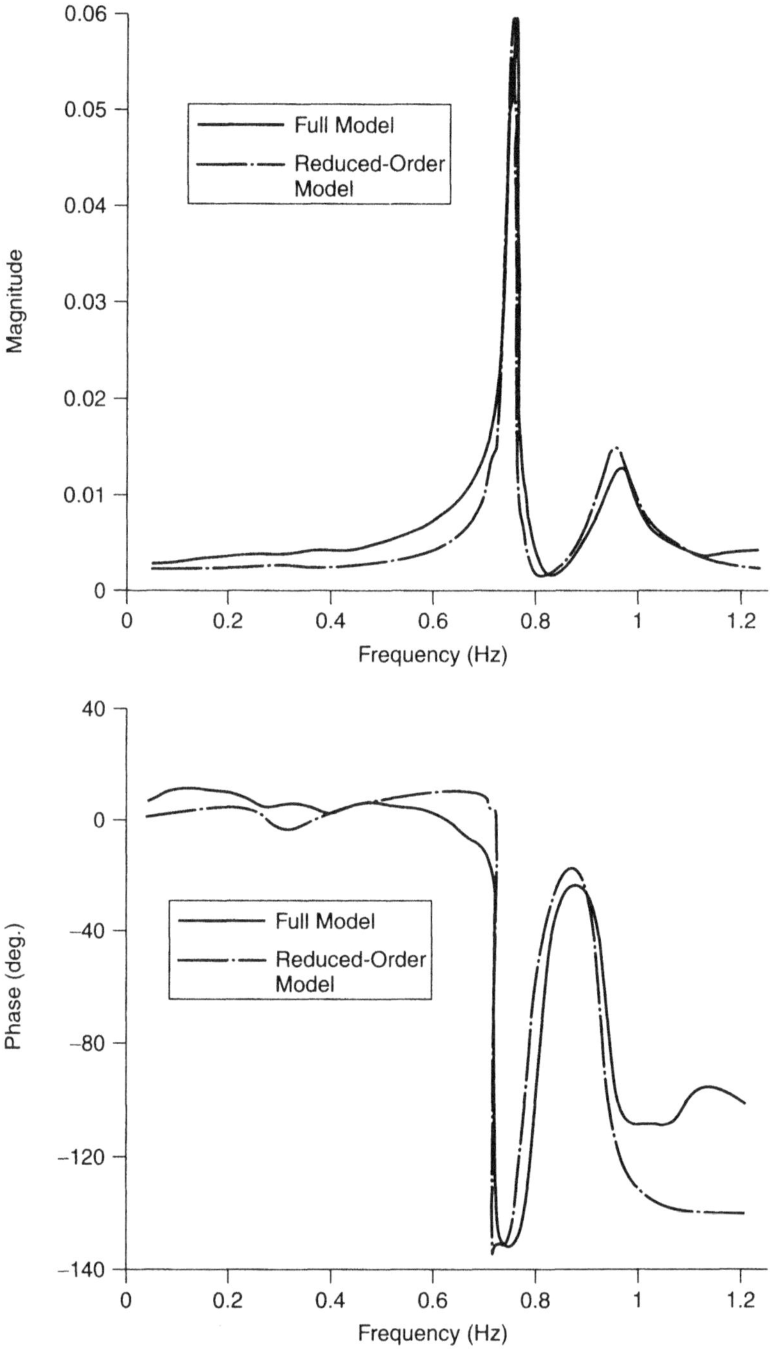

Figure 8.16 The frequency response of Edie voltage/TCSC input.

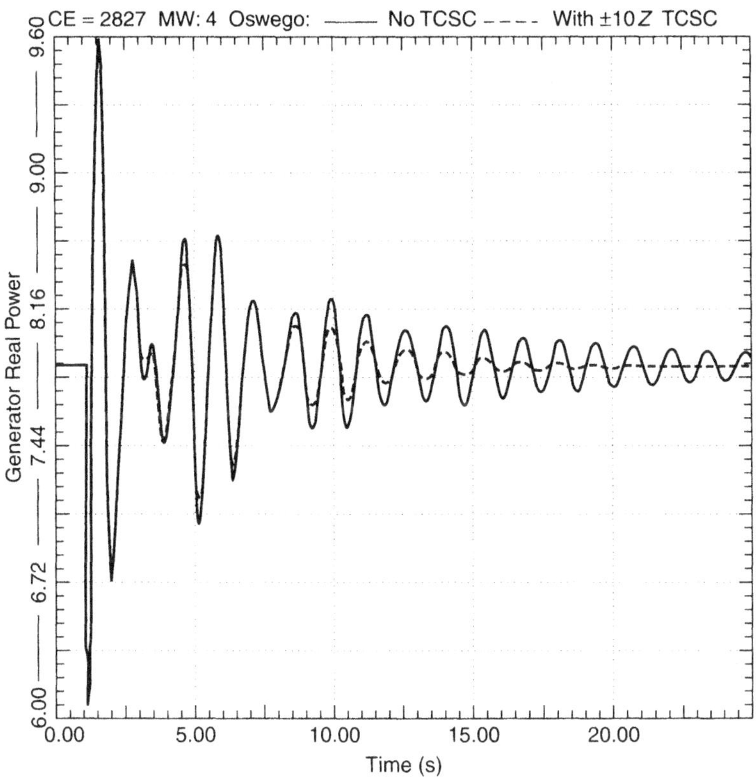

Figure 8.17 The effect of the modulated TCSC.

some cases reduce the damping slightly in the healthy system. This reduction need not be a cause for concern as long as the swing-mode damping is more than 2.5–5%.

A case study of PSDC design for a TCSC in the New York State power system is presented in ref. [22]. The system comprises 6700 buses, 1200 generating units, 5 HVDC links, and several SVCs. A TCSC compensation range of ±10% is shown to be sufficient for damping the key 0.75-Hz swing mode. Figure 8.17 shows the post-fault electrical power output of one of the major generator units with the TCSC–PSDC installed.

It is worth nothing that in the post-fault initial ten seconds, the generator oscillations are caused by other system modes that are related to TCSC. Furthermore, the TCSC response is also nonlinear because of the ±10%-limit imposition on the TCSC compensation range. It is shown from eigenvalue analysis that the TCSC–PSDC enhances the post-fault damping ratio of the critical mode to 0.1—an effect that is revealed in the latter part of the Fig. 8.17 simulation.

A succinct description of the application of the H_∞ robust-control technique for designing the TCSC damping controller is presented in refs. [23] and [24].

8.5.6 Alternative Techniques for PSDC Design

Instead of using a single modulating signal, a state feedback controller can be designed for the TCSC based on pole placement [25]. Although system poles can be assigned to desired locations irrespective of TCSC location, the corresponding controller gains may emerge as fairly large, causing prolonged saturation of the TCSCs and consequent deterioration of the controller performance. Satisfactory performance can be obtained only if the TCSC is appropriately located, that is, based on controllability considerations. In such cases, the controller gains attain acceptable values.

Modal reference algorithms, which have found application in the design of HVDC link-modulation controllers [26], [27], may also be used for designing TCSC damping controllers. Linear, robust, decentralized fixed-structure PSDCs for TCSCs have already been designed by using genetic algorithms [28].

8.5.7 Placement of the TCSC

The placement of FACTS controllers at appropiate locations is a critical issue. An optimally placed FACTS device requires a lower rating to achieve the same control objective than if it were located elsewhere [8], [9], [11]. At times, however, the FACTS controllers may need to be placed at nonoptimal locations to minimize costs, especially when land prices and environmental concerns become important.

The following conditions generally apply when considering the placement of TCSCs:

1. The TCSCs should be located in lines that experience limiting power oscillations.
2. The swing of voltages on each side of the TCSC must be within acceptable limits; otherwise, multiple sites may be necessary.
3. The control action of the TCSC in one transmission path should not cause undue power swings in a parallel path. If it does, then variable series compensation may become necessary in the parallel path.
4. Sometimes, it may be advisable to distribute the control action among multiple TCSCs rather than confining the control action to one large-rating TCSC. Doing so ensures some system reliability if one of the TCSCs should fail.

8.6 SUBSYNCHRONOUS RESONANCE (SSR) MITIGATION

Series compensation of long transmission lines may cause the following kinds of oscillations:

1. *Subsynchronous oscillations*, caused by interaction between the electrical network and the generator torsional system.

2. *Low-frequency (≈10 Hz) oscillations*, caused by interaction between the series capacitors and the shunt inductors, especially during line switchings and faults. These oscillations have large magnitudes and last for long periods because of high shunt-reactor Q-factors.
3. *Switching oscillations*, caused by the switching of lines.

The TCSCs can be employed successfully to mitigate the listed oscillations [29].

The principle of SSR mitigation by TCSCs has been obtained from the pioneering work done by Dr. N. G. Hingorani, for whom the NGH scheme of damping SSR [30]–[32] was named. This scheme involves a thyristor-controlled discharge resistor connected in shunt with the series capacitor and is installed in practical systems. The NGH scheme is described in great length in refs. [30]–[32], for which reason this book focuses on the impact the TCSC controller has in suppressing the SSR.

8.6.1 TCSC Impedance at Subsynchronous Frequencies

The TCSC impedance is purely capacitive at a nominal synchronous frequency and can be raised above the rated capacitance of the series capacitor by a factor of 2.5–3 through the appropriate firing-angle control. However, at subsynchronous frequencies, the TCSC under constant-reactance control presents a very different impedance characteristic, which greatly aids the damping of subsynchronous oscillations.

A frequency-domain EMTP study of the TCSC equivalent impedance is described in ref. [33]. The simulation is performed on a test system shown in Fig. 8.18 that comprises a single-segment TCSC placed in a transmission line having resistance R and inductance L. The system is energized by the combination of a synchronous 60-Hz voltage source and a subsynchronous 10-Hz

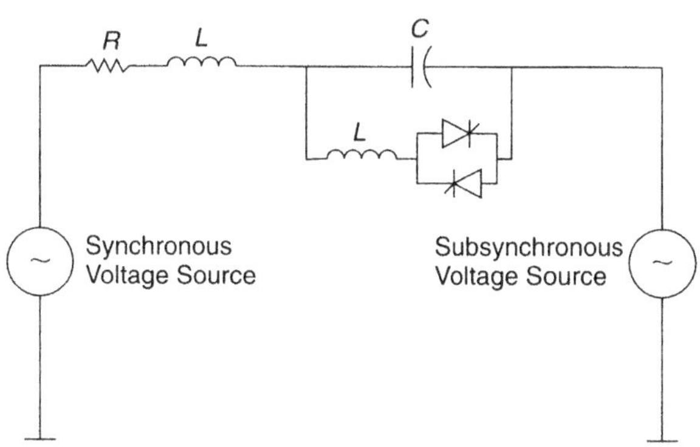

Figure 8.18 Network for the simulation study of the TCSC equivalent impedance.

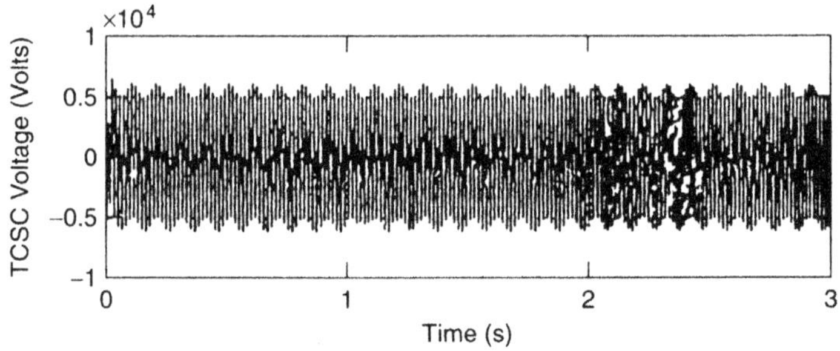

Figure 8.19 Simulated TCSC voltage.

voltage source, with the voltage magnitude of the latter much smaller than that of the former. The reactance of the TCSC fixed capacitor is 1.33 Ω.

The simulated TCSC voltage corresponding to a 70° conduction angle is shown in Fig. 8.19, and the square root of its power-density spectrum (PDS) is shown in Fig. 8.20. Two main frequency components—one at 60 Hz, the other at 10 Hz—are visible in the TCSC voltage signal. These individual-frequency components of TCSC voltage, as well as the TCSC current, are filtered through a second-order Chebyshev filter, as illustrated in Fig. 8.21. The equivalent TCSC impedance is then obtained by dividing the TCSC voltage by the TCSC current for the corresponding frequency. The equivalent impedance is calculated as $1.65\angle{-90^\circ}$ Ω at 60 Hz that, as expected, is 1.33 times the X_{order}, which has been specified as 1.2 pu. However, the TCSC equivalent impedance at 10 Hz is $3.4\angle{-8.5^\circ}$ Ω.

The equivalent impedance is similarly obtained for different frequencies corresponding to a given set of excitation-voltage magnitudes and the desired TCSC reactance X_{order}. It may be noted that because TCSC operation is nonlinear, the equivalent impedances will be functions of excitation voltages. The equivalent TCSC impedance, $Z_e(\omega)$, at different subsynchronous frequencies can be expressed as

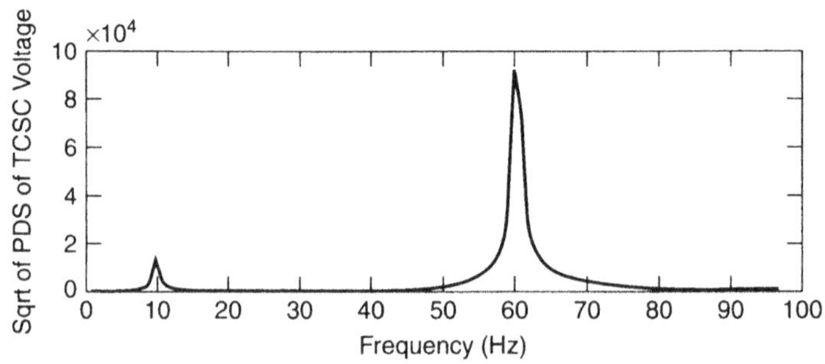

Figure 8.20 The square root (sqrt) of the power-density spectrum (PDS) of the simulated TCSC voltage.

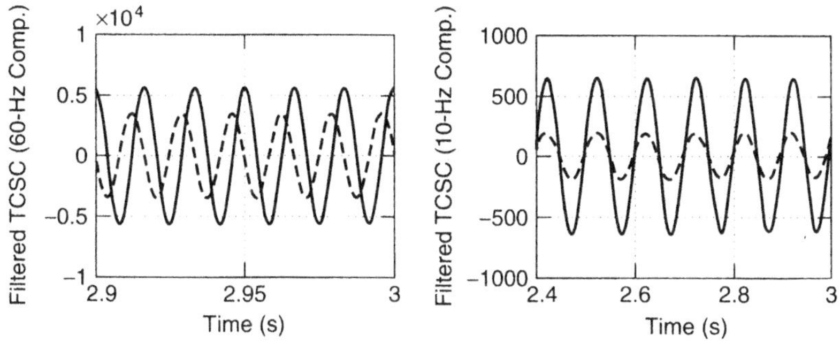

Figure 8.21 Filtered frequency components of the system responses: solid line denotes voltage in volts and dashed line denotes current in amps.

$$Z_e(\omega) = R_e(\omega) + jX_e(\omega) \tag{8.14}$$

where R_e = the equivalent resistance of the TCSC
X_e = the equivalent reactance of the TCSC

The equivalent resistance and reactance components are plotted as functions of frequency in Fig. 8.22. The TCSC presents an increasingly resistive effect as the frequency decreases below the synchronous level, implying that the TCSC offers resistive damping to subsynchronous oscillations. On the other hand, from Fig. 8.22(b), it is seen that while TCSC amplifies the capacitive reactance at the fundamental frequency according to the specified X_{order}, the equivalent TCSC capacitive reactance is reduced as the frequency is decreased below the subsynchronous frequency. In other words, the TCSC introduces an inductive effect at subsynchronous frequencies. The TCSC equivalent reactance corresponds to the reactance of the TCSC fixed capacitor at neither the nominal frequency nor the subsynchronous frequency. Thus, in the vernier mode of operation, the TCSC detunes the electrical resonant frequency with respect to the torsional mode by changing its equivalent-capacitive reactance at subsynchronous frequency. It further provides a positive, resistive damping to the subsynchronous oscillations.

It is noted from Fig. 8.22(a) that in the subsynchronous-frequency range of 28–60 Hz, the equivalent resistance increases with the X_{order}. Because the electrical frequency shown in the figure is complementary to the torsional-mode frequency (60 Hz), it can be argued that a higher X_{order} will impart a larger resistive damping to the low-frequency torsional modes in the range of 0–32 Hz.

It is seen from Fig. 8.22(b) that for a specified X_{order}, the equivalent-capacitive reactance deviates most from the fixed-capacitor reactance (or offers a maximum inductive effect) at lower electrical frequencies. This deviation implies that the TCSC will be more effective in suppressing the complementary higher-frequency torsional modes.

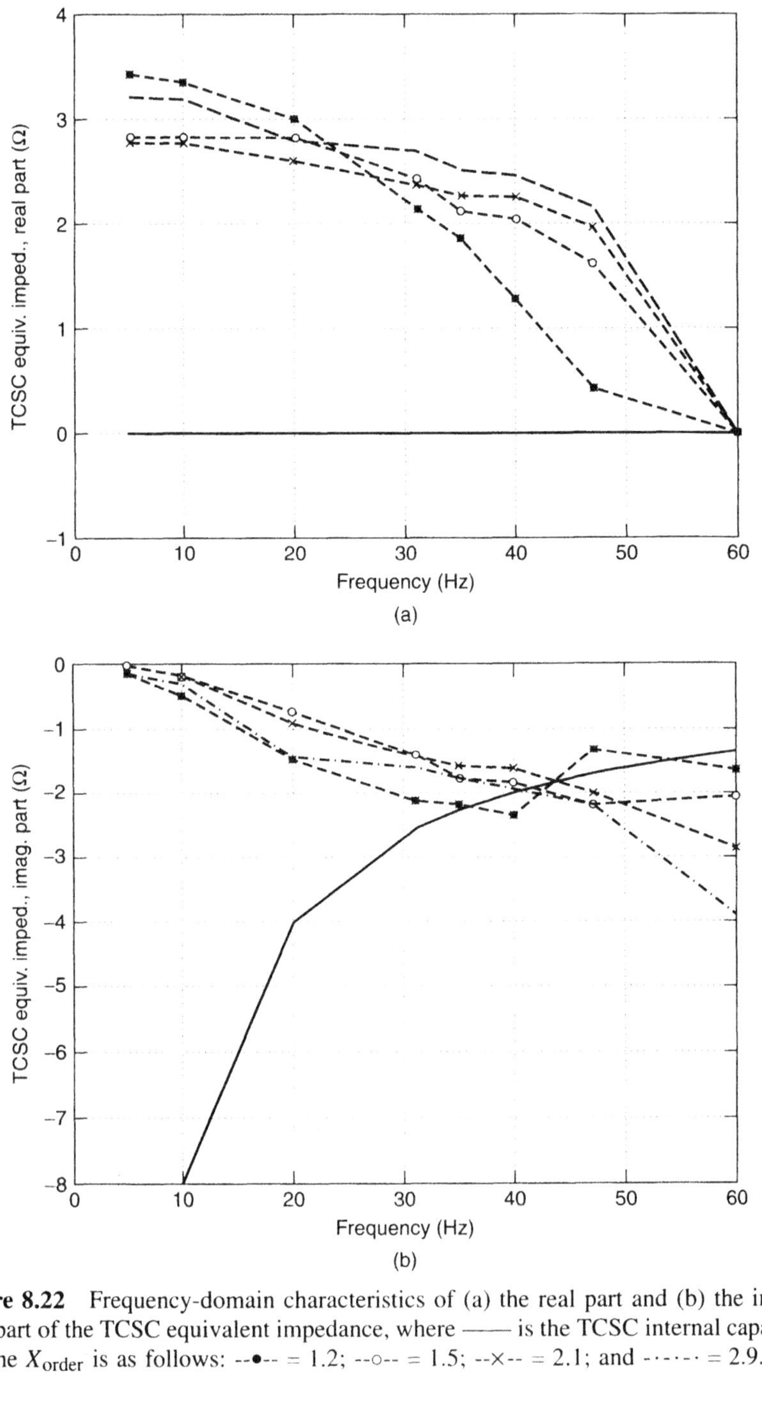

Figure 8.22 Frequency-domain characteristics of (a) the real part and (b) the imaginary part of the TCSC equivalent impedance, where —— is the TCSC internal capacitor and the X_{order} is as follows: --•-- = 1.2; --○-- = 1.5; --×-- = 2.1; and -·-·-· = 2.9.

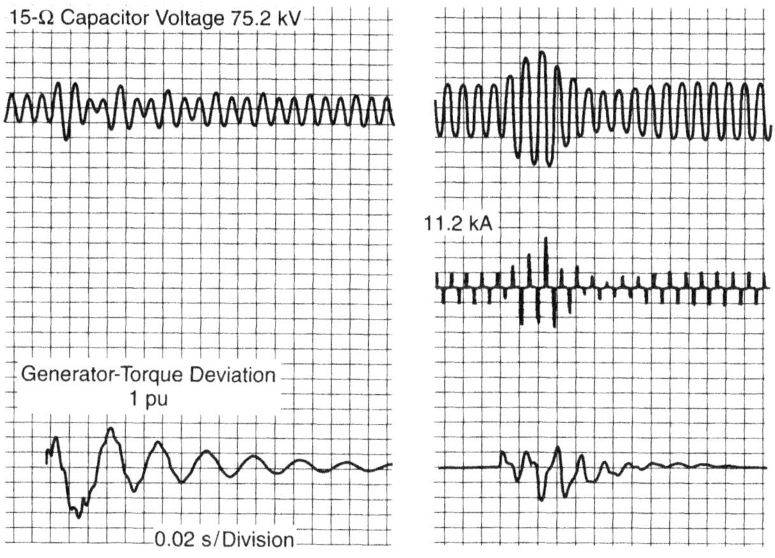

Figure 8.23 The transient torques: (a) conventional and (b) with TCSC.

The preceding study is indicative only of TCSC behavior. The exact performance may vary with differing system configurations, system parameters, TCSC operating conditions, and so forth.

An example of TCSC detuning of a transient-torque frequency is presented in ref. [34]. A series-compensated electrical network that carries power from a fossil generator to an infinite bus is considered. The total series reactance, X_L, is 203 Ω, whereas the capacitive reactance, X_c, of the conventional fixed capacitor equals 90 Ω, resulting in an electrical natural frequency of 40 Hz. A transient-torque inception scenario with the foregoing conventional capacitor is depicted in Fig. 8.23. The 15-Ω capacitor voltage exhibits a 40-Hz electrical natural frequency; the generator-torque deviation, on the other hand, undergoes a prolonged 20-Hz (60 − 40 Hz) oscillation.

A comparison is made between the performance of the conventional capacitor bank and a TCSC of same rating at synchronous frequency. The 90-Ω series-capacitive reactance is realized through two 15-Ω conventional-capacitor segments and two 15-Ω TCSC segments operating at twice the nominal reactance by the appropriate firing control (90 Ω = 2 × 15 Ω + 2 × 15 Ω × 2). The transient torque for the same disturbance is illustrated in Fig. 8.23(b). It is interesting to note that the TCSC with constant-reactance control is able to reduce the magnitude of generator transient-torque oscillation and also change the frequency of torque oscillation from 20 Hz to approximately 37 Hz. The electrical frequency corresponding to 37 Hz is 23 Hz (60 − 37 Hz), which is determined by the capacitance of the conventional capacitor 30 Ω (2 × 15 Ω) and a TCSC reactance of $2 \times 30\ \Omega$. The TCSC thus minimizes the transient torque by detuning a resonant condition.

Usually, generation and transmission projects are financially and, to a great extent, technically decoupled, so it is highly desirable to implement SSR countermeasures in the offending series-compensation equipment [35]. Thus, in a series-compensated system in which a small percent of series compensation is provided by the TCSC, the SSR effects can be mitigated successfully by enhanced control of the TCSC.

8.6.2 A Case Study

In this case study, adapted from ref. [35], a simple radial transmission system is depicted in Fig. 8.24, in which a 600-MW turbine generator supplies power over a series-compensated line. The turbine generator unit has two torsional modes: one at 18 Hz, the other at 29.8 Hz.

$$Z(j\omega) = r_g + r_s + \left(j\omega L - \frac{j}{\omega C}\right) \tag{8.15}$$

where r_g = the generator stator resistance

r_s = the transmission-line resistance

C = the capacitance of the series capacitor

$L = L_g + L_s$, the total circuit inductance

L_g = the generator inductance

L_s = the transmission-line inductance

Eigenvalues of the Fig. 8.24 system are plotted in Fig. 8.25. Path 1 in the figure corresponds to the root loci for varying series compensation of a fixed capacitor without any TCSC; path 2 corresponds to 36% series compensation and a variable-conduction angle tuned to critically excite mode 2; and path 3 in the root loci corresponds to 68% series compensation and an effective TCSC reactance tuned to mode 1. The roots correspond to the electrical system and torsional system as viewed from a reference frame attached to the generator rotor. The TCSC operates in constant-reactance control.

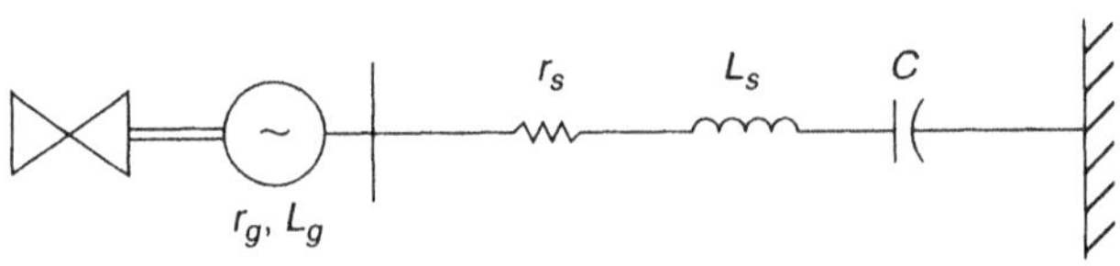

Figure 8.24 A 600-MW turbine generator supplying power over a series-compensated line.

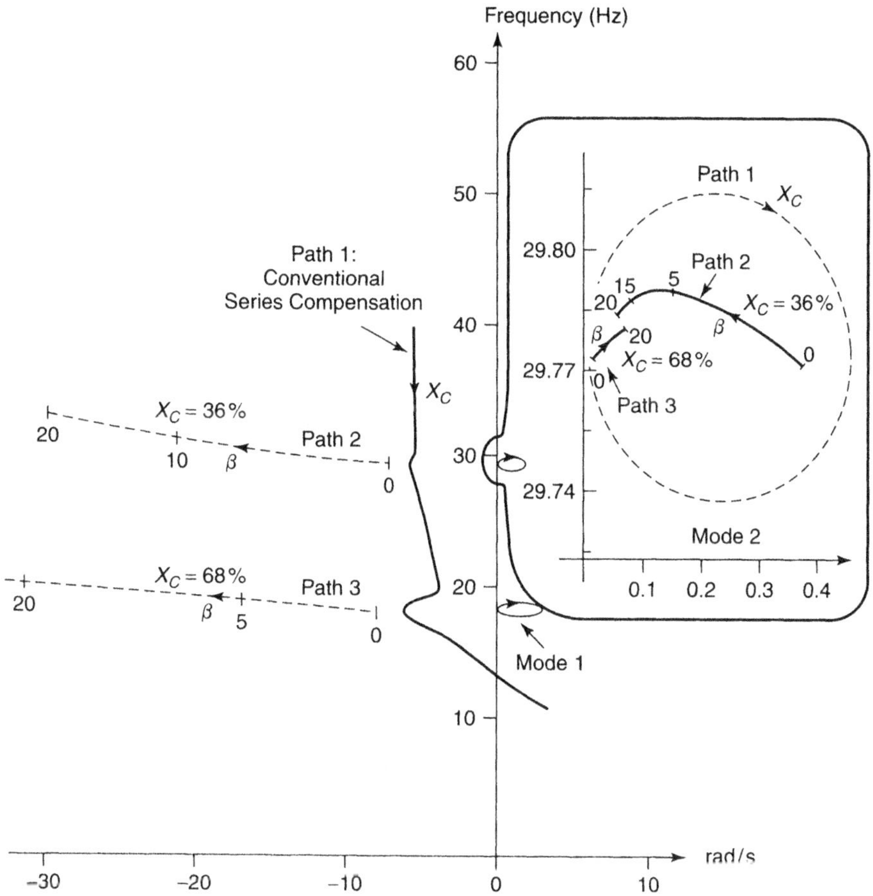

Figure 8.25 The root locus of the electric and mechanical mode versus the series compensation and conduction angle.

As the series compensation increases, the electrical-mode frequency decreases (as seen from the rotor). The damping of this mode decreases from the induction-generator effect until it finally becomes unstable beyond a certain level of series compensation—the well-understood self-excitation phenomenon.

The performance of TCSC constant-reactance control is evident from paths 2 and 3 in the root loci. Even though the TCSC is able to damp the critically destabilized mode, it is unable to stabilize the overall system, thus necessitating the application of modulated-reactance control or a PSDC that can be derived in a manner similar to a PSDC designed for damping power-swing oscillations. The subsynchronous response of the network is sensed, and a capacitor-offset voltage, generated by thyristor conduction, is modulated to produce the appropriate sideband voltages. The sideband voltages cause a countercurrent to flow

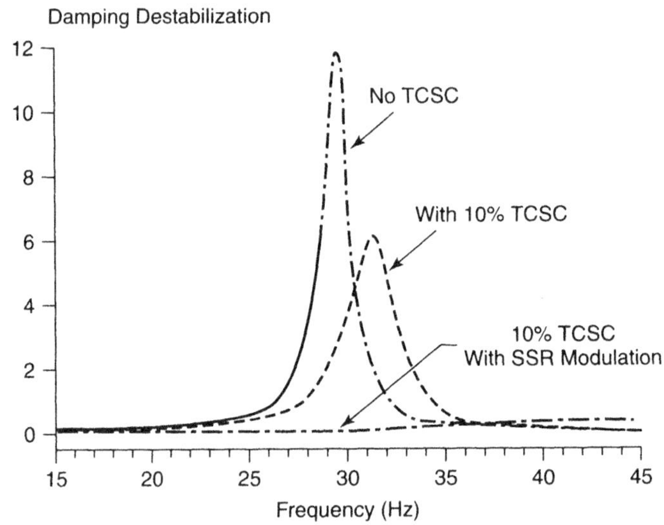

Figure 8.26 Benefit of the TCSC to small-signal stability.

into the network, equalizing the original current flow in the capacitor by means of torsional oscillations. A synthesized rotor-frequency signal is shown to be highly effective for this purpose [36].

The destabilizing, or negative, damping torque for cases lacking a TCSC with constant-reactance control, lacking a TCSC with modulated-reactance control, and even lacking a TCSC altogether is depicted in Fig. 8.26. Maximum undamping exists for the case lacking a TCSC. The negative damping is reduced by a 10% TCSC operating on constant-reactance control, athough the beneficial influence is shifted to a higher frequency. An appropriately designed SSR-modulation control for 10% TCSC successfully stabilizes all the torsional modes.

8.6.2.1 Transient-Torque Minimization During severe faults in a series-compensated system, the undischarged series capacitor causes subsynchronous currents to flow into the network and generating system. These currents result in electromagnetic torques in the generator, which may have frequency components close to the torsional-mode frequency. Very large transient-shaft torque are experienced by the torsional system; these continue until the post-fault system settles to an equilibrium state.

Damping control of the TCSC can rapidly mitigate the transient torque in the study system following a 3-phase fault, as shown in Fig. 8.27. Without the TCSC, the shaft torque persists for a long time, even when the capacitor voltage is limited to 2 pu by protective action.

8.6.2.2 Criteria for SSR Mitigation by the TCSC The following conditions are prescribed for TCSCs to damp SSR [35], [37]:

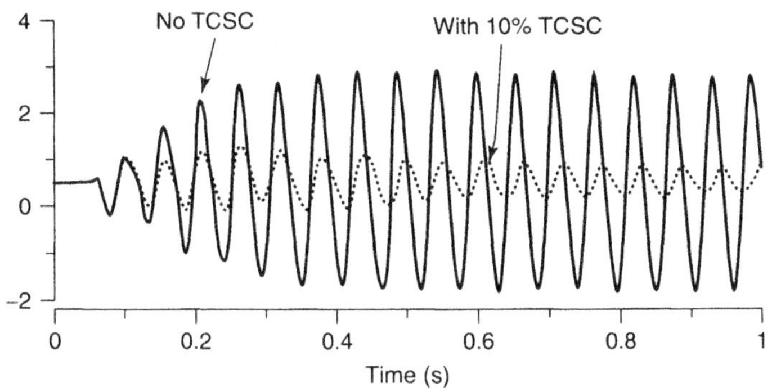

Figure 8.27 Benefit of the TCSC to damp transient-shaft torques.

1. The TCSC should be able to damp the SSR effects simultaneously on all turbine generators.
2. If the TCSC is unable to damp the SSR, the TCSC control must act to remove the capacitor from the transmission line.
3. If some remaining SSR effects are observed on turbine generators because of other series capacitors in the network, they must be identified and mitigated independently.

While achieving the aforementioned objectives, a minimum control range of the TCSC must be used, and to ensure the reliability of operation, the controller gain must not be unduly large.

Although constant-reactance control may not be adequate in limiting the SSR, other reactance-modulation schemes, based on line power or line current, (e.g., the one employing enhanced power control with a fast inner-current controller), offer robust SSR mitigation solutions [2], [6]. A constant-angle control proposed in ref. [3] also successfully damps the subsynchronous oscillations. This control strategy supersedes the constant-current control in its ability to suppress the SSR.

8.7 VOLTAGE-COLLAPSE PREVENTION

Voltage-collapse problems are a serious concern for power-system engineers and planners. Voltage collapse is mathematically indicated when the system Jacobian becomes singular. The collapse points are indicative of the maximum loadability of the transmission lines or the available transfer capability (ATC) [38]. The TCSCs can significantly enhance the loadability of transmission networks, thus obviating voltage collapse at existing power-transfer levels. While the TCSC reduces the effective line reactance, thereby increasing the power flow, it generates reactive power with increasing through-current, thus exercising a beneficial influence on the neighboring bus voltage.

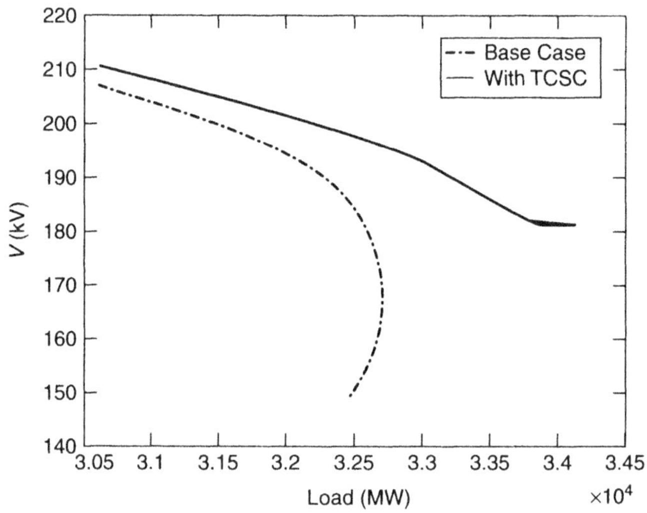

Figure 8.28 The voltage profile of the critical bus employing 50% TCSC compensation.

An application of the TCSC for the preceding purpose is presented in ref. [38] for a European system. The system faces voltage collapse or a maximum loading point corresponding to a 2120-MW increase in the net load. If a TCSC is installed to provide 50% compensation of the line experiencing the highest increase in power at the point of collapse, the maximum loadability will be enhanced to 3534 MW. The influence of the TCSC on the voltage profile of a critical bus is illustrated in Fig. 8.28.

A performance factor, f_p, is proposed in ref. [38] that indicates the maximum increase in loadability, λ_0, for a given percent of line compensation:

$$f_p = \frac{\lambda_0\ [\text{MW}]}{X_{\text{ref}}\ [\ \%\ \text{compensation of}\ X_{\text{line}}]} \tag{8.16}$$

where X_{ref} = the reactance-reference setting of the TCSC

X_{line} = the line reactance

This index can be gainfully employed to obtain the best location of the TCSC in a system. The enhancement of system loading and variation of the performance factor with TCSC compensation are depicted in Fig. 8.29.

It is suggested that TCSC reactance-modulation schemes based on line current or line power, or on the angular difference across lines, may prove unsuccessful for voltage-stability enhancement. The reason is that these controls constrain any variation in the corresponding variables that may be necessary with changing loads, thereby limiting any power-flow enhancement on the line.

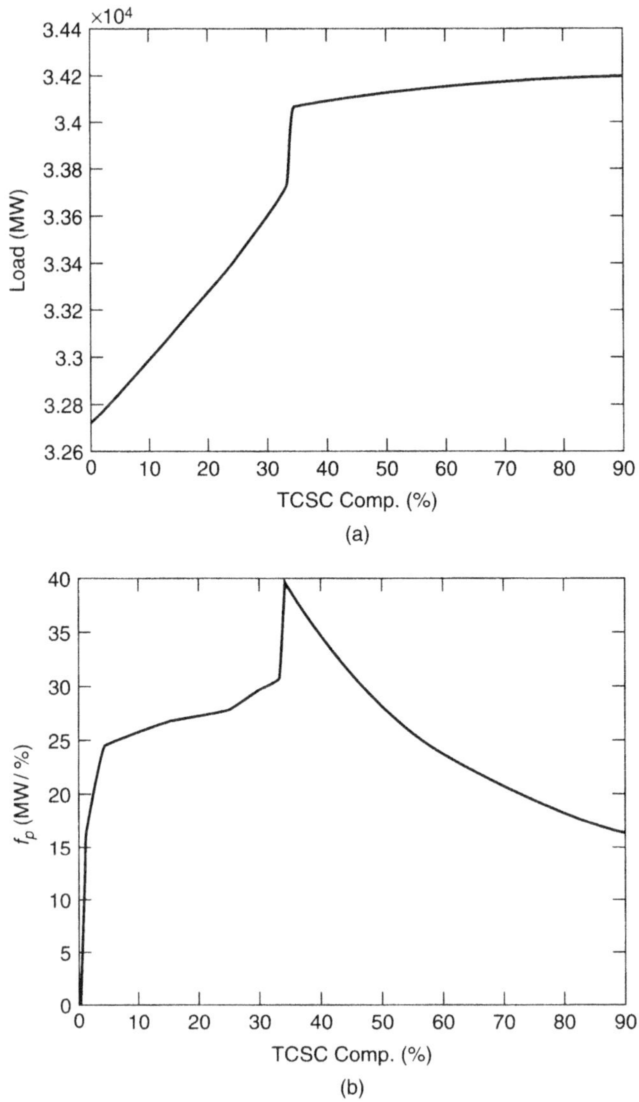

Figure 8.29 The effect of changing the TCSC compensation levels for the critical line: (a) the loading margin and (b) the performance measure, f_p.

8.8 TCSC INSTALLATIONS

This section describes the performance of two installed TCSC controllers: one, for power-oscillation damping in the Brazilian north–south inter-tie [39]–[41]; the other, for mitigating subsynchronous resonance in the Swedish Stode system [42].

Figure 8.30 The Brazilian north–south (NS) interconnection.

8.8.1 Imperatriz–Serra da Mesa TCSCs in Brazil

The north–south (NS) interconnection in Brazil, connecting the north–northeast (NNE) system and the south–southeast (SSE) system, is shown in Fig. 8.30. This interconnection comprises a single 500-kV compact transmission line of 1020-km length that links the Imperatriz substation in the N system to the Serra da Mesa substation in the S system. A single line diagram of the system is given in Fig. 8.31. The installed-generation capacities in the NNE and SSE systems are 14 GW and 48 GW, respectively, which are primarily hydroelectric. The power-transfer capacity of the inter-tie is 0–1300 MW in either direction. The 500-kV line is provided with 100% shunt compensation through the line reactors and 54% fixed-series compensation distributed on six series-capacitor banks (161 MVAR or $-23.8\ \Omega$ each) along the transmission line.

Although for purely technical reasons such an interconnection would have been expected to be installed through a dc link, it was decided that an ac inter-

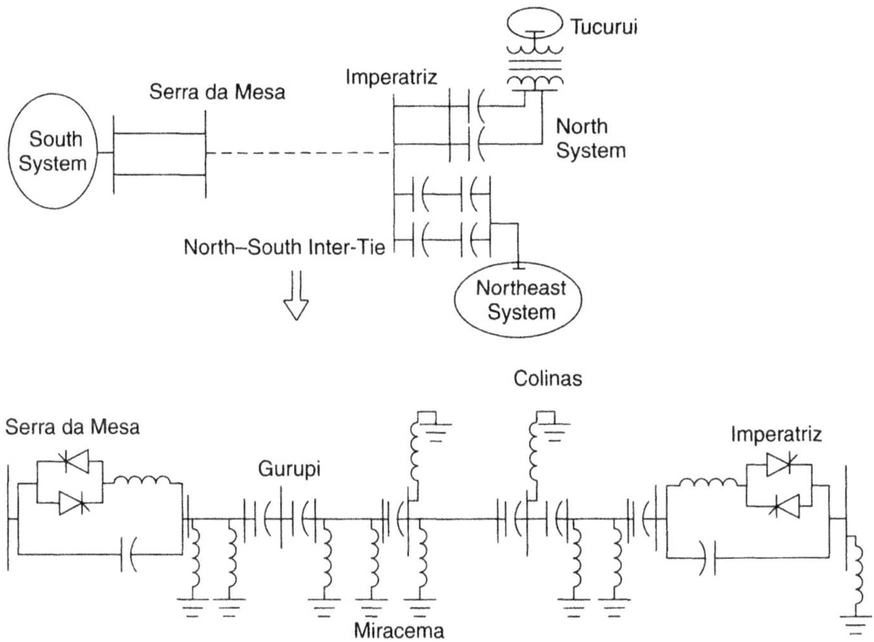

Figure 8.31 Single-line diagram of the Imperatriz–Serra da Mesa transmission line.

connection would be installed instead—for strategic and political reasons and also because of possible future ac-system expansion in the region. However, this interconnection has led to the emergence of low-frequency, poorly damped 0.18-Hz inter-area oscillation modes. The magnitude of the oscillations can be of the order of ±300 MW.

Power-system stabilizers (PSSs) have been conventionally employed to dampen electromechanical modes in the frequency range of 0.5–20 Hz. However, as the NS inter-area mode has a frequency of the order of 0.2 Hz, it is argued that the PSS solution to dampen this mode may not be feasible for the following reasons:

1. All major plants in the NE system will require modified PSSs.
2. Even the modified PSSs, which have fixed structures and parameters, may not be able to generate requisite damping for the NS mode.
3. The new frequency range that will be assigned to PSSs for damping will be extensive and possibly difficult to realize in a reliable manner.
4. The modified PSSs may adversely affect the damping of local modes in the NE system and the inter-area modes pertaining to the N–NE system.
5. The modified PSS gains cannot be made very large because of practical limits during operation at very low frequencies, thus restricting the effectiveness of the PSSs.

TABLE 8.2 The TCSC Controller Data

Parameter	Value	Unit
Maximum system voltage (phase–phase)	550	kV
Nominal reactive power	107.5	MVAR
Physical capacitive reactance (1 pu)	13.27	Ω/ph
Nominal capacitive reactance (1 · 2 pu)	15.92	Ω/ph
Nominal capacitive boost factor (X_{app}/X_c)	1.2	pu
Maximum capacitive reactance (3 pu)	39.8	Ω/ph
Reactor inductance	5.63	mH
Inductive reactance in bypass	2.5	Ω/ph
Rated line current	1500	A
Rated continuous TCSC voltage (1 pu)	23.88	kV
30-min overload current (1.5 pu)	2025	A
10-s overload current (2 pu)	3000	A
Maximum thermal short-circuit current	23	kA

It was therefore proposed that TCSCs be used for damping the N–S inter-area mode. Two TCSC banks were installed in the inter-tie to provide damping of the inter-area oscillations. One TCSC is located close to the Imperatriz substation; the other is located next to the Serra da Mesa substation. The basic TCSC parameters are listed in Table 8.2.

The reason for having two TCSCs, instead of one with twice the capacity, is based on the planned future expansion of the system when some hydroelectric stations are likely to be constructed and linked along the same route. In this scenario, it is expected that the two TCSCs, if coordinated appropriately, will result in superior system damping.

The reactance of each TCSC can be controlled from $-13.3\ \Omega$, with thyristors blocked up to $-39.8\ \Omega$ in the capacitive-vernier mode, to give a voltage boost factor of 3. In the steady state, both TCSCs operate at $-15.9\ \Omega$, providing a voltage boost factor of 1.2 and implying that during power oscillations, the effective line reactance can be modulated between 43 Ω (both TCSCs at a 3.0 pu boost) and 128 Ω (both TCSCs bypassed).

8.8.1.1 TCSC Power-Oscillation Damping (POD) Control The main objective of the TCSC installation is the damping of N–S inter-area mode. Power-flow control is not considered, nor is SSR damping, as the generators are mainly hydroelectric.

The control input to the POD controller is selected as the active power flow in the line. However, the line-current signal is found to be almost equally effective. It may be noted that the amount of line active-power variation caused by the TCSC is a function of both the change in TCSC reactance as well as the line current/loading. Thus if the TCSC control operates with a constant gain, the TCSC will be most effective in damping power oscillations during high line loadings and least effective during low line loadings. To maintain a near-uni-

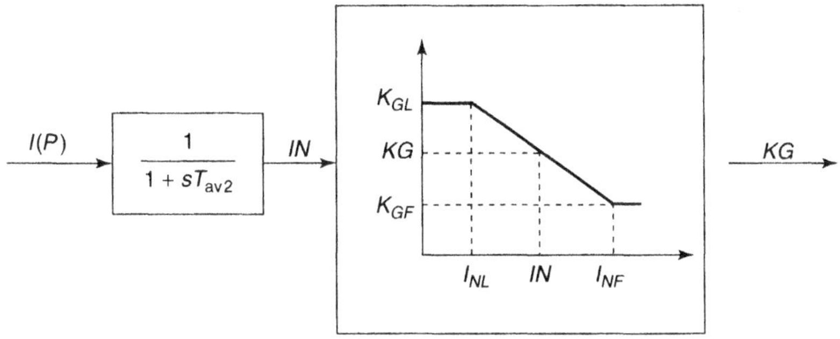

Figure 8.32 A gain-scheduling scheme.

form damping contribution over a wide range of line loadings, a gain-scheduling scheme is adopted, as shown in Fig. 8.32.

In the figure, the input, *IN*, to the gain controller is derived from the control-signal input, *I*(*P*) to the POD controller by passing it through a low-pass filter having a time constant $T_{av\,2}$, which is large enough to not interact with the power-oscillation frequency [40], [41]. A low value of line-power flow, I_{NL}, provides a high gain, K_{GL}; a high line loading, I_{NF}, provides a low gain, K_{GF}. Any intermediate level of power-flow *IN* results in an interpolated gain K_G, which is then used for power-oscillation damping.

The TCSC control system, illustrated in Fig. 8.33, provides a reactance-order output X_{order}, which essentially consists of a steady-state reactance-reference signal, X_0, added to an output signal emanating from the POD controller. Two limits, X_{max} and $X_{min}(t)$, are imposed in the limitation block on the reactance-order output X_{order}. The limit X_{max} corresponds to the reactance of the fixed capacitor, whereas the limit $X_{min}(t)$ relates to the maximum compensation level achievable by the TCSC firing control, and is determined by the maximum permissible temporary voltage across the TCSC. It is also a dynamic limit, as it is dependent on the line current, I_{line}, which is a function of time.

$$X_{min}(t) = 3X_{max} \qquad \text{if } |X_l|\,|I_{line}(t)| \leq 2V_{max} \tag{8.17}$$

$$X_{min}(t) = \frac{-2V_{max}}{|I_{line}(t)|} \qquad \text{if } |X_l|\,|I_{line}(t)| > 2V_{max} \tag{8.18}$$

where both X_{max} and $X_{min}(t)$ are negative quantities because they denote capac-

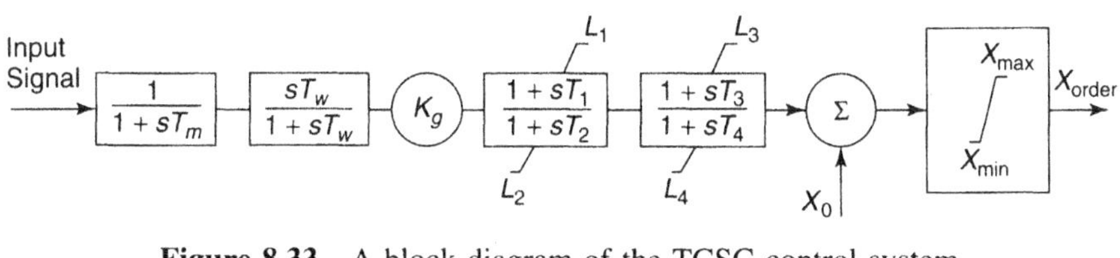

Figure 8.33 A block diagram of the TCSC control system.

TABLE 8.3 The TCSC Parameters (Base: 100 MVA, 500 kV)

Parameter	Value	Unit
K_l	Variable[a]	pu (Ω)/pu (MW)
T_m	0.05	s
T_1	1.5	s
T_2	0.3	s
T_3	2.2	s
T_4	0.3	s
T_5	2.2	s
X_{max}	–0.00531	pu (Ω)
$X_{min}(t)$	See Eqs. (8.17), (8.18)	pu (Ω)

[a]0.1 for power transfer = 1000 MW.

itive reactances, and

V_{max} = the maximum continuous voltage across the TCSC

X_l = the TCSC capacitive-reactance output before the limitation block

The POD controller comprises a measurement block with time constant T_m, a washout stage with time constant T_w, a variable gain K_G, and two lead-lag filters with time constants T_1–T_4. The washout stage eliminates the average component of the measured line-power-flow signal. The POD controller thus generates a signal proportional to the oscillatory component of power flow, which is appropriately phase-shifted. The typical values of these controller parameters are given in Table 8.3 [40].

For the dynamic-control action of the POD to not exceed the controllable range of the TCSC, the lead-lag filters are provided with non-windup limiters. These limits also ensure that the output of the lead-lag filters will not acquire a dc offset during system transients and cause the TCSC reactance to saturate on one side of the main-circuit controllable range. However, they are associated with a major discrepancy: During large power swings and whenever the POD gain is high, the limit curb the POD output signal in almost each cycle, resulting in the loss of desirable phase shift from the lead-lag filters. The effectiveness of the TCSC in damping the power oscillations is thus greatly diminished.

8.8.1.2 Phasor Estimation An ingenious alternate method of POD controller design is described in refs. [39] and [40]. The main objective of this method is to produce a TCSC reactance-order signal than can augment system damping. It is known that the line real power comprises two components:

1. the average power on the line, and
2. the oscillating power having the frequency of the inter-area mode.

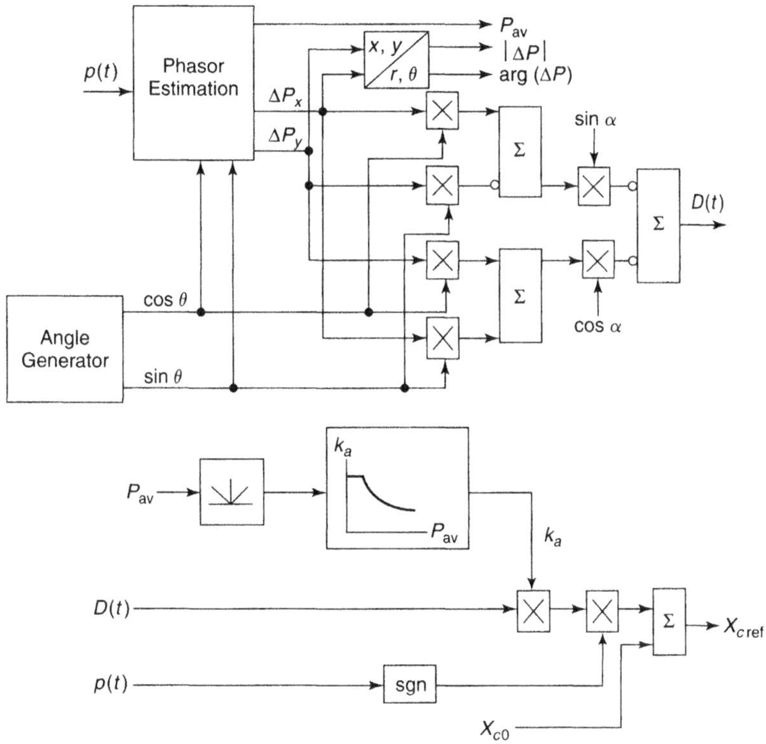

Figure 8.34 A block diagram of a phasor POD controller.

In this method, a phase-locked loop (PLL) is used to extract the instantaneous-phase angle and amplitude of the oscillating-power signal. Then, the TCSC reactance is modulated accordingly to introduce damping for these power oscillations. A block-schematic diagram of the phasor POD that performs the signal extraction is depicted in Fig. 8.34. A reference signal is generated with the same frequency as the inter-area mode. The PLL locks to this frequency and extracts the following signals:

1. the 0.2-Hz real-power oscillation;
2. the average real-power flow on the inter-tie;
3. the amplitude of the power oscillation; and
4. the phase angle of the power oscillation.

System damping can now be enhanced if the TCSC reactance order is made to lag behind the active power signal by 90° [43]. The phasor POD controller-output signal is duly corrected in sign based on the direction of power flow in the line. Even though this controller is designed to function for a fixed inter-area mode frequency, it can be adapted to operate over a wider frequency range.

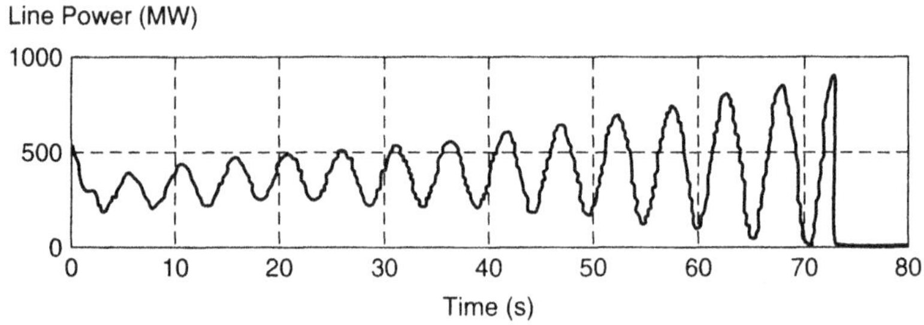

Figure 8.35 Tripping of 300-MW generator in Tucurui, with the lead-lag and phasor PODs both disabled.

8.8.1.3 Performance of Both TCSCs The performance of these POD controllers is tested in the field by tripping a 300-MW generator in Tucurui, which is considered a severe contingency. When the PODs at both the Imperatriz and Serra da Mesa substations are disabled, the foregoing contingency causes the system of each to become unstable, as shown in Fig. 8.35, which subsequently causes the tripping of the N–S interconnection.

Field tests further demonstrate that when the Imperatriz TCSC is equipped with a lead-lag filter–based POD controller, and if the POD at the Serra da Mesa TCSC is disabled, valuable phase shift gets lost from the limiter action during power oscillation in the same critical loss of generation scenario. Therefore, the TCSC capability cannot be used effectively for damping enhancement. This behavior is demonstrated in Fig. 8.36.

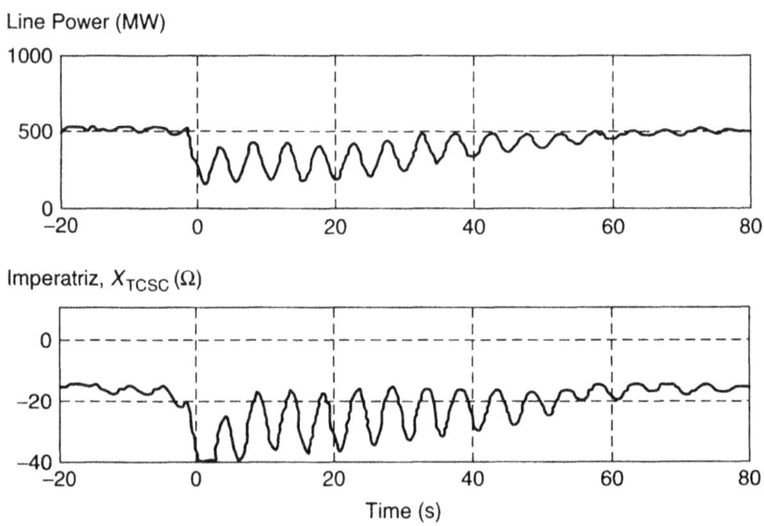

Figure 8.36 Tripping of 300-MW generator in Tucurui, with the lead-lag POD in Impertriz and the Serra da Mesa disabled.

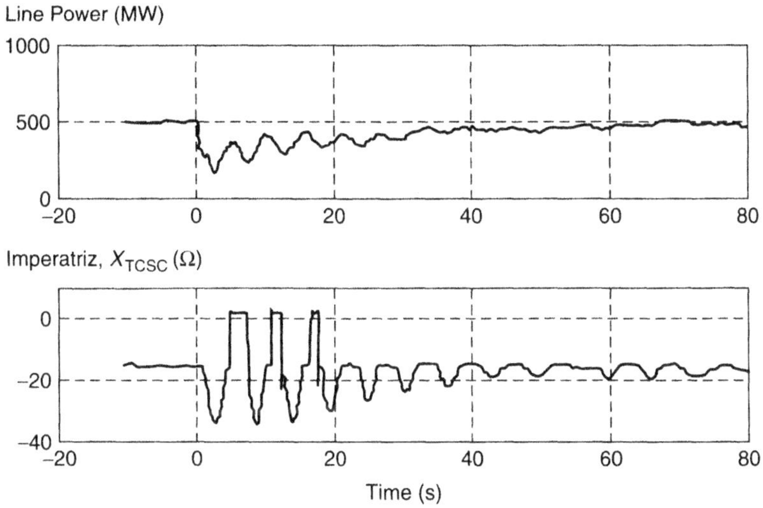

Figure 8.37 Tripping of 300-MW generator in Tucurui, with the phasor POD in Imperatriz and the Serra da Mesa disabled.

A dual-gain scheme is employed in which a high POD gain is implemented when the power-swing amplitude exceeds 30 MW and a lower steady-state gain is used when power oscillations are not as high. This damping scheme also uses a bang-bang strategy during low-power transfer situations, which involve the bypassing of TCSCs. This scheme has proven to be highly effective. The test result for the same fault scenario, with the phasor POD installed at the Imperatriz TCSC and the POD at the Serra da Mesa TCSC disabled, is depicted in Fig. 8.37.

When, however, the phasor POD controller is installed on the Imperatriz TCSC, and when the POD at the Serra da Mesa TCSC is activated, a significant damping is introduced into the system for the NS mode after the same 300-MW generator tripping in Tucurui. The damped real-power flow and corresponding TCSC apparent reactance are displayed in Fig. 8.38.

During initial system operation, the Imperatriz TCSC and the Serra da Mesa TCSC present similar behavior. Because they both use simple open-loop-impedance control, along with similarly tuned damping controllers that use local information and only do not communicate between the two TCSCs, no adverse interaction between them is anticipated.

8.8.2 Stode TCSC in Sweden

The Stode TCSC in Sweden is installed for alleviating SSR [42], [44]. In that country, some long-distance 400-kV transmission lines connected to nuclear power–generating stations are series-compensated, which has caused repeated occurrences of subsynchronous currents within the series capacitors (SCs) that

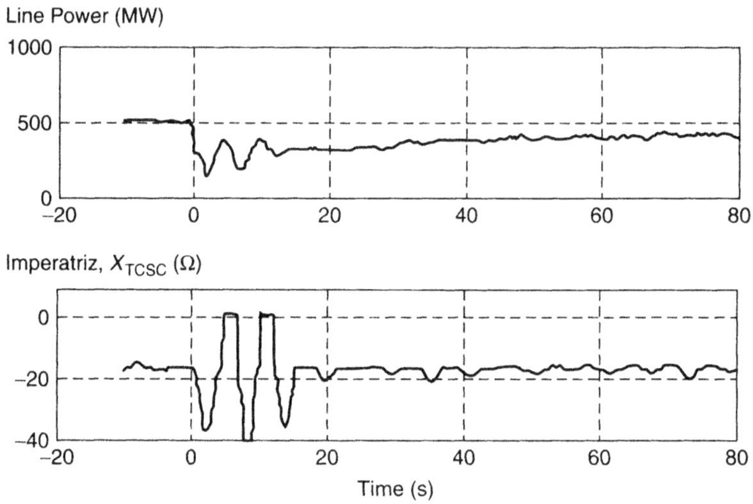

Figure 8.38 Tripping of 300-MW generation in Tucurui, with the phasor POD in Imperatriz and the Serra da Mesa enabled.

must be bypassed. One investigated means of eliminating this problem involved reducing the levels of series compensation in the lines, but doing so was found unacceptable; it substantially reduced the power transfers in other corridors, and it also increased the power losses. This unacceptability led to another means of eliminating the problem: by splitting the existing SC of the Stode unit into two segments. Of the original SC, 70% was kept intact as a fixed component; the balance (30%) segment was converted into a TCSC. The 30% balance was selected so that if any inadvertent failure of TCSC equipment occurred, the line could still operate with a reduced value of series compensation and yet provide reasonable system benefits while, at the same time, the TCSC could be bypassed. The design parameters of the Stode TCSC are given in Table 8.4 [42].

The synchronous voltage reversal (SVR) control scheme [3], [12], [32], [42], [44] is used to obtain the desired apparent impedance of TCSC. A TCSC voltage boost level of 1.2 is chosen for this scheme, which evolved from the following considerations:

TABLE 8.4 Stode TCSC Basic Design Parameters

Parameter	Value
Physical capacitor reactance	18.25 Ω/ph
Apparent TCSC reactance in the steady state	21.9 Ω/ph
Rated current	1500 A, rms continuous
Short-time current	2025 A, rms 30 min; 2250 A, rms 10 min

1. A higher boost level implies higher TCSC current that requires an increase in the capacitor bank rating.
2. A higher boost factor also raises the magnitude of thyristor current.
3. A higher boost level also increases the harmonic voltage inserted in series with the line.
4. A minimum boost level is needed to ensure that the thyristors do not need to turn on near the voltage zero crossing.

The Stode TCSC allows the degree of series compensation to remain unchanged at its original level as the SSR conditions modify to those corresponding to the case in which 30% of the SC reactance is bypassed. These conditions need not be a cause for concern.

A significant aspect of the Stode TCSC is that it demonstrates how an existing series-compensation installation can be reorganized partly by employing existing equipment and partly by employing some newly acquired apparatuses. If the TCSC component is set up on independent platforms, the line is also not required to be put out of service for a long period.

On a global level, the Stode TCSC may become a precursor for the part conversion of existing series-compensated lines into TCSC schemes.

8.9 SUMMARY

This chapter described the different controllers for TCSCs. Various applications of TCSCs were presented together with case studies. The choice of control signals for modulating TCSC reactance was discussed, along with the methodology for controller design. The objectives and performance of two recent TCSC installations were described and discussed.

REFERENCES

[1] S. Nyati, C. A. Wegner, R. W. Delmerico, R. J. Piwko, D. H. Baker, and A. Edris, "Effectiveness of Thyristor-Controller Series Capacitor in Enhancing Power System Dynamics: An Analog Simulator Study," *IEEE Transactions on Power Delivery*, April 1994, pp. 1018–1027.

[2] L. A. S. Pilotto, A. R. Carvalho, A. Bianco, W. F. Long, F. L. Alvarado, C. L. DeMarco, and A. Edris, "The Impact of Different TCSC Control Methodologies on the Subsynchronous Resonance Problem," *Proceedings of EPRI Conference on FACTS*, Washington, DC, April 1996.

[3] K. R. Padiyar, *Analysis of Subsynchronous Resonance in Power Systems*, Kluwer Academic Publishers, Boston, 1999.

[4] R. J. Piwko, C. A. Wegner, B. L. Damsky, B. C. Furumasu, and J. D. Eden, "The Slatt Thyristor-Controlled Series Capacitor Project—Design, Installation, Commissioning, and System Testing," CIGRE Paper 14-104, Paris, 1994.

[5] N. Martins, H. J. C. P. Pinto, and J. J. Paserba, "TCSC Controls for Line Power Scheduling and System Oscillation Damping—Results for a Small Example System," *Proceedings of 14th Power System Control Conference*, Trondheim, Norway, June 28–July 2, 1999.

[6] The Electric Power Research Institute (EPRI) Report TR-109969, "Analysis of Control Interactions on FACTS Assisted Power Systems," Final Report, Palo Alto, CA, January 1998.

[7] H. F. Wang, F. J. Swift, and M. Li, "The Indices for Selecting the Best Locations of PSS or FACTS-Based Stabilizers in Multimachine Power Systems: A Comparison Study," *IEE Proceedings—Generation Transmission Distribution*, Vol. 2, 1997, pp. 155–159.

[8] H. F. Wang, F. J. Swift, and M. Li, "Selection of Installing Locations and Feedback Signals of FACTS-Based Stabilizers in Multimachine Power Systems by Reduced-Order Modal Analysis," *IEE Proceedings—Generation Transmission Distribution*, No. 3, 1997, pp. 263–270.

[9] H. F. Wang, "Selection of Robust Installing Locations and Feedback Signals of FACTS-Based Stabilizers in Multimachine Power Systems," *IEEE Transactions on Power Systems*, Vol. 2, 1999, pp. 569–574.

[10] L. Ronco and F. L. Pagola, "An Eigenvalue Sensitivity Approach to Location and Controller Design of Controllable Series Capacitors for Damping Power System Oscillations," *IEEE Transactions on Power Systems*, Vol. 12, No. 4, November 1997, pp. 1660–1666.

[11] IEEE Power Engineering Society, *FACTS Applications*, Document 96TP116-0, IEEE Press, New York, 1996.

[12] Y. H. Song and A. T. Johns, Eds., *Flexible AC Transmission Systems (FACTS)*, IEEE Press, U.K., 1999.

[13] CIGRE Working Group 14.29, "Coordination of Controls of Multiple FACTS/HVDC Links in the Same System," CIGRE Technical Brochure No. 149, Paris, December 1999.

[14] L. Angquist, B. Lundin, and J. Samuelsson, "Power Oscillation Damping Using Controlled Reactive Power Compensation—A Comparison Between Series and Shunt Approaches," *IEEE Transactions on Power Systems*, Vol. 8, No. 2, May 1993, pp. 687–700.

[15] E. V. Larsen, J. J. Sanchez-Gasca, and J. H. Chow, "Concepts for Design of FACTS Controllers to Damp Power Swings," *IEEE Transactions on Power Systems*, Vol. 10, No. 2, May 1995, pp. 948–955.

[16] The Electric Power Research Institute (EPRI) WO 3789-05, "Assessment of Applications and Benefits of Phasor Measurement Technology in Power Systems," Final Report by General Electric (GE) Company, EPRI Report TR-107903, April 1997.

[17] N. A. Mittlestadt, B. Furumasu, P. Ferron, and J. J. Paserba, "Planning and Testing for Thyristor Controlled Series Capacitors," *Proceedings: FACTS Conference 2*, EPRI Report TR-101784, December 1992.

[18] M. Klein, G. J. Rogers, and P. Kundur, "A Fundamental Study of Inter-Area Oscillations in Power Systems," *IEEE Transactions on Power Systems*, Vol. 6, 1991, pp. 914–920.

[19] N. Martins, H. J. C. P. Pinto, and L. T. G. Lima, "Efficient Methods for Finding Transfer Function Zeros of Power Systems," *IEEE Transactions on Power Systems*, Vol. 7, No. 3, August 1992, pp. 1350–1361.

[20] T. S. Luor, Y. Y. Hsu, T. Y. Guo, J. T. Lin, and C. Y. Huang, "Application of Thyristor-Controlled Series Compensators to Enhance Oscillatory Stability and Transmission Capability of a Longitudinal Power System," *IEEE Transactions on Power Systems*, Vol. 14, No. 1, February 1999, pp. 179–185.

[21] B. A. Francis, *A Course in H_∞ Control Theory*, Lecture notes in the control and information Sciences, Springer-Verlag, 1987.

[22] M. Klein, L. X. Le, G. J. Rogers, S. Farrokhpay, and N. J. Balu, "H_∞ Damping Controller Design in Large Power Systems," *IEEE Transactions on Power Systems*, Vol. 10, No. 1, February 1995, pp. 158–166.

[23] Q. Zhao and J. Jiang, "TCSC Damping Controller Design Using Robust Control Theory," *International Journal of Electrical Power and Energy Systems*, Vol. 20, No. 1, January 1998, pp. 25–33.

[24] Q. Zhao and J. Jiang, "Robust SVC Controller Design for Improving Power System Damping," *IEEE Transactions on Power Systems*, Vol. 10, No. 4, November 1995, pp. 1927–1932.

[25] X. R. Chen, N. C. Pahalwaththa, U. D. Annakkage, and C. S. Kumble, "Enhancement of Power System Stability by Using Controlled Series Compensation," *Electrical Power and Energy Systems*, Vol. 18, No. 7, 1996, pp. 475–481.

[26] C. E. Grund, R. V. Pohl, and J. Reeve, "Control Design of an Active and Reactive Power HVDC Modulatikon System with Kalman Filtering," *IEEE Transactions on Power Apparatus and Systems*, Vol. 101, No. 10, October 1982.

[27] E. V. Larsen, "Control Aspects of FACTS Applications," *Proceedings of EPRI Conference on FACTS: The Future of High-Voltage Transmission*, Cincinnati, OH, November 1990.

[28] G. N. Taranto and D. M. Falcao, "Robust Decentralized Control Design Using Genetic Algorithms in Power System Damping Control," *IEEE Proceedings—General Transmission Distribution*, Vol. 145, No. 1, January 1998, pp. 1–6.

[29] K. W. Braun and K. Renz, "Application Concepts of Advanced Series Compensation System," Paper 92-SP-180, *CEA Transactions of Engineering and Operating Division*, March 1992.

[30] N. G. Hingorani, "A New Scheme for Subsynchronous Resonance Damping of Torsional Oscillations and Transient Torque, Part I," *IEEE Transactions on Power Apparatus and Systems*, Vol. PAS–100, No. 4, April 1981, pp. 1852–1855.

[31] R. A. Hedin, K. B. Stump, and N. G. Hingorani, "A New Scheme for Subsynchronous Resonance Damping of Torsional Oscillations and Transient Torque, Part II, Performance," *IEEE Transactions on Power Apparatus and Systems*, Vol. PAS–100, No. 4, April 1981, pp. 1856–1863.

[32] N. G. Hingorani and L. Gyugyi, *Understanding FACTS*, IEEE Press, New York, 1999.

[33] W. Zhu, R. Spee, R. R. Mohler, G. C. Alexander, W. A. Mittelstadt, and D. Maratukulam, "An EMTP Study of SSR Mitigation Using the Thyristor-Controlled Series Capacitor," *IEEE Transactions on Power Delivery*, Vol. 10, No. 3, July 1995, pp. 1479–1485.

[34] R. A. Hedin, S. Weiss, D. Torgerson, and L. E. Eilts, "SSR Characteristics of Alternative Types of Series Compensation Schemes," Paper 94 SM 534-8 PWRS, Presented at IEEE/PES 1994 Summer Meeting, San Francisco, 1994.

[35] E. Larsen, C. Bowler, B. Damsky, and S. Nilsson, "Benefits of Thyristor-Controlled Series Compensation," Paper 14/37/38-04, CIGRE, Paris, 1992.

[36] C. E. J. Bowler, D. H. Baker, and C. G. Moran, "FACTS and SSR-Focus on TCSC Application and Mitigation of SSR Problems," *Proceedings of EPRI Conference on FACTS*, Boston, 1992.

[37] CIGRE Working Group 14.18, "Thyristor-Controlled Series Compensation," Technical Brochure, Paris, 1996.

[38] C. A. Canizares and Z. T. Faur, "Analysis of SVC and TCSC Controllers in Voltage Collapse," *IEEE Transactions on Power Systems*, Vol. 14, No. 1, February 1999, pp. 158–165.

[39] CIGRE Task Force 38.02.16, "Impact of Interactions Among Power System Controls," CIGRE Technical Brochure No. 166, Paris, August 2000.

[40] C. Gama, R. L. Leoni, J. Gribel, R. Fraga, M. J. Eiras, W. Ping, A. Ricardo, J. Cavalcanti, and R. Tenorio, "Brazilian North–South Interconnection—Application of Thyristor-Controlled Series Compensation (TCSC) to Damp Inter-Area Oscillation Mode," CIGRE Paper 14-101, Paris, 1998.

[41] G. Gama, G. Ingestrom, and L. Angquist, "Brazilian North–South Interconnection Control Application and Experience with a TCSC," *Proceedings of the IEEE/PES Summer Meeting*, Edmonton, Alberta, July 1999.

[42] D. Holmberg, M. Danielsson, P. Halvarsson, and L. Angquist, "The Stode Thyristor-Controlled Series Capacitor," CIGRE Paper 14-105, Paris, 1998.

[43] C. Gama and M. Noroozian, "Control Strategy for Damping of Power Swings Using TCSC," *CIGRE Symposium*, Kuala Lumpur, Malaysia, 1999.

[44] L. Angquist, G. Ingestrom, and H. A. Jonsson, "Dynamical Performance of TCSC Schemes," CIGRE Paper 14-302, Paris, 1996.

CHAPTER 9

Coordination of FACTS Controllers

9.1 INTRODUCTION

Flexible ac transmission system (FACTS) controllers either extend the power-transfer capability of existing transmission corridors or enhance the stability and security margins for given power-transmission limits. Fast controls associated with FACTS controllers do provide these system improvements, but they also can interact adversely with one another. In an interconnected power system, when the controller parameters of a dynamic device are tuned to obtain the best performance, the remaining power system is generally assumed to be passive or represented by slowly varying elements. This assumption is strictly not true; hence the adjusted parameters may not prove optimal when the dynamics of the various other controllers are, in effect, found in real systems [1].

This chapter presents different scenarios when various FACTS controllers interact unfavorably with one another. Linear-control techniques for coordinating the controls of different FACTS controllers are described, and nonlinear-control design methods that can be extended for the same purpose are described as well.

9.2 CONTROLLER INTERACTIONS

An excellent discussion on controller interactions is presented in refs. [1], [2], and [3]. Controller interactions can occur in the following combinations:

1. Multiple FACTS controllers of a similar kind.
2. Multiple FACTS controllers of a dissimilar kind.
3. Multiple FACTS controllers and HVDC converter controllers.

Because of the many combinations that are possible, an urgent need arises for power systems to have the controls of their various dynamic devices coordinated. The term *coordinated* implies that the controllers have been tuned simultaneously to effect an overall positive improvement of the control scheme [2].

The frequency ranges of the different control interactions have been classified as follows [1]:

- 0 Hz for steady-state interactions
- 0–3/5 Hz for electromechanical oscillations
- 2–15 Hz for small-signal or control oscillations
- 10–50/60 Hz for subsynchronous resonance (SSR) interactions
- >15 Hz for electromagnetic transients, high-frequency resonance or harmonic resonance interactions, and network-resonance interactions

9.2.1 Steady-State Interactions

Steady-state interactions between different controllers (FACTS–FACTS or FACTS–HVDC) occur between their system-related controls. They are steady state in nature and do not involve any controller dynamics. These interactions are related to issues such as the stability limits of steady-state voltage and steady-state power; included are evaluations of the adequacy of reactive-power support at buses, system strength, and so on. An example of such control coordination may be that which occurs between the steady-state voltage control of FACTS equipment and the HVDC supplementary control for ac voltage regulation.

Load-flow and stability programs with appropriate models of FACTS equipment and HVDC links are generally employed to investigate the foregoing control interactions. Steady-state indices, such as voltage-stability factors (VSF), are commonly used. Centralized controls and a combination of local and centralized controls of participating controllers are recommended for ensuring the desired coordinated performance.

9.2.2 Electromechanical-Oscillation Interactions

Electromechanical-oscillation interactions between FACTS controllers also involve synchronous generators, compensator machines, and associated power-system stabilizer controls [1]–[3]. The oscillations include *local mode* oscillations, typically in the range of 0.8–2 Hz, and *inter-area mode* oscillations, typically in the range of 0.2–0.8 Hz. The local mode is contributed by synchronous generators in a plant or several generators located in close vicinity; the inter-area mode results from the power exchange between tightly coupled generators in two areas linked by weak transmission lines.

Although FACTS controllers are used primarily for other objectives, such as voltage regulation, they can be used gainfully for the damping of electromechanical oscillations. In a coordinated operation of different FACTS controllers, the task of damping different electromechanical modes may be assumed by separate controllers. Alternatively, the FACTS controllers can act concertedly to damp the critical modes without any adverse interaction.

Eigenvalue analysis programs are employed for determining the frequency and damping of sensitive modes.

9.2.3 Control or Small-Signal Oscillations

Control interactions between individual FACTS controllers and the network or between FACTS controllers and HVDC links may lead to the onset of oscillations in the range of 2–15 Hz (the range may even extend to 30 Hz). These oscillations are largely dependent on the network strength and the choice of FACTS controller parameters, and they are known to result from the interaction between voltage controllers of multiple SVCs [4], the resonance between series capacitors and shunt reactors in the frequency range of 4–15 Hz [5]–[6], and so forth. The emergence of these oscillations significantly influences the tuning of controller gains.

Analysis of these relatively higher frequency oscillations is made possible by frequency-scanning programs, electromagnetic-transient programs (EMTPs), and physical simulators (analog or digital). Eigenvalue analysis programs with modeling capabilities extended to analyze higher-frequency modes as well may be used [7].

9.2.4 Subsynchronous Resonance (SSR) Interactions

Subsynchronous oscillations may be caused by the interaction between the generator torsional system and the series-compensated-transmission lines, the HVDC converter controls, the generator excitation controls, or even the SVCs [8]. These oscillations, usually in the frequency range of 10–50/60 Hz, can potentially damage generator shafts. Subsynchronous damping controls have been designed for individual SVCs and HVDC links. In power systems with multiple FACTS controllers together with HVDC converters, a coordinated control can be more effective in curbing these torsional oscillations.

Analysis techniques and tools for investigating these control interactions are similar to those stated in Section 9.2.3.

9.2.5 High-Frequency Interactions

High-frequency oscillations in excess of 15 Hz are caused by large nonlinear disturbances, such as the switching of capacitors, reactors, or transformers, for which reason they are classified as electromagnetic transients. Control coordination for obviating such interactions may be necessary if the FACTS and HVDC controllers are located within a distance of about three major buses.

Instabilities of harmonics (those ranging from the 2nd to the 5th) are likely to occur in power systems because of the amplification of harmonics in FACTS controller loops. Harmonic instabilities may also occur from synchronization or voltage-measurement systems, transformer energization, or transformer saturation caused by geomagnetically induced currents (GICs).

FACTS controllers need to be coordinated to minimize or negate such interactions. The techniques employed to analyze these interactions are similar to those stated in Section 9.2.3.

9.2.6 The Frequency Response of FACTS Controllers

The composite-frequency response of a FACTS controller, together with its associated ac system, provides a good indication of the control-system stability, especially while an attempt is made to coordinate several FACTS or HVDC controllers.

A time domain–based frequency-scanning method (FSM) [2], [9] is used for obtaining the frequency responses of individual and coordinated FACTS controllers. A current source is used to inject a spectrum of frequencies at the FACTS controller bus. The local voltage developed at the bus is measured, and its harmonic content is evaluated through the use of Fourier analysis. The simulations are performed with an EMTP that has detailed models of FACTS controllers.

To avoid the operation of any system component in its nonlinear region, the magnitudes of injected harmonic currents are chosen to be quite small, thereby ensuring linearized system behavior around the operating point. In HVDC converters, an injected-current magnitude is considered sufficiently small if it does not cause a firing-angle oscillation in excess of 0.5° [9]. Two frequency-response examples of FACTS controllers [1]—one for the SVC, the other for the TCSC—are presented in the following text.

9.2.6.1 The Frequency Response of the SVC The study system considered is shown in Fig. 9.1. A ±50 MVAR SVC is connected at the midpoint of the network that connects systems 1 and 2. The frequency response is obtained for two operating points.

At the first operating point, the SVC maintains a bus voltage of 1.02 pu, with a firing angle $\alpha = 102°$ corresponding to a reactive-power absorption of 22.5 MVAR (inductive). Small-magnitude harmonic currents, I_h, are injected

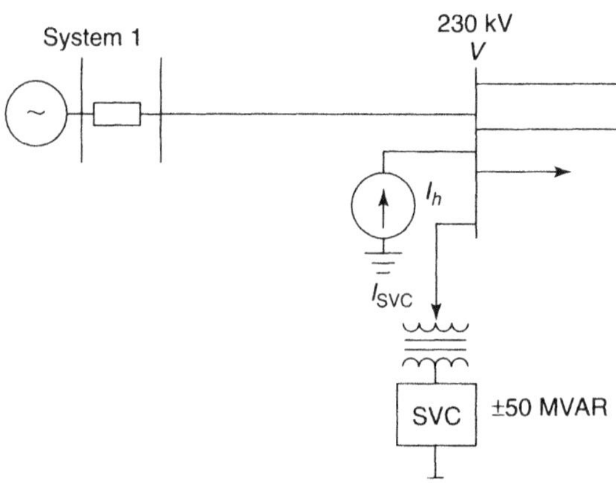

Figure 9.1 A study system for frequency scanning of the SVC.

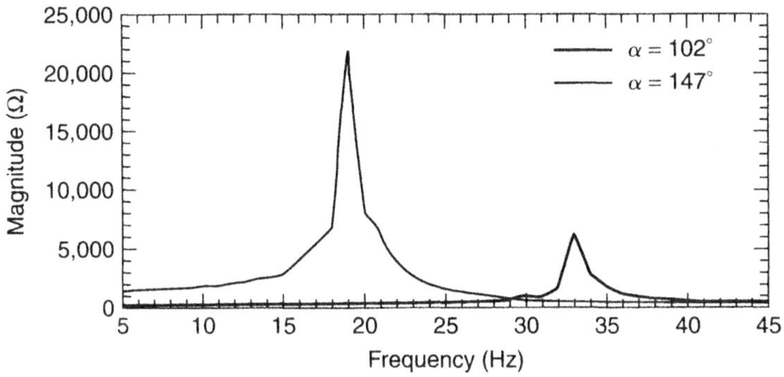

Figure 9.2 The impedance magnitude of the SVC frequency response.

at discrete frequencies ranging from 5 to 45 Hz. The corresponding impedances are computed as the ratio of the developed voltage and the injected-harmonic-disturbance-current components. The impedance magnitude and angle-frequency responses are plotted in Figs. 9.2 and 9.3, respectively. The SVC presents a parallel resonance at 33 Hz and behaves inductively from 5 to 33 Hz, becoming capacitive at resonance and tending to resume inductive behavior as the frequency is increased beyond 33 Hz.

The frequency response is obtained for the second steady state–operating point. The bus voltage is now regulated at 1.10 pu, with a thyristor firing angle $\alpha = 147^\circ$ corresponding to a reactive-power injection of 50 MVAR (capacitive). The corresponding magnitude and angle-frequency responses are, again, plotted in Figs. 9.2 and 9.3, respectively. It is seen that the resonant frequency modifies to 19 Hz and the impedance peak becomes three times that of the inductive SVC operation. The phase plot indicates that the higher the firing angle, the smaller the frequency range of inductive operation.

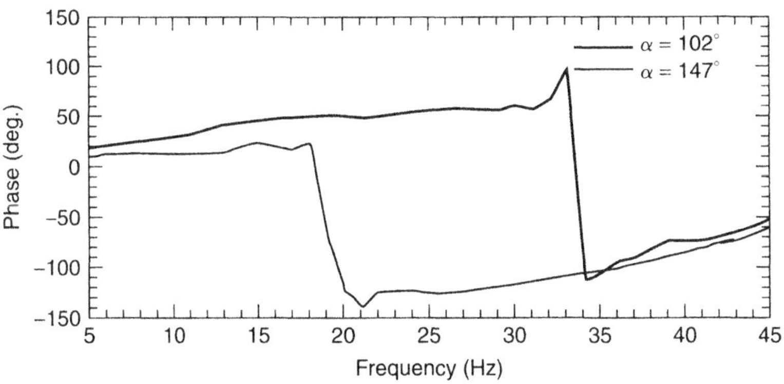

Figure 9.3 The impedance angle of the SVC frequency response.

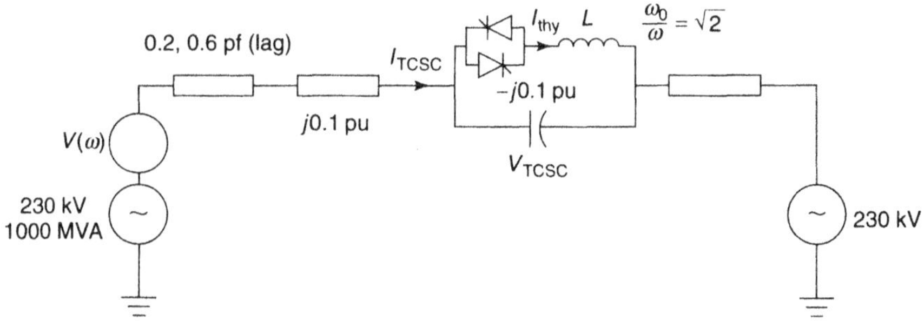

Figure 9.4 A study system for the TCSC frequency response.

9.2.6.2 The Frequency Response of the TCSC The 60-Hz test system used for evaluating the TCSC frequency response is depicted in Fig. 9.4. The frequency response is obtained for two conduction angles: 80° and 86°. The impedance magnitude and angle plots are illustrated in Fig. 9.5 and Fig. 9.6, respectively. It is evident that the TCSC-compensated system presents an inductive behavior until it reaches 45–50 Hz; thereafter, its behavior tends to become capacitive, resembling that of a pure capacitor.

9.3 SVC–SVC INTERACTION

9.3.1 The Effect of Electrical Coupling and Short-Circuit Levels

A detailed case study of control interaction between multiple SVCs in a large power system is given in refs. [3] and [10]. The interaction phenomena are investigated as functions of electrical distance (electrical coupling) between the SVCs and the short-circuit level at the SVC buses.

9.3.1.1 Uncoupled SVC Buses A simplified test system shown in Fig. 9.7 is considered for the interaction analysis performed through eigenvalue anal-

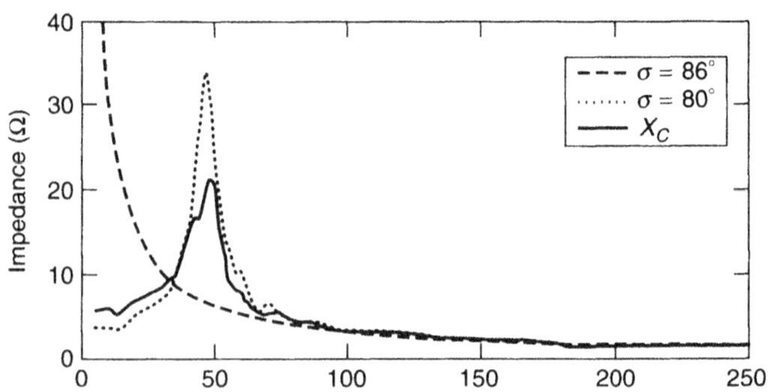

Figure 9.5 The impedance-magnitude plot of the TCSC frequency response.

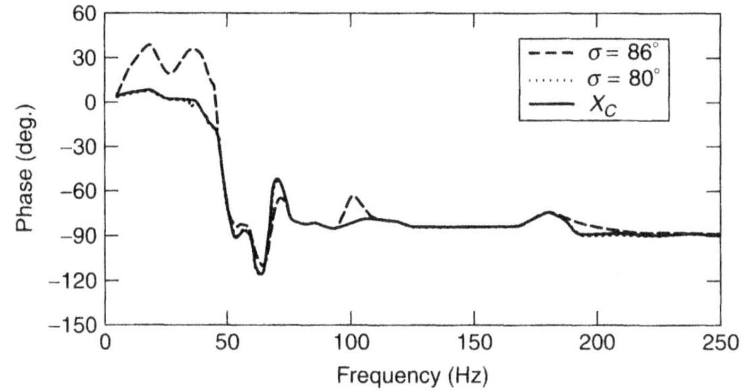

Figure 9.6 The impedance-angle plot of the TCSC frequency response.

yses and root-loci plots. All the generating units are represented by infinite buses. If the transfer reactance between buses 1 and 2 is high, making the buses electrically uncoupled, then the SVCs connected to those buses do not interact adversely. Increasing the proportional gain of SVC 1 connected to bus 1, even to the extent of making the SVC unstable, does not affect the eigenvalues of SVC 2—implying that the controller designs of SVCs can be done independently for multiple SVCs in a power system if the transfer reactance between their connecting buses is high.

9.3.1.2 Coupled SVC Buses If, however, the reactance between the two SVC buses is low, it constitutes a case of high electrical coupling between the SVCs. Here again, two possibilities exist with respect to short-circuit capacity of the region where the SVCs are installed: the SVC region with a high short-circuit capacity and the SVC region with a low short-circuit capacity.

For high short-circuit capacity conditions in the same system as Fig. 9.7, refs. [3] and [10] reveal that by increasing the proportional gain of one SVC, the eigenvalues of the other SVC are impacted very slightly. Almost no control interaction exists between the two SVCs irrespective of their electrical coupling, as long as they are in a high short-circuit-level region, that is, when the ac system is stiff. The reason for this condition is that the interlinking variable between the two SVCs is the bus voltage. Thus the controls of both SVCs can be independently designed and optimized, but if the short-circuit capacity of the SVC region is low, varying the proportional gain of SVC 1 will strongly influence the eigenvalues associated with SVC 2. It is therefore imperative that a coordinated control design be undertaken for both SVCs.

Despite simplifications in the study system and in the analysis approach, the aforementioned interaction results are general, for the phenomena investigated are independent of the number of buses, transmission lines, or generators.

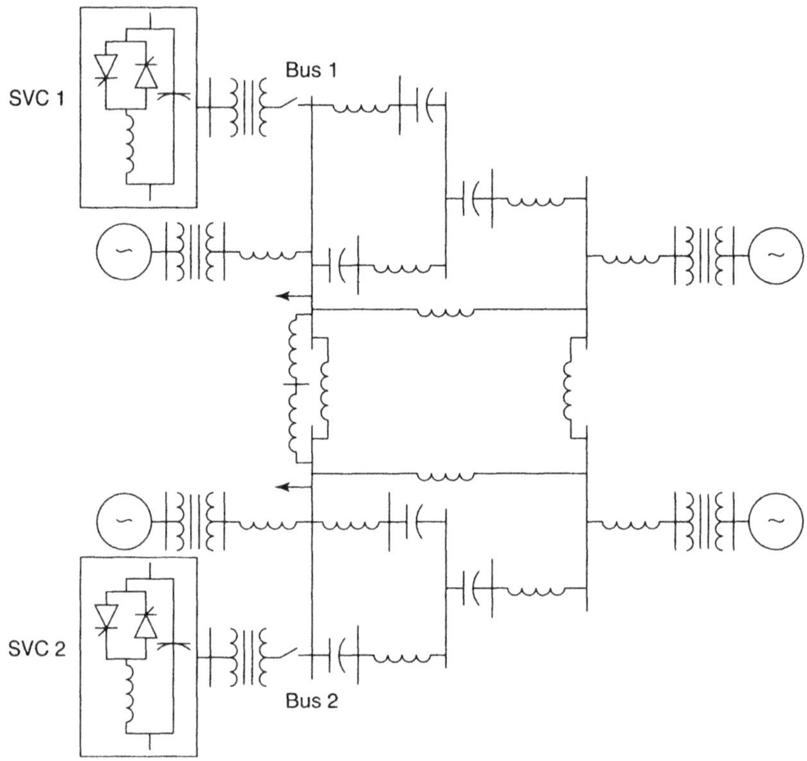

Figure 9.7 An SVC interaction-analysis network.

9.3.2 The System Without Series Compensation

In power systems with several SVCs, the maximum response rate of the SVCs is limited by the stability of individual voltage-control loops [6] during major contingencies.

9.3.2.1 Study System The Hydro-Quebec summertime power system, shown in Fig. 9.8 and studied in ref. [6], considers a possible option of 30 SVCs distributed along major transmission networks. In the study presented in ref. [6], control interaction between the SVCs was investigated for the summer transmission system and was revealed to be more crucial than that of the winter transmission system not only because of the reduced short-circuit levels but the lesser loads as well. It was noted that loads contribute to the damping of power-system oscillations.

A critical contingency that influences the SVC interactions is the loss of two lines south of La Verendrye—the *La Verendrye contingency.* The performance of SVCs is examined over a range of operating conditions spanning the zero output to the maximum capacitive output. The SVC voltage regulator is mod-

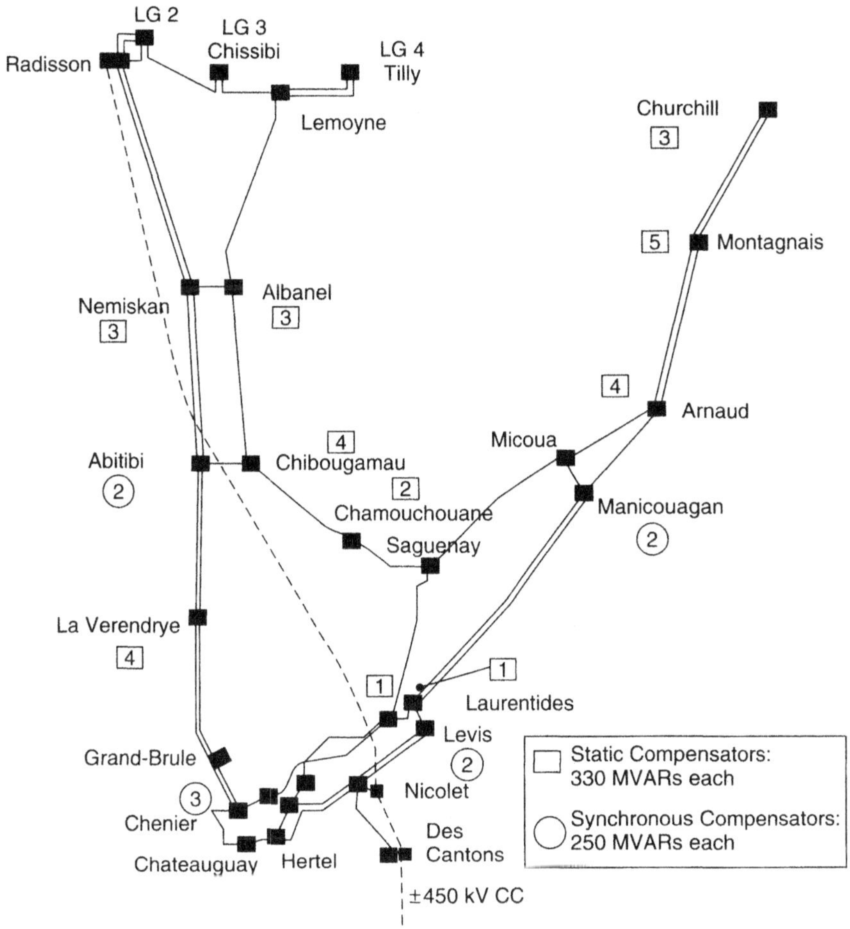

Figure 9.8 A proposed Hydro-Quebec summertime study system with shunt compensation only.

eled by the gain–time-constant representation: the gain is the inverse of the SVC slope, whereas the time constant is the response rate. On the ac side of the SVC measurement systems are notch filters (80 Hz and 96 Hz) to counteract network resonances; on the dc side are low-pass and harmonic-notch filters to obtain the pure dc equivalent of the SVC bus voltage. The interaction between SVC controllers is examined through both eigenvalue analysis and the simulation of transients by using EMTP.

The response to small-reactor switching for the La Verendrye contingency is depicted in Fig. 9.9(a). The initial conditions were selected as those likely to occur 30 cycles after this contingency. Increasing oscillations of 16-Hz frequency are noticed in the transient response because of the adverse interaction

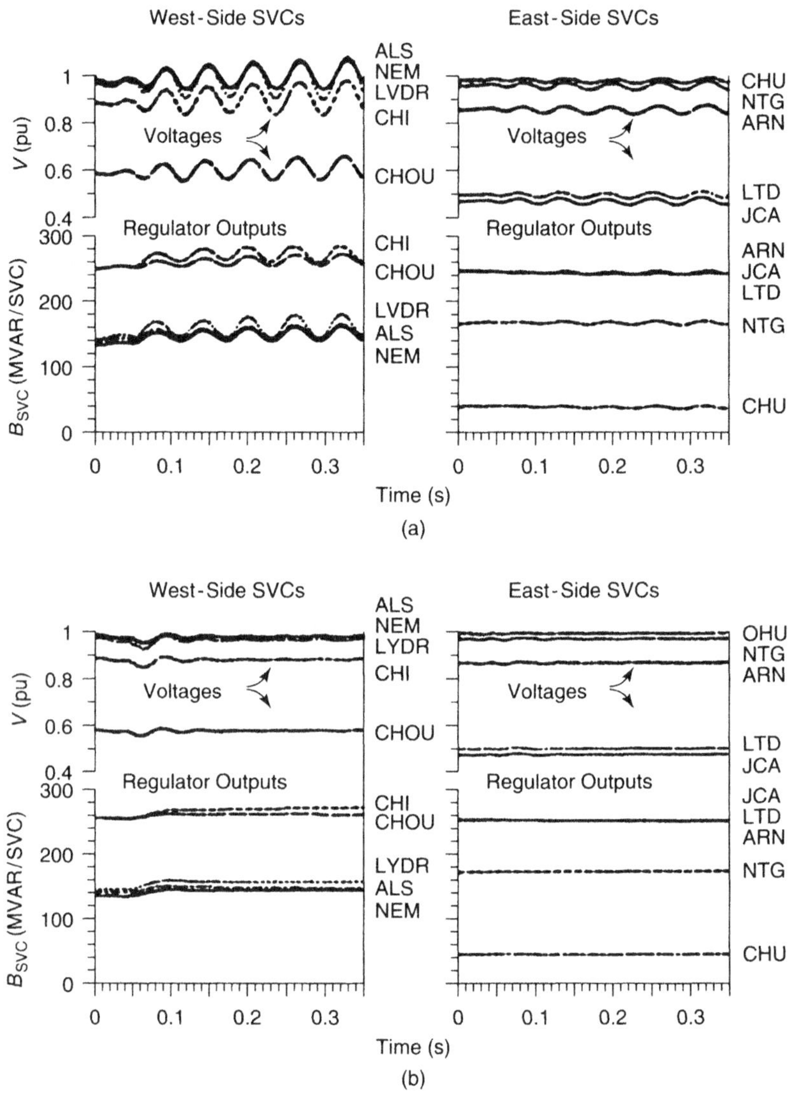

Figure 9.9 The SVC transient behavior in La Verendrye system due to "snapshot" reactor switching at Abitibi. (a) The existing SVC response rate ($T_r = 0.133$ s) and (b) the reduced SVC response rate ($T_r = 0.5$ s).

between the fast SVC controllers. An increase in the regulator time constant (the slowing down of the SVC) from 0.133 s to 0.5 s stabilizes the response as depicted in Fig. 9.9(b)—an activity that is also clearly revealed from the root loci of critical modes with varying controller-response rates, T_r, as shown in Fig. 9.10. In the figure, the 16-Hz mode is unstable for $T_r = 0.133$ s (symbol 2) but stable for $T_r = 0.5$ s (symbol 4).

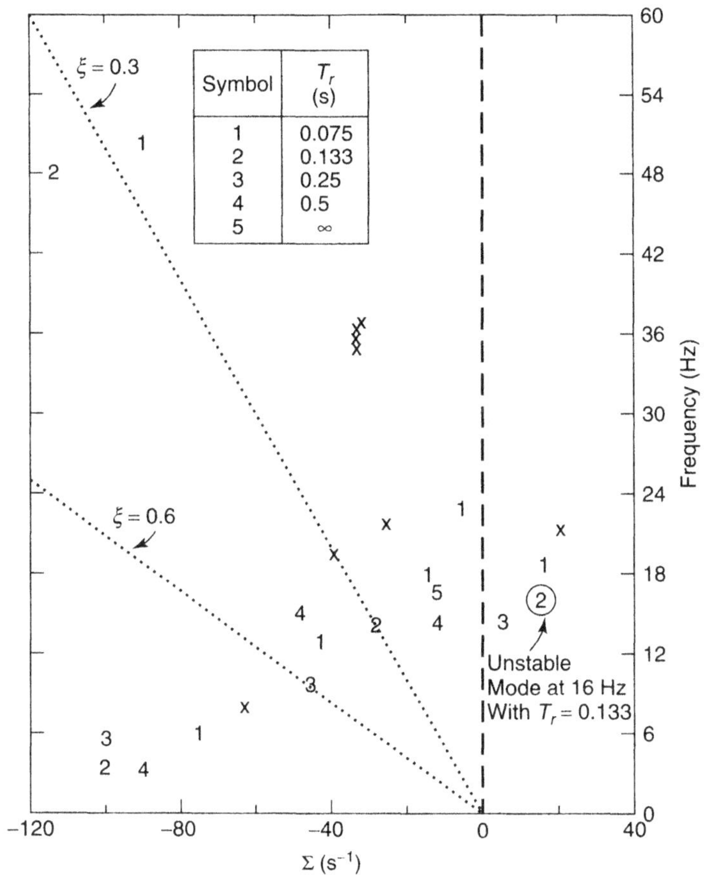

Figure 9.10 The effect of the SVC response rate on system eigenvalues in La Verendrye system stimulated by reactor switching at Abitibi.

Although Fig. 9.10 illustrates the root loci for a general operating point, Fig. 9.11 illustrates the loci for an operating point with highly capacitive SVCs (280 MVAR/SVC). The system in Fig. 9.11 is unstable, even with greatly slowed SVCs (T_r = 0.5 s), and clearly represents the worst-case scenario. To alleviate this condition, gain supervisors need to be installed on the SVCs [11]. These, however, are not without problems; even though they slow the response rate of the relevant SVCs to stabilize the oscillations, the damping of the oscillations is very low, causing an altogether poor system response after the peak of the first major system swing. This low-damping oscillation can be resolved only by the selective tripping of certain SVCs, for which, again, an elaborate evaluation of all likely contingencies must be performed and the tripping sequences developed. Obviously, therefore, a completely shunt-compensated system is not desirable from the viewpoint of voltage-regulator stability [6].

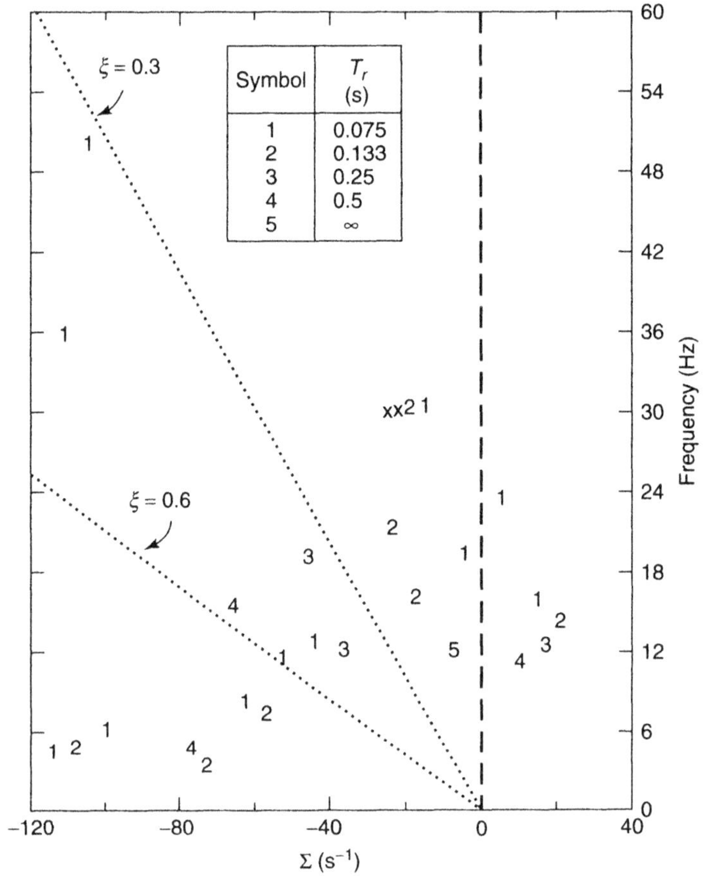

Figure 9.11 The effect on system eigenvalues for a contingency case with a 2-line loss south of La Verendrye—all SVCs at 280 MVAR.

Another study presented in ref. [6] shows that when multiple SVCs are connected on the same line, the controller mode of the SVC with the lower effective short-circuit ratio (ESCR) becomes susceptible to instability. The difference in ESCR, as viewed from the TCR terminals of each SVC, is caused by the unequal rating of the SVCs; hence a coordination in controller parameters of the different SVCs is necessary. The interactions between the voltage controllers become more prominant with high controller gains.

The use of high voltage (HV)–side instead of low voltage (LV)–side voltage as the feedback signal effectively makes the SVCs become closer electrically and more interlinked. The HV-side voltage feedback lowers both the damping and the frequency of SVC controller modes—an effect that is even greater for higher reactances of the coupling transformer.

9.3.3 The System With Series Compensation

The main problem with multiple SVCs is the presence of large shunt capacitance to the network. This capacitance interacts with the network reactance, creating a shunt-capacitance-resonant mode that, after demodulation through the measurement system, interacts adversely with the voltage-regulator mode. It is therefore desirable to series-compensate the network and use a smaller number of SVCs. A comparative case study is presented in ref. [6] for a proposed alternative Hydro-Quebec system that is mostly series-compensated. All major 735-kV lines are assumed to be 30–40% series-compensated, and a total of 13 SVCs are placed in the system as shown in Fig. 9.12.

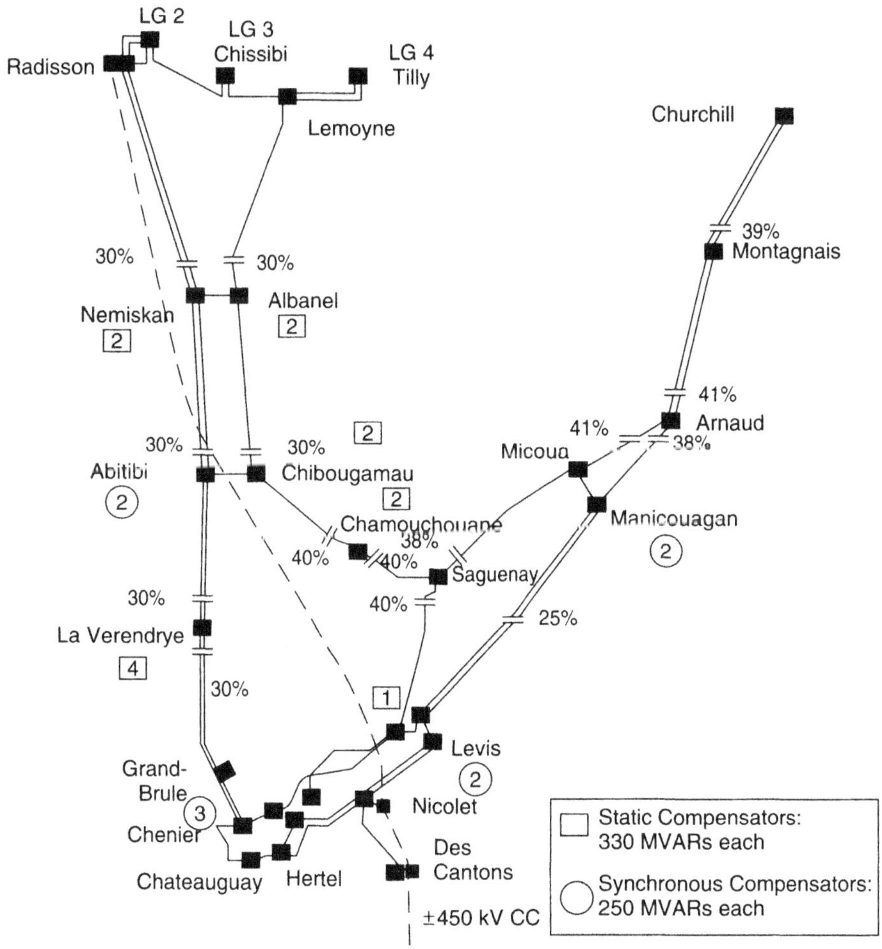

Figure 9.12 Proposed Hydro-Quebec summertime study system with both series and shunt compensation.

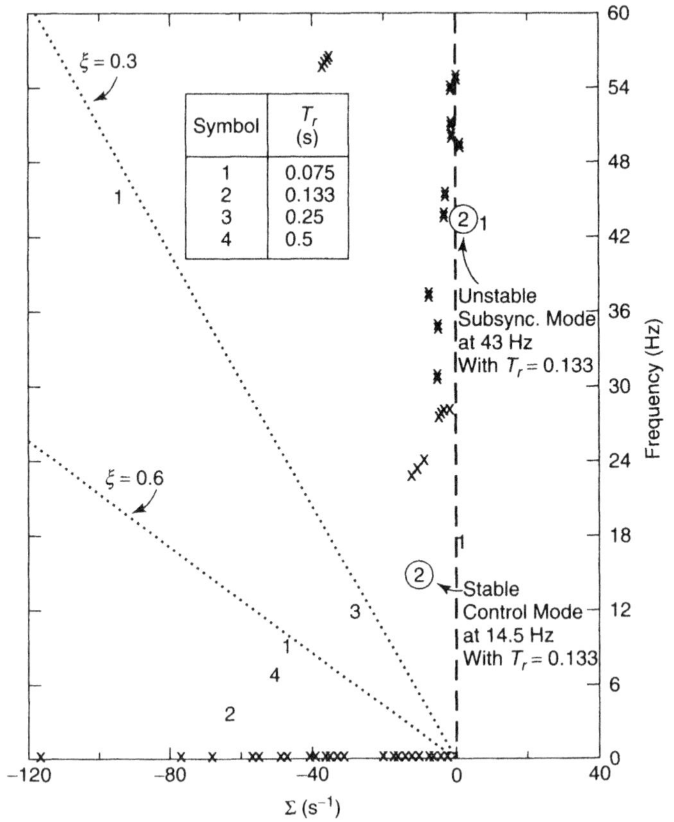

Figure 9.13 The effect of the SVC response rate on system eigenvalues for a mostly series-compensated system due to a 2-line loss south of La Verendrye.

The root loci for this system, corresponding to the La Verendrye contingency, are illustrated in Fig. 9.13. Because of the reduced shunt capacitance, with a smaller number of SVCs and a lowered line reactance from series compensation, the shunt-capacitance-resonant mode shifts to a higher mode frequency of 83 Hz that, following the demodulation effect of the measurement system, appears as a 83 − 60 = 23 Hz component that has much less interaction with the basic SVC voltage–regulator mode. In the totally shunt-compensated system, the shunt-capacitance mode is of 77 Hz, which in Fig. 9.10 appeared as 77 − 60 Hz = 17 Hz in the root loci and was indicated as 0.5 ($T_r = \infty$, i.e., no regulator response).

As with the case in Fig. 9.10, that of Fig. 9.13 must also have its SVC response rate lowered to ensure voltage-regulator stability. This reduction, however, need not be of the same extent; for instance, reducing the SVC response rate from 0.133 s to 0.25 s stabilizes the overall system.

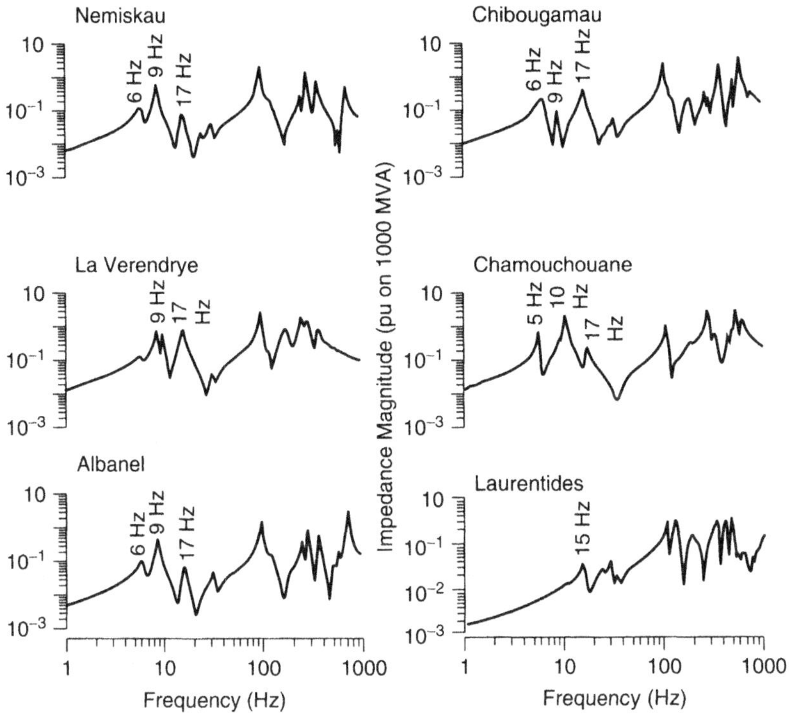

Figure 9.14 Impedance versus frequency from SVC locations for a mostly series summertime network (La Verendrye contingency).

9.3.3.1 Shunt-Reactor Resonance If shunt reactors are present in the system (which they usually are), the series compensation of the transmission line introduces additional resonant modes from the interaction of series capacitance with the inductance of shunt inductors [4], [5]. These modes are visible in the driving-point-impedance plots, as seen from the different SVC locations depicted in Fig. 9.14, and usually lie in the range of 0–20 Hz. For the study system, three major modes are indicated at 6 Hz, 9 Hz, and 17 Hz, the magnitudes of which vary with location. Also, these modes are associated with substantially low damping because of the very small line resistance and the high Q-factor of shunt reactors.

In the root loci of Fig. 9.13, the 17-Hz shunt-reactor-resonant mode is reflected as its 60-Hz complement, that is, 43 Hz (60 Hz − 17 Hz), again from the demodulation process in the measurement system. As the SVC response rate is increased (by reducing T_r), these modes become unstable.

The adverse interaction between SVCs and the shunt-reactor modes can be minimized by installing a high-pass filter with a cutoff frequency of typically 15–20 Hz on the ac-side measurement circuit. The root loci with and without a 15-Hz high-pass filter for the study system is illustrated in Fig. 9.15. It is

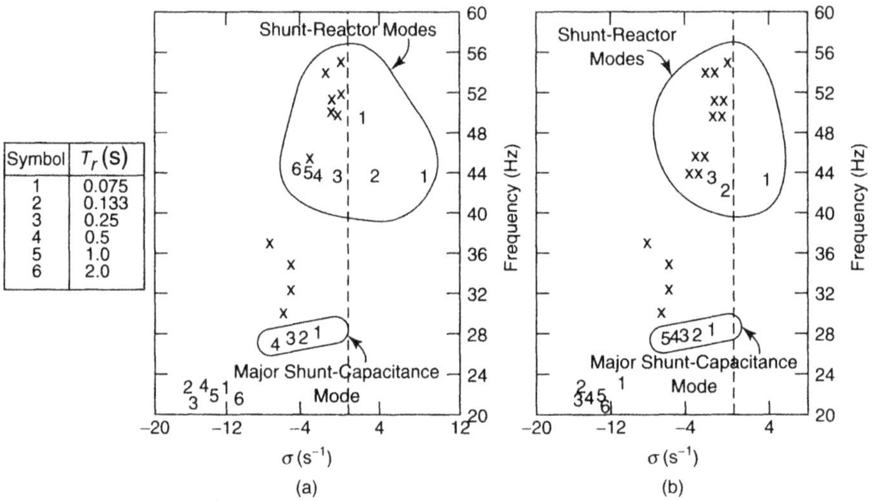

Figure 9.15 Effect of the high-pass filter on subsynchronous modes for a mostly series-compensated La Verendrye system (contingency at 280 MVAR/SVC): (a) with no high-pass filter and (b) with a 15-Hz high-pass filter.

evident that without the filter, the system would be barely stable for a response rate $T_r = 0.25$ s; however, it achieves a reasonably high damping with the filter.

The transient time responses of the system with and without the 15-Hz filter are illustrated in Fig. 9.16, which confirm the observations from root loci. The shunt-reactor mode is evidenced in the high-frequency (40–50 Hz) oscillations in the time responses.

The cutoff frequency of the high-pass filters is obtained as a trade-off between the stability of low-frequency modes and that of the shunt-reactor modes. The reason is that when the cutoff frequency is increased to stabilize the shunt-reactor modes, some lower-frequency modes become undamped.

9.3.4 High-Frequency Interactions

An example of high-frequency control interaction between two SVCs in the same electrical area is presented in ref. [1]. Figure 9.17 depicts the study system comprising two SVCs each rated +30 to −70 MVAR and operating on the voltage-control mode. Because high-frequency interactions are being analyzed, it may be noted that the transmission lines are represented as π sections. The total system is modeled by 78 state variables.

To analyze the control interaction, one SVC is modeled in detail using the generalized-switching-functions approach [12]–[14], while the second SVC is represented by a passive equivalent network with same reactive-power consumption at the operating point under consideration. In Fig. 9.18, the equivalent system used for the control design of SVC 1 is shown.

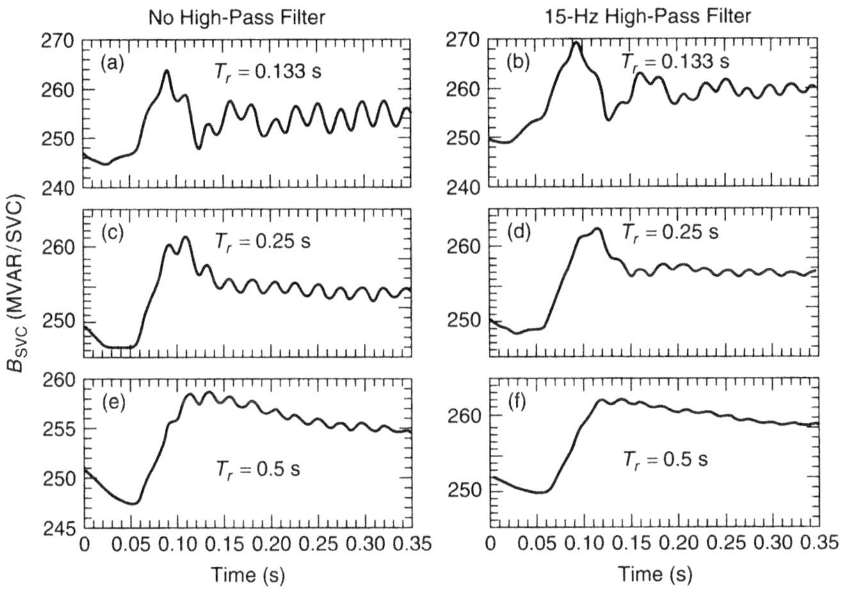

Figure 9.16 The SVC transient in a series-compensated system due to the SVC response rate and the high-pass filter (the regulator output is at Chibougamau).

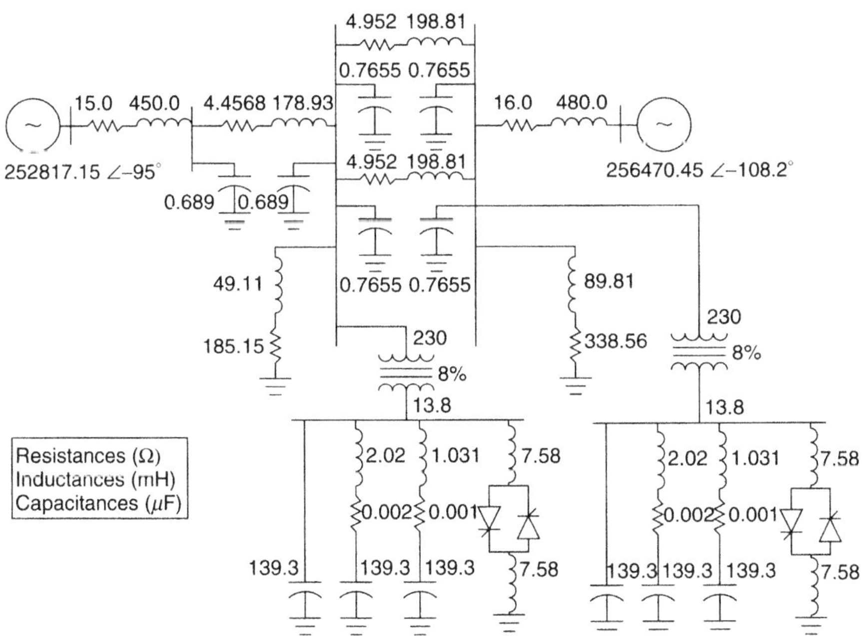

Figure 9.17 A study system for the analysis of high-frequency interaction between SVCs.

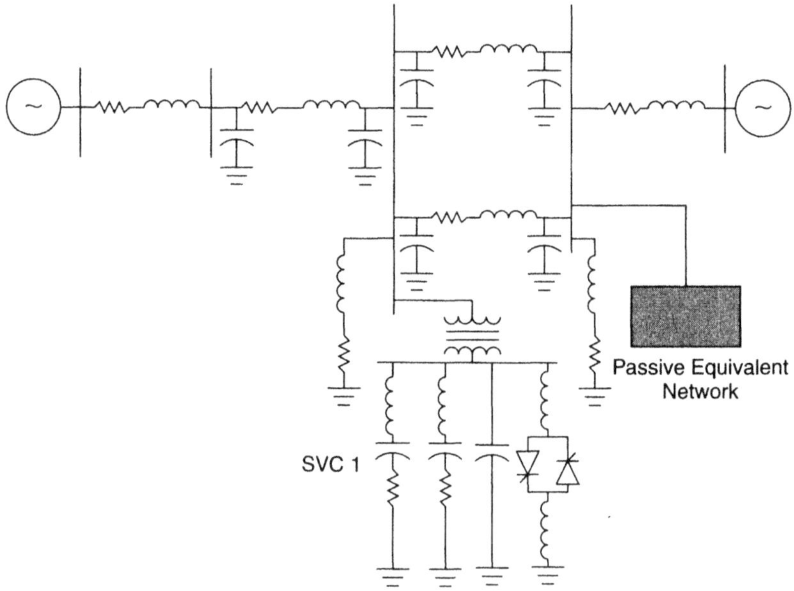

Figure 9.18 An equivalent system for the control design of SVC 1.

The SVC controller is designed to give an acceptable rise, overshoot, and settling times. Eigenvalue analysis reveals that the dominant oscillation mode has an eigenvalue of λ_{SVC1} $(= \sigma + j2\pi f) = -19 \pm j119.7$. This mode constitutes an oscillation of frequency 19.1 Hz in SVC 1's response to a 2% step in the reference-voltage input, as shown in Fig. 9.19. The system simulation is performed using a nonlinear switching function–based EMTP-type simulation.

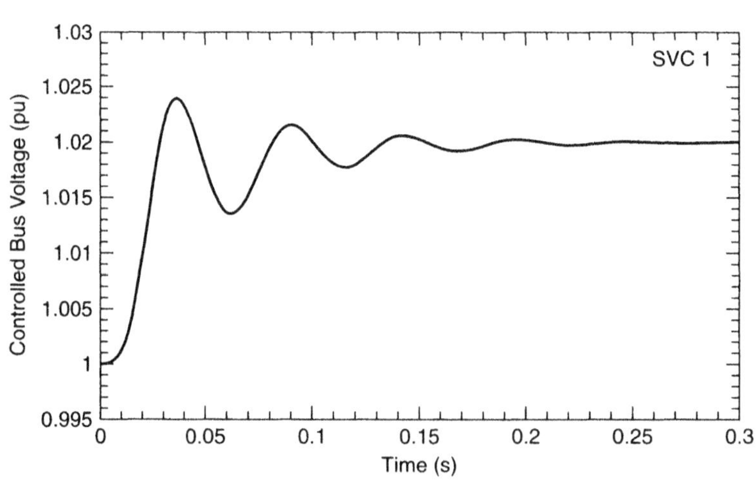

Figure 9.19 The response of SVC 1 to a step input in the reference voltage.

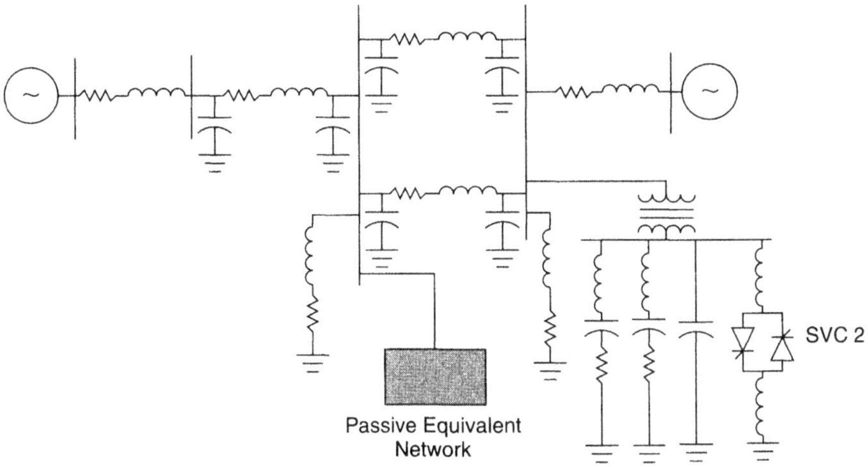

Figure 9.20 An equivalent system for the control design of SVC 2.

The SVC 2 controller is similarly designed. In this case, SVC 1 is represented by a passive equivalent network, as depicted in Fig. 9.20. The dominant mode of oscillation has an eigenvalue of $\lambda_{SVC\,2} = -19.7 \pm j118.5$ and constitutes an oscillation of 18.9 Hz in SVC 2's response to a 2% step in the reference-voltage input, as depicted in Fig. 9.21.

When both independently designed SVC controllers are together put into operation in the same electrical area shown in Fig. 9.17, an unstable operation results. The dominant oscillation mode has an associated eigenvalue of $\lambda = +3.2 \pm j129.9$, leading to an unstable oscillation of frequency 20.7 Hz in both SVCs'

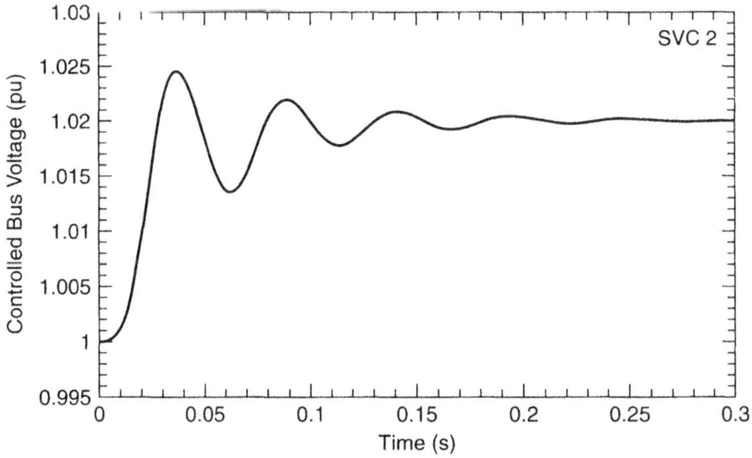

Figure 9.21 The response of SVC 2 to a step input in the reference voltage.

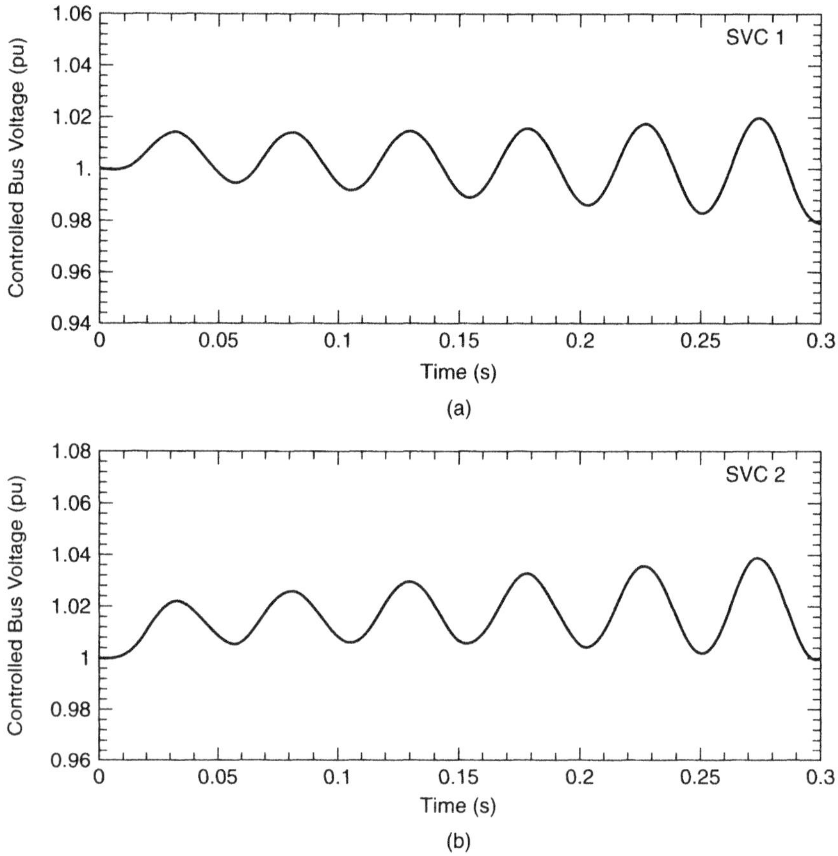

Figure 9.22 The response of independently designed SVCs to a step input in the reference voltage of SVC 2.

responses when a 2% step input is applied to the reference voltage of SVC 2, as depicted in Fig. 9.22.

This phenomenon clearly illustrates the adverse high-frequency interaction between the two independently designed SVC controllers, and brings out the need for a coordinated control design of the two SVCs. A high-frequency eigenvalue analysis program is employed, and the gains of both SVC controllers are adjusted simultaneously to stabilize the unstable mode. The response of both SVCs to a 2% reference step in SVC 2 is presented in Fig. 9.23. This response is rapid as well as stable.

The frequency response is then obtained for the transfer function relating the controlled bus voltage with the reference voltage of SVC 1. Responses are presented in Fig. 9.24 for the original SVC control (the other SVC is represented by a passive equivalent), as well as controllers designed independently and those designed coordinately. The resonant peak of the original SVC transfer

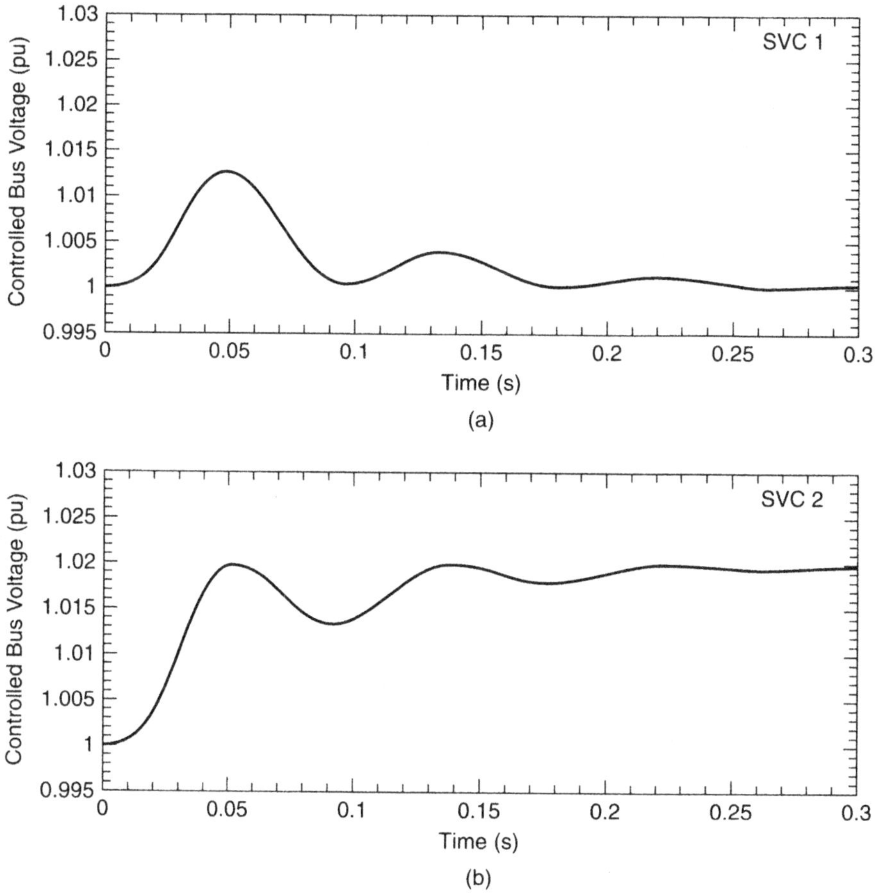

Figure 9.23 The response of coordinately designed SVCs to a step input in the SVC 2 reference voltage.

function is enhanced substantially during the independent design of SVC controllers, leading to instability. A coordinated design significantly reduces the resonant peak, thereby stabilizing the overall system.

9.3.5 Additional Coordination Features

9.3.5.1 Parallel SVCs If in a subsystem two parallel SVCs connected at the same bus exist, the slope settings in the voltage-control loops may be made comparable with the system short-circuit level to ensure an appropriate sharing of reactive-power compensation that, during any system exigency, are provided by the two SVCs.

Even though the two SVCs may be able to independently control the switching of internal shunt-capacitive filter banks and external shunt reactors on the

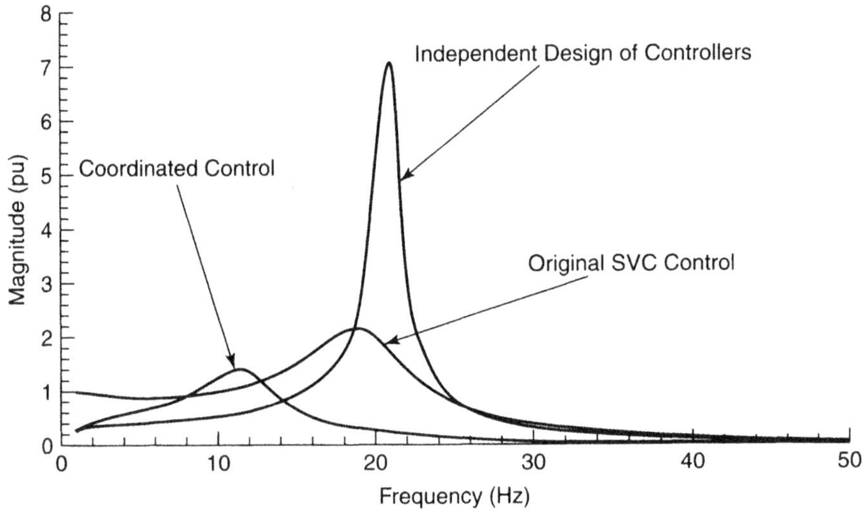

Figure 9.24 The frequency response of SVC 1.

EHV bus, a communication link may be provided between the two SVCs for coordinating this activity when both SVCs are connected to the same bus bar. This link can establish a master/slave SVC control, whereby one SVC can cyclically control the external reactors and filter banks of both SVCs.

The features described here exist in the Cape transmission SVCs of the ESKOM Cape Town transmission system in South Africa [1].

9.3.5.2 Electrically Close SVCs In some situations, SVCs do a self-determination of their voltage-controller loop gains, based on their droop setting and the system strength [1]. A SVC periodically estimates the system strength by injecting a small MVAR step into the system and measuring the corresponding variation in the bus voltage. It then uses a lookup table to select its gain corresponding to the calculated system strength. This scheme inherently assumes that only one SVC performs the voltage regulation in that electrical area.

However, if there are multiple SVCs in close electrical proximity for voltage control, the preceding gain computation may provide erroneous results. The reason is that when one SVC injects a reactive-power step, the other SVCs respond to correct the bus voltage. A communication data link is therefore established between the various SVCs to coordinate the gain selection. If the link indicates that the other SVCs are also participating, the computed gain is correspondingly reduced on all SVCs to ensure system stability. In the ESKOM system used in the Natal province of South Africa, where two SVCs are located in close vicinity the gains on both SVCs are reduced by a multiplying factor of 0.55 [1].

9.4 SVC–HVDC INTERACTION

A parametric analysis of the possible interaction between the SVC and two closely coupled HVDC links [3], [10] has revealed that there is no control interaction if the short-circuit capacity of the ac systems is high. The interrelating variable between the two controllers is the ac voltage, as the SVC and HVDC converter are connected to the same bus. Hence the controls of both can be designed independently. However, if the short-circuit level of ac system is low, strong interaction is likely to occur between the SVC and HVDC converter controls, requiring them to be designed coordinately. This control interaction may not always be construed as undesirable; after all, the SVC in this situation can be used to prevent voltage collapse at the HVDC inverter station. Nevertheless, if the gain of the SVC voltage regulator is increased substantially, the HVDC link may become unstable.

Multiple SVCs are shown to adversely interact with the HVDC when they are made to operate with fast response rates and especially so when they function in a highly capacitive output mode. A case study of interaction between multiple SVCs and the Radisson–Nicolet HVDC system in a proposed Hydro-Quebec system with extensive series compensation is illustrated in Fig. 9.25 for the root loci with and without the HVDC system.

In the figure, the root loci indicate that a high SVC response rate, $T_r = 0.075$ s, keeps the system well stabilized without the HVDC link but makes it nearly unstable when the HVDC system is inducted. However, if the SVC response rate is slower—$T_r \geq 0.25$ s, for instance—the HVDC system would not conflict with SVC controls.

From the case study of Fig. 9.25, the HVDC controls themselves are assumed to be tuned independently to provide a well-damped response without SVCs. Although the HVDC control system itself can interact adversely with the system's shunt-capacitance-resonance modes, its control is adjusted to avoid conflicts with SVCs. Furthermore, the location of the HVDC system is such that it prevents the system from interacting unfavorably with the shunt-reactor modes.

The SVCs are usually installed at HVDC converter terminals for regulating ac bus voltage and for damping any low-frequency, inter-area oscillations in the transmission system. The HVDC converters are also equipped with a reactive power controller (RPC) that controls either the ac bus voltage or the reactive-power exchange with the ac system. In addition, the RPC supervises the switching of capacitor banks, harmonic filters, and shunt reactors. The SVCs act during transients; however, the RPC is mainly a steady-state controller.

Both the RPC and the SVCs operate with different time constants and hence do not interact adversely with one another. If an SVC comprises switchable elements, such as a thyristor-switched capacitor (TSC) or a thyristor-switched reactor (TSR), the coordination established is such that the SVC initiates its switching control after the maximum and minimum voltage limits of the RPC are violated [1].

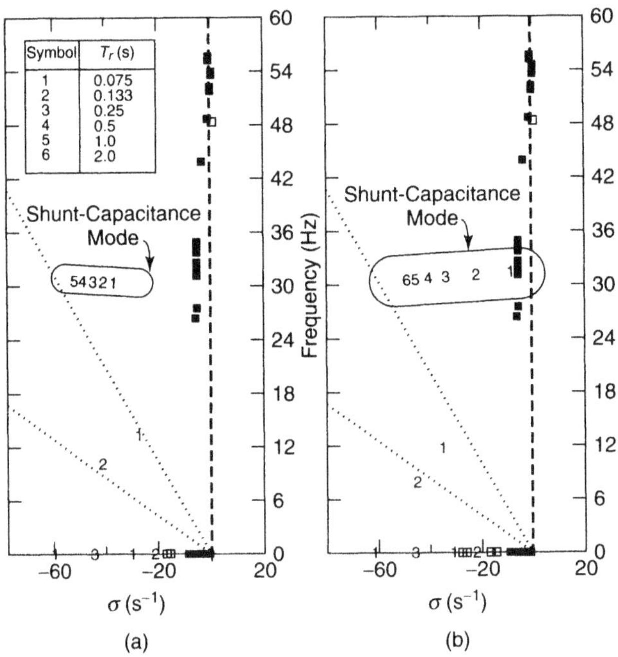

Figure 9.25 The effect of the Radisson–Nicolet HVDC on SVC interaction for a mostly series wintertime network (280 MVAR/SVC): (a) with no HVDC and (b) with HVDC.

9.5 SVC–TCSC INTERACTION

Both SVCs and TCSCs can be combined and used gainfully within power systems to enhance inter-area stability [3], [10]. The SVCs provide voltage support, whereas TCSCs contribute synchronizing torque to improve the electromechanical stability. Also, the TCSCs generate reactive power that improves the voltage stability of the system. However, the control-coordination issues of the SVCs and TCSCs must be addressed carefully so that no control conflicts arise between their fast controls.

A detailed investigation, in the form of a planning study, of SVC–TCSC interaction for the New York State transmission system is presented in ref. [15] and shown in Fig. 9.26. The SVC- and TCSC-compensated lines are connected to the same 345-kV bus at Fraser. The series-compensated line between Edic and Fraser comprises series capacitors of 45% line reactance and TCSC of 10% line reactance.

The frequency variation of system impedance, as viewed from the TCR terminals, is obtained by including in the system the TSCs together with the SVC filter branches and the fixed-series capacitors in the transmission line. Illustrated in Fig. 9.27 is a subsynchronous-system resonance of 30 Hz and a first super-

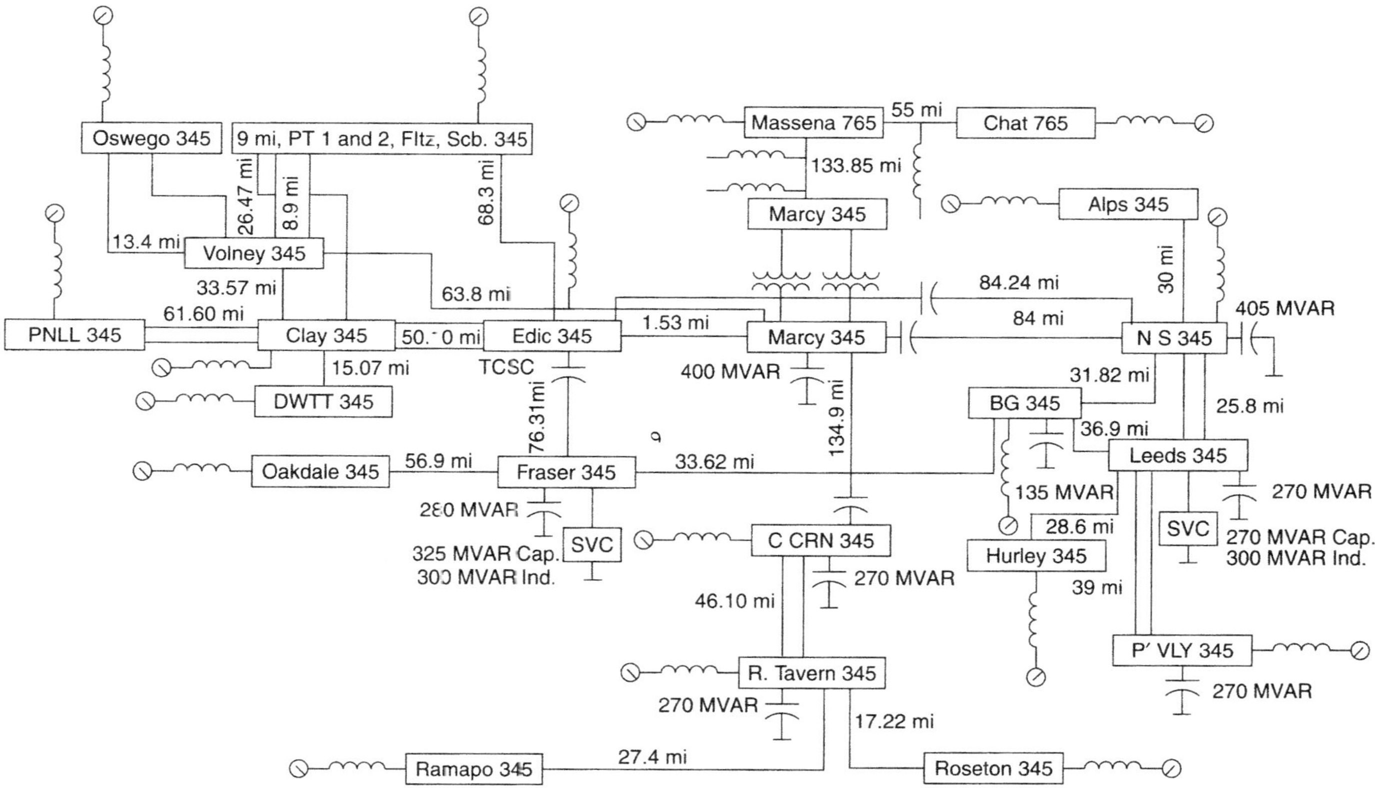

Figure 9.26 New York State 345-kV and 765-kV transmission-system model.

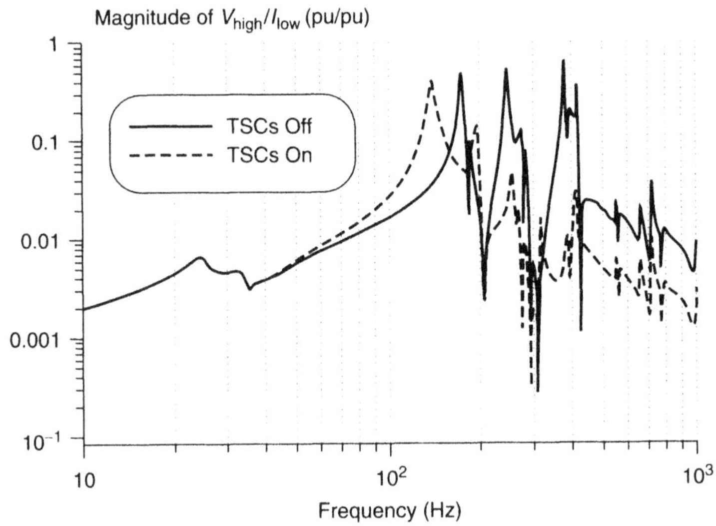

Figure 9.27 Driving-point impedance at the Fraser bus for a base-case-system model with and without TSCs.

synchronous resonance of 140 Hz, which shifts upward to 170 Hz if the TSCs are excluded.

When a power-swing damping controller (PSDC) is not considered on the TCSC, no adverse interaction between the TCSC and SVC controls is observed. The line-series capacitor excites lightly damped subsynchronous oscillations that are then damped successfully by the TCSC operating in a constant-reactance (1.1-pu reactance) mode. The variations of the TCSC voltage, the Edic bus voltage PSDC output, the Fraser SVC output, the measured voltage, and the reference-voltage signals are all depicted in Fig. 9.28. The main interaction is observed between the PSDC of TCSC and the SVC voltage controllers, as both of them are fast controllers. This interaction is examined for two PSDC input signals, each discussed in the following text.

9.5.1 Input Signal of the TCSC–PSDC With Bus Voltage

The Edic bus voltage is taken as the input signal for the PSDC. First, the signal is passed through a low-pass filter of a 1-ms time constant to remove any high-frequency noise; then, it is passed through a 2nd harmonic–notch filter. This notch filter is used to eliminate 2nd harmonics introduced by the nontransposed transmission lines.

The responses of the SVC and TCSC (with the voltage-input PSDC) are indicated in Fig. 9.29. As soon as the TCSC is brought into circuit at 0.02 s, poorly damped oscillations are excited at a frequency that is complementary to the network-series-resonant frequency of 30 Hz, that is, at 60 Hz − 30 Hz = 30

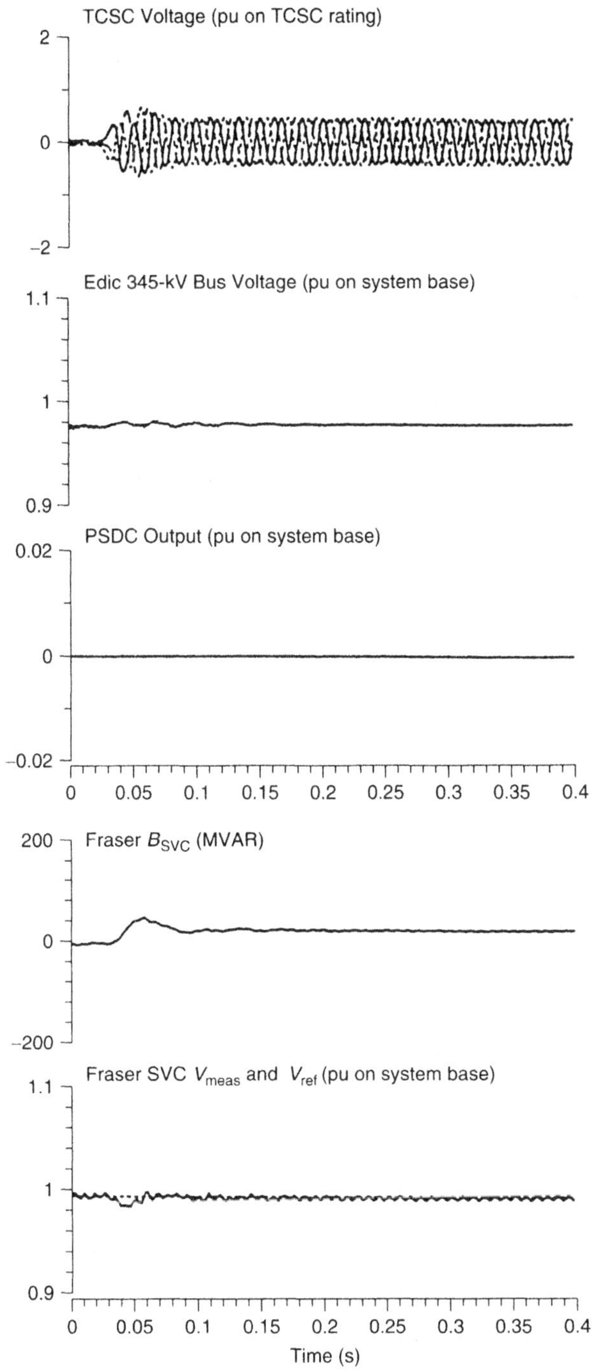

Figure 9.28 The TCSC and SVC response to TCSC insertion without a PSDC.

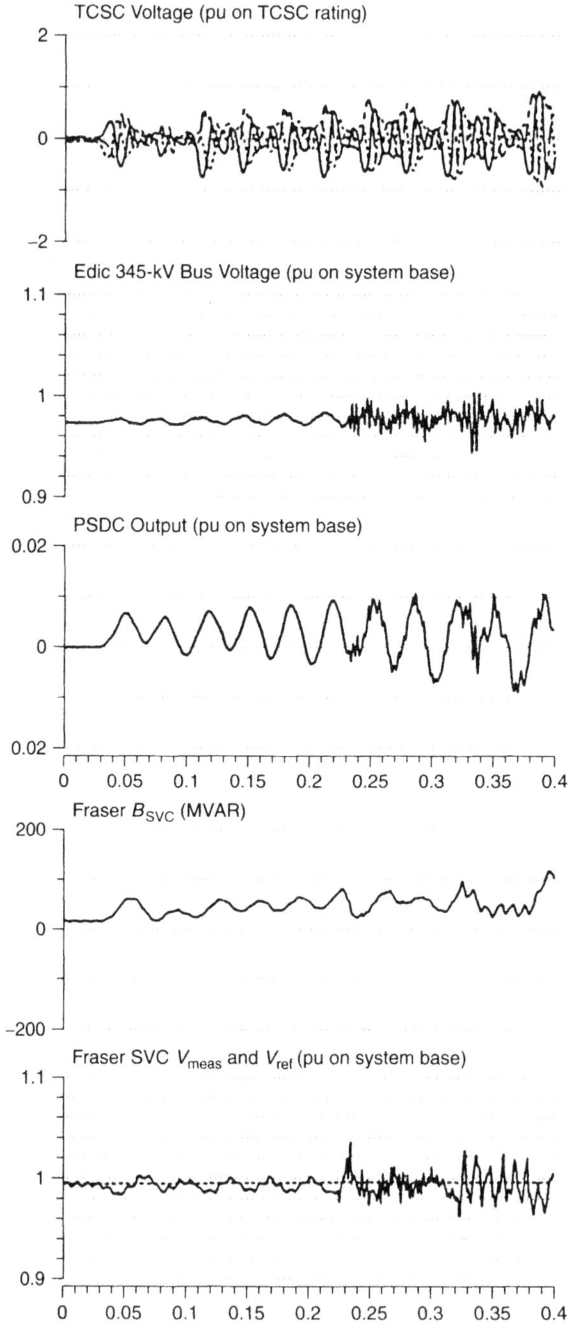

Figure 9.29 The TCSC and SVC response to TCSC insertion with a base voltage-input PSDC.

Hz. This activity demonstrates an adverse interaction between the TCSC and the series-compensated line resonance. The SVC voltage control exacerbates this interaction, as discussed in ref. [5]. This interaction can be avoided by appropriate filtering of the undesirable high-frequency signals. A double-low-pass filter with a cutoff frequency near 4 Hz (40-ms time constant) is included in the PSDC transfer function, which then modifies to

$$G(s) = \frac{X_{\mathrm{TCSC}}(\mathrm{pu})}{V_{\mathrm{Edic}}(\mathrm{pu})} = \frac{K(s0.21)(1 + s0.57)}{(1 + s0.21)(1 + s0.04)^2(1 + s0.5)} \tag{9.1}$$

The new responses of the SVC and TCSC are illustrated in Fig. 9.30. It is interesting to note that although the 30-Hz oscillations are damped out, 10-Hz new oscillations start appearing in the response. These oscillations persist but do not increase in magnitude, indicating the stability limit with the voltage-input PSDC. To impart a sufficient gain margin in the system, the PSDC gain is reduced by half.

9.5.2 Input Signal of the TCSC–PSDC With a System Angle

The angular difference, or *system angle*, between synthesized remote bus voltages is examined as an alternative PSDC input signal to obtain an improved stability limit. This signal is computed from local measurements of bus voltages, line current, and synthesizing impedance that are approximately equivalent to the short-circuit impedance. The synthesizing impedances, however, are modified to minimize potential control interaction, as discussed in Chapter 8.

The TCSC and SVC responses with the modified PSDC transfer function given in Eq. (9.1) are depicted in Fig. 9.31. The PSDC gain with the system-angle input is reduced by a factor of 4, as compared to that of the voltage-input PSDC, to ensure compatibility in responses. This reduction occurs because the peak-to-peak amplitude of the system angle is four times the peak-to-peak amplitude of the Edic bus voltage input. Both the 10-Hz and 30-Hz oscillations are well stabilized with the system-angle-input PSDC.

Thus it is seen that the appropriate filtering of high-frequency signals that lie outside the desired TCSC control bandwidth mitigates the adverse interaction between the TCSC and PSDC, the series-compensated line, and the SVC.

9.5.3 High-Frequency Interactions

Extensive investigations on high-frequency-control interactions between the SVC and TCSC in a complex power system that has several FACTS controllers and HVDC links embedded, as shown in Fig. 9.32, are presented in refs. [1], [3], and [10]. In the system of Fig. 9.32, the HVDC links are modeled by equivalent constant-current loads. First, TCSC 1 is optimized independently, ignoring the presence of SVC 3 that is connected to the same bus. The optimized response of this TCSC to a current-order step is obtained through EMTP simulation, as

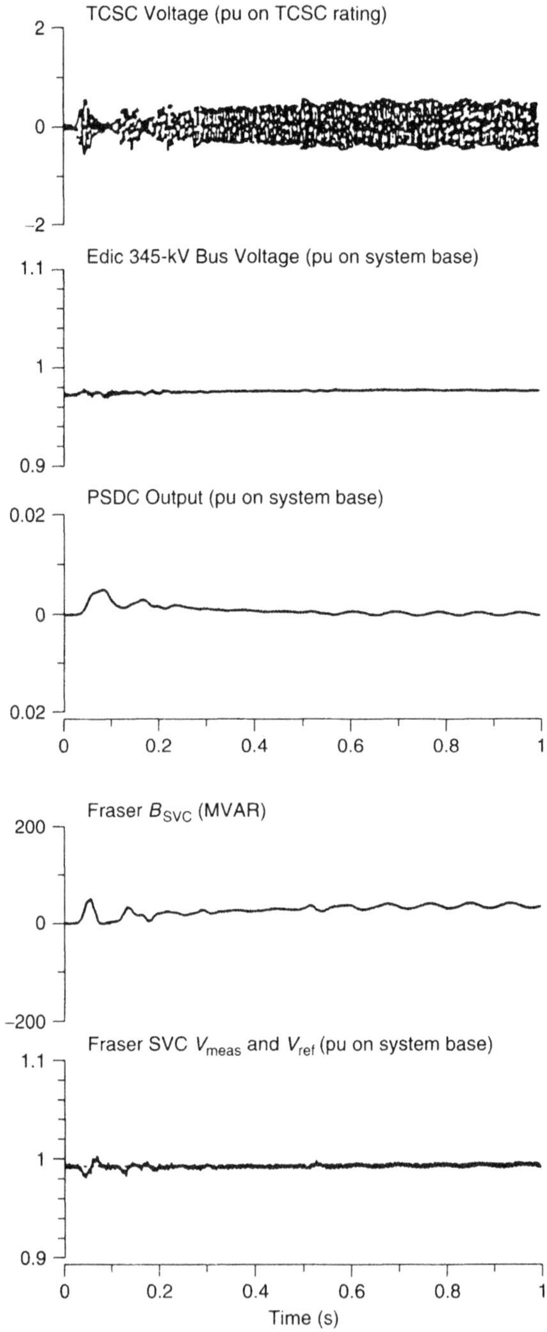

Figure 9.30 The TCSC and SVC response to TCSC insertion with a modified voltage-input PSDC.

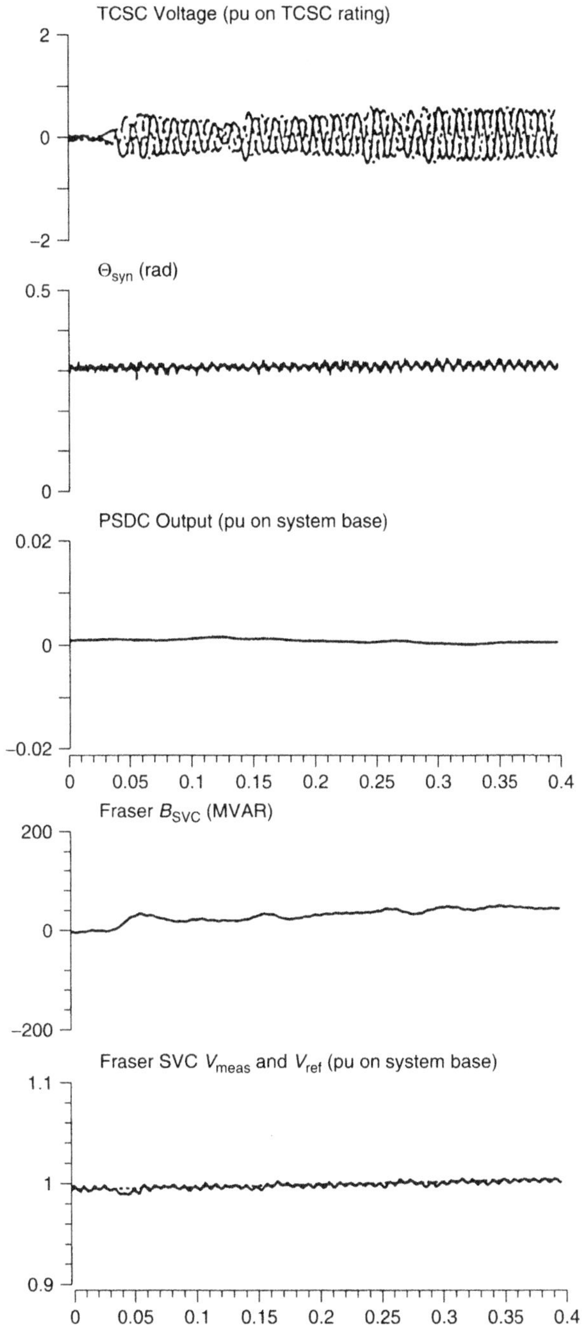

Figure 9.31 The TCSC and SVC response to TCSC insertion with an angle-input PSDC.

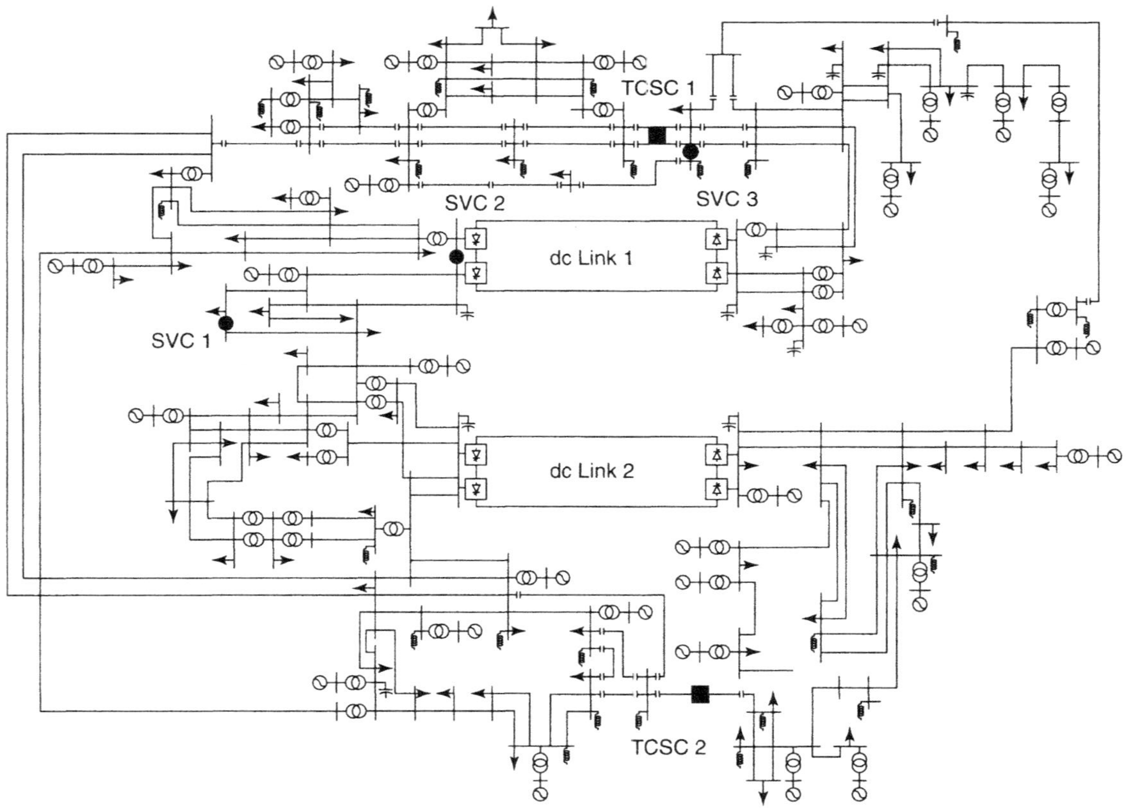

Figure 9.32 A multi-infeed, multi-FACTS study network.

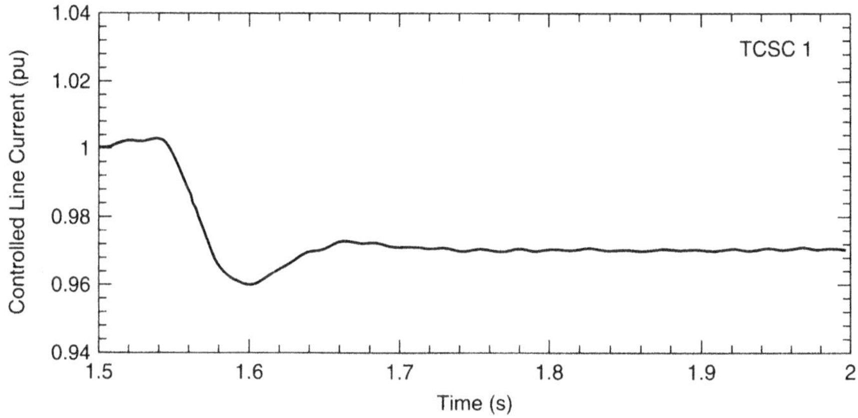

Figure 9.33 An independently optimized TCSC controller.

shown in Fig. 9.33. Next, the voltage controller of SVC 3 is optimized independently, with TCSC 1 not considered. The well-damped response of SVC 3 to a voltage-reference step input is shown in Fig. 9.34. When the independently optimized SVC 3 and TCSC 1 are considered connected, the responses of both the SVC and the TCSC to a reference-voltage step input to SVC 3 are as depicted in Fig. 9.35. These responses, obtained through EMTP simulation, indicate an unstable 44-Hz high-frequency oscillation from the adverse interaction between the two controllers in the same electrical area.

This study makes the need for a simultaneous global optimization of both controllers clearly evident [3], [16]. A procedure is suggested in ref. [3] whereby the SVC controller is first optimized with a TCSC controller, which is subsequently optimized with the optimized SVC controller in place. The proce-

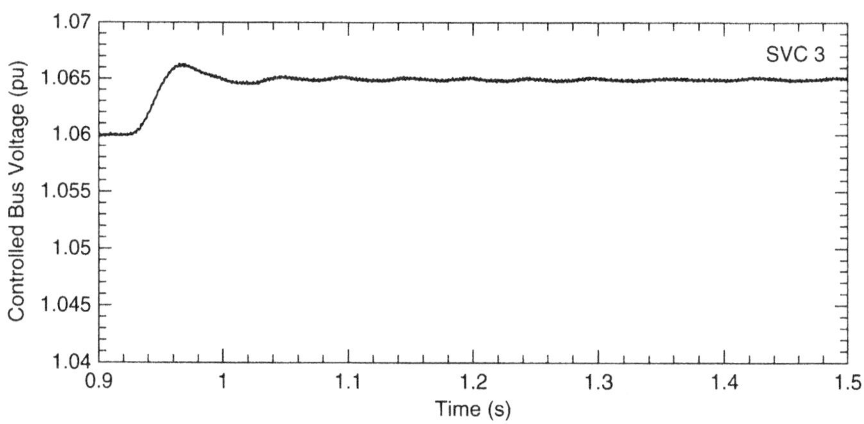

Figure 9.34 An independently optimized SVC controller.

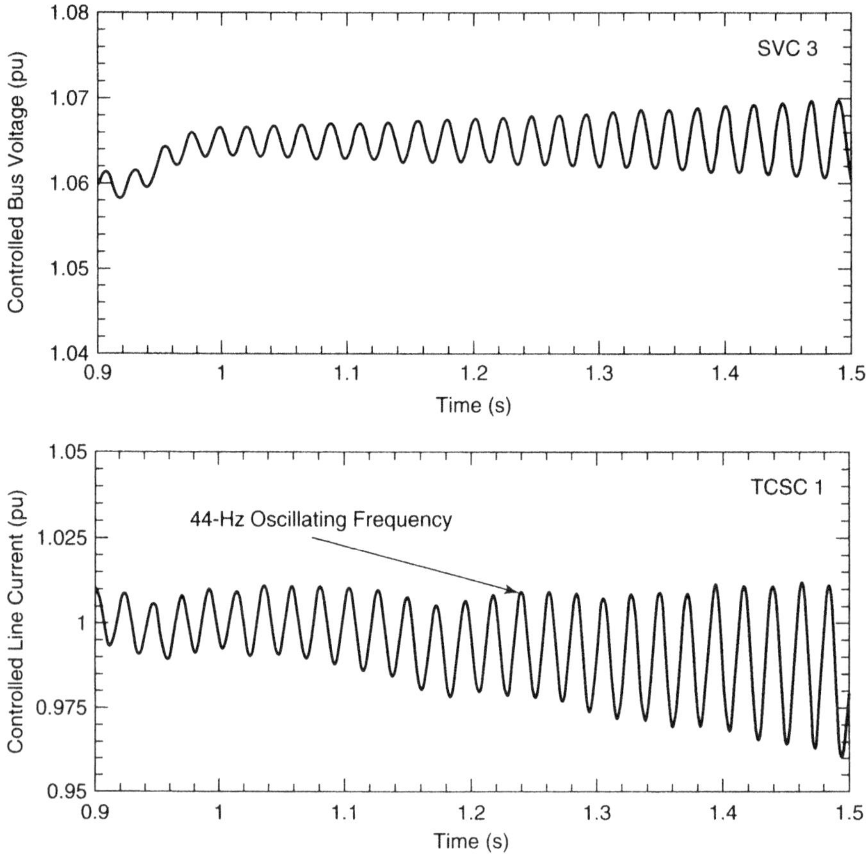

Figure 9.35 The combined operation of the independently designed SVC and TCSC controllers.

dure's objective is to obtain a fast rise time with minimum overshoot and settling time. The influence of the TCSC controller optimization is then checked on the SVC response; if necessary, the SVC controller is optimized once again and the procedure repeated. This process usually converges rather quickly.

After adopting the foregoing global-coordinated optimization process for determining the TCSC and SVC controller parameters, the responses of both controllers to an SVC 3 reference-voltage step input are obtained and illustrated in Fig. 9.36 [3]. The response of both controllers is very fast and stable.

It is important to note that the high-frequency interaction of Fig. 9.36 was detected through EMTP simulation. Conventional eigenvalue analysis–based programs model their transmission lines through algebraic equations, not through differential equations; hence the high-frequency "self-modes" of these FACTS controllers in large power systems do not become visible, for which reason EMTP-based simulations become necessary. Such simulations do not

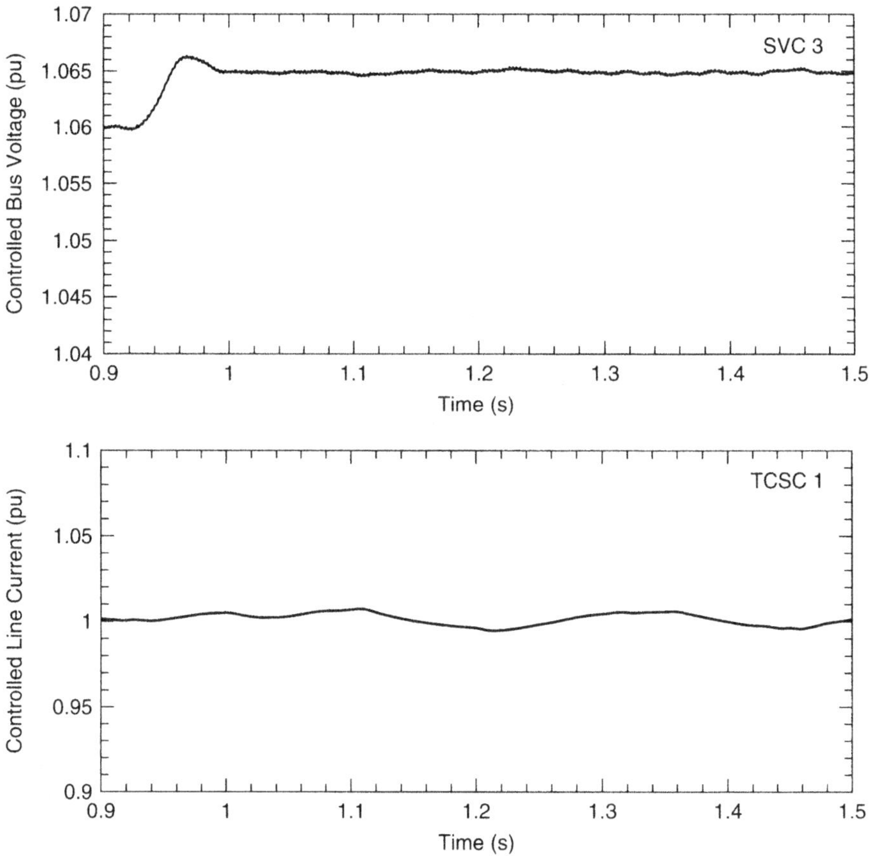

Figure 9.36 The combined operation of the coordinated designed SVC and TCSC controllers.

undermine the utility of algebraic modeling of transmission lines in conventional eigenvalue analysis–based programs, for they are adequate to examine the low-frequency 0.1- to 2-Hz electromechanical-mode stability.

9.6 TCSC–TCSC INTERACTION

9.6.1 The Effect of Loop Impedance

Control interaction between multiple TCSCs is investigated in refs. [1], [3], and [10]. Two cases of electrical coupling between the TCSCs are examined in the following text: low-loop impedance and high-loop impedance.

9.6.1.1 Low-Loop Impedance In the low-loop impedance scenario of Fig. 9.37, varying the gain of the TCSC 2 controller strongly influences not only its

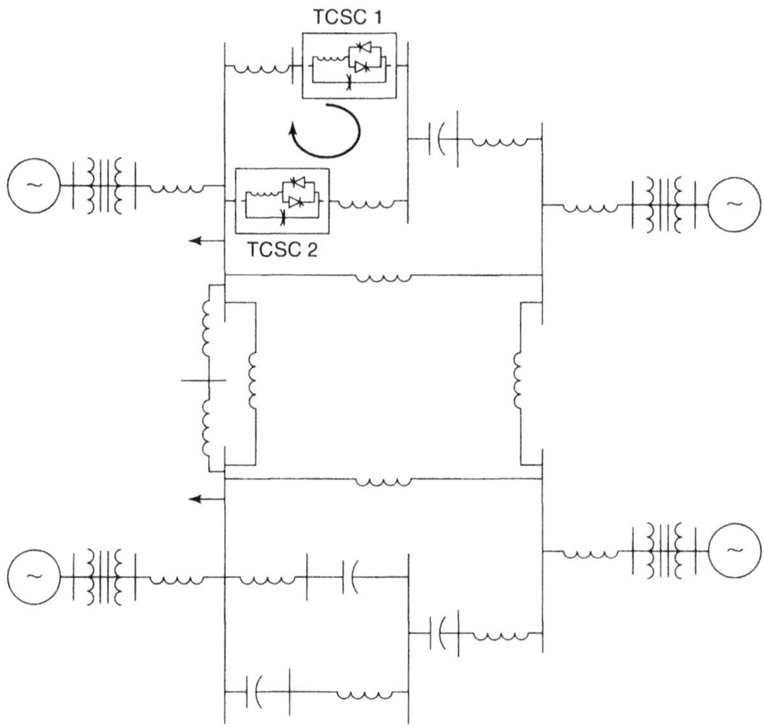

Figure 9.37 A low-loop impedance configuration with coupled TCSCs.

own eigenvalues but also those associated with TCSC 1. Thus a strong interaction occurs between the two TCSCs. The interconnecting variable between the two TCSCs is the ac line current; hence, a coordinated design of both controllers must be made to obviate any adverse interaction.

9.6.1.2 High-Loop Impedance In the high-loop impedance scenario of Fig. 9.38, even a large variation in the controller gain of TCSC 2 does not affect the eigenvalues of TCSC 1. This lack of effect reveals that no control interaction exists between the two TCSCs, for which reason they can be optimized independently.

9.6.2 High-Frequency Interaction

The potential interaction in the high-frequency range between two TCSCs located in the same electrical area is demonstrated through the test system shown in Fig. 9.39 [1]. Both TCSCs have similar ratings and operate on constant-current control with a current reference of 800 A. The total system is modeled by 90 state variables, with the TCSCs represented through the generalized switching functions [1], [12]–[14].

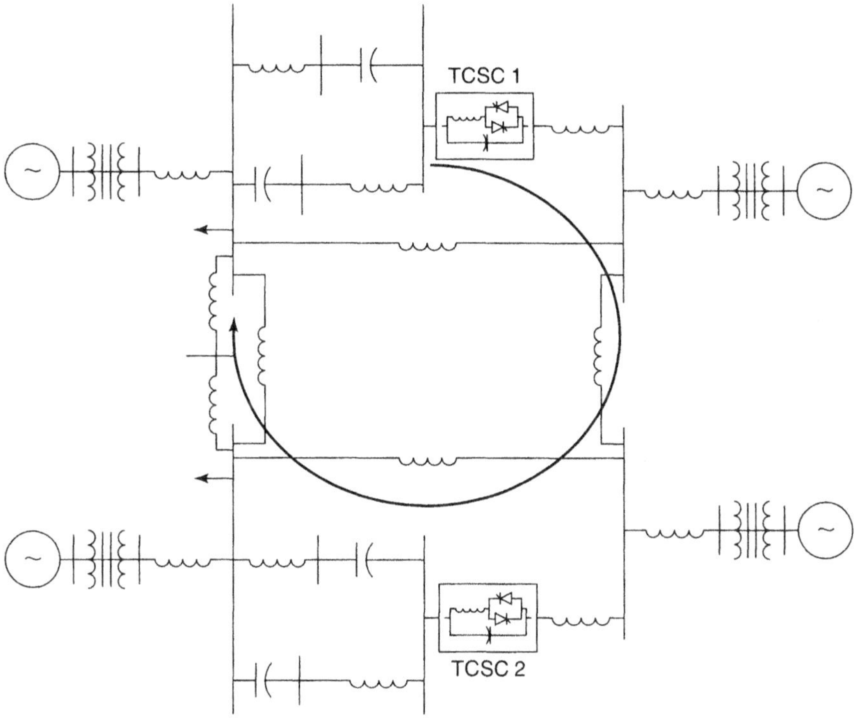

Figure 9.38 A high-loop impedance configuration with uncoupled TCSCs.

First, the controller of TCSC 1 is designed independently using the system depicted in Fig. 9.40, where TCSC 2 is replaced by a passive equivalent network with same level of series compensation for the relevant operating point. The controller is designed to give small rise time and settling time with acceptable overshoot. The equivalent system presents a dominant mode associated with eigenvalue $\lambda_{TCSC\,1} = -5 \pm j41$ and reflected as a 6.5-Hz oscillation during the TCSC response to a reference-current step, as illustrated in Fig. 9.41.

Next, the TCSC 2 controller is designed independently by replacing TCSC 1 with an equivalent passive circuit. A careful examination reveals that both TCSCs are connected to exactly the same networks; hence the same dynamic performance is obtained when both TCSCs are designed independently. However, when both TCSCs are designed as such and operated simultaneously, the response of TCSC 1 to its reference-current step will show growing oscillations at a 6.4-Hz frequency, as shown in Fig. 9.42, which is attributed to the dominant-mode eigenvalue $\lambda = \pm 2 \pm j40$. Therefore a need arises for a coordinated design of both TCSC current controllers; such design stabilizes the system. In this situation, the response of TCSC 1 to a +1% reference-current step is illustrated in Fig. 9.43.

The TCSC controller designs for the three control cases—the original TCSC

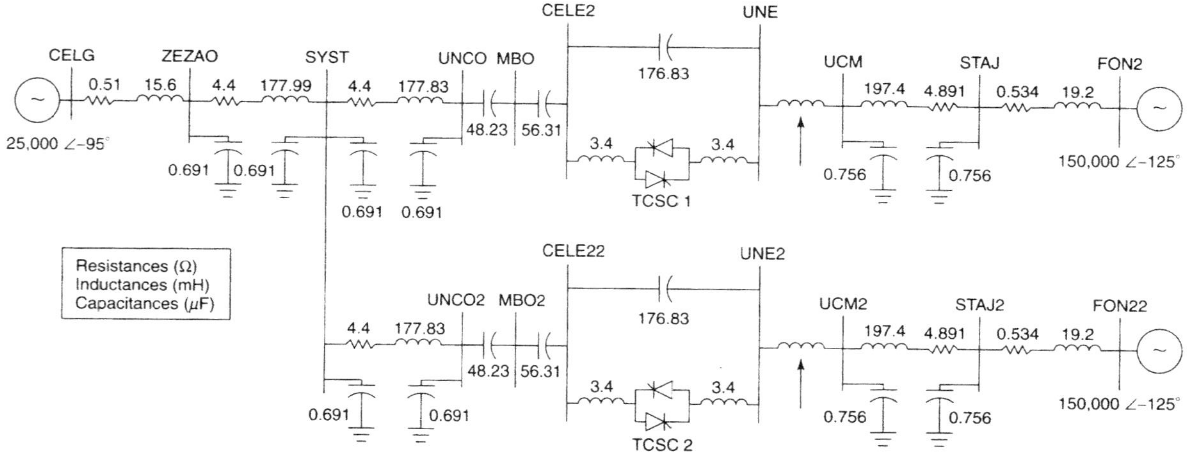

Figure 9.39 A test system for analysis of the high-frequency interaction between TCSCs.

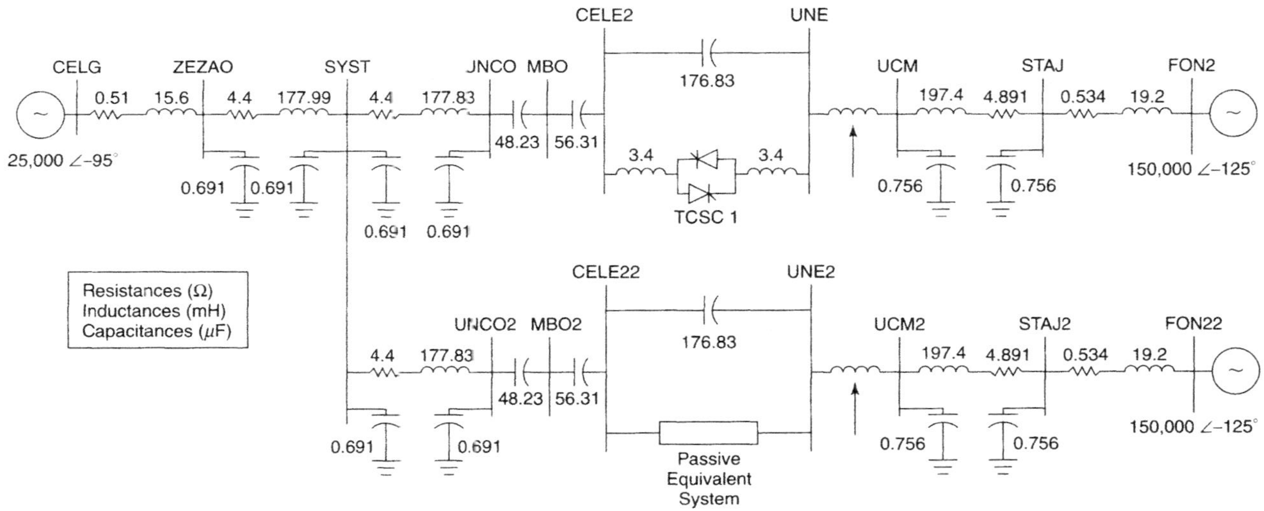

Figure 9.40 An equivalent system for the control design of TCSC.

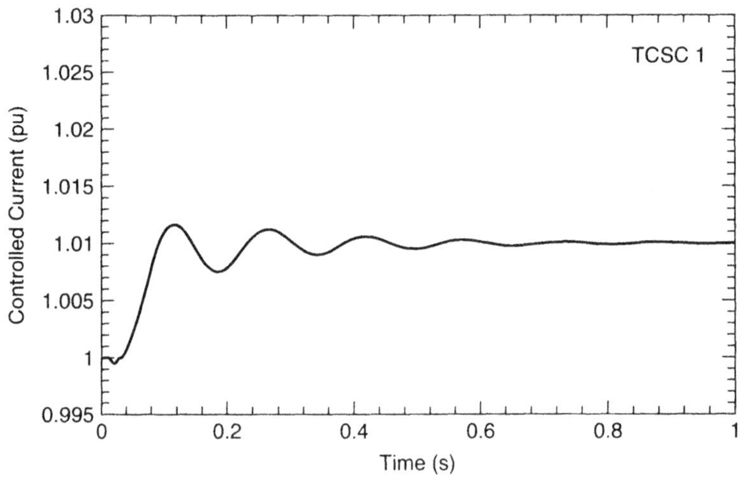

Figure 9.41 The response of TCSC 1 to the reference-current step input.

control, the joint operation of independently designed controllers, and the coordinated control—are now examined through frequency-response studies. The frequency response is obtained for a transfer function relating the controlled current with the reference of TCSC 1 in each of the aforementioned cases. This response is plotted in Fig. 9.44. In this figure, the stable, coordinated control results in the lowest magnitude of resonant peak. The control coordination may be based on an appropriate index, such as the optimal damping factor, or any index that provides a rapid response with minimum overshoot.

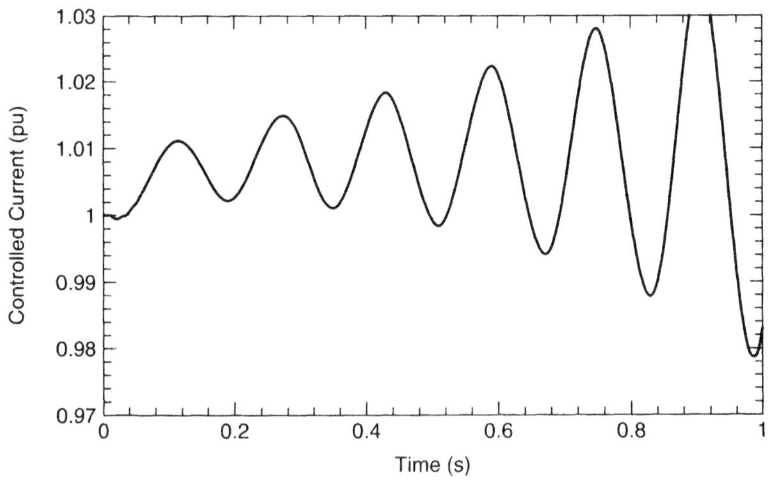

Figure 9.42 The response of TCSC 1 to the reference-current step input, with both TCSCs modeled.

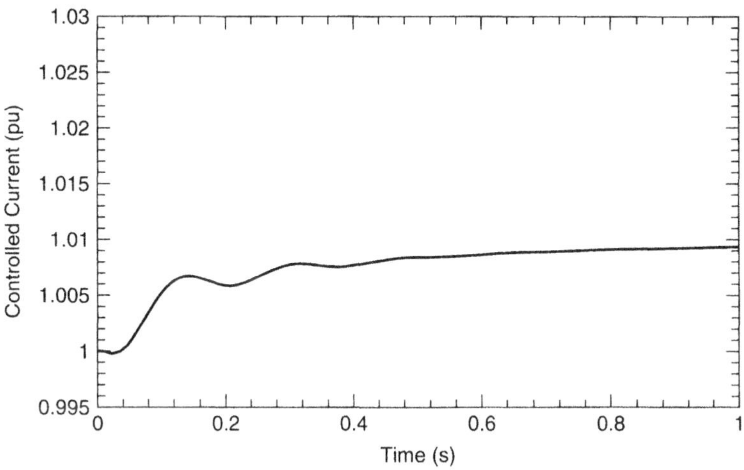

Figure 9.43 The response of TCSC 1 to the reference-current step input with coordinated control of both TCSCs.

9.7 PERFORMANCE CRITERIA FOR DAMPING-CONTROLLER DESIGN

The damping criteria for both local- and inter-area modes of rotor oscillation under normal and post–fault-operating conditions are specified in the following list. (See also refs. [2] and [17].)

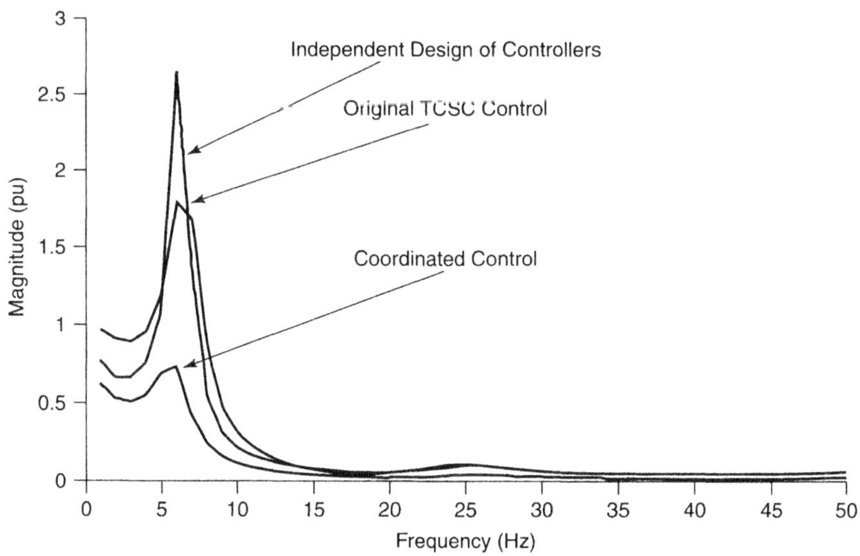

Figure 9.44 The frequency response of TCSC 1.

TABLE 9.1 Comparison of Settling Times for Typical Local-Area and Inter-Area Modes

	Settling Time		
Mode Type	Criterion 1 $\xi > 0.03$–0.05	Criterion 2 $T_H = 5$ s	Criterion 3 $T_S = 10.20$ s
Local-Area: $\omega = 8$ rad/s	8.3–7.5 s	21.6 s	10–20 s
Inter-Area: $\omega = 2$ rad/s	50–30 s	21.6 s	10–20 s

1. For all i-electromechanical modes specified by $-\alpha_i \pm j\omega_i$, the following damping ratio should be greater than a minimum level of 0.03–0.05:

$$\xi_i \left(= \frac{\alpha}{\sqrt{\alpha^2 + \omega^2}} \right) \tag{9.2}$$

2. The halving time $T_H(s)$ (i.e., the time that the oscillation amplitude corresponding to the rotor mode takes to decay to within half the initial value $= 0.6931/\alpha$) for the most lightly damped mode should be lower than a specified value, typically 5 s.
3. The electromechanical oscillations should have a settling time $T_s(s)$ (i.e., the time taken to decay to within ±5% of the final value) of 10–20 s.

A comparison of the settling times of the local- and inter-area modes based on the preceding three criteria is presented in the Table 9.1. (See also ref. [2].) It is noted that the three criteria result in different settling time requirements. Also, the settling time is highly different for the local- and inter-area modes, even if they have the same criterion 1 of the preceding list. This discrepancy is attributed to the fact that these damping criteria were enumerated years ago when only power system stabilizers (PSSs) were available to damp both the local- and inter-area oscillations. Although PSSs are highly effective in damping the local modes of related generators, their capability of damping the inter-area modes—even in the tuned state—is limited for the following reasons:

1. lower participation factors associated with PSSs;
2. adverse interaction between PSSs; and
3. substantial damping contribution provided by only several PSSs, with worsening of the overall system damping if an outage of some of these PSSs occurs.

The selection of the aforementioned damping criteria, therefore, appears to have been based only on the capabilities of the PSSs available for damping the local- and inter-area modes. In recent years, however, properly located FACTS controllers have proven greatly effective in damping the inter-area modes. Thus it is advised that the following stricter damping criteria now be introduced for normal and post-fault conditions (see also ref. [2]):

1. For local-area modes having an angular frequency $\omega \geq 4$ rad/s, the damping ratio should be $\xi > 0.08$.
2. For inter-area modes having an angular frequency $\omega < 4$ rad/s, the real part of the mode should be $\alpha > 0.2 - 0.25$ nepers/s. This will result in corresponding settling times of $15 - 12$ s.

It may be noted that although most controllers are designed using linear-analysis techniques, the power systems in which they are applied are highly nonlinear. Hence the damping levels obtained in linearized design are not witnessed in transient-stability studies that incorporate the effect of limiters and other nonlinear effects. The damping criteria must therefore incorporate enough leavage to account for the nonlinearities in power systems.

9.8 COORDINATION OF MULTIPLE CONTROLLERS USING LINEAR-CONTROL TECHNIQUES

The essential design features of multiple FACTS controllers that can ensure secure operation with sufficient damping over a wide range of power-system operating conditions are discussed elaborately in ref. [2]. The term *coordination* does not imply centralized control; rather, it implies the simultaneous tuning of the controllers to attain an effective, positive improvement of the overall control scheme. It is understood that each controller relies primarily on measurements of locally available quantities and acts independently on the local FACTS equipment.

9.8.1 The Basic Procedure for Controller Design

The controller-design procedure involves the following steps:

1. derivation of the system model;
2. enumeration of the system-performance specifications;
3. selection of the measurement and control signals;
4. coordination of the controller design; and
5. validation of the design and performance evaluation.

9.8.1.1 Derivation of the System Model First, a reduced-order nonlinear-system model must be derived for the original power system. This model should retain the essential steady-state and dynamic characteristics of the power system [18]. Then, the model is linearized around an operating point to make it amenable to the application of linear-control design techniques. If a controller must be designed for damping electromechanical oscillations, a further reduced-linear model is selected [19] that exhibits the same modal characteristics over the relevant narrow range of frequencies as the original system.

In situations where linearized-system models may not be easily obtainable, identification techniques are employed [20] to derive simple linear models from time-response information.

9.8.1.2 Enumeration of the System-Performance Specifications

The damping controller is expected to satisfy the following criteria [2].

1. It should help the system survive the first few oscillations after a severe system disturbance with an adequate safety margin. This safety factor is usually specified in terms of bus-voltage levels that should not be violated after a disturbance.
2. A minimum level of damping must be ensured in the steady state after a disturbance.
3. Potentially deleterious interactions with other installed controls should be avoided or minimized.
4. Desired objectives over a wide range of system-operating conditions should be met (i.e., it should be robust).

9.8.1.3 Selection of the Measurement and Control Signals

The choice of appropriate measurement and control signals is crucial to controller design. The signals must have high observability and controllability of the relevant modes to be damped, and furthermore, the signals should only minimally affect the other system modes. The selection of these signals is usually based on system-modal magnitudes, shapes, and sensitivties—all of which can be obtained from small-signal-stability analysis.

9.8.1.4 Controller Design and Coordination

The FACTS controller structures are usually chosen from industry practice. Typically, the controller-transfer function, $H_j(s)$, of controller j is assumed to be

$$H_j(s) = k_j G_j(s) = k_j \frac{sT_W}{1 + sT_W} \left(\frac{1 + s\tau_1}{1 + s\tau_2} \right)^p \frac{1}{(1 + sT_1)(1 + sT_2) \cdots (1 + sT_n)} \tag{9.3}$$

This transfer function consists of a gain, a washout stage, and a pth-order lead-lag block, as well as low-pass filters. Alternatively, it can be expressed as

$$H_j(s) = k_j G_j(s) = k_j \left[k_0 \frac{(s + \cdots + b_m s^m}{1 + a_1 s + \cdots + a_n s^n} \right], \qquad m \leq n \tag{9.4}$$

Although the basic structure of different controllers is assumed as from the preceding text, the coordination of controllers involves the simultaneous selection of gains and time constants through different techniques. Doing so permits the system-operating constraints and damping criteria to be satisfied over a wide range of operating conditions.

The coordination techniques may use linearized models of the power system and other embedded equipments, capitalizing on the existing sparsity in system representation. This model may be further reduced by eliminating certain algebraic variables yet still retaining the essential system behavior in the frequency range of interest.

Eigenvalue analysis–based controller-optimization and -coordination techniques are applicable to power systems typically with a thousand states—occurring when full modal analysis must be performed. However, sometimes a limited number of electromechanical modes must be damped; hence the eigenvalue analysis of a selected region can be performed even for relatively larger power systems.

In the case of large systems, procedures are employed that automate the tuning and coordination of controllers.

9.8.1.5 Validation of the Design and Performance Evaluation Even though the controller design is performed on the simplified system model, the performance of the controller must still be established by using the most detailed system model. The controller should meet the specifications over a wide range of operating conditions and consider all credible contingencies. This validation is generally performed with nonlinear time-domain simulations of the system.

9.8.2 Controller Coordination for Damping Enhancement

This technique for the coordination of controllers to improve the damping of electromechanical modes is based on the damping-torque contribution of each FACTS controller, as well as that of any other controller present in the system—PSS, HVDC, and so on. The damping-torque contribution of a controller is related to the left shift that it introduces in the relevant electromechanical mode.

The relative effectiveness of each controller can thus be measured in terms of the attained left-shift magnitude in the relevant mode for a given change in the controller-transfer function gain after the interaction between the controllers has been accounted for. This study results in a *controller-damping contribution diagram*, illustrating the damping contribution of each controller. It is also possible to infer from this study which individual supplementary-feedback signal or combinaiton of signals is most effective for a controller in augmenting the mode damping. Furthermore, the adverse effects of a controller on any mode other than the electromechanical one for which it was designed to damp are determined from this study as well.

Certain assumptions are made in the use of this technique:

1. All controllers in the system, including FACTS, have the transfer function of the type $k_j G_j(s)$, as given in Eq. (9.3).
2. The component $G_j(s)$ in the transfer function is responsible for causing the left shift in the electromechanical mode.
3. The gain k_j in the transfer function decides the magnitude of left shift in the mode of interest.

The controller interactions can then be examined with respect to changes in the controller gains, and the exercise of coordinating the controllers is simplified to that of coordinating the gains k_j of different controllers.

Some constraints, however, are imposed in selecting the gains on individual controllers. The gains of the damping control loop in a FACTS controller must be reasonably low to not interfere with the main control loop of the controller. It may be recalled that the primary reason why FACTS controllers are installed may not be for damping enhancement in all cases.

The gains on the different PSSs, too, should be kept relatively small to minimize the following:

1. the operating-limit influence in the stabilizer, the automatic voltage regulator (AVR), and the excitation systems.
2. the reactive-power oscillations in generators for small contingencies.

The damping-enhancement controller-coordination technique provides useful insight and is simple to implement. It can be applied manually to power systems where the number of controllers to be coordinated (including FACTS) is small and/or the number of electromechanical modes to be damped is low [21]. However, the manual process can be highly time-consuming and susceptible to errors when

1. the total controllers to be coordinated is substantial, and
2. the coordination of controllers must satisfy a wide range of performance specifications to ensure robustness.

A need then arises for using automated techniques for controller coordination. The control-coordination technique described here is automated by formulating it as a linear-programming problem [22]. A weighted sum of gains k_j of all controllers is minimized and is subject to the following constraints:

1. The left shift of all electromechanical modes is greater than or equal to the specified desired values.
2. The change in various mode frequencies is less than certain limits. Because the controllers may be unable to impart pure damping—and also because of the controllers' interactions—the modal frequencies can become changed, possibly causing a deterioration of the synchronizing

torque, which must be restricted. If large mode-frequency changes are unavoidable, the various parameters of controller $G_j(s)$ must be adjusted.

3. The magnitude of gains associated with the coordinated controllers should be less than specified values.

9.8.3 Linear Quadratic Regulator (LQR)–Based Technique

The LQR technique is one of optimal control that can be used to coordinate the controllers with the overall objective of damping low-frequency inter-area modes during highly stressed power-system operations [19], [23]. The system model is first linearized and later reduced to retain the modal features of the main system over the frequency range of interest. The control-system specifications are laid out as described previously. Appropriate measurement and control signals are selected, based on observability and controllability considerations, to have only a minimal interaction with other system modes.

Using a projective-controls approach [24], the control-coordination method involves formulating an LQR problem to determine a full-state-feedback controller in which a quadratic performance index is minimized. An output-feedback controller is then obtained, based on the reduced eigenspace of the full-state solution. The dominant modes of the full-state-feedback system are retained in the closed-loop system with output feedback. The order of the controller and the number of independent measurements influence the number of modes to be retained. The output-feedback solution results in the desired coordinated control.

The performance of coordinated controls is later tested and evaluated through time-domain simulation of the most detailed model of the nonlinear system. An application example is described in refs. [2] and [23].

9.8.4 Constrained Optimization

Constrained-optimization techniques for control coordination use control structures generally used in industry, but they may or may not use robustness criteria explicitly in the design process. In such a case controller robustness must be verified separate from the design process.

9.8.4.1 Techniques Without Explicit Robustness Criteria

Pole placement is one technique [25] in which the robustness requirement is not explicitly considered. In this method, the critical electromechanical modes are assigned a priori new locations that are placed deeper into the left half of the s plane. The controller parameters are then selected to result in these assigned pole locations.

Let us consider that the jth controller is to be tuned. The characteristic equation for the closed-loop system can be expressed as

$$1 - F(s)G_j(s) = 0 \tag{9.5}$$

where $F(s)$ = the transfer function between output y_i and input u_i as obtained from the state-space model of the linearized power system

$G_j(s)$ = the transfer function of the FACTS damping controller

Let the critical mode, which is also a pole of the closed-loop system, be assigned a location λ. If $G_j(\lambda)$ is expressed as $e + jf$, then

$$e + jf = \frac{1}{F(\lambda)} \tag{9.6}$$

The parameters of the controller structure are then obtained by comparison with Eq. (9.6). This procedure is applied sequentially to all the controllers to be tuned, thus completing one iteration. At each step, the tuning process of a controller considers the interactions with other controllers. The iterative process is continued until convergence is achieved, which is indicated when the largest absolute difference between the assigned eigenvalues in any two iterations becomes less than the specified tolerance.

This method can be applied easily to large systems, as the inherent sparsity in system matrices can be used to the advantage of such systems.

9.8.4.2 Techniques With Explicit Robustness Criteria In techniques that explicitly consider the robustness requirements [26], any changes in the system operating points caused by line switching, load variations, or contingencies are incorporated as changes in the matrix elements of the linearized systems. These matrix-element variations are called *admissible uncertainties*. The robust-control design relates to the determination of a state-feedback matrix that can maintain stability of the closed-loop system for any admissible uncertainty. This method is somewhat difficult to apply to large systems, as it relies totally on a linearized-system model.

9.8.5 Nonlinear-Constrained Optimization of a Selective-Modal-Performance Index

In the nonlinear-constrained optimization technique, the objective function is called a *modal-performance index*, which relates exclusively to system damping enhancement [27]. This index computes the energy enveloped by the damped sinusoidal-signal responses that arises from a given initial condition. A large value of this index implies a scanty level of damping; a minimum value, on the other hand, implies an optimally damped closed-loop system.

The optimization process uses the closed-loop state-space model and is computationally efficient irrespective of the controller structures. Because some oscillation modes may not be controllable from certain locations or be completely noncontrollable by certain stabilizing controllers, they are not considered in the optimization process; hence the term *selective*-modal-performance index.

9.8.6 Global Coordination Using Nonlinear-Constrained Optimization

In the global-coordination technique, the parameters (both gain and time constants) of the damping controllers of multiple FACTS controllers are coordinated globally by using nonlinear-constrained optimization [2]. It is usually observed in power systems that if the FACTS controllers have large ratings, their damping action on the real-power oscillations causes substantial reactive-power oscillations. To restrict oscillations—in both the real and the reactive power—the optimization problem is stated as follows:

Minimize

$$F = \int_0^\infty [\Delta P^{\mathrm{T}} I \Delta P + K_Q \Delta Q^{\mathrm{T}} I \Delta Q]\, dt \tag{9.7}$$

subject to

$$V_{\min} \le V_{\mathrm{bus}\,k} \le V_{\max}, \qquad k = 1, 2, m \tag{9.8}$$

where for n FACTS controllers,

$\Delta P = [\Delta P_1, \Delta P_2, ..., \Delta P_n]^{\mathrm{T}}$
= the n locally measured real-power-flow oscillations

$\Delta Q = [\Delta Q_1, \Delta Q_2, ..., \Delta Q_n]^{\mathrm{T}}$
= the n locally measured reactive-power oscillations

$V_{\mathrm{bus}\,k}$ = the voltage magnitude of the kth bus

$V_{\max}, V_{\min}$ = the upper and lower limits of the bus-voltage magnitude, respectively

K_Q = the weighting factor

If the primary control objective is to damp active-power oscillations, the weighting factor K_Q is assigned a small value, typically 0.2. However, if reactive-power swings must also be restricted, K_Q is assigned a higher magnitude.

The foregoing optimization scheme results in robust controllers having a significant damping influence on both large and small disturbances. An example of the global coordination of damping controllers of an SVC and a TCSC is presented in ref. [2]. The transfer function, $F_D(s)$, of both damping controllers is assumed to be of the form

$$F_D(s) = K \cdot \frac{sT_1}{1 + sT_2} \cdot \frac{1 + sT_3}{1 + sT_4} \tag{9.9}$$

Although T_2 and T_4 are chosen a priori, parameters K, T_1, and T_3 are determined for both controllers through the optimization procedure.

9.8.7 Control Coordination Using Genetic Algorithms

Genetic algorithms are optimization techniques based on the laws of natural selection and natural genetics [28] that recently have been applied to the control design of power systems [29], [30]. These techniques provide robust, decentralized control design and are not restricted by problems of nondifferentiability, nonlinearity, and nonconvexity, all of which are often limiting in optimization exercises.

Genetic-algorithm techniques use the linearized state-space model of the power system. The objective function is defined as the sum of the damping ratios of all the modes of interest. This sum is evaluated over several likely operating conditions to introduce robustness. A minimum damping level is specified for all the modes; the other constraints include limits on the gain and time constants of the damping controllers assumed to be from a fixed structure, as given in Eq. (9.3). The optimization problem is therefore stated as follows:

Maximize

$$F = \sum_{i=1}^{m} \left[\sum_{j=1}^{n} (\xi_i) \right] \tag{9.10}$$

subject to the following constraints:

$$\begin{aligned} k_{j\,\min} &\le k_j \le k_{j\,\max} \\ \tau_{1\,\min} &\le \tau_1 \le \tau_{1\,\max} \\ \tau_{2\,\min} &\le \tau_2 \le \tau_{2\,\max} \\ \xi_{\min} &\le (\xi_j)_i \end{aligned} \tag{9.11}$$

where n = the number of modes to be damped

m = the number of different possible operating conditions

k_j = the gain of the controller

τ_1, τ_2 = the time constants of the lead-lag blocks

ξ = the damping ratio of the closed-loop eigenvalue

This maximization yields the gain k_j and the time constants τ_1, τ_2 for all the controllers for a prespecified order p of the lead-lag blocks. The time constant T_W of the washout filter is assumed to be adequately large and known a priori. Likewise, the time constants $T_1, T_2, \ldots, T_n$ of the low-pass filters are selected beforehand.

The foregoing optimization problem involves a computation of eigenvalues of a large system matrix, which is usually difficult to solve with conventional techniques. An advantage of genetic-algorithm techniques is that the parameter limits can be varied during the optimization, making the techniques computationally efficient. An application of these techniques to two large power systems is presented in ref. [2]. Use of high-performance computation (parallel processing) is recommended for reducing the computational time associated with the application of these techniques.

9.9 COORDINATION OF MULTIPLE CONTROLLERS USING NONLINEAR-CONTROL TECHNIQUES

Even though rotor-angle stability is essentially a nonlinear-control problem, the theoretical concepts of linear-control theory can be applied to solve many issues related to this stability problem, as described in previous sections of this chapter as well as in ref. [31]. Recently, however, several nonlinear-control techniques have been applied for the design of FACTS controllers [31]. These techniques are likely to yield greatly improved controllers, as they include the effects of system nonlinearities. Some of these methods are described briefly in the following text.

One nonlinear-control technique in which the system nonlinearities are expressed as system changes constituting a function of time is the *adaptive control* [32]–[34]. If the number of controller parameters to be optimized is not too large, a *cost-penalty function* technique can be used, which is based on nonlinear simulation [35]. An effective technique commonly used for enhancing transient stability during large disturbances is the *discontinuous control* or *bang-bang control* [36], [37]. Another nonlinear-control technique is the *normal forms*, which includes the effects of higher-order terms in Taylor's series to represent power systems—especially during high-power transfers [38], [39]. For damping low-frequency oscillations, FACTS controllers can be designed using the *dissipation* technique, which is based on the concept that passive systems always absorb energy [40]. For designing the controls of FACTS controllers in large power systems, the *energy*, or *Lyapunov*, technique can be used [41], [42]. For stability enhancement, *nonlinear fuzzy* and *neural net* techniques are presently being researched [43].

In the future, these techniques may be extended for coordination of FACTS controllers. One possible approach could be to first do a "coarse" coordination using linear-control techniques, followed by a "fine" coordination employing the nonlinear-control methods.

9.10 SUMMARY

This chapter described the various interactions that can potentially occur between the different FACTS controllers, as well as between FACTS and HVDC controllers, in power systems. The likely interactions have been classified into different frequency ranges. The damping criteria for the design of controllers was enumerated. Linear-control techniques were presented for coordinating the FACTS controllers. Some nonlinear-control techniques for FACTS controller design and coordination were also indicated.

REFERENCES

[1] CIGRE Working Group 14.29, "Coordination of Controls of Multiple FACTS/HVDC Links in the Same System," CIGRE Technical Brochure No. 149, Paris, December 1999.

[2] CIGRE Task Force 38.02.16, "Impact of Interactions Among Power System Controls," CIGRE Technical Brochure No. 166, Paris, August 2000.

[3] The Electric Power Research Institute (EPRI) Report TR-109969, "Analysis of Control Interactions on FACTS Assisted Power Systems," Final Report, Palo Alto, CA, January 1998.

[4] M. Parniani and M. R. Iravani, "Voltage Control Stability and Dynamic Interaction Phenomena of Static Var Compensators," Presented at IEEE/PES 1995 Winter Meeting.

[5] E. V. Larsen, D. H. Baker, A. F. Imece, L. Gerin-Lajoie, and G. Scott, "Basic Aspects of Applying SVCs to Series-Compensated AC Transmission Lines," *IEEE Transactions on Power Delivery*, Vol. 5, No. 3, July 1990, pp. 1466–1473.

[6] L. Gerin-Lajoie, G. Scott, S. Breault, E. V. Larsen, D. H. Baker, and A. F. Imece, "Hydro-Quebec Multiple SVC Application Control Stability Study," *IEEE Transactions on Power Delivery*, Vol. 5, No. 3, July 1990, pp. 1543–1551.

[7] L. A. S. Pilotto, W. W. Ping, and M. Szechtman, "Converter Control Interactions on Multi-Infeed HVDC Systems," CIGRE International Colloquium on HVDC and FACTS, Montreal, September 1995.

[8] P. Kundur, *Power System Stability and Control*, McGraw-Hill, EPRI Power Systems Engineering Series, 1994.

[9] X. Jiang and A. M. Gole, "A Frequency Scanning Method for the Identification of Harmonic Instabilities in HVDC Systems," *IEEE Transactions on Power Delivery*, Vol. 10, No. 4, October 1995.

[10] L. A. S. Pilotto, W. W. Ping, A. R. Carvalho, W. F. Long, F. L. Alvarado, C. L. De Marco, and A. Edris, "Coordinated Design of FACTS Controllers to Enhance Power System Dynamic Performance," Presented at International Colloquium on HVDC and FACTS, Johannesburg, South Africa, September 1997.

[11] J. Belanger, G. Scott, T. Anderson, and S. Torseng, "Gain Supervisor for Thyristor-Controlled Shunt Compensators," CIGRE Paper 38-01, August 1984.

[12] L. Gyugyi and B. R. Pelly, *Static Power Frequency Changers: Theory, Performance, and Application*, John Wiley and Sons, New York, 1976.

[13] L. A. S. Pilotto, "Advanced Modelling of AC/DC Systems," DSc thesis, Federal University of Rio de Janeiro, Brazil, April 1994.

[14] J. E. R. Alves, "Thyristor-Controlled Reactor Modelling Based on Generalized Switching Functions," DSc candidacy paper, Federal University of Rio De Janeiro, Brazil, March 1997.

[15] K. Clark, B. Fardanesh, and R. Adapa, "Thyristor-Controlled Series Compensation Application Study—Control Interaction Considerations," *IEEE Transactions on Power Delivery*, Vol. 10, No. 2, April 1995, pp. 1031–1037.

[16] X. Z. Lei, D. Povh, and K. Renz, "Improving Power Swing Damping by Coordinated FACTS Controls," Presented at CIGRE International Colloquium on HVDC and FACTS, Johannesburg, South Africa, September 1997.

[17] CIGRE Task Force 38.01.07, "Analysis and Control of Power System Oscillations," CIGRE Technical Brochure No. 111, Paris, December 1996.

[18] J. H. Chow, R. Galarza, P. Accari, and W. W. Price, "Inertial and Slow Coherency Aggregation Algorithms for Power System Dynamic Model Reduction," *IEEE Transactions on Power Systems*, Vol. 10, No. 2, May 1995, pp. 680–685.

[19] J. J. Sanchez-Gasca and J. H. Chow, "Power System Reduction to Simplify the Design of Damping Controllers for Inter-Area Oscillations," *IEEE Transactions on Power Systems*, Vol. 11, No. 3, August 1996, pp. 1342–1349.

[20] J. R. Smith, J. F. Hauer, and D. J. Trudnowski, "Transfer Function Identification in Power Systems Applications," *IEEE Transactions on Power Systems*, Vol. 8, No. 3, August 1993, pp. 1282–1290.

[21] M. J. Gibbard, P. Pourbeik, and D. J. Vowles, "Coordination of Multiple, Distributed Stabilizers in Interconnected Power Systems," *Proceedings of the CONTROL '97*, Sydney, Australia, October 20–22, 1997, pp. 458–463.

[22] P. Pourbeik and M. J. Gibbard, "Simultaneous Coordination of Power System Stabilizers and FACTS Device Stabilizers in a Multimachine Power System for Enhancing Dynamic Performance," *IEEE Transactions on Power Systems*, Vol. 13, No. 2, May 1998, pp. 473–479.

[23] J. J. Sanchez-Gasca, "Coordinated Control of Two FACTS Devices for Damping Inter-Area Oscillations," *IEEE Transactions on Power Systems*, Vol. 13, No. 2, May 1998, pp. 428–434.

[24] D. Arnautovic and J. Medanic, "Design of Decentralized Multivariable Excitation Controller in Multimachine Power Systems by Projective Controls," *IEEE Transactions on Energy Conversion*, Vol. 2, No. 4, December 1987, pp. 598–604.

[25] S. Elangovan and C. M. Lim, "Efficient Pole-Assignment Method for Designing Stabilizers in Multimachine Power Systems," *IEE Proceedings*, Vol. 134, No. 6, November 1987, pp. 383–394.

[26] A. S. Bazanella, A. Fischman, A. S. Silva, J. M. Dion, and L. Dugard, "Coordinated Robust Controllers in Power Systems," *Proceedings of the IEEE Stockholm Power Tech.*, Stockholm, Sweden, June 1995, pp. 256–261.

[27] J. B. Simo, I. Kamwa, G. Trudel, and S. A. Tahan, "Validation of a New Modal Performance Measure for Flexible Controller Design," *IEEE Transactions on Power Systems*, Vol. 11, No. 2, May 1996, pp. 819–828.

[28] D. E. Goldberg, "Genetic Algorithms in Search, Optimization, and Machine Learning," Addison-Wesley, Reading, MA, 1989.

[29] P. Ju, E. Handschin, and F. Reyer, "Genetic Algorithm–Aided Controller Design With Application to SVC," *IEEE Proceedings on Generation, Transmission, and Distribution*, Vol. 143, No. 3, 1996, pp. 258–262.

[30] A. L. B. do Bomfim, G. N. Taranto, and D. M. Falco, "Simultaneous Tuning of Power System Damping Controllers Using Genetic Algorithms," *IEEE Transactions on Power Systems*, Paper PE-233-PWRS 0-03-1999.

[31] CIGRE Task Force 38.02.17, "Advanced Angle Stability Controls," CIGRE Technical Brochure No. 155, Paris, April 2000.

[32] J. R. Smith, D. A. Pierre, I. Sadighi, M. H. Nehrir, and J. F. Hauer, "A Supplementary Adaptive Var Unit Controller for Power System Damping," *IEEE Transactions on Power Systems*, Vol. 4, No. 3, August 1989, pp. 1017–1023.

[33] J. Reeve and M. Sultan, "Gain Scheduling Adaptive Control Strategies for HVDC Systems to Accommodate Large Disturbances," *IEEE Transactions on Power Systems*, Vol. 9, No. 1, February 1994, pp. 366–372.

[34] X. Zhou and J. Liang, "Nonlinear Adaptive Control of the TCSC to Improve the Performance of Power Systems," *IEEE Proceedings on Generation, Transmission, and Distribution*, Vol. 146, No. 3, May 1999, pp. 301–305.

[35] P. S. Dolan, J. R. Smith, and W. A. Mittelstadt, "Study of TCSC Optimal Damping Control Parameters for Different Operating Conditions," *IEEE Transactions on Power Systems*, Vol. 10, No. 4, November 1995, pp. 1972–1978.

[36] D. N. Kosterev and W. J. Kolodziej, "Bang-Bang Series Capacitor Transient Stability Control," *IEEE Transactions on Power Systems*, Vol. 10, No. 2, May 1995, pp. 915–929.

[37] J. Chang and J. H. Chow, "Time-Optimal Control of Power Systems Requiring Multiple Switchings of Series Capacitors," *IEEE Transactions on Power Systems*, Vol. 13, No. 2, May 1998, pp. 367–373.

[38] G. Jang, V. Vittal, and W. Kliemann, "Effects of Nonlinear Modal Interactions on Control Performance: Use of Normal Forms Technique in Control Design, Part I: General Theory and Procedure," *IEEE Transactions on Power Systems*, Vol. 13, No. 2, May 1998, pp. 401–407.

[39] G. Jang, V. Vittal, and W. Kliemann, "Effects of Nonlinear Modal Interactions on Control Performance: Use of Normal Forms Technique in Control Design, Part II: Case Studies," *IEEE Transactions on Power Systems*, Vol. 13, No. 2, May 1998, pp. 408–413.

[40] A. M. Stankovic, P. C. Stefanov, G. Tadmor, and D. J. Sobajic, "Dissipativity as a Unifying Control Design Framework for Suppression of Low-Frequency Oscillations in Power Systems," *IEEE Transactions on Power Systems*, Vol. 14, No. 1, February 1999, pp. 192–199.

[41] M. A. Pai, *Energy Function Analysis for Power System Stability*, Kluwer Academic Publishers, 1989.

[42] J. J. Gronquist, W. A. Sethares, F. L. Alvarado, and R. H. Lasseter, "Power Oscillation Damping Control Strategies for FACTS Devices Using Locally Measurable Quantities," *IEEE Transactions on Power Systems*, Vol. 10, No. 3, August 1995, pp. 1598–1605.

[43] T. Hiyama, W. Hubbi, and T. H. Ortmeyer, "Fuzzy Logic Control Scheme With Variable Gain for Static Var Compensator to Enhance Power System Stability," *IEEE Transactions on Power Delivery*, Paper PE-375-PWRS-0-2-1998.

CHAPTER 10

Emerging FACTS Controllers

10.1 INTRODUCTION

With the ever-increasing complexities in power systems across the globe and the growing need to provide stable, secure, controlled, economic, and high-quality power—especially so in the derugulated environment—it is envisaged that FACTS controllers will play a critical role in power systems in the future. Major FACTS controllers—SVCs and TCSCs, for instance—have already been discussed in previous chapters. A host of other highly versatile, effective FACTS controllers exist that are fast emerging in power systems [1]–[8], some of which have already been installed in the field. These controllers include the static synchronous compensator (STATCOM) [9]–[23], the thyristor-controlled phase-shifting transformer (TCPST) [24]–[27], the static synchronous series compensator (SSSC) [28]–[31], the unified power-flow controller (UPFC) [32]–[47], the interphase power controller (IPC) [2], [3], [7], [8], the thyristor-controlled braking resistor (TCBR) [2], [7], [8], [48], the thyristor-controlled voltage limiter (TCVL) [7], [8], [48], the battery energy storage system (BESS), and the superconducting magnetic energy storage (SMES) [1], [7], [8]. In Appendix D is a list of different FACTS controllers and their definitions as defined by the IEEE.

This chapter presents the operating principles and applications of several highly versatile controllers—the STATCOM, SSSC, and UPFC—without any prejudice to other FACTS controllers. These three FACTS controllers are second-generation types that are based on nonthyristor devices such as gate turn-offs (GTOs) and insulated-gate bipolar transistors (IGBTs).

10.2 THE STATCOM

The STATCOM (or SSC) is a shunt-connected reactive-power compensation device that is capable of generating and/or absorbing reactive power and in which the output can be varied to control the specific parameters of an electric power system. It is in general a solid-state switching converter capable of generating or absorbing independently controllable real and reactive power at its output terminals when it is fed from an energy source or energy-storage device at its input terminals. Specifically, the STATCOM considered in this chapter is a voltage-source converter that, from a given input of dc voltage, produces a set of 3-phase ac-output voltages, each in phase with and coupled to the corresponding ac system voltage through a relatively small reactance (which is provided by either an interface reactor or the leakage inductance of a coupling transformer). The dc voltage is provided by an energy-storage capacitor.

A STATCOM can improve power-system performance in such areas as the following:

1. The dynamic voltage control in transmission and distribution systems;
2. the power-oscillation damping in power-transmission systems;
3. the transient stability;
4. the voltage flicker control; and
5. the control of not only reactive power but also (if needed) active power in the connected line, requiring a dc energy source.

Furthermore, a STATCOM does the following:

1. it occupies a small footprint, for it replaces passive banks of circuit elements by compact electronic converters;
2. it offers modular, factory-built equipment, thereby reducing site work and commissioning time; and
3. it uses encapsulated electronic converters, thereby minimizing its environmental impact.

A STATCOM is analogous to an ideal synchronous machine, which generates a balanced set of three sinusoidal voltages—at the fundamental frequency—with controllable amplitude and phase angle. This ideal machine has no inertia, is practically instantaneous, does not significantly alter the existing system impedance, and can internally generate reactive (both capacitive and inductive) power [9]–[11].

The Tennessee Valley Authority (TVA) installed the first ±100-MVA STATCOM in 1995 at its Sullivan substation. The application of this STATCOM is expected to reduce the TVA's need for load tap changers, thereby achieving savings by minimizing the potential for transformer failure. This STATCOM aids in resolving the off-peak dilemma of overvoltages in the Sullivan substation area while avoiding the more labor- and space-intensive installation of an additional transformer bank. Also, this STATCOM provides instantaneous control—and therefore increased capacity—of transmission voltage, providing the TVA with greater flexibility in bulk-power transactions, and it also increases the system reliability by damping grids of major oscillations in this grid [12], [13].

To summarize, a STATCOM controller provides voltage support by generating or absorbing reactive power at the point of common coupling without the need of large external reactors or capacitor banks.

10.2.1 The Principle of Operation

A STATCOM is a controlled reactive-power source. It provides the desired reactive-power generation and absorption entirely by means of electronic processing of the voltage and current waveforms in a voltage-source converter (VSC). A single-line STATCOM power circuit is shown in Fig. 10.1(a), where a VSC is connected to a utility bus through magnetic coupling. In Fig. 10.1(b), a STATCOM is seen as an adjustable voltage source behind a reactance—meaning that capacitor banks and shunt reactors are not needed for reactive-power generation and absorption, thereby giving a STATCOM a compact design, or small footprint, as well as low noise and low magnetic impact.

The exchange of reactive power between the converter and the ac system can be controlled by varying the amplitude of the 3-phase output voltage, E_s, of the converter, as illustrated in Fig. 10.1(c). That is, if the amplitude of the output voltage is increased above that of the utility bus voltage, E_t, then a current flows through the reactance from the converter to the ac system and the converter generates capacitive-reactive power for the ac system. If the amplitude of the output voltage is decreased below the utility bus voltage, then the current flows from the ac system to the converter and the converter absorbs inductive-reactive

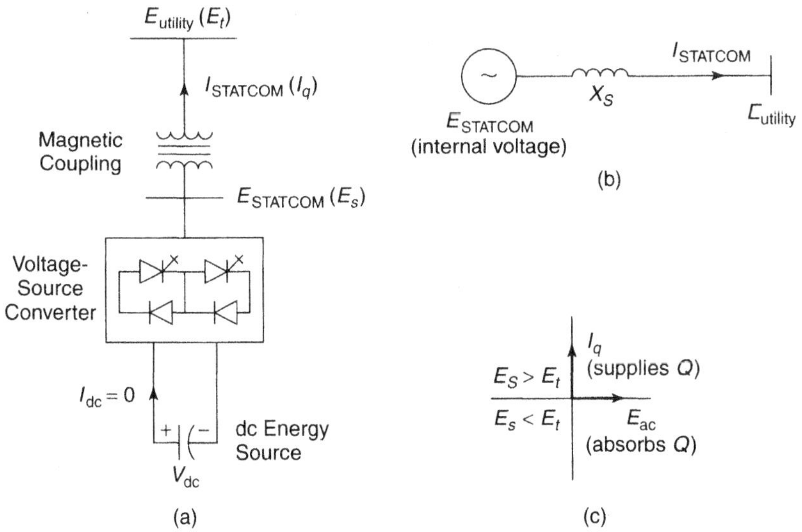

Figure 10.1 The STATCOM principle diagram: (a) a power circuit; (b) an equivalent circuit; and (c) a power exchange.

power from the ac system. If the output voltage equals the ac system voltage, the reactive-power exchange becomes zero, in which case the STATCOM is said to be in a floating state.

Adjusting the phase shift between the converter-output voltage and the ac-system voltage can similarly control real-power exchange between the converter and the ac system. In other words, the converter can supply real power to the ac system from its dc energy storage if the converter-output voltage is made to lead the ac-system voltage. On the other hand, it can absorb real power from the ac system for the dc system if its voltage lags behind the ac-system voltage.

A STATCOM provides the desired reactive power by exchanging the instantaneous reactive power among the phases of the ac system. The mechanism by which the converter internally generates and/or absorbs the reactive power can be understood by considering the relationship between the output and input powers of the converter. The converter switches connect the dc-input circuit directly to the ac-output circuit. Thus the net instantaneous power at the ac-output terminals must always be equal to the net instantaneous power at the dc-input terminals (neglecting losses) [11].

Assume that the converter is operated to supply reactive-output power. In this case, the real power provided by the dc source as input to the converter must be zero. Furthermore, because the reactive power at zero frequency (dc) is by definition zero, the dc source supplies no reactive power as input to the converter and thus clearly plays no part in the generation of reactive-output power by the converter. In other words, the converter simply interconnects the three output terminals so that the reactive-output currents can flow freely among them. If the terminals of the ac system are regarded in this context, the converter establishes a circulating reactive-power exchange among the phases. However, the real power that the converter exchanges at its ac terminals with the ac system must, of course, be supplied to or absorbed from its dc terminals by the dc capacitor.

Although reactive power is generated internally by the action of converter switches, a dc capacitor must still be connected across the input terminals of the converter. The primary need for the capacitor is to provide a circulating-current path as well as a voltage source. The magnitude of the capacitor is chosen so that the dc voltage across its terminals remains fairly constant to prevent it from contributing to the ripples in the dc current. The VSC-output voltage is in the form of a staircase wave into which smooth sinusoidal current from the ac system is drawn, resulting in slight fluctuations in the output power of the converter. However, to not violate the instantaneous power-equality constraint at its input and output terminals, the converter must draw a fluctuating current from its dc source. Depending on the converter configuration employed, it is possible to calculate the minimum capacitance required to meet the system requirements, such as ripple limits on the dc voltage and the rated-reactive-power support needed by the ac system.

The VSC has the same rated-current capability when it operates with the capacitive- or inductive-reactive current. Therefore, a VSC having a certain

MVA rating gives the STATCOM twice the dynamic range in MVAR (this also contributes to a compact design). A dc capacitor bank is used to support (stabilize) the controlled dc voltage needed for the operation of the VSC.

The reactive power of a STATCOM is produced by means of power-electronic equipment of the voltage-source-converter type. The VSC may be a 2-level or 3-level type, depending on the required output power and voltage [2], [3]. A number of VSCs are combined in a multi-pulse connection to form the STATCOM. In the steady state, the VSCs operate with fundamental-frequency switching to minimize converter losses. However, during transient conditions caused by line faults, a pulse width–modulated (PWM) mode is used to prevent the fault current from entering the VSCs [2], [3]. In this way, the STATCOM is able to withstand transients on the ac side without blocking.

10.2.2 The *V-I* Characteristic

A typical *V-I* characteristic of a STATCOM is depicted in Fig. 10.2. As can be seen, the STATCOM can supply both the capacitive and the inductive compensation and is able to independently control its output current over the rated maximum capacitive or inductive range irrespective of the amount of ac-system voltage. That is, the STATCOM can provide full capacitive-reactive power at any system voltage—even as low as 0.15 pu.

The characteristic of a STATCOM reveals another strength of this technology: that it is capable of yielding the full output of capacitive generation almost independently of the system voltage (constant-current output at lower voltages). This capability is particularly useful for situations in which the STATCOM is needed to support the system voltage during and after faults where voltage collapse would otherwise be a limiting factor.

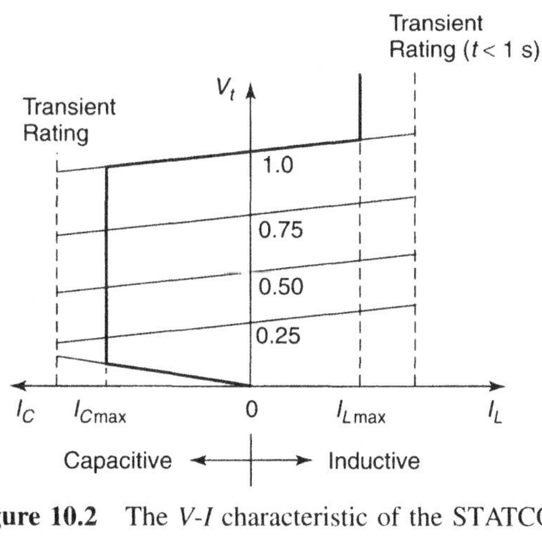

Figure 10.2 The *V-I* characteristic of the STATCOM.

Figure 10.2 also illustrates that the STATCOM has an increased transient rating in both the capacitive- and the inductive-operating regions. The maximum attainable transient overcurrent in the capacitive region is determined by the maximum current turn-off capability of the converter switches. In the inductive region, the converter switches are naturally commutated; therefore, the transient-current rating of the STATCOM is limited by the maximum allowable junction temperature of the converter switches.

In practice, the semiconductor switches of the converter are not lossless, so the energy stored in the dc capacitor is eventually used to meet the internal losses of the converter, and the dc capacitor voltage diminishes. However, when the STATCOM is used for reactive-power generation, the converter itself can keep the capacitor charged to the required voltage level. This task is accomplished by making the output voltages of the converter lag behind the ac-system voltages by a small angle (usually in the 0.1°–0.2° range). In this way, the converter absorbs a small amount of real power from the ac system to meet its internal losses and keep the capacitor voltage at the desired level. The same mechanism can be used to increase or decrease the capacitor voltage and thus, the amplitude of the converter-output voltage to control the var generation or absorption.

The reactive- and real-power exchange between the STATCOM and the ac system can be controlled independently of each other. Any combination of real-power generation or absorption with var generation or absorption is achievable if the STATCOM is equipped with an energy-storage device of suitable capacity, as depicted in Fig. 10.3. With this capability, extremely effective control strategies for the modulation of reactive- and real-output power can be devised to improve the transient- and dynamic-system-stability limits.

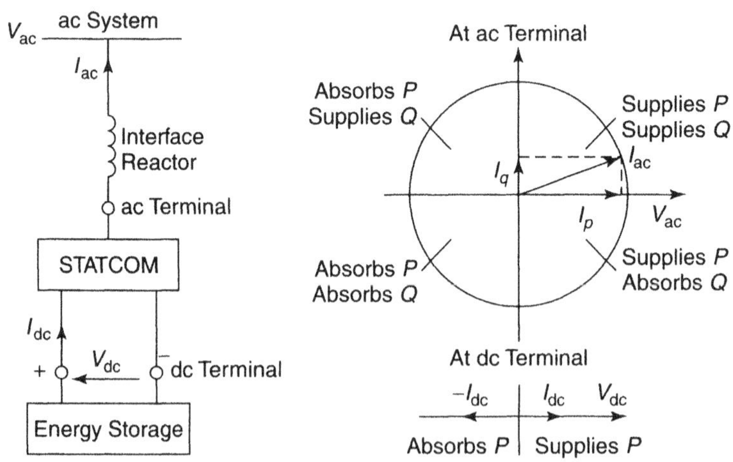

Figure 10.3 The power exchange between the STATCOM and the ac system.

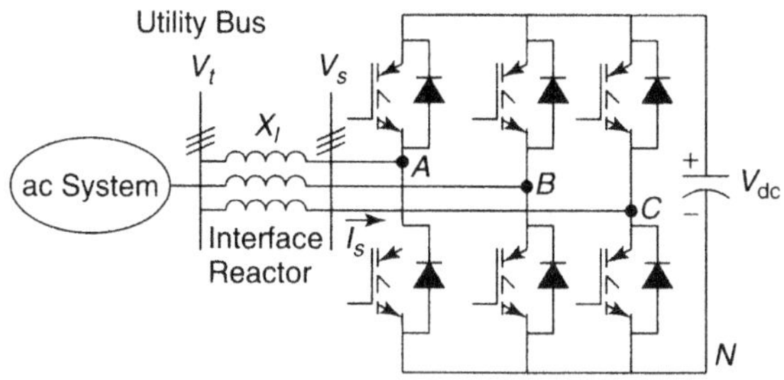

Figure 10.4 An elementary 6-pulse VSC STATCOM.

10.2.3 Harmonic Performance

An elementary 6-pulse VSC STATCOM is shown in Fig. 10.4, consisting of six self-commutated semiconductor switches (IGBT, IGCT, or GTO) with antiparallel diodes. In this converter configuration, IGBTs constitute the switching devices. With a dc-voltage source (which may be a charged capacitor), the converter can produce a balanced set of three quasi-square voltage waveforms of a given frequency by connecting the dc source sequentially to the three output terminals via the appropriate converter switches.

The power quality embraces issues such as voltage flicker, voltage dip, and voltage rise, as well as harmonic performance and high-frequency noise. Power-electronic devices distort voltage and current waveforms in a power network, influencing power facilities and customer equipment in a diverse manner. Harmonic currents induce abnormal noise and parasitic losses, and harmonic voltages cause a loss of accuracy in measurement instruments and the faulty operation of relays and control systems. Electromagnetic noise, caused by the noise of the high-frequency electromagnetic waves emitted from power-electronic circuits, affects electronic devices used in business and industry and often induces interfering voltage in communication lines. The corrective measure generally recommended for mitigating harmonics and high-frequency noise is to limit their generation at the source.

In principle, the STATCOM-output voltage wave is a staircase-type wave synthesized from the dc-input voltage with appropriate combinations of converter switches. For example, the 6-pulse converter shown in Fig. 10.4 is operated typically with either a 120° or 180° conduction sequence for converter switches. For a 180° conduction sequence, three switches conduct at a time; for a 120° conduction sequence, two switches conduct at a time. Figure 10.5 shows the 3-step staircase-line voltage, v_{ab}, along with the fundamental component, V_{fund}, for a conduction sequence of 180°.

The line voltage v_{ab}, in terms of its various frequency components, can be described by the following Fourier-series equation:

$$v_{ab} = a_0 + \sum_{h=1}^{\infty} a_h \cos(h\omega t) + \sum_{h=1}^{\infty} b_h \sin(h\omega t) \tag{10.1}$$

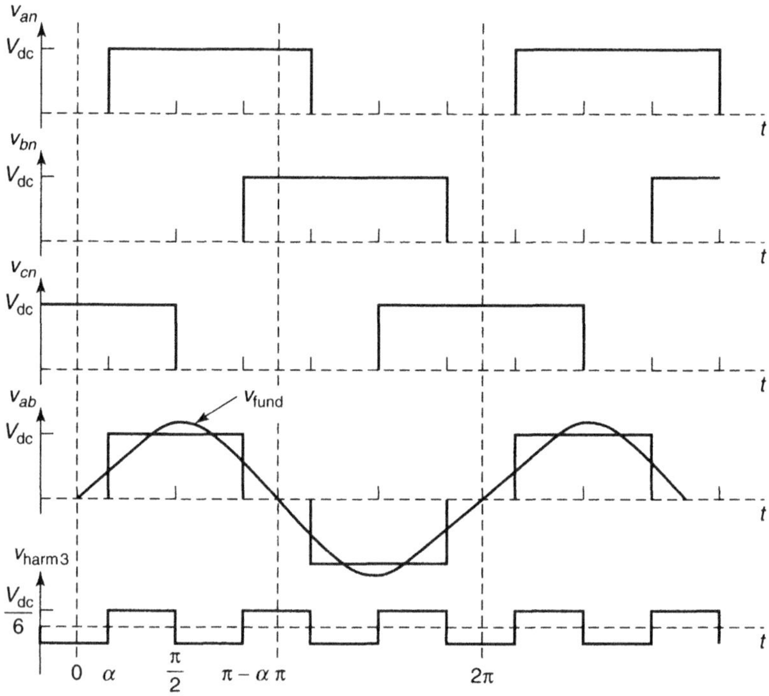

Figure 10.5 An ac line-voltage output of a 6-pulse voltage-source inverter for a 180° conduction sequence.

where coefficients a_0, a_h, and b_h can be determined by considering one fundamental period of v_{ab}. The v_{ab} waveform is symmetrical, so the average voltage $a_0 = 0$. It also has odd-wave symmetry; therefore, $a_h = 0$. The coefficient b_h is determined as

$$b_h = \frac{2}{\pi}\int_0^{\pi} V_{dc}\sin(h\omega t)\, d(\omega t) \tag{10.2}$$

$$b_h = \frac{2}{\pi}\int_{\alpha}^{\pi-\alpha} V_{dc}\sin(h\omega t)\, d(\omega t) = \frac{4V_{dc}}{\pi h}\cos(h\alpha) \tag{10.3}$$

Therefore

$$v_{ab} = \sum_{h=1,3,5}^{\infty} \frac{4V_{dc}}{\pi h}\cos(h\alpha)\sin(h\omega t) \tag{10.4}$$

For 180° conduction sequence, $\alpha = 30°$; hence the triplen harmonics are zero

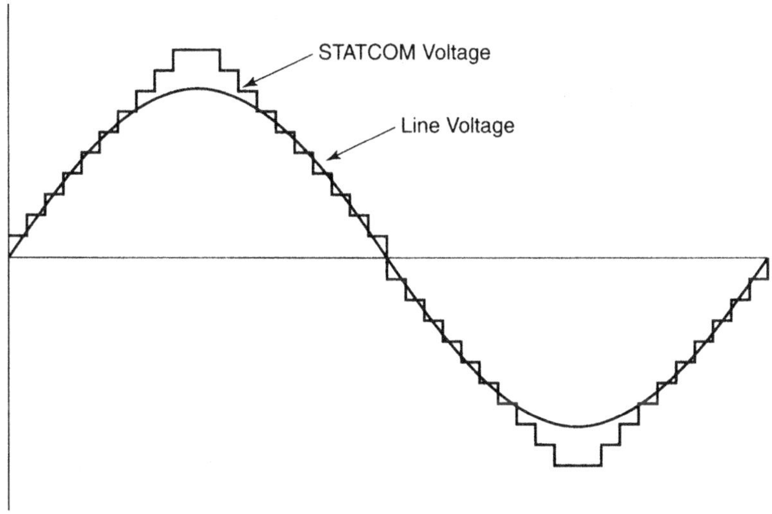

Figure 10.6 The output voltage of a 48-pulse STATCOM that generates reactive power.

in the line voltage, as seen from Eq. (10.4). It is also noted that the converter has harmonic components of frequencies $(6k \pm 1)f_0$ in its output voltage and $6kf_0$ in its input current, where f_0 is the fundamental-output frequency and $k = 1, 2, 3, \ldots$ As is evident, the high harmonic content in the output voltage makes this simple converter impractical for power-system applications.

To reduce harmonic generation, various converter configurations and converter-switching techniques are suggested in the literature. For example, the first installed commercial STATCOM [13] has a 48-pulse converter configuration so that the staircase ac-line output-voltage waveform has 21 steps, as shown in Fig. 10.6, and approaches an ideal sinusoidal waveform with a greatly reduced harmonic content. Switching strategies, such as selective harmonic-elimination techniques, also aid in limiting harmonic generation at its source.

10.2.4 Steady-State Model

A STATCOM is always connected in shunt with the ac system through some magnetic coupling, namely, the coupling transformer or interface reactor. A typical STATCOM connection is shown in Fig. 10.7; it consists of a VSC using either a GTO or IGBT as a switching device, and a capacitor, C_s, on the dc side as an energy-storage device. The resistance, R_p, in parallel with C_s represents both the capacitor losses and switching losses. The STATCOM is connected to the ac system through magnetic coupling, represented by leakage inductance, L_s, and resistance, R_s [14]. The STATCOM improves the desired power-system performance, including dynamic compensation, mitigating the SSR by modulating the reactive power at the common-coupling point, and so forth.

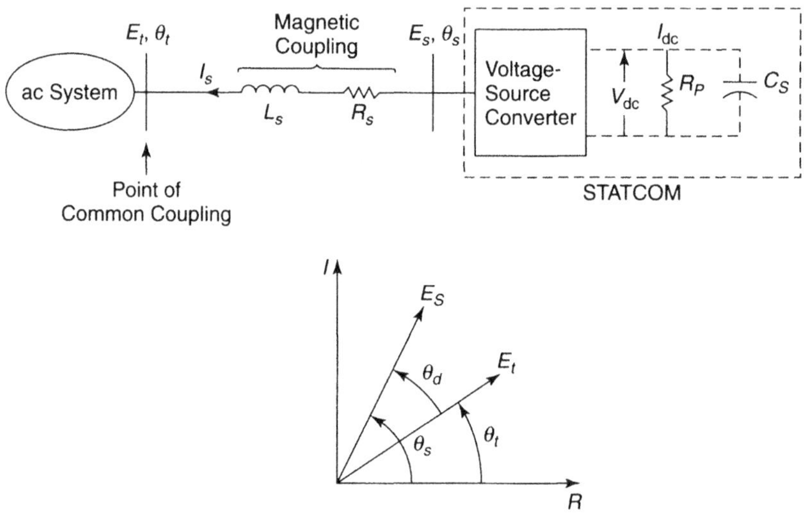

Figure 10.7 A typical STATCOM connection to the ac system.

The first-order differential equations for the ac-side circuit of the STATCOM circuit in Fig. 10.7 can be written as

$$\frac{di_{sa}}{dt} = \frac{1}{L_s}(-R_s i_{sa} + e_{sa} - e_{ta})$$

$$\frac{di_{sb}}{dt} = \frac{1}{L_s}(-R_s i_{sb} + e_{sb} - e_{tb})$$

$$\frac{di_{sc}}{dt} = \frac{1}{L_s}(-R_s i_{sc} + e_{sc} - e_{tc}) \tag{10.5}$$

These equations are converted on *R-I* frame of reference (the synchronously rotating frame of reference) [49] as follows:

$$\begin{bmatrix} \dot{I}_{sR} \\ \dot{I}_{sI} \end{bmatrix} = \begin{bmatrix} \frac{-R_s}{L_s} & \omega_0 \\ -\omega_0 & \frac{-R_s}{L_s} \end{bmatrix} \begin{bmatrix} I_{sR} \\ I_{sI} \end{bmatrix} + \frac{1}{L_s}\begin{bmatrix} E_{sR} - E_{tR} \\ E_{sI} - E_{tI} \end{bmatrix} \tag{10.6}$$

The STATCOM dc-side-circuit equation can be written as

$$\frac{dV_{\mathrm{dc}}}{d_t} = \frac{-1}{C_s}\left(I_{\mathrm{dc}} + \frac{V_{\mathrm{dc}}}{R_p}\right) \tag{10.7}$$

The instantaneous powers at the ac and dc terminals of the converter are equal, giving the following power-balance equation:

$$V_{dc}I_{dc} = \tfrac{3}{2}(E_{sR}I_{sR} + E_{sI}I_{sI}) \tag{10.8}$$

where the constant 3/2 is the reference-frame transformation constant. Based on the phasor diagram given in Fig. 10.7, E_{sR} and E_{sI} can be defined as follows:

$$\begin{aligned} E_{sR} &= E_s \cos\theta_s = K_{cs}V_{dc}\cos\theta_s \\ E_{sI} &= E_s \sin\theta_s = K_{cs}V_{dc}\sin\theta_s \end{aligned} \tag{10.9}$$

where K_{cs} is the constant relating the ac and dc voltage. For example, in a 12-pulse VSC, $K_{cs} = 2\sqrt{6}\pi$. Therefore, Eq. (10.8) becomes

$$V_{dc}I_{dc} = \tfrac{3}{2}(K_{cs}V_{dc}\cos\theta_s I_{sR} + K_{cs}V_{dc}\sin\theta_s I_{sI}) \tag{10.10}$$

or

$$I_{dc} = \tfrac{3}{2}(K_{cs}\cos\theta_{sR} + K_{cs}\sin\theta_s I_{sI}) \tag{10.11}$$

Substituting the value of I_{dc} in Eq. (10.7),

$$\frac{dV_{dc}}{d_t} = \frac{-1}{C_s}\left(\frac{3}{2}K_{cs}\cos\theta_s I_{sR} + \frac{3}{2}K_{cs}\sin\theta_s I_{sI} + \frac{V_{dc}}{R_p}\right) \tag{10.12}$$

Also, substituting the values of E_{sR} and E_{sI} from Eq. (10.9) into Eq. (10.6),

$$\begin{bmatrix} \dot{I}_{sR} \\ \dot{I}_{sI} \end{bmatrix} = \begin{bmatrix} \dfrac{-R_s}{L_s} & \omega_0 \\ -\omega_0 & \dfrac{-R_s}{L_s} \end{bmatrix} \begin{bmatrix} I_{sR} \\ I_{sI} \end{bmatrix} + \frac{1}{L_s}\begin{bmatrix} K_{cs}V_{dc}\cos\theta_s - E_{tR} \\ K_{cs}V_{dc}\sin\theta_s - E_{tI} \end{bmatrix} \tag{10.13}$$

From Eqs. (10.12) and (10.13), the state-space model in the *R-I* frame for the STATCOM circuit in Fig. 10.7 can be written as follows:

$$\begin{bmatrix} \dot{I}_{sR} \\ \dot{I}_{sI} \\ \dot{V}_{dc} \end{bmatrix} = \begin{bmatrix} -\dfrac{R_s}{L_s} & \omega_0 & \dfrac{K_{cs}\cos\theta_s}{L_s} \\ -\omega_0 & -\dfrac{R_s}{L_s} & \dfrac{K_{cs}\sin\theta_s}{L_s} \\ -\dfrac{1.5K_{cs}\cos\theta_s}{C_s} & -\dfrac{1.5K_{cs}\cos\theta_s}{C_s} & -\dfrac{1}{R_pC_s} \end{bmatrix} \begin{bmatrix} I_{sR} \\ I_{sI} \\ V_{dc} \end{bmatrix} + \begin{bmatrix} -\dfrac{1}{L_s} & 0 \\ 0 & -\dfrac{1}{L_s} \\ 0 & 0 \end{bmatrix} \begin{bmatrix} E_{tR} \\ E_{tI} \end{bmatrix} \tag{10.14}$$

The steady-state solution of the STATCOM circuit—represented by Eq. (10.14) using $R_s = 0.01$ pu, $X_s = 0.15$ pu, $R_p = 128$ pu, $C_s = 0.013$ pu, and $K_{cs} = 2\sqrt{6}\pi$ [14]—is plotted in Fig. 10.8 as a function of the phase-difference angle, θ_d. In this plot, I_{sR} and I_{sI} are, respectively, the active and reactive components of the STATCOM current, I_s. The reactive-power output from STATCOM is controlled by varying θ_d. It should be noted that I_{sI} varies almost linearly with θ_d, and the range of θ_d for controlling I_{sI} within ± 1 pu is very small.

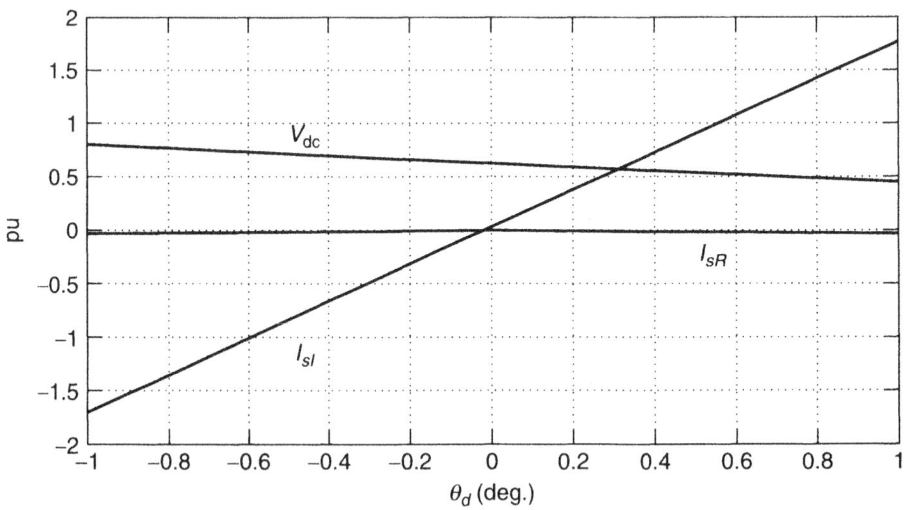

Figure 10.8 The steady-state characteristics of a STATCOM.

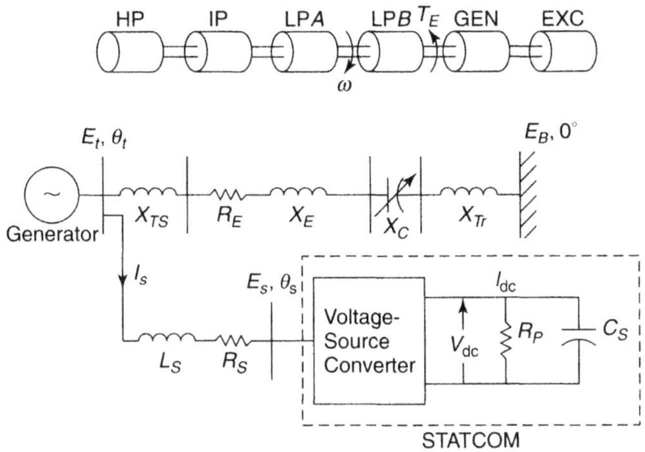

Figure 10.9 The IEEE First Benchmark System with a STATCOM for SSR damping studies.

10.2.5 SSR Mitigation

10.2.5.1 A Study System The mitigation of SSR is demonstrated on the IEEE First Benchmark System [50] with a STATCOM connected at the generator terminals, as shown in Fig. 10.9. To develop a mathematical model for the studied system, the electric network equations as they are given in ref. [51] are used. The STATCOM controller is modeled as described in the next section. The generator is represented by its d- and q-axis equivalent circuits, with one damper winding and field winding on the d-axis and two damper windings on the q-axis [49]. The mechanical system, as shown in Fig. 10.9, comprises six masses: one high-pressure turbine (HP), one intermediate-pressure turbine (IP), two low-pressure turbines (LPA and LPB), one generator (GEN), and one exciter (EXC). The mechanical damping is assumed to be zero for all masses to represent the worst damping conditions. The generator is equipped with a static-excitation system. The static-excitation and governor-system configurations are taken from ref. [52].

Combining the equations of the mechanical, excitation, and governor systems, as well as the generator- and capacitor-compensated transmission line, results in a set of twenty-seventh-order nonlinear-differential equations without the STATCOM. These differential equations are then linearized around an operating point to obtain the linear-differential equations, which are used to obtain the system eigenvalues. The six-mass model has five torsional modes in addition to an electromechanical mode (mode 0). The IEEE First Benchmark System is characterized by four unstable torsional modes, each of which has its largest SSR interaction at a certain value of the series compensation, X_C [51].

It should be noted that the definition of the series compensation percent used is

TABLE 10.1 Eigenvalues of the IEEE First Benchmark System

	% Compensation Level			
Mode	32%	46.5%	59.25%	72.25%
0	$0.168 \pm j10.08$	$0.113 \pm j10.99$	$0 \pm j11.85$	$-0.31 \pm j12.75$
1	$-0.001 \pm j99.37$	$0.006 \pm j99.70$	$0.044 \pm j100$	$6.53 \pm j98.79$
2	$0.001 \pm j127.05$	$0.006 \pm j127.11$	$1.21 \pm j126.98$	$0.004 \pm j126.90$
3	$0.01 \pm j160.81$	$2.29 \pm j160.49$	$0.017 \pm j160.25$	$0.002 \pm j160.41$
4	$2.65 \pm j203.01$	$0.015 \pm j202.58$	$0.002 \pm j202.73$	$-0.001 \pm j202.80$
5	$1E\text{-}6 \pm j298.18$	$-1E\text{-}5 \pm j298.18$	$-1E\text{-}5 \pm j298.18$	$-1E\text{-}5 \pm j298.18$
Electrical	$-5.5 \pm j203.19$	$-4.76 \pm j160.41$	$-3.04 \pm j126.88$	$-6.96 \pm j98.40$

$$\% \text{ Compensation} = \frac{X_C}{0.14 + 0.5 + 0.06} \times 100 \tag{10.15}$$

where 0.14 = the sending-end transformer reactance
0.5 = the transmission-line reactance
0.06 = the equivalent system reactance at an infinite bus

Without the STATCOM, the real part of the eigenvalues corresponding to the various torsional frequencies vary in magnitude and acquire peak instability at different levels of series compensation. For each of the critical compensation levels at which the real part of the eigenvalue is maximum (indicating worst damping), the corresponding electrical- and torsional-mode eigenvalues of the system without the STATCOM are listed in Table 10.1. It can be seen that modes 1, 2, 3, and 4 become peak-unstable at compensation levels of 72.25%, 59.25%, 46.5%, and 32%, respectively. The objective is to damp all these torsional modes by employing the STATCOM at the generator terminal.

10.2.5.2 STATCOM Performance The STATCOM controller used for the IEEE First Benchmark System is shown in Fig. 10.10. The main function of a

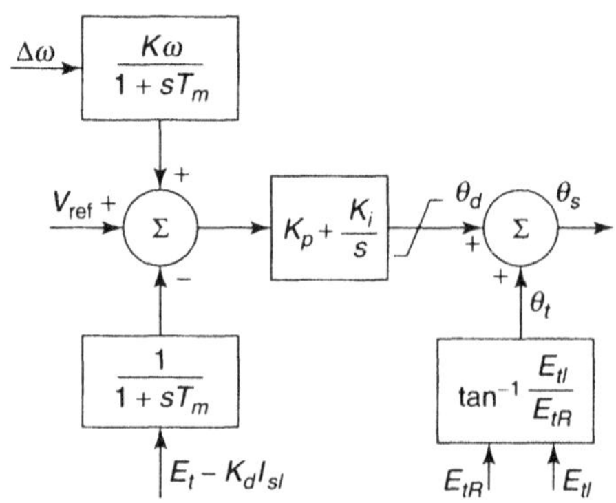

Figure 10.10 A STATCOM controller for damping the SSR.

TABLE 10.2 Eigenvalues of the IEEE First Benchmark System With Only the STATCOM Voltage Controller

Mode	% Compensation Level			
	32%	46.5%	59.25%	72.25%
0	−3.99 ± j2.92	−4.338 ± j2.91	−4.773 ± 2.74	−5.176 ± j2.12
1	−0.198 ± j99.64	−0.154 ± j100.08	0.754 ± j100.65	1.326 ± j99.44
2	−0.015 ± j127.08	0.037 ± j127.14	0.114 ± j127.04	0.059 ± j127.01
3	0.034 ± j160.91	0.354 ± j160.65	0.158 ± j160.50	0.09 ± j160.56
4	0.505 ± j203.03	0.144 ± j202.86	0.667 ± j202.91	0.04 ± j202.94
5	−1E-5 ± j298.18	−1E-5 ± j298.18	−1E-5 ± j298.18	−1E-5 ± j298.18
Electrical	−27.97 ± 193.66	−28.28 ± j144.54	−27.46 ± j107.88	−25.76 ± j76.97

STATCOM, as with the conventional SVC, is to regulate the transmission-line voltage at the point of connection. When a STATCOM having only a voltage controller is incorporated into the power system, the electrical- and torsional-mode eigenvalues for all the critical compensation levels are as they appear in Table 10.2. It can be seen that the STATCOM equipped with only a voltage controller is not sufficient to damp all the torsional modes of the IEEE First Benchmark System. Thus a need exists for an additional control signal along with the STATCOM voltage controller.

The principal strategy in controlling an SSR-oscillation-damping STATCOM is to use simple stabilizing signals [16], [51]. It is known that the generator speed contains components of all the turbine modes; consequently, if the generator speed is used to control a STATCOM, all the torsional modes—in addition to the mode corresponding to the generator mass—will be affected [51]. Therefore, the auxiliary signal employed is the generator-speed deviation. Listed in Table 10.3 for all critical compensation levels are the eigenvalues of the electrical- and torsional-modes of the system having a STATCOM equipped with a voltage controller and an auxiliary speed-deviation feedback stabilizer. All the torsional modes are seen to be stable. For the IEEE First Benchmark System, all other system eigenvalues (those not listed in Table 10.3) are also found to be stable.

TABLE 10.3 Eigenvalues of the IEEE First Benchmark System With the STATCOM Voltage Controller and the Auxiliary Stabilizer

Mode	% Compensation Level			
	32 %	46.5 %	59.25 %	72.25 %
0	−3.348 ± j2.62	−3.398 ± j2.7	−3.484 ± j2.8	−3.716 ± j2.92
1	−1.004 ± j99.14	−1.442 ± j99.25	−2.256 ± j100.50	−0.284 ± j101.2
2	−0.125 ± j127.02	−0.21 ± j127.08	−0.062 ± j127.23	−0.001 ± j127.14
3	−0.718 ± j160.73	−0.215 ± j161.33	−0.018 ± j160.97	−0.041 ± j160.86
4	−0.383 ± j204.22	−0.093 ± j203.39	−0.18 ± j203.25	−0.218 ± j203.21
5	−1E-5 ± j298.18	−1E-5 ± j298.18	−1E-5 ± j298.18	−1E-5 ± j298.18
Electrical	−25.37 ± j192.85	−25.49 ± j144.83	−23.40 ± j108.56	−23.34 ± j76.88

The STATCOM controller model shown in Fig. 10.10 has three gains: the proportional gain, K_P; the integral gain, K_i; and the speed-deviation-feedback gain, K_ω. The objective is to design a single controller, which can be used to damp all SSR modes at all series-compensation levels. Eigenvalue analysis is used to obtain the range of K_P, K_i, and K_ω for which the system is stable. It is found that the system is stable for $-6.25 \le K_P \le -1$, $-132 \le K_i \le 0$, and $4 \le K_\omega \le 11$. From these K_P, K_i, and $K_\bullet$ values, final gain parameters are selected by carrying out a step-response test on the system and ensuring that the system settling time and the generator overshoot speed is low. Thus the final values of K_P, K_i, and K_ω obtained are -1, -1.25, and +8, respectively.

10.2.6 Dynamic Compensation

Several configurations of STATCOM have been proposed in the literature (see refs. [2], [3], [9], [11], and [17]–[22]), many of which have also been implemented in practical applications. This section focuses on one such configuration and its advantages: the binary voltage source inverter (BVSI).

10.2.6.1 A Multilevel VSC–Based STATCOM The harmonic contamination of the power-system network by the addition of STATCOM into the power system can be reduced by employing multilevel VSC configurations [17], [18]. The multilevel converters usually synthesize a staircase-type voltage wave from several levels of dc-voltage sources (typically capacitor-voltage sources). The multilevel VSC schemes studied and tested so far include the *diode clamp*, the *flying capacitor*, and the *cascaded, separate dc-source converter* types [17]. Many separate dc-source converter topologies are suggested in refs. [19]–[22]. Multilevel converters can reach high voltages and reduce harmonic distortion because of their structure. To increase the voltage rating, many single-phase full-bridge converters (FBCs) can be connected in series, automatically leading to a desirable reduction of harmonic distortion. However, the need to balance capacitor voltages, the complexity of switching, and the size of the capacitors all limit the number of levels that can be practically employed.

Figure 10.11 shows the 3-phase star-connected arrangement of the separate dc-source, 3-level binary VSC commonly referred to as a BVSI [19]. It consists of three single-phase FBCs, each with its own dc source, connected in series. However, the magnitude of each dc source is in binary proportion of V_{dc}, $2V_{dc}$, and $4V_{dc}$, where V_{dc} is chosen to get the desired fundamental ac-voltage output for a normalized 1-pu modulation index. The switches are turned on and off to generate a 15-step ac-voltage output over one fundamental cycle. In general, n-level BVSI would produce a $(2^{n+1} - 1)$–step ac-voltage output versus a $(2n + 1)$–step output generated by a conventional n-level, separate dc-source VSC configuration.

Figure 10.12 illustrates the various voltages in the 3-level BVSI STATCOM. The resulting ac-phase voltage, v_{av}, and the fundamental-output voltage, v_a, of the 3-level converter are also shown in Fig. 10.12. The output-phase voltage is

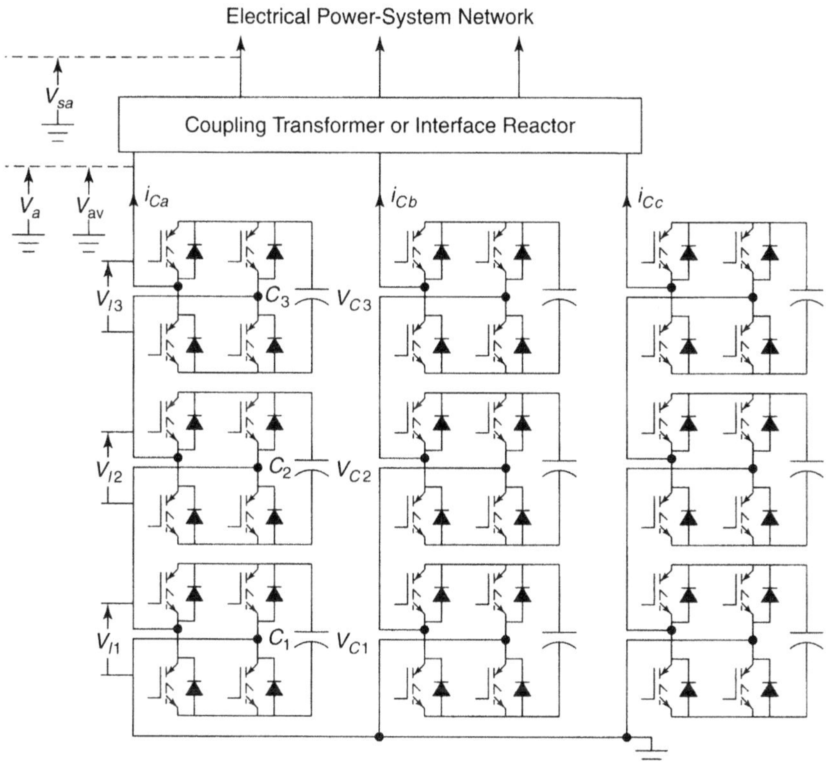

Figure 10.11 The 3-phase, star-connected 3-level BVSI.

given by

$$v_{\mathrm{av}} = v_{l1} + v_{l2} + v_{l3} \tag{10.16}$$

where

$$\begin{aligned} v_{l1} &= +V_{C1}, 0, -V_{C1} \quad \text{and} \quad V_{C1} = V_{\mathrm{dc}} \\ v_{l2} &= +V_{C2}, 0, -V_{C2} \quad \text{and} \quad V_{C2} = 2V_{\mathrm{dc}} \\ v_{l3} &= +V_{C3}, 0, -V_{C3} \quad \text{and} \quad V_{C3} = 4V_{\mathrm{dc}} \end{aligned} \tag{10.17}$$

The phase voltage given by Eq. (10.16) is obtained by varying the voltage output of each FBC level in Eq. (10.17) by appropriately switching various devices and their combinations. From Fig. 10.12, the a-phase converter-output voltage for n-level BVSI is given by

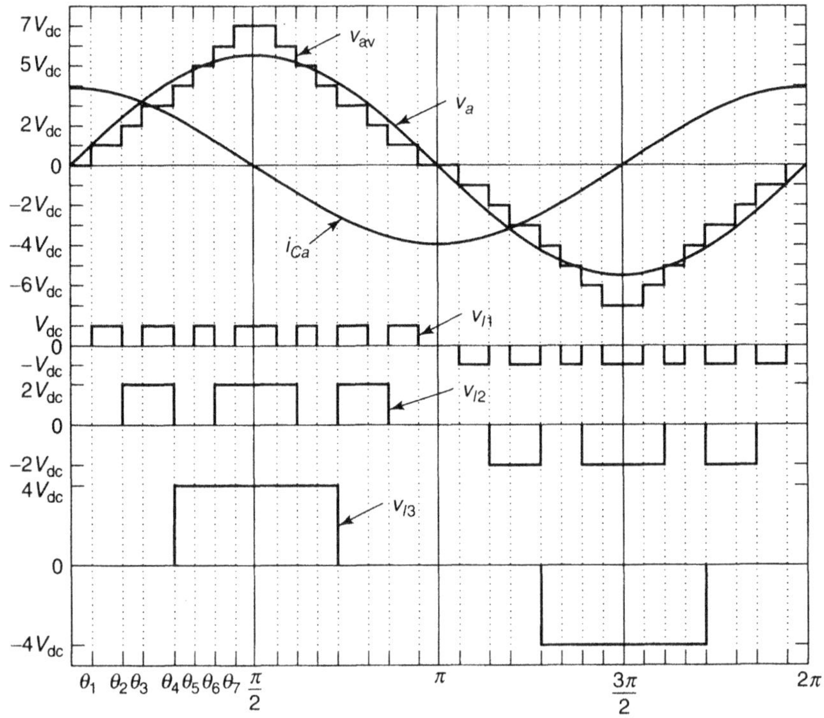

Figure 10.12 Typical voltages of the 3-level BVSI.

$$v_{av} = \frac{4V_{dc}}{(2k-1)\pi} \sum_{k=1}^{\infty} \sum_{i=1}^{2^n-1} \cos[(2k-1)\theta_i]\sin[(2k-1)\omega t] \tag{10.18}$$

From Eq. (10.18), the fundamental root mean square (rms) voltage, V_a, the harmonic rms voltage, V_{2k-1}, and the maximum fundamental-phase voltage, $V_{a\max}$, can be determined as

$$V_a = \frac{4V_{dc}}{\sqrt{2}\pi} \sum_{i=1}^{2^n-1} \cos\theta_i \tag{10.19}$$

and

$$V_{2k-1} = \frac{4V_{dc}}{\sqrt{2}(2k-1)\pi} \sum_{i=1}^{2^n-1} \cos[(2k-1)\theta_i] \qquad \text{for } k = 1, 2, 3, \ldots \tag{10.20}$$

and

$$V_{a\max} = \frac{28V_{\mathrm{dc}}}{\sqrt{2}\pi} \qquad \text{for a 3-level BVSI} \tag{10.21}$$

10.2.6.2 A Selective Harmonic-Elimination Modulation (SHEM) Technique To reduce the harmonics injected into the power system by the BVSI, a selective harmonic-elimination modulation (SHEM) technique for this converter configuration has been developed.

Equation (10.18) shows that the BVSI output voltage contains odd harmonic voltages in addition to fundamental voltage. Equations (10.19) and (10.20) show that the harmonic voltages and fundamental voltage can be changed by varying either the dc voltage, V_{dc}, or the modulation angles, θ_i. Varying the dc voltage is not economical in practical applications because it needs a rectifier. Here, the modulation angles are varied to control the fundamental-output voltage as well as to minimize the harmonic distortion.

The phase voltage, v_{av}, of the 3-level BVSI is plotted in Fig. 10.12, where seven steps are introduced into the otherwise square wave-converter output to control the magnitude of the fundamental voltage and to eliminate or minimize any of the six harmonics. On a half-cycle basis, each step provides one degree of freedom, that is, having (2^{n-1}) steps per half-cycle in an n-level BVSI provides control of the fundamental and elimination or minimization of (2^{n-2}) harmonics. Using Eqs. (10.19) and (10.21), the modulation index, MI, or the amplitude-modulation ratio is defined as

$$MI = \frac{V_a}{V_{a\max}} = \frac{1}{7}\sum_{i=1}^{7} \cos\theta_i \tag{10.22}$$

The 5th, 7th, 11th, 13th, 17th, and 19th harmonics are minimized or completely eliminated. From Eq. (10.20),

$$V_h = \sum_{i=1}^{7} \cos(h\theta_i) = 0 \qquad \text{for } h = 5, 7, 9, 11, 13, 17, 19 \tag{10.23}$$

The seven nonlinear equations, (10.22) and (10.23), are solved by the Newton–Raphson numerical technique to derive switching instants (angles) of BVSI switches.

10.2.6.3 Capacitor-Voltage Control Figure 10.13 shows the average waveforms of the phase voltage, v_{av}, the fundamental phase voltage, v_a, the system bus voltage, v_{sa} (to which the converter is connected), and the current, i_{Ca}, flowing through the capacitors. The capacitor current either leads or lags the phase voltage by 90°, depending on whether the converter acts capacitively or inductively. The average charge on each dc capacitor over every half-cycle is equal to zero, and because of this symmetric charge flow, the voltages on all the dc capacitors remain theoretically balanced [18]. However, because of

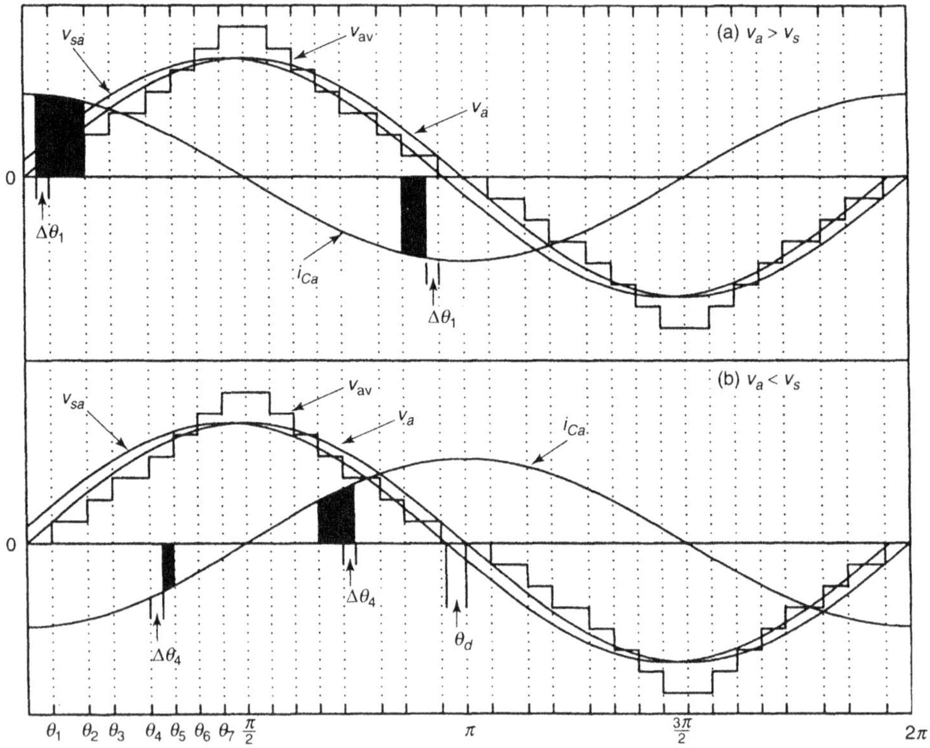

Figure 10.13 The modulating charge and discharge of capacitors to control their voltages.

the power losses in the converter and an unbalanced operation of the 3-phase system, the capacitor voltages may drift away from their set levels.

In the BVSI of Fig. 10.11, the set voltage levels are V_{dc}, $2V_{dc}$, and $4V_{dc}$. In principle, the charge and hence the voltage across each capacitor is maintained by prolonging or shortening the current flow through it, which is achieved by shifting the switching patterns [18], [21]–[22]. The direction of phase shift depends on whether the converter is producing leading var or absorbing lagging var. Figure 10.13(a) depicts the extra charging of capacitor C_1 to raise its falling voltage when the converter is producing leading var. In the positive half-cycle of i_{Ca}, C_1 conduction is increased by $\Delta\theta_1$, whereas in the negative half-cycle it is reduced by the same amount. Reflecting this behavior are shaded areas in which the positive half-cycle is larger than that in the negative half-cycle, indicating extra charging of C_1.

Figure 10.13(b) illustrates the extra charging of capacitor C_3 when the converter is absorbing lagging var. Here, also, the positive half-cycle conduction of C_3 is more than its negative half-cycle conduction. Similarly, θ_2 is modulated to adjust the conduction of capacitor C_2. The required power to charge

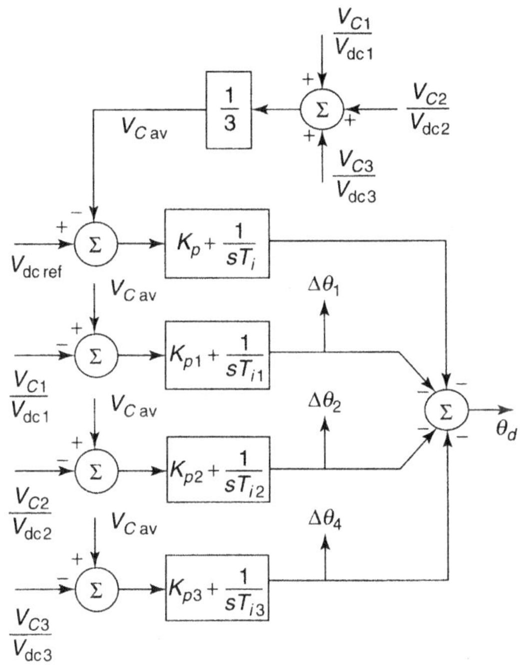

Figure 10.14 Capacitor-voltage control.

the capacitors is drawn from the system by controlling the phase-lag angle θ_d between v_a and v_{sa}, as shown in Fig. 10.13.

The capacitor-voltage control scheme employed is shown in Fig. 10.14. One such controller on each phase maintains individual capacitor voltages in binary proportion as it adjusts the phase lag between the converter-phase voltage and the system-phase voltage to draw a sufficient amount of active power from the system required to keep the capacitors charged. The net active power drawn is equal to the switching losses and capacitor losses.

In Fig. 10.14, the phase-lag angle θ_d adjusts the net active-power flow into the converter. The angles $\Delta\theta_1$, $\Delta\theta_2$, and $\Delta\theta_4$ are used to prolong or shorten the conduction of capacitors C_1, C_2, and C_3, respectively, by shifting the switching patterns $\Delta\theta_1$, $\Delta\theta_2$, and $\Delta\theta_4$ in order, as shown in Fig. 10.14. It is found that normally $\Delta\theta_1$, $\Delta\theta_2$, $\Delta\theta_4$, and θ_d are each less than 1°, meaning that such unequal capacitor conduction is not likely to affect the 3-phase system balance.

10.2.6.4 STATCOM Performance With BVSI, the dynamic compensation of the ac system at the point of common coupling is achieved by meeting the reactive-power demands in transient and in steady-state conditions. To generate the required reactive power from the converter, the modulation index *MI* is varied and the gate-firing angles θ_i are calculated for each *MI* value. When *MI* is $MI_{\min}$, the converter absorbs the rated reactive power (inductive STATCOM),

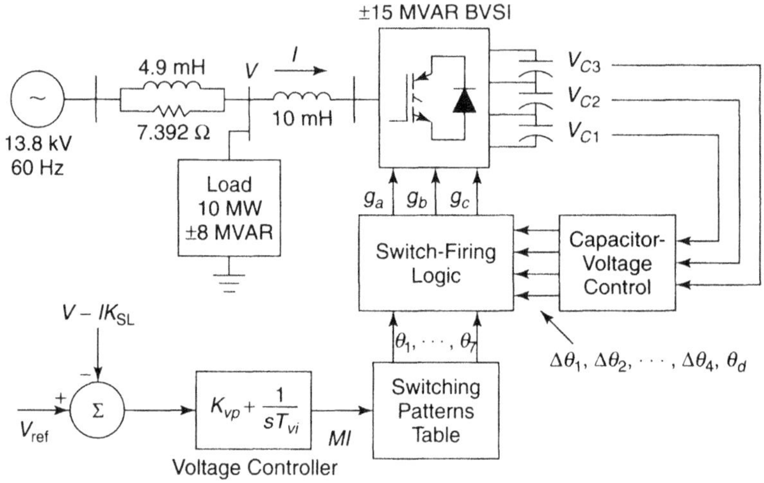

Figure 10.15 Distribution-system compensation using a BVSI STATCOM.

whereas when MI is MI_{max}, the converter generates the rated reactive power (capacitive STATCOM).

Figure 10.15 shows the configuration of a distribution system studied for examining the effectiveness of the BVSI to regulate the load-bus voltage under various operating conditions. The distribution system considered is 13.8 kV, 60 Hz with a short-circuit capacity of 100 MVA and $X/R = 4$. A 3-phase, 3-level, 15-step ±15 MVAR BVSI is connected to the load bus through a 10-mH interface reactor. The load considered is 10 MW ± 8 MVAR.

A STATCOM Voltage Controller for Dynamic Compensation The desired voltage at the load bus is maintained at 1 pu for varying operating conditions by dynamically providing the reactive-power requirement through the BVSI. To regulate the load-bus voltage, a proportional–integral (PI) controller is employed that contains a feedback signal derived from the voltage at the load bus, V, and the reactive current, I, flowing into the converter weighted with the V-I characteristic droop, K_{SL}, as indicated in Fig. 10.15. The output of this PI controller comprises the modulation index MI, and depending on its value, the gate-firing angles (of $\theta_1 \cdots \theta_7$) to turn on various switches are selected from the precomputed switching-patterns table. The capacitor-voltage controller (see Section 10.2.6.3) generates $\Delta\theta_1$, $\Delta\theta_2$, $\Delta\theta_4$, and θ_d. These angles, together with the gate-firing angles, are processed through a switch-firing logic algorithm to modulate the conduction period of all capacitors to control their voltages in a binary proportion. The switch-firing logic generates the appropriate gate-firing pulses of g_a, g_b, and g_c. The following circuit and controller parameters are considered in the transient simulation:

1. The ac-voltage rating (line-to-line, rms): 13.8 kV, 60 Hz
2. The MVAR rating: ±15 MVAR
3. The normalized per-unit modulation index: 0.631 pu
4. The undermodulation (lagging var): 0.51–0.63 pu
5. The overmodulation (leading var): 0.632–0.886 pu
6. The dc-voltage regulation factor (ϵ): ±5%
7. The dc-voltage rating (V_{C1}, V_{C2}, V_{C3}): 2, 4, 8 kV
8. The dc capacitors (C_1, C_2, C_3): 6.5, 2.57, 1.6 mF
9. The K_{p1}, K_{p2}, K_{p3}, K_p (leading var): 12, 10, 12, 10
10. The K_{p1}, K_{p2}, K_{p3}, K_p (lagging var): −12, −10, −12, 10
11. The T_{p1}, T_{p2}, T_{p3}, T_p: 1, 1, 1.5, 0.8 s
12. The voltage controller (K_{vp}, T_{vi}, K_{SL}, V_{ref}): 10 s, 1 s, 1%, 1 pu

Transient Simulation The operation of the 3-level BVSI for distribution-system dynamic compensation, described in Fig. 10.15, is evaluated using PSCAD/EMTDC [53] for various disturbances and different system-operating conditions. This simulation illustrates the effectiveness of the BVSI in reactive-power compensation, as well as the suitability of the capacitor-voltage controller in maintaining the capacitor voltages in binary proportion and the suitability of the voltage controller in regulating load-bus voltage during abnormal conditions and varied operating environments. Also, it shows that the power-quality impairment from STATCOM operation can be minimized by the proper configuration of STATCOM, that is, multilevel and/or high-pulse converter configurations.

As an example case, Fig. 10.16 illustrates the operation of the BVSI when it generates leading var and absorbs lagging var. Initially, the capacitor voltages are maintained at 2-, 4-, and 8-kV levels, the BVSI is in a floating state, and the load-bus voltage is 1 pu.

At 0.12 s, a 10-MW, $+j8$–MVAR inductive load is added. To maintain the load-bus voltage at 1 pu, increased capacitive-reactive-power demand at the load bus is met by the converter. The voltage controller responds to the changed operating condition and adjusts the modulation index, and the converter generates 11-MVAR capacitive-reactive power. Figure 10.16 also shows the variations in the capacitor voltages, V_{C1}, V_{C2}, and V_{C3}; the rms load-bus voltage, V_l; the BVSI-bus voltage, V_s; and instantaneous load-bus line voltages, v_{ab}, v_{bc}, and v_{ca}. It is seen that after the initial disturbance, the capacitor voltages adjust to the predisturbance binary levels, and V_l is regulated to 1 pu.

At 0.4 s, a 16-MVAR capacitor bank is switched in at the load bus, changing the load to 10 MW, $-j8$ MVAR capacitive. It is seen that the converter is able to switch successfully from capacitive to inductive operation and the load-bus voltage is regulated rapidly to 1 pu. However, the capacitor voltages are maintained at newly reduced steady-state values but still remain in a binary proportion. In principle, the variation in the dc-link voltages does not inherently affect the ability of the converter to produce the designated sinusoidal-output

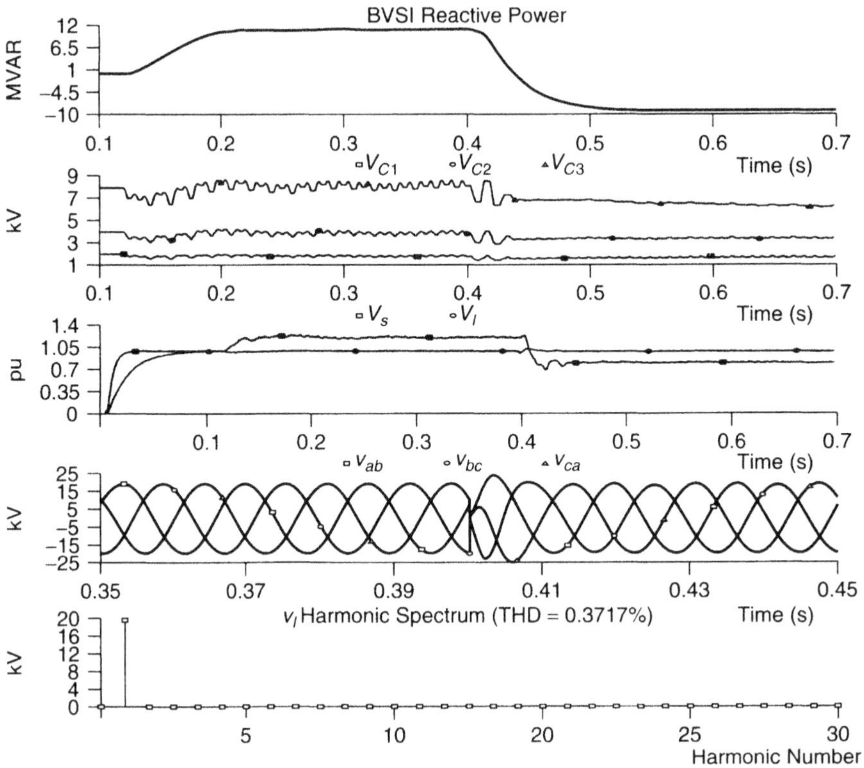

Figure 10.16 The addition of inductive (at 0.12 s) and capacitive (at 0.4 s) loads.

voltages at the desired single synchronous frequency by appropriately controlling the (instantaneous) magnitude of the converter-output voltage [28]. Here, too, the reduced dc voltages do not affect the effectiveness of the BVSI in any manner as long as they remain in a binary proportion and are not reduced substantially. If the capacitor voltages drift from their binary proportions, the harmonic distortion would increase.

The line-voltage harmonic spectrum is shown in Fig. 10.16. The total harmonic distortion (THD) with a 60-Hz base frequency and the first 100 harmonics considered is 0.3717% at 0.6499 s. The THD values at a different time instant are shown in Table 10.4. These very low THD values indicate that a SHEM algorithm with precomputed switching patterns improves the system harmonic profile significantly.

Figure 10.17 illustrates real-power flow from an ac system into the STATCOM to meet its losses. The real power drawn by the STATCOM is very low when there are no disturbances; the STATCOM draws more power to meet increased losses during transient conditions. This figure also shows $\Delta\theta_{1a}$, $\Delta\theta_{2a}$, and $\Delta\theta_{4a}$ angles used to modulate capacitor-switching periods to regulate their

TABLE 10.4 The THD During STATCOM Operation in the MVAR Absorption-Supply Modes

No. of Cycles After a Load Change at Which the THD Is Measured	Inductive/Capacitive Loads	
	Time	THD
5	0.2033 s	1.2496%
	0.4833 s	0.5488%
10	0.2866 s	1.0945%
	0.5666 s	0.2857%
15	0.3699 s	1.1043%
	0.6499 s	0.3717%

voltages, which vary from $-1°$ to $+1°$. The STATCOM bus-phase voltages—v_{av}, v_{bv}, and v_{cv}—distinctly show the 15-step output.

10.3 THE SSSC

The SSSC, sometimes called the S^3C, is a series-connected synchronous-voltage source that can vary the effective impedance of a transmission line by injecting a voltage containing an appropriate phase angle in relation to the line current. It has the capability of exchanging both real and reactive power with the transmission system [1]–[8], [28]–[31]. For instance, if the injected voltage is in phase with the line current, then the voltage would exchange real power. On the other hand, if a voltage is injected in quadrature with the line current, then reactive power—either absorbed or generated—would be exchanged. The SSSC emerges as a potentially more beneficial controller than the TCSC because of its ability to not only modulate the line reactance but also the line resistance in consonance with the power swings, thereby imparting enhanced damping to the generators that contribute to the power oscillations.

The SSSC comprises a multi-phase VSC with a dc-energy storage controller, as shown in Fig. 10.18(a). Here, the controller is connected in series with the transmission line. The operating modes of the SSSC are illustrated graphically in Fig. 10.18(b).

10.3.1 The Principle of Operation

A series capacitor compensates the transmission-line inductance by presenting a lagging quadrature voltage with respect to the transmission-line current. This voltage acts in opposition to the leading quadrature voltage appearing across the transmission-line inductance, which has a net effect of reducing the line inductance. Similar is the operation of an SSSC that also injects a quadrature voltage, V_C, in proportion to the line current but is lagging in phase:

$$\overline{V_C} = -jkX\overline{I_L} \tag{10.24}$$

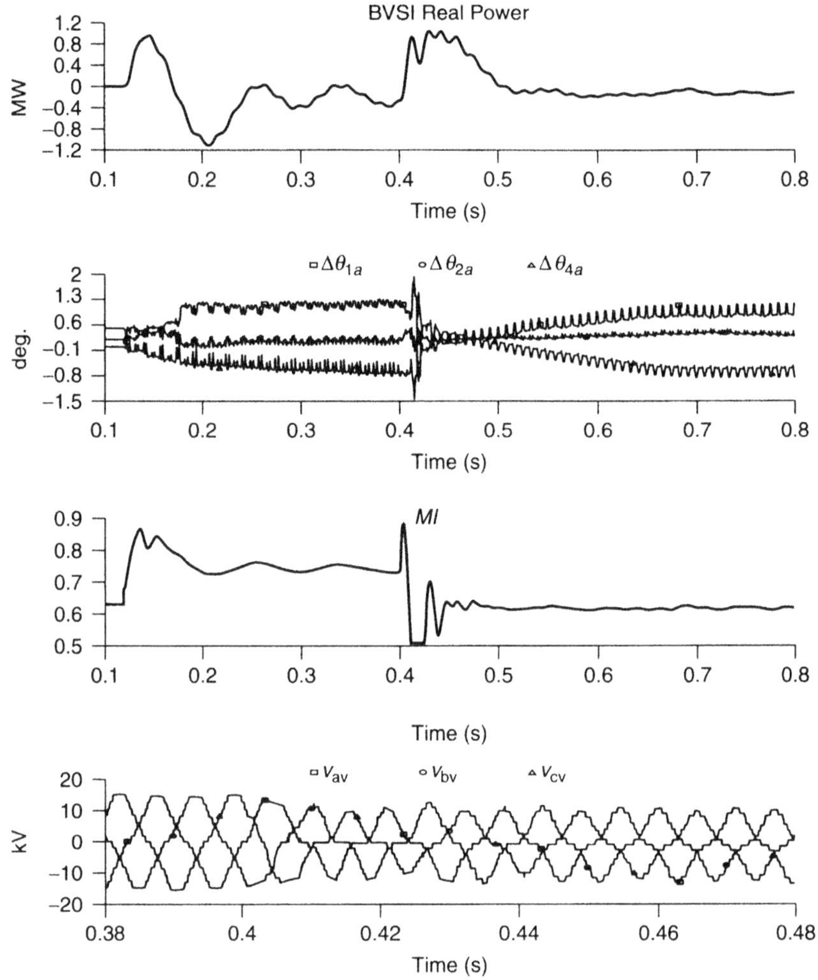

Figure 10.17 Typical STATCOM real-power consumption.

where $\overline{V_C}$ = the injected compensating voltage
$\overline{I_L}$ = the line current
X = the series reactance of the transmission line
k = the degree of series compensation

The current in a line compensated at its midpoint by the SSSC is expressed as [11], [20]:

$$I_L = \frac{2V \sin \delta/2}{X} + \frac{V_C}{X} \tag{10.25}$$

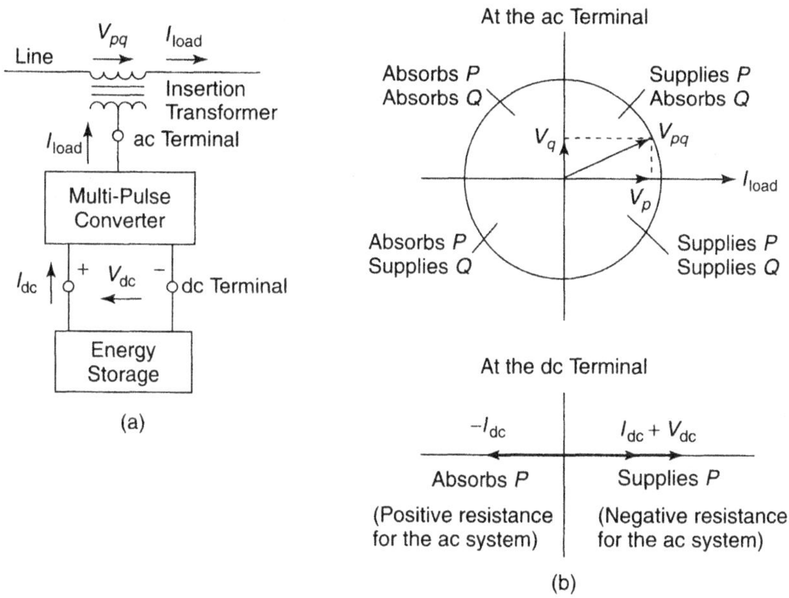

Figure 10.18 (a) Generalized series-connected synchronous-voltage source employing a multi-pulse converter with an energy-storage device; (b) the different operating modes for real- and reactive-power exchange.

where V = the magnitude of voltage (assumed to be the same) at the two ends of the transmission line

δ = the angular difference across the line

The corresponding line-power flow is then expressed as

$$P = VI_L \cos(\delta/2) \tag{10.26}$$

or

$$P = \frac{V^2 \sin\delta}{X} + \frac{VV_C}{X} \cos(\delta/2) \tag{10.27}$$

A series-compensation scheme using the SSSC is depicted in Fig. 10.19. Normally, the SSSC-output voltage lags behind the line current by 90° to provide effective series compensation. In addition, the SSSC can be gated to produce an output voltage that leads the line current by 90°, which provides additional inductive reactance in the line. This feature can be used for damping power swings and, if the converter has adequate rating, for limiting short-circuit currents.

A typical SSSC controller connected in a transmission line is shown in Fig. 10.20 [29]. This controller comprises a VSC in which its coupling transformer

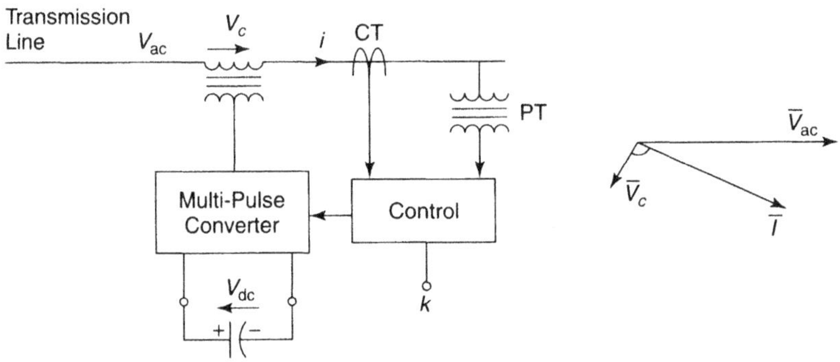

Figure 10.19 A synchronous-voltage source employing a multi-phase dc/ac converter that is operated as a series-capacitive compensator.

is connected in series with the transmission line. The valve-side voltage rating is higher than the line-side voltage rating of the coupling transformer to reduce the required current rating of the gate turn-off (GTO) thyristor valves. The valve-side winding is delta-connected to provide a path for 3rd harmonics to flow. Solid-state switches are provided on the valve side to bypass the VSC during periods of very large current flow in the transmission line or when the VSC is inoperative. The basic dc voltage for conversion to ac is provided by the capacitor, and the dc/ac conversion is achieved by pulse width–modulation techniques [27], [54]. The dc-capacitor rating is chosen to minimize the ripple in the dc voltage. An MOV is installed across the dc capacitor to limit its voltage and provide protection to the valves.

10.3.2 The Control System

A typical SSSC control system is depicted in Fig. 10.21 [11]. It accomplishes the following functions:

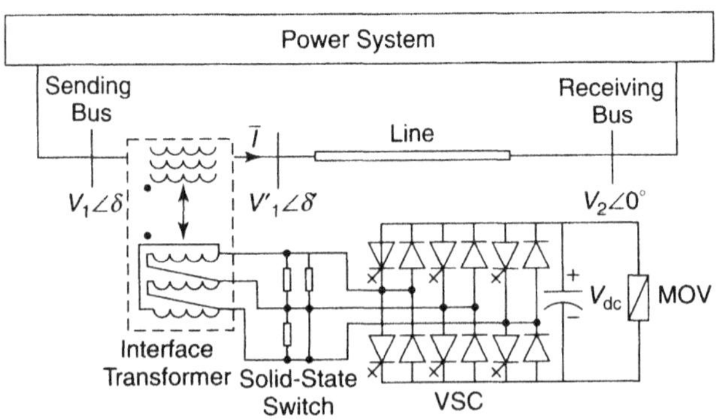

Figure 10.20 A line compensated with an SSSC.

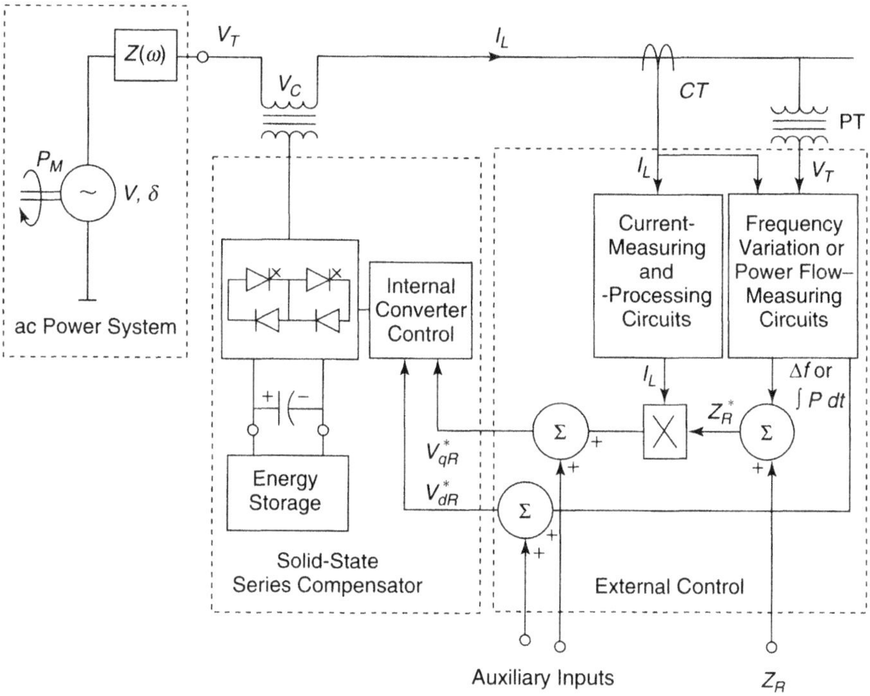

Figure 10.21 A basic control scheme for the solid-state series compensator to control (reactive and real) line impedance and improve system stability.

1. The introduction of desired series-reactive compensation (capacitive or inductive).
2. The damping of power-swing oscillations and enhancement of transient stability.
3. The control of current in the SSSC-compensated line.

The line current, I_L, and the SSSC terminal voltage, V_T, are measured together with the bus frequency or the line-power flow, which can either be measured directly or calculated from I_L and V_T measurements. The desired SSSC reactance is set by a reactance reference, Z_R.

The SSSC acts as a voltage source in synchronism with the ac-system voltage—the magnitude and phase of which can be controlled by voltage-reference inputs of V_{dr}^* and V_{qr}^*. The signal V_{qr}^* regulates the SSSC-output voltage component in quadrature with the line current. It thus determines the amount of reactive compensation (capacitive or inductive) introduced in the transmission line. The reactance reference Z_R is modulated with bus frequency or line-power signals to generate Z_R^*, which when multiplied with the rms line current I_L results in the signal V_{qr}^*. The bus frequency or line-power signals are representative of the generator-rotor oscillations, $(d\delta/dt)$, as follows:

1. If $(d\delta/dt) > 0$—that is, the generators accelerate—the signal V^*_{qr} controls the SSSC output to increase the power drawn from the generator, thereby decreasing its kinetic energy. This action is achieved by enhancing the series-capacitive compensation provided by SSSC.
2. If $(d\delta/dt) < 0$—that is, the generators decelerate—the signal V^*_{qr}, effectively inserts an inductive reactance in the line to decrease the power transmitted from the generator.

The signal V^*_{dr} determines the magnitude of the SSSC-output voltage component that is in phase (or out of phase) with the line current. It thus controls the real-power exchange between the SSSC and the system and is particularly effective when the SSSC is equipped with an energy-storage device such as the one shown in the Fig. 10.21. To damp power swings, the modulation signal representing $(d\delta/dt)$ is used to modulate V^*_{dr}, as follows:

1. If $(d\delta/dt) > 0$—that is, the generators accelerate—the signal V^*_{dr} controls the SSSC to absorb real power from the system. It effectively introduces a positive (apparent) resistance in the transmission network.
2. If $(d\delta/dt) < 0$—that is, the generators decelerate—the signal V^*_{dr} instructs the SSSC to inject real power into the system. This action is equivalent to the introduction of a negative (virtual) resistance in the transmission circuit.

The SSSC can also be used to balance the currents in parallel lines. This objective is attained by implementing an additional current loop in which the measured current is compared with the desired reference value. The error signal is then employed to modulate Z_R to effect the required current flow in the compensated line.

10.3.3 Applications

10.3.3.1 Power-Flow Control A case study that illustrates the effectiveness of the SSSC in providing line-power control is presented in ref. [29]. Figure 10.22 depicts a two-area 500-kV test system interconnected by a 600-km double-circuit line. Each circuit is compensated by fixed capacitors to the extent of 70.5% line reactance, which translates to a net series compensation of 46.6% if the area impedances are also considered.

The SSSC is inserted in series with the line. The coupling transformer is rated at 28.9 kV (line side) and 57.7 kV (valve side) per phase. The dc side of the GTO-based VSC is rated at 75 kV (which can be increased to 85 kV) by the MOV across the dc capacitor. The SSSC can thus inject a maximum phase voltage of 28.9 kV, corresponding to a line-to-line voltage of 50 kV.

The SSSC response to changes in the line-power reference is shown in Fig. 10.23. Initially, the SSSC is blocked and bypassed with the dc capacitor of the

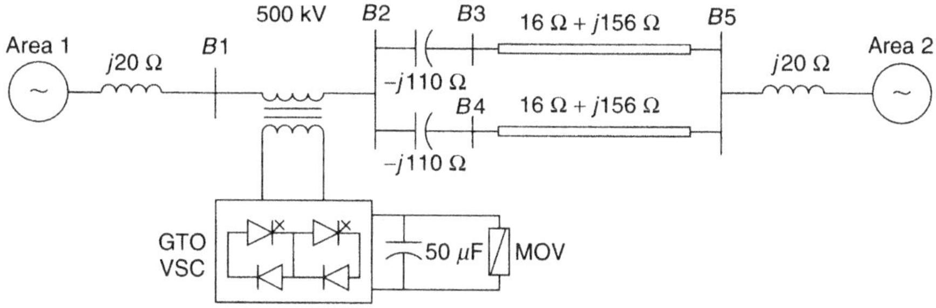

Figure 10.22 A test system.

VSC uncharged. At this instant, the line-power flow is 2015 MW, corresponding to an angular difference of 30° between the two areas. At $t = 0.1$ s, the SSSC is deblocked; the line-power reference setting is first increased to 2400 MW, then reduced in steps to a final value of 1600 MW. It is seen from Fig. 10.23 that even a small SSSC that is rated at 10% of the line voltage can provide effective control of 800 MW of line power, or 40% of the line power. The total harmonic distortion with the switching scheme of Fig. 10.23 is about 1.5%. Even in unbalanced-voltage scenarios, the SSSC is shown to operate satisfactorily.

10.3.3.2 SSR Mitigation An important aspect of the SSSC is that because it does not introduce a physical capacitor in the line, it does not cause SSR. However, it assists in the damping of subsynchronous oscillations caused by other series capacitors inserted in the transmission network. This damping is

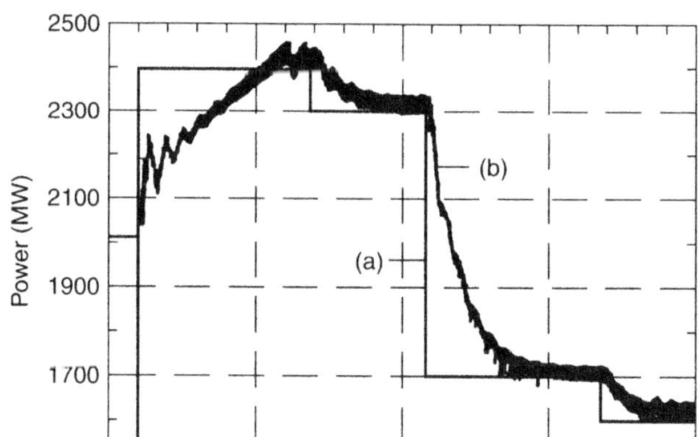

Figure 10.23 The system response to power-reference change: (a) line-power reference and (b) line power.

achieved by introducing subsynchronous voltages of appropriate magnitude, frequency, and phase to negate the effect of original subsynchronous currents. The subsynchronous voltages can be generated together with the fundamental-frequency voltage.

The influence of the SSSC in suppressing SSR in a modified IEEE First Benchmark System [50] is described in refs. [38] and [55]. The line is assumed to be compensated by 50%. Three cases are considered for line compensation:

1. Total of 50% series compensation by the fixed capacitor.
2. Total of 35% compensation by a fixed capacitor and 15% by an SSSC.
3. Total of 50% compensation by an SSSC.

Eigenvalue analysis is used to show that both cases 1 and 2 of the preceding list reduce the undamping of the concerned torsional mode. The SSSC of case 2 damps the torsional oscillations effectively with a damping controller. It may be noted that the SSSC, as a very expensive controller, is not used alone to compensate the line. A practical compensating scheme involves the use of a relatively small-rating SSSC in conjunction with a fixed-series capacitor.

For the case-study system, depicted in Fig. 10.24 is the response of the SSSC in case 2 to a step increase in the HP turbine's mechanical-torque output. The generator-rotor-angle oscillations decay very rapidly, and in the GEN–LP shaft torques, and the subsynchronous oscillations are also stabilized—although slowly. Moreover, the SSSC closely regulates the dc-capacitor voltage.

10.4 THE UPFC

10.4.1 The Principle of Operation

The UPFC is the most versatile FACTS controller developed so far, with all encompassing capabilities of voltage regulation, series compensation, and phase shifting. It can independently and very rapidly control both real- and reactive-power flows in a transmission line [1]–[8], [32]–[47]. It is configured as shown in Fig. 10.25 and comprises two VSCs coupled through a common dc terminal. One VSC—converter 1—is connected in shunt with the line through a coupling transformer; the other VSC—converter 2—is inserted in series with the transmission line through an interface transformer. The dc voltage for both converters is provided by a common capacitor bank. The series converter is controlled to inject a voltage phasor, V_{pq}, in series with the line, which can be varied from 0 to $V_{pq\,\max}$. Moreover, the phase angle of V_{pq} can be independently varied from 0° to 360°. In this process, the series converter exchanges both real and reactive power with the transmission line. Although the reactive power is internally generated/absorbed by the series converter, the real-power generation/absorption is made feasible by the dc-energy–storage device—that is, the capacitor.

The shunt-connected converter 1 is used mainly to supply the real-power demand of converter 2, which it derives from the transmission line itself. The

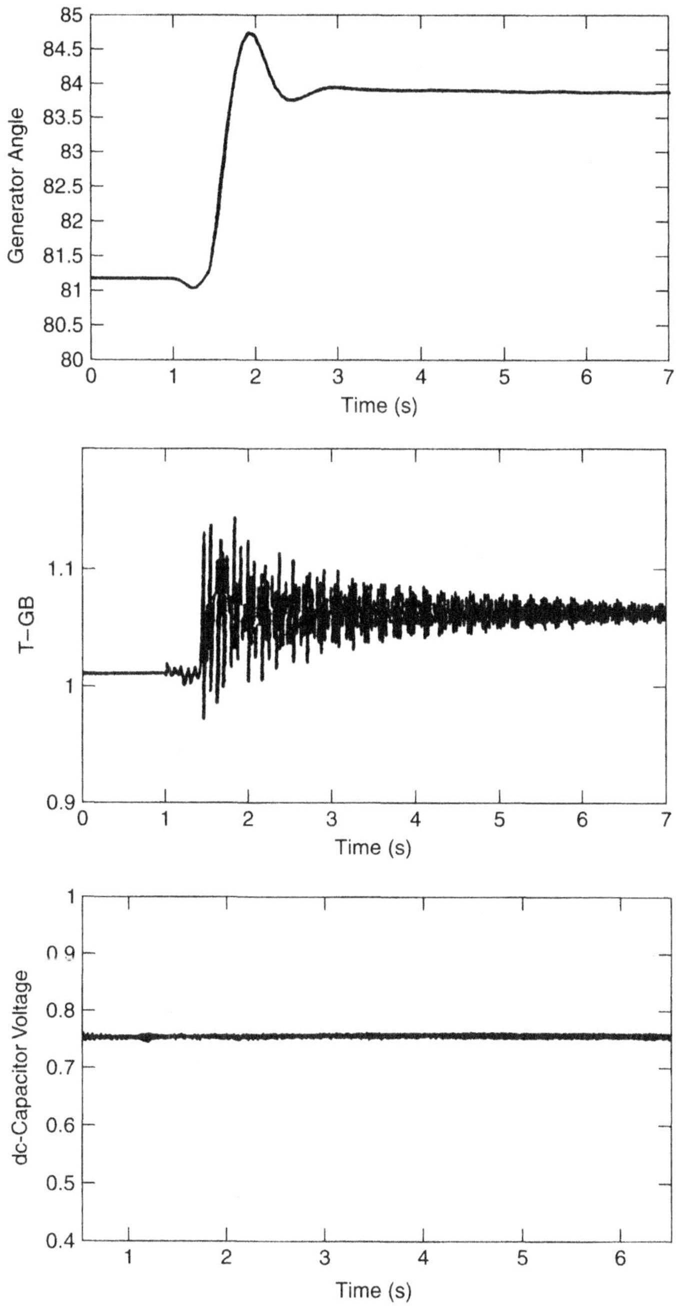

Figure 10.24 The SSSC response for a step change in the mechanical input of an HP turbine.

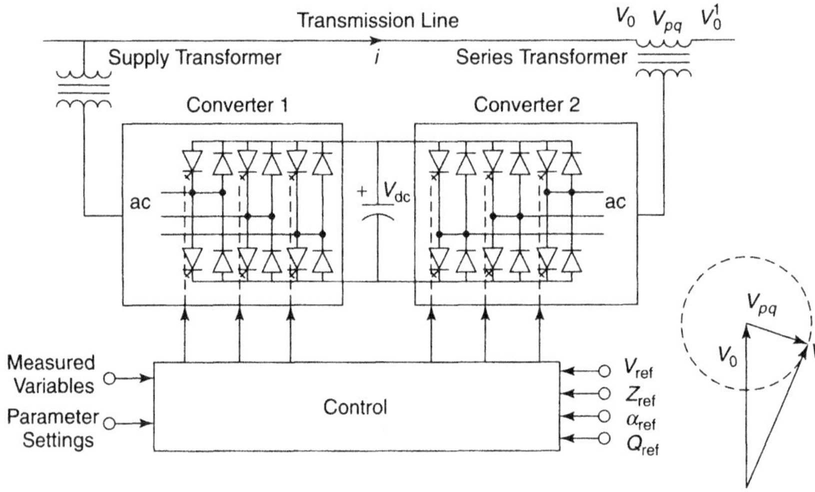

Figure 10.25 The implementation of the UPFC using two "back-to-back" VSCs with a common dc-terminal capacitor.

shunt converter maintains constant voltage of the dc bus. Thus the net real power drawn from the ac system is equal to the losses of the two converters and their coupling transformers. In addition, the shunt converter functions like a STATCOM and independently regulates the terminal voltage of the interconnected bus by generating/absorbing a requisite amount of reactive power.

The concepts of various power-flow control functions by use of the UPFC are illustrated in Figs. 10.26(a)–(d). Part (a) depicts the addition of the general

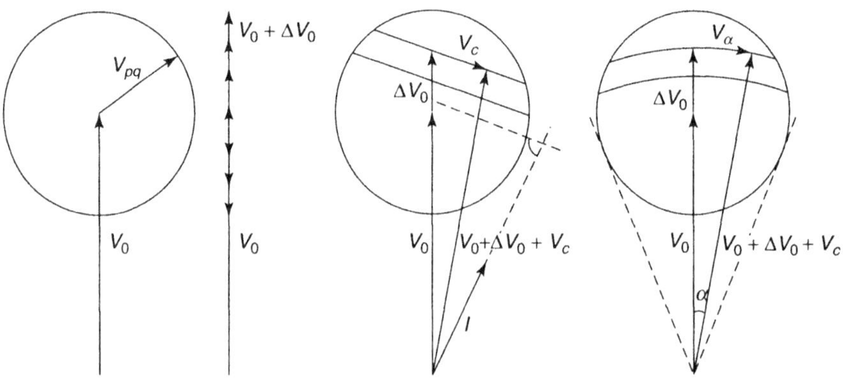

Figure 10.26 A phasor diagram illustrating the general concept of series-voltage injection and attainable power-flow control functions: (a) series-voltage injection; (b) terminal-voltage regulation; (c) terminal-voltage and line-impedance regulation; and (d) terminal-voltage and phase-angle regulation.

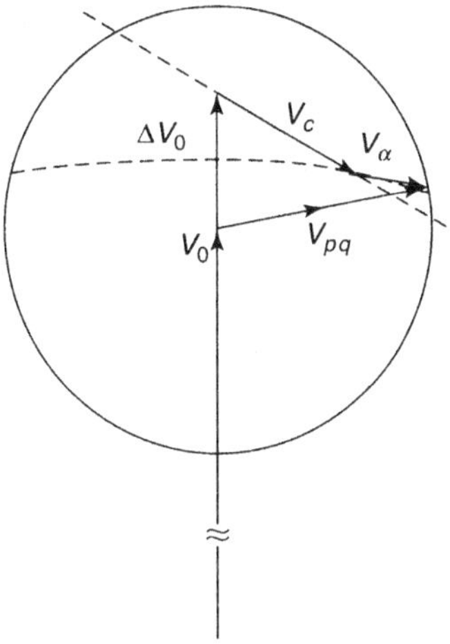

Figure 10.27 A phasor diagram illustrating the simultaneous regulation of the terminal voltage, line impedance, and phase angle by appropriate series-voltage injection.

voltage phasor V_{pq} to the existing bus voltage, V_0, at an angle that varies from 0° to 360°. Voltage regulation is effected if V_{pq} (= ΔV_0) is generated in phase with V_0, as shown in part (b). A combination of voltage regulation and series compensation is implemented in part (c), where V_{pq} is the sum of a voltage-regulating component ΔV_0 and a series compensation providing voltage component V_c that lags behind the line current by 90°.

In the phase-shifting process shown in part (d), the UPFC-generated voltage V_{pq} is a combination of voltage-regulating component ΔV_0 and phase-shifting voltage component V_α. The function of V_α is to change the phase angle of the regulated voltage phasor, $V_0 + \Delta V$, by an angle α. A simultaneous attainment of all three foregoing power-flow control functions is depicted in Fig. 10.27. The controller of the UPFC can select either one or a combination of the three functions as its control objective, depending on the system requirements.

The UPFC operates with constraints on the following variables [37]:

1. the series-injected voltage magnitude;
2. the line current through series converter;
3. the shunt-converter current;
4. the minimum line-side voltage of the UPFC;
5. the maximum line-side voltage of the UPFC; and
6. the real-power transfer between the series converter and the shunt converter.

The theoretical concepts of UPFC-control design and equipment rating are well-explained in refs. [2]–[8] and [32]–[47]; hence they are not described here.

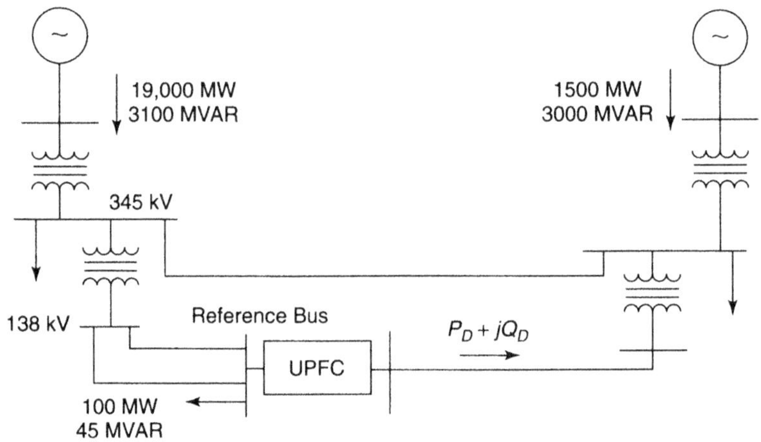

Figure 10.28 A case-study system.

However, some control-application case studies are described here to illustrate the capabilities of the UPFC.

10.4.2 Applications

A case study for power-flow control and oscillation damping in the two-area system discussed in ref. [37] is presented in Fig. 10.28. The two areas exchange power via two transmission lines of unequal power-transfer capacity—one operating at 345 kV, the other at 138 kV. Although the 345-kV line is 100 mi long, the 138-kV system is composed of two parallel 60-mi-long lines feeding a load and a single 40-mi-long line leading to the other area.

The power-transmission capability is determined by the transient-stability considerations of the 345-kV line. The UPFC is installed in the 138-kV network. A 3-phase-to-ground fault is applied on the 345-kV line for four cycles, and the line is disconnected after the fault. The maximum stable power flow possible in the 138-kV line without the UPFC is shown in Fig. 10.29 to be 176 MW. However, the power transfer with the UPFC can be increased 181 MW (103%) to 357 MW. Although this power can be raised further by enhancing the UPFC rating, the power increase is correspondingly and significantly lower than the increase in the UPFC rating, thereby indicating that the practical limit on the UPFC size has been attained.

The UPFC also provides very significant damping to power oscillations when it operates at power flows within the operating limits. The UPFC response to a 3-phase-line-to-ground fault cleared after four cycles, leaving the 345-kV line in service, is illustrated in Fig. 10.30. Because the 345-kV line remains intact, the oscillation frequency changes from that shown in Fig. 10.29.

Dramatic enhancements in power-oscillation damping with the use of the UPFC are also reported in the planning study of the Mead–Phoenix project

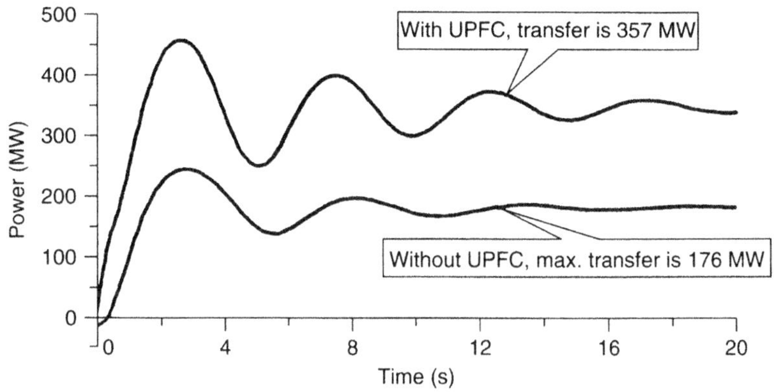

Figure 10.29 Power-transfer capability with the UPFC.

[34] and also with a modulation controller–equipped UPFC in a simple system [38]. The application of UPFC for transient-stability improvement [42], [56] is presented in the next section.

10.5 COMPARATIVE EVALUATION OF DIFFERENT FACTS CONTROLLERS

The various functions achievable by different FACTS controllers described in this book are listed in Table 10.5 [57]. A specific task can be performed by several FACTS controllers, so their effectiveness in achieving the same tasks can vary substantially.

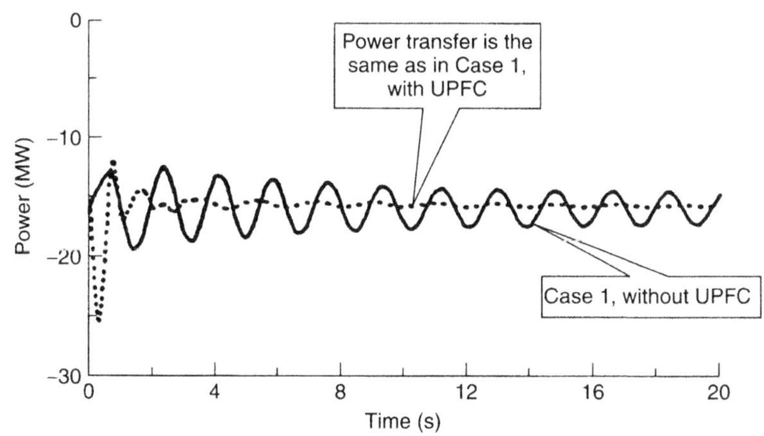

Figure 10.30 A comparison of oscillation damping.

TABLE 10.5 Capabilities of Different FACTS Controllers

Controller	Voltage Control	Transient Stability	Damping-Power Oscillations	Reactive-Power Compensation	Power-Flow Control	SSR Mitigation
BESS	X		X			
SMES		X	X			
SSSC	X	X	X	X	X	X
STATCOM	X	X	X	X		
SVC	X	X	X	X		
TCPST		X	X		X	X
TCSC	X	X	X		X	X
TSBR		X	X			X
TSSC	X	X	X		X	
UPFC	X	X	X	X	X	X

10.5.1 Performance Comparison

The relative efficacy of different FACTS controllers in achieving a major objective in power systems—namely, the enhancement of transient stability—is investigated in ref. [56]. Four FACTS controllers are evaluated and compared in this study: the static var compensator (SVC), the static synchronous compensator (STATCOM), the thyristor-controlled series capacitor (TCSC), and the unified power-flow controller (UPFC). The comparison also includes the high initial response (HIR) brushless generator excitation system. The effects of different controllers are expressed in terms of the critical fault-clearing time (CFCT).

The evaluation is performed on a small 2-generator, 5-bus system depicted in Fig. 10.31. Faults are applied at the indicated bus and cleared after the CFCT to ensure generator synchronism after fault removal. The ratings of all the controllers are kept the same as the following:

- ±50 MVAR SVC
- ±50 MVAR STATCOM

Backup Generation	General Comments
X	• Uninterruptible power supply. • Primarily used for load leveling and spinning reserve.
X	• Potentially useful for many future applications, including stabilizing systems, serving as a standby power source for critical loads, and reducing governor ripple.
	• Can deliver desired var even at depressed voltages, thus giving better voltage support than an SVC. • Has a faster response than an SVC.
	• Provides continuous control if used as a TCR with an FC bank. • Otherwise, TSR and TSC use provides discrete control.
	• Essentially a phase-shifting transformer with a thyristor-control circuit that facilitates high-speed changes in the phase angle and hence the power flow.
	• Can operate in a vernier mode that allows continuously varying of impedance from the inductive to the capacitive.
	• Can be used to effectively damp power swings.
	• Essentially a discrete version of a TCSC (i.e., series capacitors are switched in and out by thyristor-control circuitry).
	• Essentially a STATCOM–SSSC combination connected back-to-back by a dc link, thus combining the capability of power control (through phase) and voltage control (through reactive power); hence the term *unified*.

- +50 MVAR TCSC (100% controllable)
- UPFC with ±50 MVA STATCOM, ±50 MVA series element

The control systems used for the SVC, STATCOM, TCSC, and UPFC are illustrated in Figs. 10.32, 10.33, 10.34, and 10.35, respectively. The controller parameters are listed along with the gains and time constants of different controllers in each case are selected with only consideration of maximizing the CFCT. Issues such as power-oscillation damping are not considered. The transient-stability simulations were conducted with the EPRI's ETMSP package [58].

The CFCT obtained for the different controllers are compared in the bar chart depicted in Fig. 10.36. This study demonstrates that all FACTS controllers have a significantly higher beneficial impact on transient stability than HIR excitation systems. Among the shunt controllers, the STATCOM performs better than SVC because of its better characteristics in the low-voltage region. The TCSC is more effective than the shunt controllers, as it offers greater controllability of the power flow in the line. The UPFC is by far the best controller, as it provides

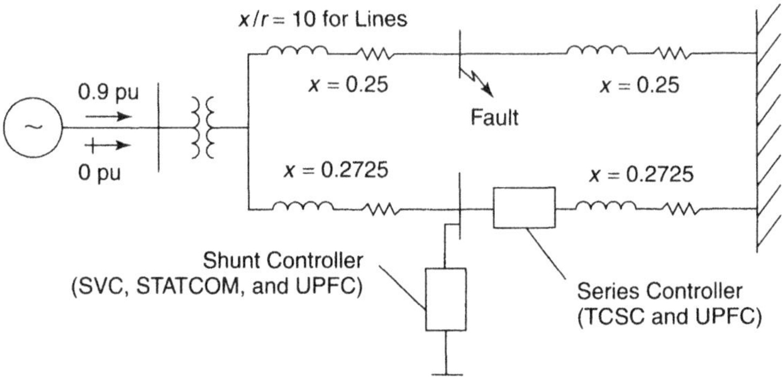

Figure 10.31 A study system for transient-stability evaluation.

independent control over the bus voltage and the line real- and reactive-power flows.

This study on a simplistic system gives only an indication of the relative performances of different FACTS controllers. The actual quantitative benefits that accrue from these controllers in realistic systems depend on several other factors, including interconnections and the range of operating conditions.

10.5.2 Cost Comparison

An elaborate cost comparison of different FACTS controllers is presented in Table 10.6 [59], [4]. These cost figures have been estimated from various EPRI reports and information obtained from a vender involved in the installation of STATCOM and UPFC. The UPFC cost is equivalent to that of two STATCOMs—one in shunt, the other in series. The cost of conventional series and shunt capacitors and power angle–regulating (PAR) transformers are also included for sake of completeness. Although HVDC links do not fall in the category of FACTS equipment, the average turnkey costs for HVDC links of different capacities are also indicated. This information may be of benefit while deciding one of two options: using HVDC or using FACTS for attaining similar power-system performance enhancements.

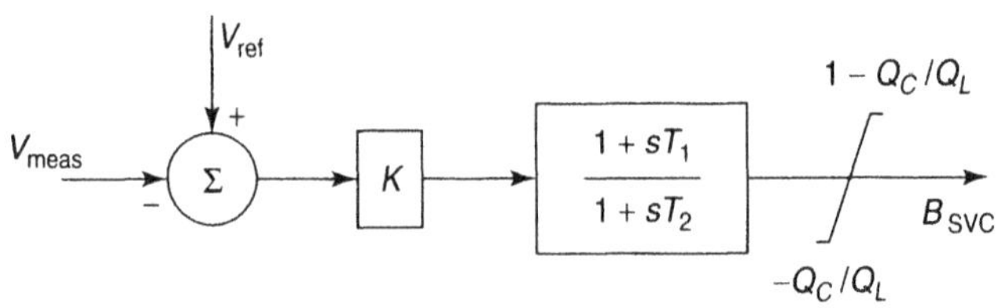

Figure 10.32 An SVC control system: $K = 5.0$, $T_1 = 0.05$, $T_2 = 0.05$, $Q_C = 50$, and $Q_L = 100$.

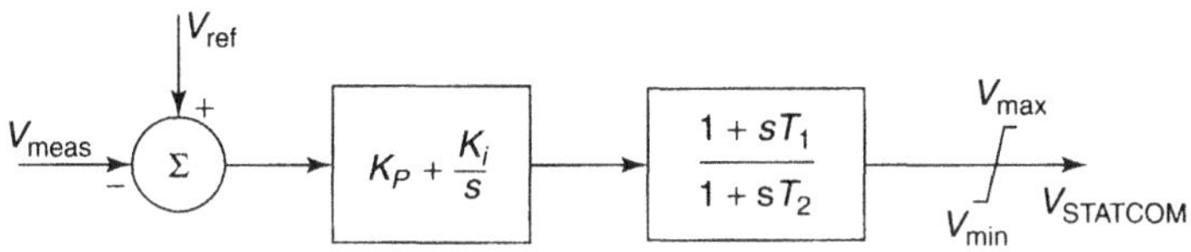

Figure 10.33 A STATCOM PI controller for voltage regulation: $K_P = 0.05$, $K_i = 1.0$, $T_1 = 0$, and $T_2 = 0$.

The total installed cost of a FACTS or HVDC project mainly comprises two components: direct costs and indirect costs. The direct costs encompass those for construction material and labor related to the installation. Indirect costs encompass those associated with licensing and permits, legal expenses, finance, project administration, insurance, taxes, interest, and so on.

As an example, the costs are presented for the Chester SVC, which is rated for −125/+425 MVAR [59]. This SVC is installed on a 345-kV transmission line to strengthen the ac system linked to the New England Phase II HVDC Project. For this project, the SVC vendor costs constituted 86% of the direct costs, whereas the indirect costs constituted 14% of the direct costs. The total installed cost for the Chester SVC included these costs, as well as the cost of ac-system reinforcements. The breakdown of the total installed cost for different items of the SVC project is presented in Table 10.7 [59].

The final selection of a specific FACTS controller is based on a technoeconomic analysis [4]. The choice is a function of the objectives to be achieved, the performance of each controller in attaining the performance specifications, and the total costs involved.

10.6 FUTURE DIRECTION OF FACTS TECHNOLOGY [60]–[62]

The technology behind phase-controlled thyristor-based FACTS controllers has been present for several decades and is therefore considered mature. More utilities are likely to adopt this technology in the future, even as the newer, more promising switch-mode GTO-based FACTS technology is fast emerging. Fore-

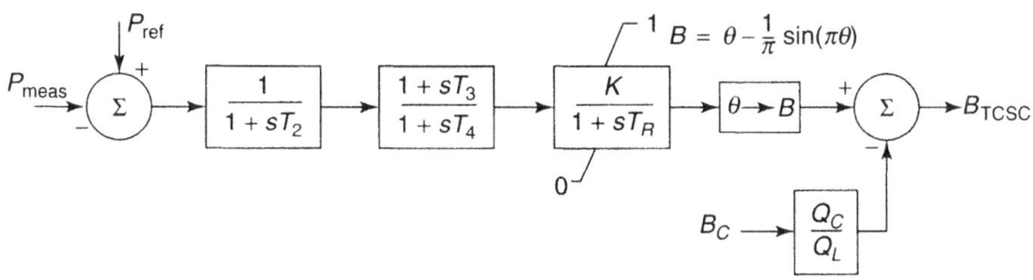

Figure 10.34 The TCSC power-flow controller (100% variable B): $T_2 = 0.02$, $T_3 = 0.2$, $T_4 = 2.0$, $T_R = 0$, $Q_C = 2869$, and $Q_L = 666$.

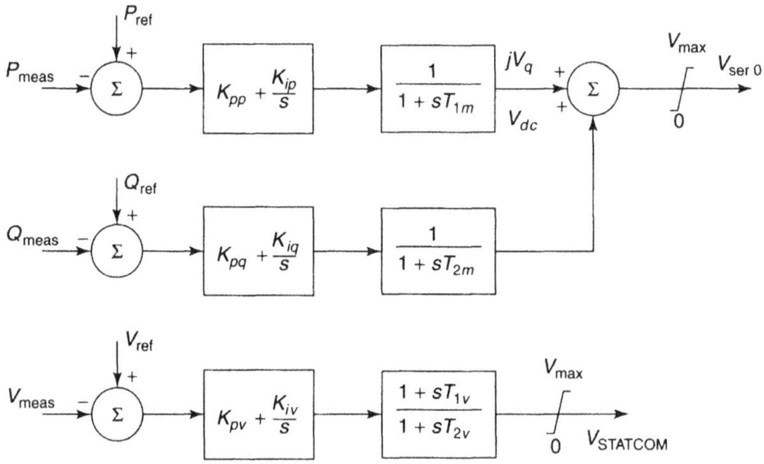

Figure 10.35 The UPFC control system with both series real- and reactive-power flow control and shunt control for bus-voltage regulation: $K_{pp} = 0.05$, $K_{ip} = 5.0$, $T_{1m} = 0.01$, $T_{2m} = 0.01$, $K_{pq} = 0.05$, $K_{iq} = 1.2$, $K_{pv} = 0.05$, $K_{iv} = 0.5$, $T_{1v} = 0$, and $T_{2v} = 0$.

seen in the near future is the application of a hybrid technology involving both thyristors and GTOs; for instance, STATCOM-compensated HVDC converters that perform better than SVC-compensated HVDC links.

The second generation of FACTS controllers, such as the STATCOM, the SSSC, and the UPFC, use switch-mode GTO-based VSC configurations. Novel GTO-centered topologies are being researched and are expected to evolve into another mature family of FACTS controllers. Recent advances in silicon power-switching devices that significantly increase their power ratings will contribute even further to the growth of FACTS technology.

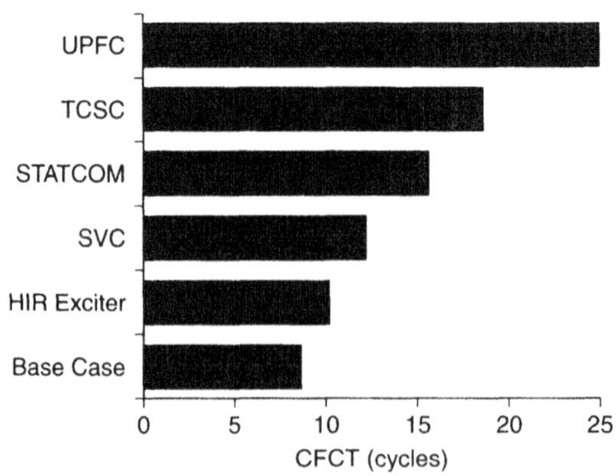

Figure 10.36 Comparative evaluation of the CFCT.

TABLE 10.6 Cost Estimates for FACTS Controllers and HVDC Links

Controller	Cost
Shunt capacitor	$8/kVAR
Conventional series capacitor	$20/kVAR
Conventional PAR transformer	$20/kVAR
SVC	$40/kVAR–controlled part
TCSC	$40/kVAR–controlled part
STATCOM	$50/kVAR
UPFC series portion	$50/kW series-power flow
UPFC shunt portion	$50/kVAR–controlled part
HVDC Links	
Back-to-Back, 200 MW	$108/kW/terminal
±250 kV, 500 MW	$145/kW/terminal
±350 kV, 1000 MW	$107/kW/terminal
±500 kV, 3000 MW	$75/kW/terminal

10.6.1 The Role of Communications

The ongoing restructuring of power systems poses new stability problems because of frequently changing power-transfer patterns [61]. The FACTS controllers will play an important role in facing this challenge. Indeed, FACTS controllers will be required to ensure global stability objectives. At such a time, reliance on only locally measured signals may not be adequate. In addition, the requirement that all supplementary damping-control signals be only locally derived may make the placement of FACTS controllers difficult task from the viewpoint of robustness [63], [64]. Hence it is increasingly felt that reliable, globally obtained signals might be made available to the controllers to supplement local signals. The selected remote signals that demonstrate superior observability of the relevant modes will be transmitted primarily through communication channels; it may not be all that unreliable to do so.

Digital fiber-optic communications hold much promise in this respect [65], [66]. In the future, power systems are likely to see a wider application of distributed measurement technology, using the global-positioning system and accurate phasor-measurement units for increases in overall system stability [61]. Wide-area control-information systems are anticipated to permeate the application of FACTS controllers [60]. This is because the control interactions between FACTS–HVDC–ac controllers proliferate to large areas because of the closely interlinked systems.

10.6.2 Control-Design Issues

In the emerging deregulated power systems, FACTS controllers will provide the following benefits at existing or enhanced levels of reliability [67]:

TABLE 10.7 Total Installed Cost of the Chester, Maine, –125/+425–MVAR SVC

Item Description	% Total Project
Thyristor valves	16.39
345/24–kV transformers	19.82
ac-Switchyard equipment	22.48
Station auxiliary power	5.47
Control and protections	8.47
Station costs	23.04
ac-Transmission tap line	2.88
ac-Network reinforcements	1.45

1. Balancing the power flow over a wide range of operating conditions (including contingencies), thereby using the power-system network most efficiently.
2. Balancing the power flow in parallel networks operating at different voltage levels.
3. Alleviating unwanted loop flow in large, integrated power systems.
4. Mitigating inter-area power oscillations.
5. Obviating or delaying the construction of new transmission facilities by significantly enhancing the power-transfer capacity of existing transmission corridors.

The delivery period of FACTS technology is much less than that of the installation of new transmission lines, thus making the technology highly attractive.

The FACTS controllers will be burdened with the responsibility of operating reliably over an expansive range of unpredictable power-flow scenarios. Hence control systems for these FACTS controllers may have to be designed not just on the basis of linear-control techniques, but also by using intelligent, adaptive digital controllers based on information obtained from wide-area measurement networks.

For systems using FACTS controllers, aiming for high levels of damping may not be a safe design goal for wide-area control. Adequate damping over the largest realistic range of operating conditions may be a more desirable criterion to fulfill [62]. The coordination of multiple FACTS controllers in the same system as well as in the adjacent systems must be investigated extensively and implemented to ensure the security of power-system operation.

10.7 SUMMARY

In this chapter, three important second-generation FACTS controllers—the STATCOM, the SSSC, and the UPFC—were described briefly. The essential features of their control systems were mentioned, and some application examples for each controller were presented. One primary task for any utility is to select a specific FACTS controller for a given objective; the comparison given in this chapter of the technical capabilities and costs of various FACTS controllers might facilitate this selection. The likely future direction of FACTS technology, especially in restructured power systems, was discussed as well.

REFERENCES

[1] IEEE Power Engineering Society/CIGRE, *FACTS Overview*, Publication 95TP108, IEEE Press, New York, 1995.

[2] N. G. Hingorani and L. Gyugyi, *Understanding FACTS*, IEEE Press, New York, 1999.

[3] Y. H. Song and A. T. Johns, Eds., *Flexible AC Transmission Systems (FACTS)*, IEE Press, London, 1999.

[4] IEEE Power Engineering Society, *FACTS Applications*, Publication 96TP116-0, IEEE Press, New York, 1996.

[5] W. H. Litzenberger, Ed., "An Annotated Bibliography of High-Voltage Direct-Current Transmission, 1989–1991," Published by the Bonneville Power Administration (BPA) and the Western Area Power Administration, Portland, OR, 1992.

[6] W. H. Litzenberger, Ed., "An Annotated Bibliography of High-Voltage Direct-Current Transmission and Flexible AC Transmission (FACTS) Devices, 1991–1993," Published by the Bonneville Power Administration (BPA) and the Western Area Power Administration, Portland, OR, 1994.

[7] W. H. Litzenberger and R. K. Varma, Eds., "An Annotated Bibliography of High-Voltage Direct-Current Transmission and FACTS Devices, 1994–1995," Published by the Bonneville Power Administration (BPA) and the U.S. Department of Energy, Portland, OR, 1996.

[8] W. H. Litzenberger, R. K. Varma, and J. D. Flanagan, Eds., "An Annotated Bibliography of High-Voltage Direct-Current Transmission and FACTS Devices, 1996–1997," Published by the Electric Power Research Institute (EPRI) and the Bonneville Power Administration (BPA), Portland, OR, 1998.

[9] L. Gyugi et al., "Advanced Static Var Compensator Using Gate Turn-off Thyristors for Utility Applications," CIGRE Paper 23-203, 1990.

[10] S. Mori and K. Matsuno, "Development of a Large Static Var Generator Using Self-Commutated Converters for Improving Power System Stability," *IEEE Transactions on Power Systems*, Vol. 8, No. 1, February 1993, pp. 371–373.

[11] L. Gyugyi, "Dynamic Compensation of AC Transmission Lines by Solid-State Synchronous Voltage Sources," *IEEE Transactions on Power Delivery*, Vol. 9, No. 2, April 1994, pp. 904–911.

[12] C. Schauder, M. Gernhardt, E. Stacey, T. Lemak, L. Gyugyi, T. Cease, and A. Edris, "TVA STATCOM Project: Design, Installation, and Commissioning," CIGRE Paper 14-106, 1996.

[13] C. D. Schauder et al., "Development of a ±100 MVAR Static Condenser for Volt-

age Control of Transmission Lines," *IEEE Transactions on Power Delivery*, Vol. 10, No. 3, July 1995, pp. 1486–1493.

[14] C. D. Schauder and H. Mehta, "Vector Analysis and Control of Advanced Static Var Compensators," *IEEE Proceedings–C*, Vol. 140, No. 4, 1993, pp. 299–306.

[15] E. Larsen, N. Miller, S. Nilsson, and S. Lindgreen, "Benefits of GTO-Based Compensating System for Electric Utility Applications," *IEEE Transactions on Power Delivery*, Vol. 7, No. 4, 1992, pp. 2056–2063.

[16] K. V. Patil, J. Senthil, J. Jiang, and R. M. Mathur, "Application of STATCOM for Damping Torsional Oscillations in Series Compensated AC Systems," *IEEE Transactions on Energy Conversion*, Vol. 13, No. 3, September 1998, pp. 237–243.

[17] J.-S. Lai and F. Z. Peng, "Multilevel Converter—A New Breed of Power Converters," *IEEE Transactions on Industry Applications*, Vol. 32, No. 3, May/June 1996, pp. 509–517.

[18] F. Z. Peng, J.-S. Lai, J. W. McKeever, and J. Vancoevering, "A Multilevel Voltage-Source Converter with Separate DC Sources for Static Var Generation," *IEEE Transactions on Industry Applications*, Vol. 32, No. 5, September/October 1996, pp. 1130–1138.

[19] K. V. Patil, R. M. Mathur, J. Jiang, and S. H. Hosseini, "Distributed System Compensation Using a New Binary Multilevel Voltage Source Converter," *IEEE Transactions on Power Delivery*, Vol. 14, No. 2, April 1999, pp. 459–464.

[20] J. B. Ekanayake and N. Jenkins, "A Three-Level Advanced Static Var Compensator," *IEEE Transactions on Power Delivery*, Vol. 11, No. 1, January 1996, pp. 540–545.

[21 C. Hochgraf and R. H. Lasseter, "A Transformerless Static Synchronous Compensator Employing a Multilevel Converter," *IEEE Transactions on Power Delivery*, Vol. 12, No. 2, April 1997, pp. 881–887.

[22] C. J. Hatziadoniu and F. E. Chalkiadakis, "A 12-Pulse Static Synchronous Compensator for the Distribution System Employing the 3-Level GTO Converter," *IEEE Transactions on Power Delivery*, October 1997, pp. 1830–1835.

[23] I. A. Erinmez and A. M. Foss, Eds., "Static Synchronous Compensator (STATCOM)," Working Group 14.19, CIGRE Study Committee 14, Document No. 144, August 1999.

[24] M. R. Iravani and D. Maratukulam, "Review of Semiconductor-Controlled (Static) Phase Shifters for Power System Applications," *IEEE Transactions on Power Systems*, Vol. 9, No. 4, 1994, pp. 1833–1839.

[25] M. R. Iravani, P. L. Dandeno, K. H. Nguyen, D. Zhu, and D. Maratukulam, "Applications of Static Phase Shifters in Power Systems," Paper 94WM068-7 PWRD, Presented at IEEE/PES 1994 Winter Meeting, New York, January–February, 1994.

[26] S. Nyati, M. Eitzmann, J. Kappenman, D. Van House, N. Mohan, and A. Edris, "Design Issues for a Single Core Transformer Thyristor-Controlled Phase Angle Regulator," *IEEE Transactions on Power Delivery*, Vol. 10, No. 4, October 1995, pp. 2013–2019.

[27] B. T. Ooi, S. Z. Dai, and F. D. Galiana, "A Solid-State PWM Phase-Shifter," *IEEE Transactions on Power Delivery*, Vol. 8, April 1993, pp. 573–579.

[28] L. Gyugyi, C. D. Schauder, and K. K. Sen, "Static Synchronous Series Compen-

sator: A Solid-State Approach to the Series Compensation of Transmission Lines," *IEEE Transactions on Power Delivery*, Vol. 12, No. 1, January 1997, pp. 406–417.

[29] C. J. Hatziadoniu and A. T. Funk, "Development of a Control Scheme for Series-Connected Solid-State Synchronous Voltage Source," *IEEE Transactions on Power Delivery*, Vol. 11, No. 2, April 1996, pp. 1138–1144.

[30] K. K. Sen, "SSSC—Static Synchronous Series Compensator: Theory, Modelling, and Applications," *IEEE Transactions on Power Delivery*, Vol. 13, No. 1, January 1998, pp. 241–246.

[31] L. Gyugyi, C. D. Schauder, and K. K. Sen, "Static Synchronous Series Compensator: A Solid-State Approach to the Series Compensation of Transmission Lines," *IEEE Transactions on Power Delivery*, Vol. 12, No. 1, January 1997.

[32] CIGRE Task Force 14-27, "Unified Power Flow Controller," CIGRE Technical Brochure, 1998.

[33] L. Gyugyi, "A Unified Power Flow Control Concept for Flexible AC Transmission Systems," *IEE Proceedings–C*, Vol. 139, No. 4, July 1992.

[34] L. Gyugyi, C. D. Schauder, S. L. Williams, T. R. Reitman, D. R. Torgerson, and A. Edris, "The Unified Power Flow Controller: A New Approach to Power Transmission Control," *IEEE Transactions on Power Delivery*, Vol. 10, No. 2, April 1995.

[35] C. D. Schauder, L. Gyugyi, M. R. Lund, D. M. Hamai, T. R. Reitman, D. R. Torgerson, and A. Edris, "Operation of the Unified Power Flow Controller (UPFC) Under Practical Constraints," *IEEE Transactions on Power Delivery*, Vol. 13, No. 2, April 1998.

[36] M. Rahman et al., "UPFC Application on the AEP System: Planning Considerations," *IEEE Transactions on Power Systems*, Vol. 12, No. 4, November 1997.

[37] J. Bian, D. G. Ramey, R. J. Nelson, and A. Edris, "A Study of Equipment Sizes and Constraints for a Unified Power Flow Controller," *Proceedings IEEE T&D Conference*, 1996.

[38] K. R. Padiyar and A. M. Kulkarni, "Control Design and Simulation of Unified Power Flow Controller," *IEEE Transactions on Power Delivery*, 1998.

[39] B. A. Renz et al., "AEP Unified Power Flow Controller Performance," *IEEE Transactions on Power Delivery*, Paper 1998.

[40] B. A. Renz et al., "World's First Unified Power Flow Controller on the AEP System," CIGRE Paper No. 14-107, 1998.

[41] K. K. Sen and E. J. Stacey, "UPFC—Unified Power Flow Controller: Theory, Modelling, and Applications," Presented at the IEEE/PES 1998 Winter Meeting, New York.

[42] R. Mihalic, P. Zunko, and D. Povh, "Improvement of Transient Stability Using Unified Power Flow Controller," *IEEE Transactions on Power Delivery*, Vol. 11, No. 1, 1996, pp. 485–491.

[43] J. Y. Liu and Y. H. Song, "Comparison Studies of Unified Power Flow Controller With Static Var Compensators and Phase Shifters," *Electric Machines and Power Systems*, Vol. 27, No. 3, 1999.

[44] M. Noorozian, L. Angquist, M. Ghandhari, and G. Andersson, "Use of UPFC for

Optimal Power Flow Control," *IEEE Transactions on Power Delivery*, Vol. 12, No. 4, 1997, pp. 1629–1634.

[45] X. Lombard and P. G. Therond, "Control of Unified Power Flow Controller: Comparison of Methods on the Basis of a Detailed Numerical Model," *IEEE Transactions on Power Systems*, Vol. 12, No. 2, May 1997, pp. 824–830.

[46] S. Arabi and P. Kundur, "A Versatile FACTS Device Model for Power Flow and Stability Simulations," *IEEE Transactions on Power Systems*, Vol. 11, No. 4, November 1996, pp. 1944–1950.

[47] A. Nabavi-Niaki and M. R. Iravani, "Steady-State and Dynamic Models of Unified Power Flow Controller (UPFC) for Power System Studies," *IEEE Transactions on Power Systems*, Vol. 4, 1996, pp. 1937–1943.

[48] A. Edris et al., "Proposed Terms and Definitions for Flexible ac Transmission System (FACTS)," *IEEE Transactions on Power Delivery*, Vol. 12, No. 4, October 1997, pp. 1848–1853.

[49] P. Kundur, *Power System Stability and Control*, McGraw-Hill, New York, 1994.

[50] IEEE SSR Task Force, "First Benchmark Model for Computer Simulation of Subsynchronous Resonance," *IEEE Transactions on Power Apparatus and Systems*, Vol. 96, No. 5, 1977, pp. 1565–1570.

[51] A. E. Hammad and M. El-Sadek, "Application of a Thyristor-Controlled Var Compensator for Damping of Subsynchronous Oscillations in Power System," *IEEE Transactions on Power Apparatus and Systems*, Vol. 103, No. 1, 1984, pp. 198–212.

[52] Y. N. Yu, *Electric Power System Dynamics*, Academic Press, New York, 1983.

[53] *PSCAD/EMTDC Power Systems' Simulation Software, User's Manual*, Manitoba HVDC Research Center, Winnipeg, Manitoba, 1996.

[54] Z. Zhang, J. Kuang, X. Wang, and B. Ooi, "Force Commutated HVDC and SVC Based on Phase-Shifted Multi-Converter Modules," *IEEE Transactions on Power Delivery*, Vol. 8, No. 2, April 1993, pp. 712–718.

[55] K. R. Padiyar, *Analysis of Subsynchronous Resonance in Power Systems*, Kluwer Academic Publishers, Boston, 1999.

[56] J. Bian, R. Nelson, D. Ramey, T. Rietman, and J. Hill, "Transient Stability Enhancement With FACTS Controllers," *Proceedings of IEEE Sixth International Conference AC and DC Transmission*, London, May 1996.

[57] The Electric Power Research Institute (EPRI) Report, "Guide for Economic Evaluation of Flexible AC Transmission Systems (FACTS) in Open Access Environments," EPRI TR 108500, Final Report Prepared by (GE) Company, Schenectady, NY, August 1997.

[58] *Extended Transient Mid-Term Stability Program Manual*, EPRI TR-102004-V3RI, The Electric Power Research Institute, Palo Alto, CA, May 1994.

[59] J. Van Coevering, J. P. Stovall, R. L. Hauth, P. J. Tatro, B. D. Railing, and B. K. Johnson, "The Next Generation of HVDC—Needed R&D, Equipment Costs, and Cost Comparisons," *Proceedings of EPRI Conference on Future of Power Delivery*, Washington, DC, April 1996.

[60] CIGRE Working Group 14.29, "Coordination of Controls of Multiple FACTS/HVDC Links in the Same System," CIGRE Technical Brochure No. 149, Paris, December 1999.

[61] CIGRE Task Force 38.02.17, "Advanced Angle Stability Controls," CIGRE Technical Brochure No. 155, Paris, April 2000.

[62] CIGRE Task Force 38.02.16, "Impact of Interactions Among Power System Controls," CIGRE Technical Brochure No. 166, Paris, August 2000.

[63] J. F. Hauer, "Robust Damping Controls for Large Power Systems," *IEEE Control Systems Magazine*, January 1989, pp. 12–19.

[64] CIGRE Task Force 38.01.07, "Analysis and Control of Power System Oscillations," CIGRE Technical Brochure No. 111, Paris, December 1996.

[65] J. F. Hauer, D. J. Trudnowski, G. J. Rogers, W. A. Mittelstadt, W. H. Litzenberger, and J. M. Johnson, "Keeping an Eye on Power System Dynamics," *IEEE Computer Applications in Power*, 1997, pp. 50–54.

[66] J. F. Hauer, W. A. Mittelstadt, W. H. Litzenberger, C. Clemans, D. Hamai, and P. Overholt, "Wide Area Measurements for Real-Time Control and Operation of Large Electric Power Systems: Evaluation and Demonstration of Technology for the New System," Report prepared for the U.S. Department of Energy by the Bonneville Power Administration (BPA) and the Western Area Power Administration, Portland, OR, April 1999. (This report and its attachments are available from the BPA on CD-ROM.)

[67] D. Povh, "FACTS Controller in Deregulated Systems," *SEPOPE '98 Power Systems Symposium*, Rio de Janeiro, Brazil, May 1998.

APPENDIX A

Design of an SVC Voltage Regulator

A.1 STUDY SYSTEM

The study system shown in Fig. A.1(a) consists of a synchronous generator supplying power to a 230-kV, 50-Hz infinite bus over a 600-km transmission line rated at 400 kV. The transmission line is compensted at its midpoint by an SVC of FC–TC configuration.

The dynamic range of an SVC is determined on the basis of the reactive-power requirement at the SVC bus to control the voltage under steady-state conditions. An optimal load-flow program that minimizes the total transmission loss is used to obtain the SVC rating as 200-MVA inductive to 300-MVA capacitive at 400 kV.

A proportional–integral (PI) controller is selected as the SVC voltage regulator, $G(s)$, for voltage-control applications.

$$G(s) = -\left(K_P + \frac{K_I}{s}\right) \tag{A.1}$$

The study-system data are as follows:

The Generator (on Its Own Base)

$S_n = 1110$ MVA; $V_n = 22$ kV; $f = 50$ Hz; $pf = 0.9$; $R_a = 0.0036$ pu; $x_l = 0.21$ pu; $R_0 = 0$; $x_0 = 0.195$ pu; $T'_{d0} = 6.66$ s; $T'_{q0} = 0.44$ s; $T''_{d0} = 0.032$ s; $T''_{q0} = 0.057$ s; $x_d = 1.933$ pu; $x_q = 1.743$ pu; $x'_d = 0.467$ pu; $x'_q = 1.144$ pu; $x''_d = 0.312$ pu; $x''_q = 0.312$ pu; $H = 3.22$ s; $D = 0$

The Automatic Voltage Regulator (IEEE Type 1 Rotating Exciter)

$T_R = 0$ s; $K_A = 400$ pu; $T_A = 0.02$ s; $K_E = 1$ pu; $T_E = 1$ s; $K_F = 0.06$ pu; $T_F = 1$ s; $S'_E = 0$

The power-system stabilizer (PSS) and governor control are not considered.

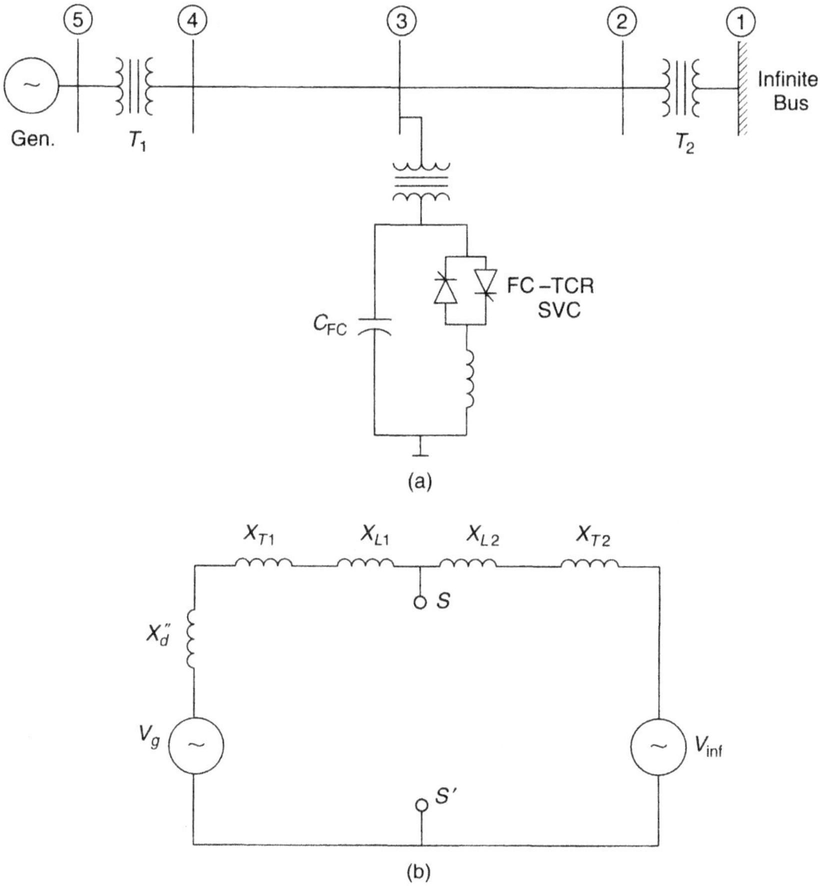

Figure A.1 (a) The single-machine infinite-bus (SMIB) study system and (b) an equivalent circuit diagram.

The Transmission Line

Resistance (R) = 0.055 Ω/phase/mi

Reactance (X_L) = 0.52 Ω/phase/mi

Susceptance (B_C) = 5.92×10^{-6} ℧/phase/mi

SVC (6-Pulse Operation)

$T_M = 2.4$ ms; $T_d = 5$ ms; $T_Y = 1.667$ ms; slope $K_{SL} = 3\%$

The design of the SVC voltage regulator is undertaken by two methods: the *method of system gain* [A.1] and *eigenvalue analysis* [A.2], [A.3].

A.2 METHOD OF SYSTEM GAIN

The equivalent circuit of the system is shown in Fig. A.1(b). The line-resistance and line-charging capacitances are neglected. The base quantities are 400 kV and 500 MVA in the line.

$$X_{\text{base}} = \frac{400^2}{500} = 320\ \Omega$$

$$X_{L1} = X_{L2} = 0.52\left(\frac{300}{1.6}\right)\left(\frac{1}{320}\right) = 0.305 \text{ pu}$$

$$X_{T1} = X_{T2} = 0.15\left(\frac{400^2}{1110^2}\right)\left(\frac{1}{320}\right) = 0.0676 \text{ pu}$$

$$X_d'' = 0.312\left(\frac{500}{1110}\right) = 0.1405 \text{ pu}$$

As viewed from SS',

$$\begin{aligned} X_N &= j(0.1405 + 0.0676 + 0.305) \,||\, (0.305 + 0.0676) \\ &= j\,\frac{0.5131 \cdot 0.3726}{0.5131 + 0.3726} \\ &= j0.216 \text{ pu} \end{aligned}$$

For the fixed capacitor,

$$X_c = -j\,\frac{500}{300} = -j1.667 \text{ pu}$$

The equivalent effective impedance, including the effect of fixed capacitor, is given by

$$X_{TH} = \frac{X_N X_C}{X_N + X_C} = j0.248$$

Thus the system gain of $K_N = V_0 X_{TH} = 1.05 \cdot 0.248 = 0.26$ pu. Because only a single-circuit transmission line is considered, the K_N corresponds to the weakest network state. Hence $K_N = K_{N\,\max}$.

The proportional gain of the controller is obtained as

$$K_P = -\frac{1}{2(K_{SL} + K_{N\max})}$$

$$= -\frac{1}{2(0.01 + 0.26)} = -1.85$$

$$K_1 = -\frac{K_P}{T_Y} = -\frac{1.85}{0.005} = -370$$

These values of K_P, K_I and the susceptance output of the voltage regulator correspond to a system base of 500 MVA. However, if the system base is chosen as 100 MVA, the modified values of K_P, K_I and the susceptance output can be obtained as given in the text that follows.

Let the subscripts n and o refer to 100-MVA and 500-MVA systems, respectively. The kV base for the two systems is identical at 400 kV. Then,

$$B_n \text{ [in pu]} = B_o \text{ [in pu]} \cdot \frac{\text{MVA}_o}{\text{MVA}_n}$$

$$= \frac{500}{100} \cdot B_o = 5B_o$$

A susceptance of $5B_o$ can be achieved by multiplying the gains K_P and K_I by the factor $5(= \text{MVA}_o/\text{MVA}_n)$. Thus the gains expressed on a system base of 100 MVA are

$$K_{Pn} = 5 \times 1.85 = 9.25 \text{ pu}$$
$$K_{In} = 5 \times 370 = 1850 \text{ pu}$$

These values of optimal gains have been obtained on very simplistic assumptions without considering the dynamics of the generator.

Now, we obtain the optimal gains by the method of eigenvalue analysis, which considers a detailed model of the system.

A.3 EIGENVALUE ANALYSIS

The voltage controller is designed at an operating point corresponding to the generator output of 500 MW. An optimal load flow that minimizes transmission losses results in the following operating data:

The generator bus: $P_g = 500$ MW; $Q_g = -39.9$ MVA; $|V_g| = 1.05$ pu; $\angle\theta_g = 38.9^\circ$

The SVC bus: $Q_{SVC} = 6.19$ MVA; $|V_{SVC}| = 1.05$ pu; $\angle\theta_{SVC} = 19.58^\circ$

The infinite bus: $P_{inf} = -471.24$ MW; $Q_{inf} = -18.14$ MVA; $|V_{inf}| = 1$ pu; $\angle\theta_{inf} = 0.0^\circ$

The power system is linearized about the operating point, and state equations are developed for the following two cases:

1. *The simplified generator model*, in which the generator is represented by a constant-voltage source behind the subtransient reactance.
2. *The detailed generator model*, in which the generator is modeled as a dependent current source in shunt with the subtransient reactance [A.3]. The automatic voltage regulator (AVR) is also included. The SVC controller is chosen to be of the PI configuration described in Eq. (A.1).

The root loci with varying integral gain K_I and proportional gain K_P, each taken one at a time, are shown in Figs. A.2(a) and (b), respectively, for a simplified generator model; the same loci for a detailed generator model are shown in Figs. A.3(a) and (b), respectively. The root loci are obtained for a generator-operating power tranfer of 500 MW. The gains K_P and K_I are expressed on a 100-MVA base.

It is observed from the root loci with varying K_I (with $K_P = 0$) displayed in Fig. A.2(a) that a negative value of K_I or a large positive value of K_I (>4000) renders the system unstable. In both K_I values, the SVC controller mode becomes destabilized. When the generator dynamics are modeled as shown in Fig. A.3(a), the lower end on the stable range of K_I is imposed by the generator-rotor mode that tends to become unstable if K_I is decreased from 200. The upper limit of K_I continues to be dictated by the SVC controller mode, and its stability range is determined to be $180 < K_I < 3500$ for $K_P = 0$.

From the variation of roots with K_P (with $K_I = 1200$) shown in Fig. A.2(b), it is seen that the stability range of K_P is limited by an SVC controller mode in the negative direction ($K_P < -12$) and by the TCR mode in the positive direction ($K_P > 6$). When the generator dynamics are considered, the stability range is governed by similar considerations. The stability range for K_P is obtained as $-14 < K_P < 5$ for $K_I = 1200$.

A.3.1 Step Response

Having obtained the range of stability of parameters K_P and K_I, the step-response studies are performed for the linearized system with the simplified generator model for a step change in the reference voltage V_{ref}. The rise time (to 90% of the steady-state value), settling time (to ±5% of the final value), and peak overshoot in the SVC bus voltage are obtained. The step-response data with varying K_I and K_P are compiled in Tables A.1 and A.2, respectively.

The actual responses for $K_I = 700$, 900, 1200, and 1800 when $K_P = 0$ are displayed in Figs. A.4(a), (b), (c), and (d), respectively. The positive step change in the reference voltage ΔV_{ref} is equal to 0.01 pu. From Table A.1, it is observed that a variation of K_I from 700 to 2500 improves the rise time but worsens the settling time. Also, although the response is damped for smaller magnitudes of K_I, it tends to become oscillatory with increasing K_I. The best value of K_I is

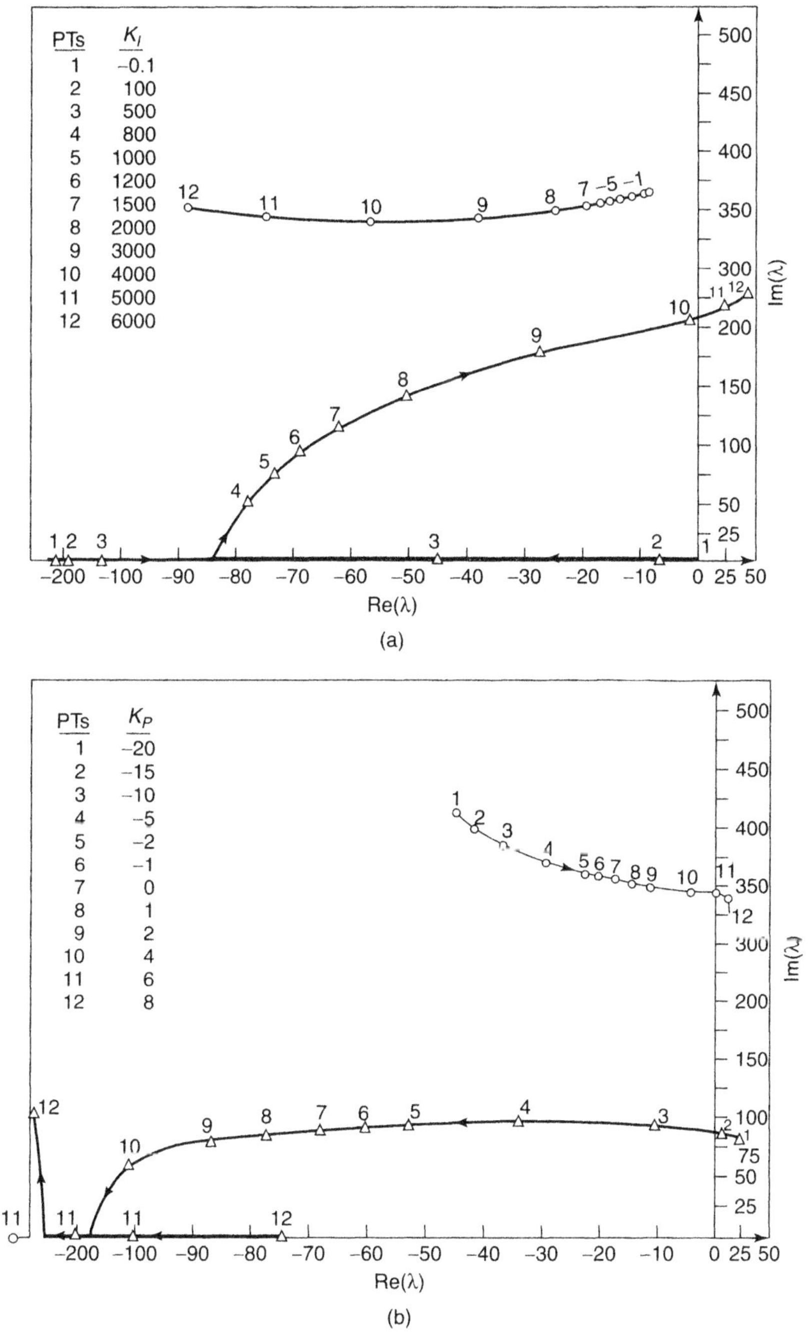

Figure A.2 (a) The root loci of a simplified model with varying K_I ($K_P = 0$) and (b) the root loci of a simplified model with varying K_P ($K_I = 1200$).

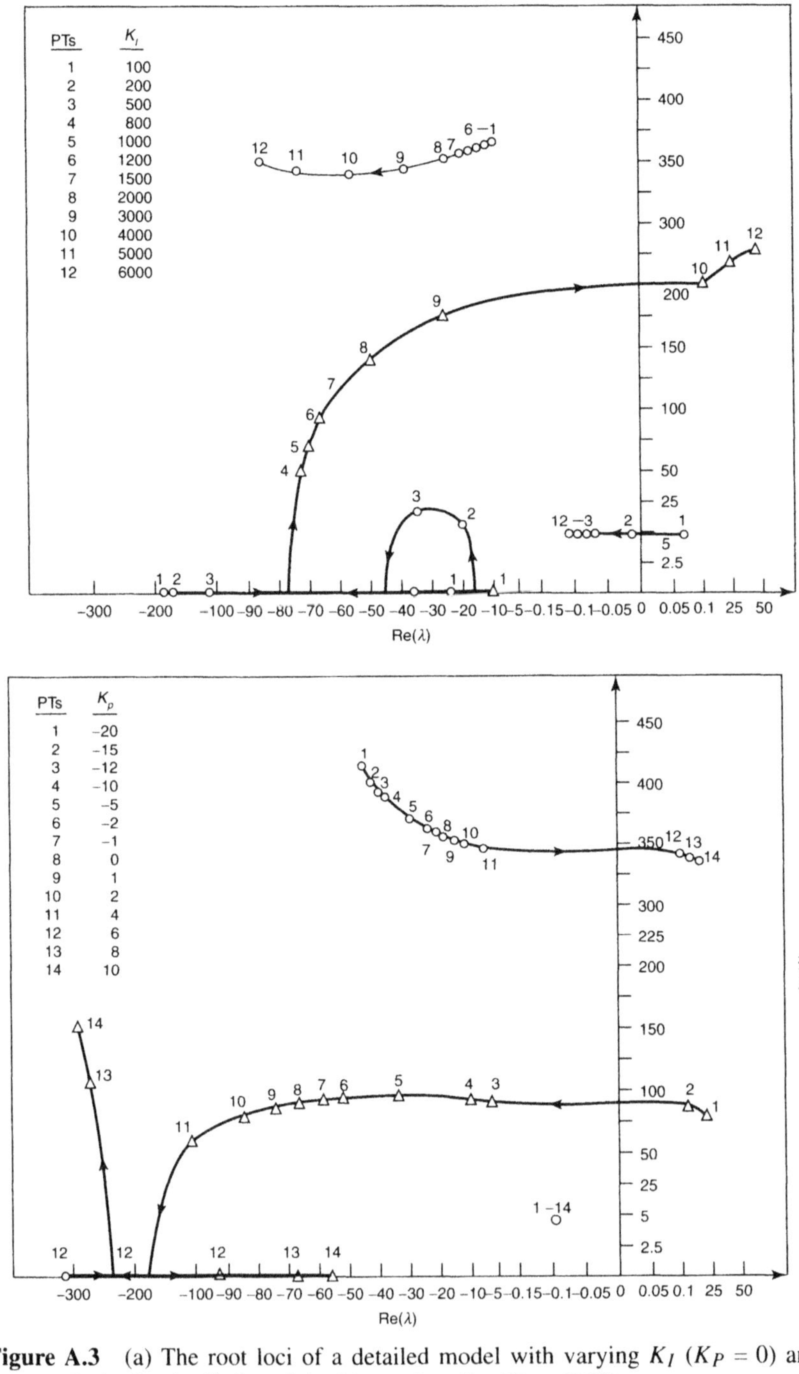

Figure A.3 (a) The root loci of a detailed model with varying K_I ($K_P = 0$) and (b) the root loci of a detailed model with varying K_P ($K_I = 1200$).

TABLE A.1 Step-Response Data With Varying K_I ($K_P = 0$; $K_{SL} = 0.01$; $P_g = 500$ MW; Integration Step = 200 μs; and $\Delta V_{ref} = 0.01$ pu)

SVC No.	K_I	Rise Time (ms)	Settling Time (ms)	Peak Overshoot (pu)
1	700	43.6	48	0.02
2	900	30	54.8	0.06
3	950	29	55.4	0.07
4	1000	28	55.4	0.09
5	1050	26.8	55.4	0.11
6	1100	24.8	55.2	0.13
7	1150	21.2	55	0.14
8	1200	19.6	54.4	0.16
9	1250	18.6	54.8	0.17
10	1300	17.8	63.6	0.18
11	1350	17.2	64.2	0.19
12	1500	16	65	0.22
13	1800	14.4	74	0.32
14	2000	14	80	0.46
15	2500	12.4	98	0.90

chosen as 1200, which results in an acceptably small rise time, the minimum possible settling time, and a reasonable overshoot of 16%.

Table A.2 shows that increasing magnitudes of K_P in both the positive and the negative direction affect the rise time and settling time detrimentally. For large negative values of K_P, a high value of peak overshoot is observed, whereas for large positive values of K_P, the response is very oscillatory and poorly damped even though the overshoot is small. Figures A.5(a) and (b) illustrate the step responses for $K_P = -3$ and $+3$ when $K_I = 1200$.

Considering the minimum settling time and acceptable peak overshoot, the best value of proportional gain K_P is selected as +1. It is interesting to note that the response is not very different in the range $0 \leq K_P \leq 1$. The best value of K_P can also be chosen as zero that reduces the controller to a purely integral type for this example case.

The step-response studies show that for the foregoing system, the integral gain K_I is essentially the one that controls the speed of response. It is further seen that the best values of K_P and K_I obtained from the detailed eigenvalue analysis do not correlate with those values obtained from the simplified method of system gain presented in Section A.2.

The optimization of voltage-regulator parameters is based on step-response studies using a simplified generator model. However, the long-term step response with optimized parameters using a detailed generator model is illustrated in Fig. A.6. The long-term oscillations visible in the response correspond to the rotor-mode oscillations.

TABLE A.2 Step-Response Data with Varying K_P (K_I = 1200; K_{SL} = 0.01; P_g = 500 MW; Integration Step = 200 μs; and ΔV_{ref} = 0.01 pu)

SVC No.	K_P	Rise Time (ms)	Settling Time (ms)	Peak Overshoot (pu)
1	−10	24.2	188	0.98
2	−5	21.2	82.6	0.44
3	−3	20.4	68.4	0.3
4	−2	20	65.6	0.22
5	−1.5	19.8	64.2	0.22
6	−1	19.8	54.2	0.2
7	−0.5	19.6	54.4	0.18
8	0	19.6	54.4	0.16
9	0.5	19.4	54.6	0.15
10	1	19.4	54.8	0.14
11	1.5	19.4	62.4	0.13
12	2	26	80	0.13
13	3	27.2	143	0.13
14	4	15.6	200	0.14

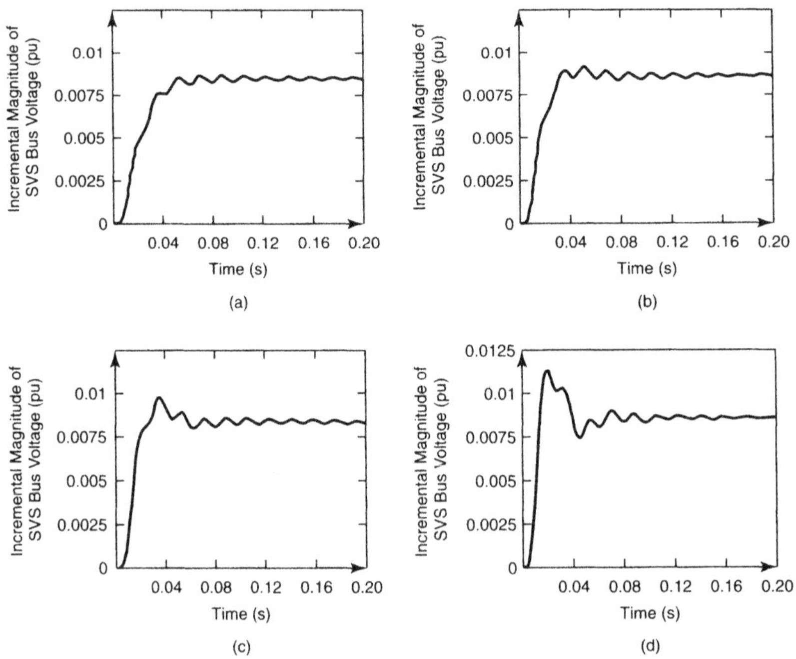

Figure A.4 The step responses of a simplified model with varying K_I: (a) K_I = 700; (b) K_I = 900; (c) K_I = 1200; and (d) K_I = 1800.

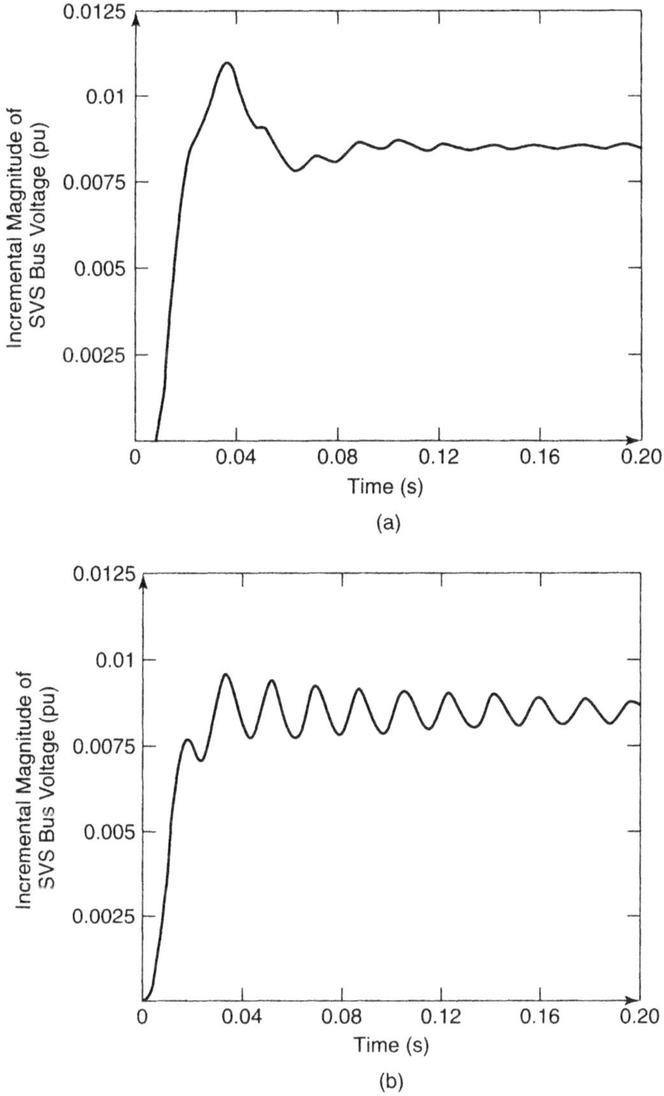

Figure A.5 The step responses of a simplified model with varying K_P: (a) $K_P = -3$ and (b) $K_P = 3$.

A.3.2 Power-Transfer Studies

The power-transfer limits are determined both with and without the optimized SVC by using eigenvalue analysis. As mentioned in the previous section, the SVC voltage regulator has been optimized for a generator power transfer of 500 MW. The power limit obtained without the SVC is 474 MW, with the limit imposed by

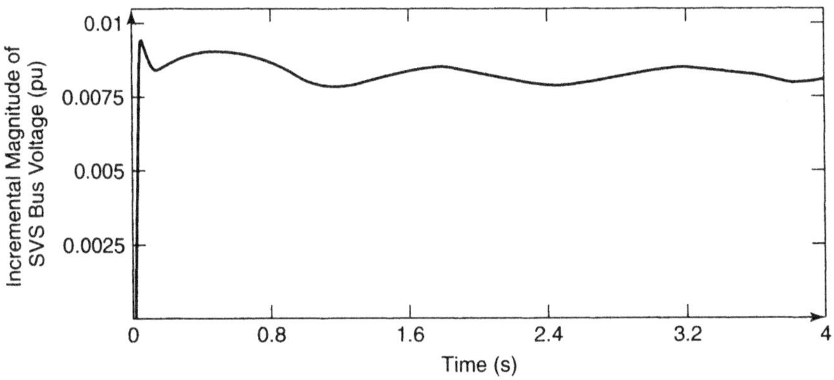

Figure A.6 The step response of a detailed model with $K_I = 1200$ ($K_P = 0$).

a rotor mode of frequency 4.36 rad/s. On the other hand, the maximum transmissible power with the optimized SVC is 610 MW, with the limit again imposed by a rotor mode—in this case, one of frequency 4.58 rad/s. The SVC with pure voltage control thus improves the power transfer in the line by 29%.

A.4 SIMULATOR STUDIES

The eigenvalue analyses pertain to the prediction of system behavior under small disturbances by using linearized models. It is therefore desirable to validate these results and investigate the performance of the SVC under large disturbances by using detailed simulator studies, wherein the various system nonlinearities and harmonic generation from the SVC thyristor-switching action are also considered.

The study system is simulated on the ABB's HVDC simulator that has elaborate models of the synchronous generator, the SVC, and the ac lines. Only integral control is modeled on the simulator.

A.4.1 Step-Response Studies

The step-response studies are conducted for two reasons: (1) to investigate the range of voltage-regulator integrator gain, K_I, to ensure stable system operation, and (2) to determine the optimal value of gain K_I.

The stability range of K_I is determined by long-term oscillations in the SVC response. Figures A.7(a)–(g) illustrate the susceptance-signal output from the voltage regulator in response to a negative step input in the reference voltage V_{ref}. These responses correspond to $K_I = 94.4$, 198.7, 632, 1415, 2831, 4246, and 6321, respectively. The numerical values of K_I are constrained by the discrete gain-knob settings of the controller.

It is noted that for $K_I = 94.4$, the susceptance-order signal B_{ref} begins to self-

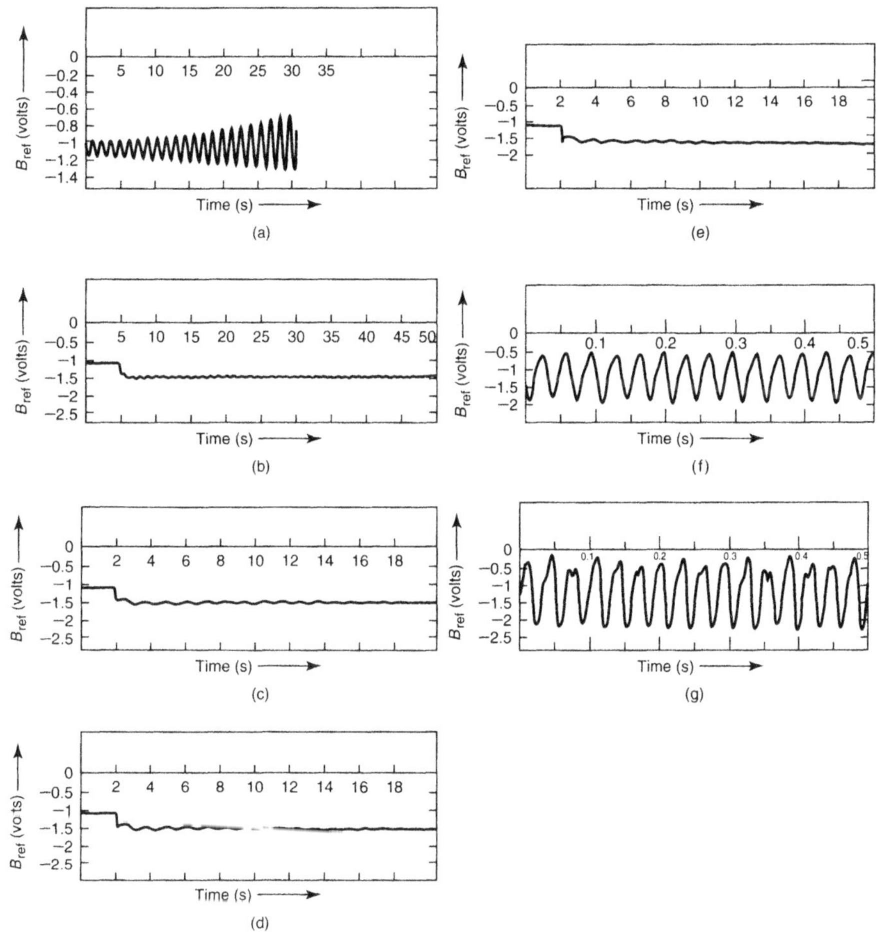

Figure A.7 The long-duration step response of the SVC as obtained on the simulator: (a) susceptance-order B_{ref} for $K_I = 94.4$ without the application of step input; (b) $K_I = 198.7$; (c) $K_I = 632$; (d) $K_I = 1415$; (e) $K_I = 2831$; (f) susceptance-order B_{ref} for $K_I = 4246$ without the application of step input; and (g) susceptance-order B_{ref} for $K_I = 6321$ without the application of step input.

oscillate even without the application of a step input. Moreover, the oscillations in B_{ref} have a frequency of 0.74 Hz (4.65 rad/s), which tends to increase over time. This increase implies that $K_I = 94.4$ falls in the unstable range of K_I. For $K_I = 198.7$, the oscillations in B_{ref} following the application of the step input diminish in magnitude, although very slowly. However, the steady state is reached in a relatively smaller time period (typically 8–12 s) when K_I is set at values 632, 1415, and 2831.

These low-frequency oscillations are also observed in the step response obtained from the linearized-system model of Fig. A.6. They are caused by the generator-rotor mode and take a long time to decay because the mechanical damping is neglected. If the synchronous machine is modeled as a constant-voltage source behind a reactance, such long-term oscillations are not observed.

As K_I is increased to 4246, the susceptance B_{ref} starts showing sustained oscillations of large magnitude, even without the application of a step input. The frequency of such oscillations is observed as 29.5 Hz (185.4 rad/s). For the next higher gain-knob setting corresponding to $K_I = 6321$, these oscillations start growing over time. However, this growth is not visible from Fig. A.6, for the plot is taken over a short time period. As measured from the plot, the frequency of these oscillations is 32.3 Hz (202.9 rad/s).

From a study of the foregoing step responses involving long-term oscillations, it is concluded that the stability boundary for K_I lies between 94.4 and 198.7 on the lower side and near 4246 on the higher side. It may be noted that because the system is nonlinear, even sustained oscillations for $K_I = 4246$ can cause instability in the corresponding linearized model. The lower-side instability is caused by rotor-mode oscillations, whereas the higher-side instability is caused by a mode of frequency 185.4 rad/s. Also, it is concluded that for stable operation, the integral gain K_I must lie between 632 and 2831.

The SVC responses are now investigated in a time period that is much shorter than that considered in the example of the preceding paragraph. The rise time and percent of peak overshoot are obtained for different values of K_I; they are based on which optimal magnitudes of integral gain are determined.

Figures A.8(a)–(d) display the step responses in which K_I = 1120, 1415, 2239, and 3330, respectively, for the same 5% negative step input. The corresponding response data obtained from these plots are listed in Table A.3. It is pertinent to note that measurement errors of 10% in the rise time and of 5–10% in the peak overshoot are likely in the values given in Table A.3. It is observed that for higher values of K_I, the rise time decreases as the peak overshoot increases. The optimal value of K_I that gives a fast response and an acceptable peak overshoot (<20%) is determined to be 1120.

A.4.2 Power-Transfer Limits

The power-transfer capacity of the study system is investigated both with and without the optimized SVC. Without the SVC, it is observed that as soon as the generator power exceeds the 521.7-MW level, it ceases to remain stable and starts oscillating with a frequency of 0.73 Hz (4.59 rad/s) and a magnitude that keeps increasing with time—marking the onset of system instability. Figure A.9(a) shows the generator-power signals at a stable level of 400 MW and 522 MW when the power oscillations are set up.

With the optimized SVC, when the machine power surpasses the level of 648 MW, it starts oscillating with a continuously increasing magnitude and at a frequency of 0.75 Hz (4.71 rad/s). Figure A.9(b) displays a stable power level of 500 MW and oscillations at 649 MW. The oscillation frequencies of 4.59 rad/s (without the SVC) and 4.71 rad/s (with the SVC) correspond to the rotor mode, thus

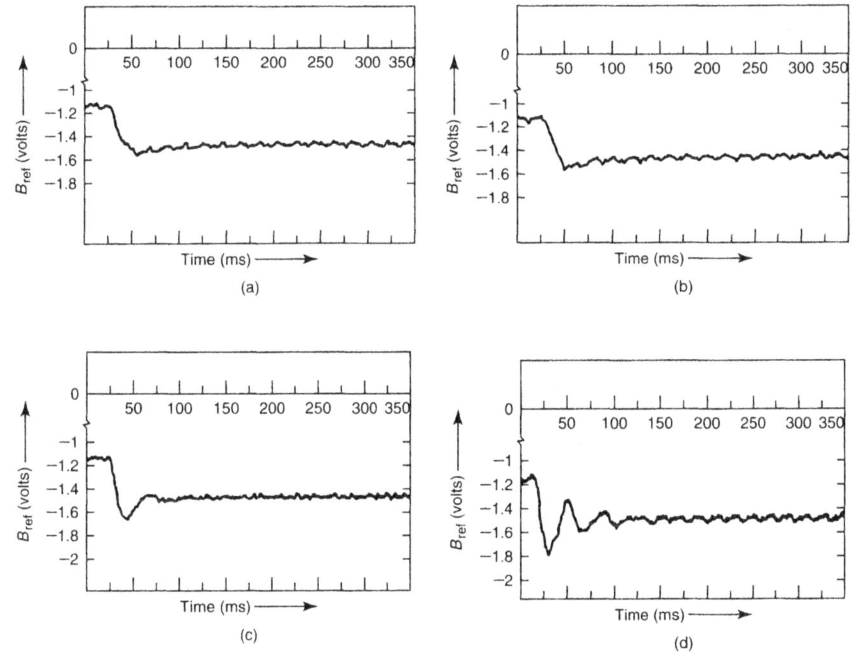

Figure A.8 The short-duration step response of the SVC as obtained on the simulator: (a) $K_I = 1120$; (b) $K_I = 1415$; (c) $K_I = 2239$; and (d) $K_I = 3330$.

confirming that the rotor mode is indeed the one that limits the power-carrying capability of the transmission system both with and without the SVC.

A.5 A COMPARISON OF PHYSICAL SIMULATOR RESULTS WITH ANALYTICAL AND DIGITAL SIMULATOR RESULTS USING LINEARIZED MODELS

Presented in Table A.4 is a comparison of results obtained from physical simulators with results predicted from eigenvalue analysis and digital simulation

TABLE A.3 Response Data With Varying K_I as Obtained on the Simulator

SVC No.	K_I	Rise Time (ms)	% Peak Overshoot
1	1120	19	18
2	1415	16.3	32.3
3	2239	7.5	71
4	3330	5	100

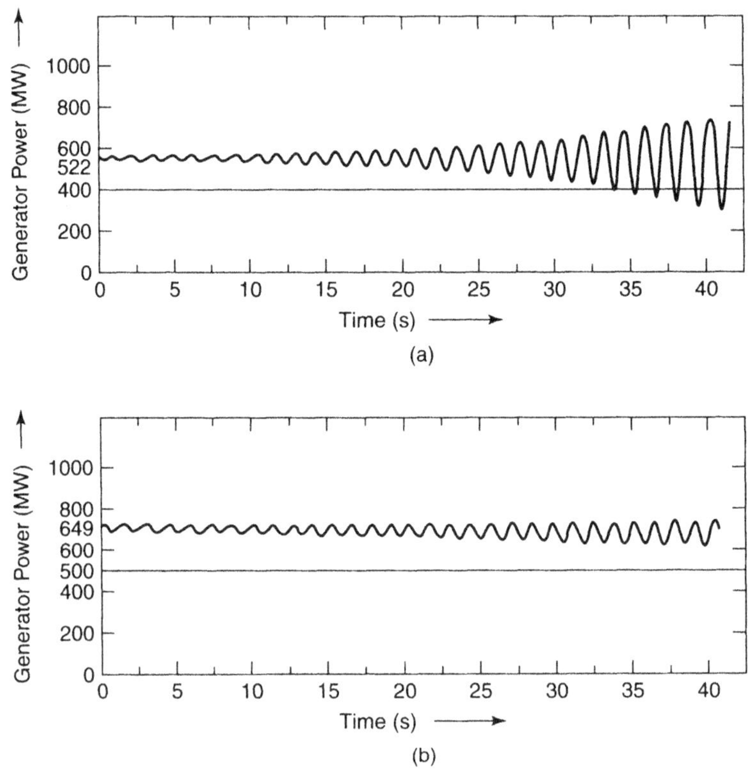

Figure A.9 The power-transfer limits obtained on the simulator: (a) without the SVC and (b) with the SVC.

using linearized models. It is seen that the stability ranges of K_I as obtained from both methods are quite similar. The exact stability boundary for both small and large values of K_I could not be obtained in the simulator because of the discrete nature of the K_I variation. The frequencies of spontaneous oscillations in the B_{ref} signal observed in the simulator for the two values of K_I lying beyond the stability range are also compared with the critical-mode frequencies for the same values of K_I obtained from eigenvalue analysis. A good correlation is observed between the magnitudes of these frequencies.

The optimal values of the integrator gain setting obtained from the two methods are also similar. Because it is not possible to vary K_I continuously in the simulator, the value $K_I = 1120$ is the best that could be obtained by changing the gain knob. It is noted from the step-response data corresponding to the simplified machine model given in Table A.2 that there is little variation in the response characteristics in the 1100–1200 K_I range.

The optimal responses obtained from both methods are observed to be similar regarding the rise time and the peak overshoot. However, the response obtained on the simulator is slightly faster. It is noted that the power limits obtained on

TABLE A.4 Comparison of Results From Physical Simulation and Analytical Prediction

SVC. No.	Item	Physical Simulation	Analytical Prediction
1	Stability range of SVC integral gain, K_I	$198.7 \leq K_I \leq 4246$	$200 \leq K_I \leq 4400$
2	Critical mode frequency (rad/s)		
	$K_I = 94.4$	4.7	4.6
	$K_I = 4246$	185.4	195.5
3	Optimal K_I	1120	1200
4	Optimal step response		
	Rise time (ms)	18.1	21
	Peak overshoot (%)	18.2	15
5	Power limit without SVC		
	MW	521.7	474
	Critical-mode frequency (rad/s)	4.59	4.36
6	Power limit with optimized SVC		
	MW	648	610.2
	Critical-mode frequency (rad/s)	4.71	4.58

the simulator are higher than those computed from eigenvalue analysis. The differences in power with and without the optimized SVC are observed to be 6% and 9%, respectively. A close correlation is also witnessed in the frequencies obtained from both methods for the rotor mode that is critical in limiting the stable transfer of power across the transmission line both with and without the SVC.

Thus the design of the SVC voltage regulator as performed through linearized analysis is validated from detailed simulation studies. This validation can also be performed through digital simulation using EMTP/EMTDC programs.

REFERENCES

[A.1] H. E. Schweickardt, G. Romegialli, and K. Reichert, "Closed Loop Control of Static Var Sources (SVC) on EHV Transmission Lines," Paper A 78 135-6, Presented at the IEEE/PES 1978 Winter Meeting, New York, January 29–February 3, 1978.

[A.2] K. R. Padiyar and R. K. Varma, "Damping Torque Analysis of Static Var System Controllers," *IEEE Transactions on Power Systems*, Vol. 6, No. 2, pp. 458–464, May 1991.

[A.3] K. R. Padiyar and R. K. Varma, "Concepts of Static Var System Control for Enhancing Power Transfer in Long Transmission Lines," *Electric Machines and Power Systems*, Vol. 18, No. 4–5, July–October 1990, pp. 337–358.

APPENDIX B

Transient-Stability Enhancement in a Midpoint SVC-Compensated SMIB System

Consider the single-machine infinite-bus (SMIB) system depicted in Fig. 6.1(b), in which the transmission line is compensated at the midpoint by an SVC. Applying Kirchoff's current law at the SVC bus, we have

$$\frac{V_1\angle\delta - V_m\angle(\delta/2)}{j(X/2)} + \frac{V_2\angle 0 - V_m\angle(\delta/2)}{j(X/2)} - V_m\angle(\delta/2)\cdot jB_s = 0 \tag{B.1}$$

or

$$V_m\angle(\delta/2) = \frac{(V_1\angle\delta + V_2\angle 0)}{X_e} \tag{B.2}$$

where

$$X_e = 2 - B_s\,\frac{X}{2} \tag{B.3}$$

Equating the real and imaginary component of Eq. (B.3), we have

$$V_m \sin\frac{\delta}{2} = \frac{1}{X_e}\,V_1 \sin\delta \tag{B.4}$$

$$V_m \cos\frac{\delta}{2} = \frac{1}{X_e}\,(V_1\cos\delta + V_2) \tag{B.5}$$

Because the SVC maintains constant voltage, V_m, at the bus, Eqs. (B.4) and (B.5) are solved to give

$$V_m = \frac{1}{X_e}\sqrt{(V_1^2 + V_2^2 + 2V_1V_2\cos\delta)} \tag{B.6}$$

The power transfer, P, over the SVC-compensated line is given by

$$P = \frac{V_m V_2}{X/2} \sin \frac{\delta}{2} \tag{B.7}$$

Substituting from Eq. (B.5) in Eq. (B.7), we get

$$P = \frac{V_1 V_2}{X_T} \sin \delta \tag{B.8}$$

where

$$X_T = \frac{X X_e}{2} \tag{B.9}$$

which establishes that X_T is the equivalent transfer reactance between the generator and the infinite bus. The synchronizing torque of the SVC-compensated system is now expressed as

$$K_s = \frac{\partial P_E}{\partial \delta} \tag{B.10}$$

Differentiating Eq. (B.8),

$$K_s = \frac{2V_1 V_2 \cos \delta}{X \cdot X_e} - \frac{2V_1 V_2 \sin \delta \cdot \partial X_e / \partial \delta}{X \cdot X_e^2} \tag{B.11}$$

Differentiating Eq. (B.6),

$$\frac{\partial X_e}{\partial \delta} = \frac{-V_1 V_2 \sin \delta}{V_m^2 X_e} \tag{B.12}$$

Substituting Eq. (B.12) in (B.11),

$$K_s = \frac{2V_1 V_2 \cos \delta}{X \cdot X_e} + \left(\frac{V_1 V_2 \sin \delta}{X_m X_e} \right)^2 \frac{2}{X} \cdot \frac{1}{X_e} \tag{B.13}$$

or

$$K_s = \frac{V_1 V_2}{X_T} \cos \delta + \left(\frac{V_1 V_2 \sin \delta}{X_m X_T} \right)^2 \frac{X^2}{4X_T} \tag{B.14}$$

The increase in synchronizing torque by the SVC is therefore given by

$$\Delta K_s = K_s - K_{uc} \tag{B.15}$$

where K_s = the synchronizing torque of the uncompensated system given by (6.17)

Substituting Eqs. (B.13) and (B.17) in (B.15), we obtain

$$\Delta K_s = \frac{V_1 V_2}{2X_e} \cdot B_s \cos \delta + \left(\frac{V_1 V_2 \sin \delta}{V_m X_e} \right)^2 \frac{2}{X \cdot X_e} \tag{B.16}$$

APPENDIX C

Approximate Multimodal Decomposition Method for the Design of FACTS Controllers

C.1 INTRODUCTION

This appendix presents a general technique of FACTS controller design used for damping various power-swing modes simultaneously, as described by Larsen et al. [C.1]. It is based on well-established concepts of power-system stabilizer (PSS) design [C.2], [C.3] and assesses the influence of the FACTS controller on overall system damping by examining the controller's incremental effect on each mode of interest. Certain assumptions, however, are made:

1. the low-frequency modes have small damping, and
2. the designed control does not substantially impact the frequency and mode shape of the relevant mode.

In a single-machine system [C.2], the rotor-inertial mode is modeled in terms of the damping and synchronizing torques. A similar approach is adopted in the multimodal decomposition as well.

Let the linearized-system model be represented by the following state equations:

$$\dot{x} = Ax + Bu \tag{C.1}$$

$$y = Cx + Du \tag{C.2}$$

where $x =$ the rearranged state vector

$= [\Delta\delta_1 \Delta\delta_2 \cdots \Delta\delta_n \Delta\omega_{g1} \Delta\omega_{g2} \cdots \Delta\omega_{gn} z^{\mathrm{T}}]^{\mathrm{T}}$

$\Delta\delta_i =$ the rotor angle of the ith generator

$\Delta\omega_i =$ the angular speed of the ith generator

$z =$ the vector of all other system state variables

$u =$ the control-input variables

$y =$ the output-measurement variables

The system matrices A and B emerge in the following form:

$$A = \begin{bmatrix} 0 & \omega_b I & 0 \\ A_{21} & A_{22} & A_{23} \\ A_{31} & A_{32} & A_{33} \end{bmatrix}; \qquad B = \begin{bmatrix} 0 \\ B_2 \\ B_3 \end{bmatrix} \tag{C.3}$$

where I = the identity matrix

ω_b = the system-base frequency (rad/s)

It is noted that the effect of the swing modes is embedded in the matrices A_{21} and A_{22}. The multimodal transformation involves diagonalizing the matrix A_{21} using the following relationship:

$$V^{-1}A_{21}V = \Lambda \tag{C.4}$$

where V = the matrix of the right eigenvectors of A_{21}

A_{21} = an $n \times n$ matrix in which n is the number of generators in the system (Incidentally, the matrix A_{21} is significantly smaller—about ten times so—than the total system matrix, A.)

The system-state equations (C.1) and (C.2) are subject to the following modal transformation:

$$x_m = U^{-1}x \tag{C.5}$$

where

$$U = \begin{bmatrix} V & 0 & 0 \\ 0 & V & 0 \\ 0 & 0 & I \end{bmatrix} \tag{C.6}$$

$$\begin{aligned} x_m &= U^{-1}AUx_m + U^{-1}Bu \\ &= A_m x_m + B_m u \end{aligned} \tag{C.7}$$

$$\begin{aligned} y &= CUx_m + Du \\ &= C_m x_m + Du \end{aligned} \tag{C.8}$$

where

$$A_m = \begin{bmatrix} 0 & \omega_b I & 0 \\ \Lambda & A_{m22} & A_{m23} \\ A_{m31} & A_{m32} & A_{m33} \end{bmatrix} \tag{C.9}$$

Such a transformation allows an independent study of each swing mode. The contribution of each swing mode can be later combined to obtain the overall system behavior. However, the analysis and the controller-design procedure enunciated in the forthcoming text correspond to a single swing mode presumed to be the dominant swing mode.

C.2 MODAL ANALYSIS OF THE *i*TH SWING MODE, λ_i

The state variables of the transformed state vector are rearranged so that the modal angle, $\Delta\delta_{mi}$, and modal speed, $\Delta\omega_{mi}$, corresponding to the swing mode, λ_i, become the first and second state variables, respectively. Rearranged, the state equation can then be written as

$$\begin{bmatrix} \Delta\dot{\delta}_{mi} \\ \Delta\dot{\omega}_{mi} \\ \dot{z}_{mi} \end{bmatrix} = \begin{bmatrix} 0 & \omega_b & 0 \\ -K_{mi} & -d_{mi} & -A_{d23} \\ A_{d31} & A_{d32} & A_{d33} \end{bmatrix} \begin{bmatrix} \Delta\delta_{mi} \\ \Delta\omega_{mi} \\ z_{mi} \end{bmatrix} + \begin{bmatrix} 0 \\ -B_{d2} \\ B_{d3} \end{bmatrix} u \tag{C.10}$$

$$y = [C_{d1} \quad C_{d2} \quad C_{d3}] \begin{bmatrix} \Delta\delta_{mi} \\ \Delta\omega_{mi} \\ z_{mi} \end{bmatrix} + Du \tag{C.11}$$

where K_{mi} = the approximate modal-synchronizing coefficient
d_{mi} = the approximate modal-damping coefficient
z_{mi} = the balance of the state variables

The foregoing form of the state equation is similar to the single-mode system described in ref. [C.2]. Hence the system equations (C.10) and (C.11) can be represented by the block diagram depicted in Fig. C.1, using the following appropriate transfer functions:

$$s\Delta\omega_{mi}(s) = -(\omega_b/s)K_{mi}(s)\Delta\omega_{mi}(s) - K_{ci}(s)U(s) \tag{C.12}$$

$$Y(s) = K_{oi}(s)\Delta\omega_{mi}(s) + K_{ILi}(s)U(s) \tag{C.13}$$

The different transfer functions used in the preceding equations are as follows:

- $K_{mi}(s)$ = the modal transfer function relating the modal angle to the modal torque, and

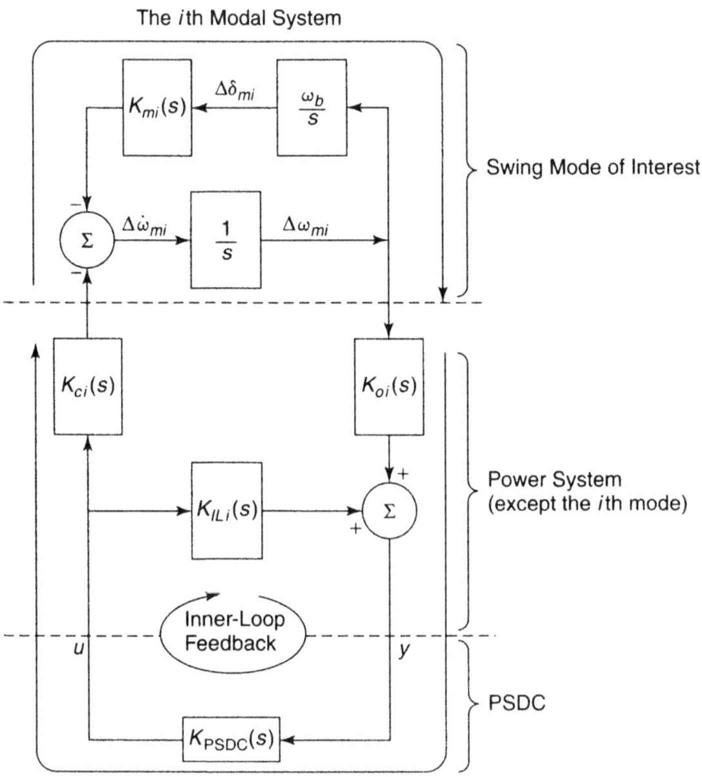

Figure C.1 Multimodal-decomposition block diagram.

$$= k_{mi} + (s/\omega_b)d_{mi} + A_{d23}(sI - A_{d33})^{-1}[A_{d31} + (s/\omega_b)A_{d32}]$$

- $K_{ci}(s)$ = the modal-controllability transfer function relating the control input, u, to $\Delta\dot{\omega}_{mi}$, and
$$= A_{d23}(sI - A_{d33})^{-1}B_{d3} + B_{d2}$$
- $K_{oi}(s)$ = the modal-observability transfer function relating the modal speed to the specific measurement signal, y, and
$$= C_{d3}(sI - A_{d33})^{-1}[A_{d31}(\omega_b/s) + A_{d32}] + [(\omega_b/s)C_{di} + C_{d2}]$$
- $K_{ILi}(s)$ = the inner-loop gain transfer function relating the control signal u to the specific measurement signal y for constant δ_{mi} and ω_{mi}, and
$$= C_{d3}(sI - A_{d33})^{-1}B_{d3} + D$$

All four transfer functions are complex when computed at $s = j\omega$. The magnitude and phase data of each provide valuable information for the design of power-swing damping controllers (PSDCs). An analysis of the transfer functions—$K_{mi}(s)$, $K_{ci}(s)$, $K_{oi}(s)$, and $K_{ILi}(s)$—is presented in the text that follows.

C.2.1 Effect of the Damping Controller

The effective control action, $K_{ei}(s)$, is expressed by

$$K_{ei}(s) = K_{ci}(s) \frac{K_{\text{PSDC}}(s)}{1 - K_{ILi}(s)K_{\text{PSDC}}(s)} K_{oi}(s) \tag{C.14}$$

It is assumed that the dominant swing mode, λ_i, is lightly damped or that its real part is negligible. Then,

$$\lambda_i \approx j\omega_i \tag{C.15}$$

For this dominant swing mode, it can be shown that

$$\omega_b K_{mi}(j\omega_i) \approx \omega_i^2 \tag{C.16}$$

Following the analysis presented in ref. [C.1], the characteristic polynomial of the closed-loop system with the PSDC is given by

$$s^2 + K_{ei}(s)s + \omega_b K_{mi}(s) = 0 \tag{C.17}$$

Let

$$K_{ei}(j\omega_i) = K_{re} + jK_{im} \tag{C.18}$$

Also, let the PSDC modify the swing mode from

$$s = j\omega_i \quad \text{to} \quad s = -\Delta\sigma_i + j(\omega_i + \Delta\omega_i) \tag{C.19}$$

Substituting Eqs. (C.18) and (C.19) in Eq. (C.17) gives

$$\{-\Delta\sigma_i + j(\omega_i + \Delta\omega_i)\}^2 + (K_{re} + jK_{im})\{-\Delta\sigma_i + j(\omega_i + \Delta\omega_i)\} + \omega_i^2 = 0 \tag{C.20}$$

Neglecting higher-order incremental terms, we get

$$2\Delta\omega_i\omega_i + K_{im}\omega_i \approx 0 \tag{C.21}$$

$$-2\Delta\sigma_i\omega + K_{re}\omega_i \approx 0 \tag{C.22}$$

Solving Eqs. (C.20), (C.21) and (C.22) results in the effective contribution of the PSDC on the dominant swing mode as

$$\Delta\lambda_i \approx -K_{ei}(j\omega_i)/2 \tag{C.23}$$

$$\Delta\sigma_i \approx K_{re/2} = \text{real}\,\{K_{ei}(j\omega_i)/2\} \tag{C.24}$$

$$\Delta\omega_i \approx -K_{im/2} = -I_m\{K_{ei}(j\omega_i)/2\} \tag{C.25}$$

C.3 IMPLICATIONS OF DIFFERENT TRANSFER FUNCTIONS

C.3.1 Controllability

The influence of the PSDC on a given ith swing mode is described by the controllability function, $K_{ci}(s)$. For $s = j\omega_i$, $K_{ci}(j\omega_i)$ indicates how effectively the ith mode can be controlled by the control signal u. If more than one control signal is employed, $K_{ci}(s)$ assumes the form of a vector-transfer function; the effective control action is dependent directly on $K_{ci}(s)$.

The controllability function $K_{ci}(s)$ is one of network structure and loads. Also, $K_{ci}(s)$ has different values for different swing modes, implying that a thyristor-controlled series capacitor (TCSC) with a given control input u may influence one swing mode beneficially and influence another adversely. An appropriate control signal or a combination of control signals must therefore be chosen to avoid such an effect.

In addition, $K_{ci}(s)$ also serves as a useful signal for deciding the placement of a TCSC. For instance, a large value of $K_{ci}(s)$ at a given site suggests that a TCSC-type PSDC can be used more effectively there.

C.3.2 Observability

The effectiveness of the measured signal y for damping control is obtained from the observability function, $K_{oi}(s)$. At $s = j\omega_i$, $K_{oi}(j\omega_i)$ expresses the modal content of the ith swing mode in y. The effective control action, $K_{ei}(s)$, is also a direct function of $K_{oi}(s)$. Because $K_{oi}(s)$ is defined mainly with respect to modal speed, any measurements corresponding to machine speeds result in real observability gains. On the other hand, power-flow signals demonstrate an integral characteristic that gives an almost 90° lag with respect to speed signals.

In conclusion, an effective input signal for damping control should have a high value of $K_{oi}(j\omega_i)$ for a given mode.

C.3.3 The Inner loop

A highly desirable feature of a damping controller is that the controller output must have a minimal direct effect on its own input (i.e., the component of measurement signal y that relates to control input u). The controller influence on y must be largely through the dominant swing mode of interest only—an effect that is expressed by the inner-loop transfer function, $K_{ILi}(s)$.

C.4 DESIGN OF THE DAMPING CONTROLLER

The TCSC-type PSDC is designed using three key indices: the *controller-phase*, *maximum damping influence*, and *natural phase influence* indices.

C.4.1 The Controller-Phase Index (CPI)

If the controller gain is so small that in Eq. (C.14) $K_{ILi}(j\omega_i) \ll 1$, the effective control action then becomes directly proportional to the PSDC gain. Equation (C.23) can then be written as

$$\Delta\lambda_i = -k_{ei}(j\omega_i)K_{\text{PSDC}}(j\omega_i)K_{oi}(j\omega_i)/2 \tag{C.26}$$

The PSDC can impart a pure damping influence if the phase of $\Delta\lambda_i$ is made zero.

A CPI for the ith mode is defined as

$$\text{CPI}(i) = -\angle K_{ci}(j\omega_i) + \angle K_{oi}(j\omega_i) \tag{C.27}$$

Ideally, pure damping can be achieved if the phase of $K_{\text{PSDC}}(j\omega_i)$ is made equal to CPI(i). The actual influence on the ith swing mode is expressed through Eqs. (C.24)–(C.26) as

$$\Delta\sigma_i \approx |K_{\text{PSDC}}(j\omega_i)||K_{ci}(j\omega_i)||K_{oi}(j\omega_i)| \cos(\Delta\varphi)/2 \tag{C.28}$$

$$\Delta\omega_i \approx |K_{\text{PSDC}}(j\omega_i)||K_{ci}(j\omega_i)||K_{oi}(j\omega_i)| \sin(\Delta\varphi)/2 \tag{C.29}$$

where $\Delta\varphi = \angle K_{\text{PSDC}}(j\omega_i) - \text{CPI}(i)$
= the difference between the actual achieved phase and the desired pure damping phase for the damping controller

To achieve damping for all relevant swing modes, the selected measurement signal must have nearly constant CPI in the same quadrant over the range of swing-mode frequencies—typically, 0.3–2 Hz.

Having selected an appropriate measurement signal, one can choose (based on personal experience) the appropriate time constants of lead-lag transfer-function PSDC(s). More important than phase matching is choosing a suitable bandwidth so that high-frequency modes do not influence the performance of the PSDC. Introducing some phase lag at the swing-mode frequency may be acceptable, for doing so enhances the synchronizing torque between the machines.

C.4.2 The Maximum Damping Influence (MDI) Index

Even though a high PSDC gain is desirable for enhanced damping, it cannot be increased beyond a specific limit; otherwise, the inner-loop gain will supersede the influence of the PSDC, as evident from Eq. (C.14). A gain margin of 10 dB is chosen for the inner loop, restricting the maximum PSDC gain to

$$K_{\text{PSDC max}} = \frac{1}{\sqrt{10}K_{ILi}(j\omega_i)} \tag{C.30}$$

The maximum eigenvalue shift attainable with the specified maximum PSDC gain of Eq. (C.30) is defined as the MDI index for each swing mode λ_i; by using Eq. (C.26), it is expressed as

$$\mathrm{MDI}(i) = \frac{|K_{ci}(j\omega_i)||K_{oi}(j\omega_i)|}{2\sqrt{10}|K_{ILi}(j\omega_i)|} \tag{C.31}$$

The MDI concept implies that a preferred measured signal must have a minimum inner-loop gain to ensure a high PSDC gain for damping improvement.

C.4.3 The Natural Phase Influence (NPI) Index

Even though the observability and controllability of a selected signal vary with the operating condition, it is desirable to have a fixed controller for damping the dominant swing modes. An effective, robust signal should therefore have a large MDI index and a constant CPI over the expected range of operating conditions. This is, however, difficult to achieve in real systems, especially when more than one swing mode must be damped. A new index—the NPI—relates to the inner-loop phase and is based on the influence of $K_{ILi}(s)$ on the system zeroes.

The frequency response of the closed-loop transfer function between control signal u and measurement y displays peaks and troughs that correspond to the poles and zeroes, respectively, of this transfer function. The poles relate to the eigenvalues of swing modes in which damping must be improved [C.4], [C.5].

If the controller gain is made infinitely large, all swing modes λ_i move to their neighboring zeroes, z_i, in the s plane, attaining the damping levels of those zeroes. This activity constitutes the "natural" behavior of closing the loop with the measurement signal. The path followed by the swing modes is influenced by the phase compensation of the controller.

When $K_{\mathrm{PSDC}} \to \infty$, then $\lambda_i \to z_i$, and the effective control action given in Eq. (C.14) modifies to

$$K_{ei}(s) = -\frac{K_{ci}(s)K_{oi}(s)}{K_{ILi}(s)} \tag{C.32}$$

The natural change in the ith swing mode λ_i is then given by Eqs. (C.23) and (C.32) as

$$\Delta\lambda_i = z_i - \lambda_i \approx \frac{-K_{ci}(j\omega_i)K_{oi}(j\omega_i)}{2K_{ILi}(j\omega_i)} \tag{C.33}$$

The NPI index for each swing mode λ_i is defined as

$$\mathrm{NPI}(i) = -\{\mathrm{CPI}(i) + \angle K_{ILi}(j\omega_i)\} \tag{C.34}$$

Substituting from Eq. (C.27),

$$\text{NPI}(i) = +\angle K_{ci}(j\omega_i) + \angle K_{oi}(j\omega_i) - \angle K_{ILi}(j\omega_i) \tag{C.35}$$

If NPI(i) = 0°, a pure, positive damping effect will be provided; however, if NPI(i) = −90°, a pure, positive synchronizing influence will be imparted. As a compromise to achieve a natural tendency for both damping- and synchronizing-torque enhancements, NPI(i) must lie in the range 0° > NPI(i) > −90°.

To summarize, the PSDC control design must be based on the following criteria:

1. The selected measurement signal must, as much as possible, ensure that the CPI remains nearly constant in the same quadrant for the range of frequencies spanned by all dominant swing modes.
2. The inner-loop gain $K_{ILi}(j\omega_i)$ should be small for the selected measurement signal to ensure a high MDI index.
3. The NPI index for each swing mode λ_i must lie in the range −90° < NPI(i) < 0°.

REFERENCES

[C.1] E. V. Larsen, J. J. Sanchez-Gasca, and J. H. Chow, "Concepts for Design of FACTS Controllers to Damp Power Swings," *IEEE Transactions on Power Systems*, Vol. 10, No. 2, May 1995, pp. 948–955.

[C.2] F. P. DeMello and C. Concordia, "Concepts of Synchronous Machine Stability as Affected by Excitation Control," *IEEE Transactions on Power Applications and Systems*, Vol. PAS–88, 1969, pp. 316–329.

[C.3] E. V. Larsen and D. A. Swann, "Applying Power System Stabilizers, Parts I–III," *IEEE Transactions on Power Apparatus and Systems*, Vol. PAS–100, 1981, pp 3017–3046.

[C.4] J. H. Chow and J. J. Sanchez-Gasca, "Pole Placement Design of Power System Stabilizers," *IEEE Transactions on Power Systems*, Vol. 4, 1989, pp. 271–277.

[C.5] J. F. Hauer, "Reactive Power Control as a Means for Enhanced Inter-Area Damping in the Western Power System—A Frequency Domain Perspective Considering Robustness Needs," *Application of Static Var Systems for System Dynamic Performance*, IEEE Publication 87TH01875-5-PWR, 1987.

[C.6] N. Martins and L. T. G. Lima, "Eigenvalue and Frequency Domain Analysis of Small Signal Electromechanical Stability Problems," *Application of Static Var Systems for System Dynamic Performance*, IEEE Publication 87TH01875-5-PWR, 1987.

APPENDIX D

FACTS Terms and Definitions

All definitions given in this appendix are reproduced from ref. [D.1].

D.1 DEFINITIONS OF BASIC TERMS

Flexibility of electric power transmission The ability to accommodate changes in the electric transmission system or operating conditions while maintaining sufficient steady-state and transient margins.

Flexible ac transmission system (FACTS) Alternating-current transmission systems incorporating power electronic–based and other static controllers to enhance controllability and increase power transfer capability.

FACTS controller A power electronic–based system and other static equipment that provide control of one or more ac transmission system parameters.

D.2 DEFINITIONS OF FACTS CONTROLLER TERMS

The following terms and definitions are arranged alphabetically.

Battery-energy–storage system (BESS) A chemical-based energy-storage system using shunt-connected switching converters to supply or absorb energy to or from an ac system which can be adjusted rapidly.

Interphase power controller (IPC) A series-connected controller of active and reactive power consisting, in each phase, of inductive and capacitive branches subjected to separately phase-shifted voltages. The active and reactive power can be set independently by adjusting the phase shifts and/or the branch impedances using mechanical or electronic switches. In the particular case where the inductive and capacitive impedances form a conjugate pair, each terminal of the IPC is a passive current source dependent on the voltage at the other terminal.

Static condenser (STATCON) This term is deprecated in favor of the static synchronous compensator (SSC or STATCOM).

Static synchronous compensator (SSC or STATCOM) A static synchronous generator operated as a shunt-connected static var compensator whose capacitive or inductive output current can be controlled independent of the ac system voltage.

Static synchronous generator (SSG) A static, self-commutated switching power converter supplied from an appropriate electric energy source and operated to produce a set of adjustable multiphase output voltages, which may be coupled to an ac power system for the purpose of exchanging independently controllable real and reactive power.

Static synchronous series compensator (SSSC or S^3C) A static synchronous generator operated without an external electric energy source as a series compensator whose output voltage is in quadrature with, and controllable independently of, the line current for the purpose of increasing or decreasing the overall reactive voltage drop across the line and thereby controlling the transmitted electric power. The S^3C may include transiently rated energy-storage or energy-absorbing devices to enhance the dynamic behavior of the power system by additional temporary real power compensation, to increase or decrease momentarily, the overall real (resistive) voltage drop across the line.

Static var compensator (SVC) A shunt-connected static var generator or absorber whose output is adjusted to exchange capacitive or inductive current so as to maintain or control specific parameters of the electrical power system (typically bus voltage).

Static var generator or absorber (SVG) A static electrical device, equipment, or system that is capable of drawing controlled capacitive and/or inductive current from an electrical power system and thereby generating or absorbing reactive power. Generally considered to consist of shunt-connected, thyristor-controlled reactor(s) and/or thyristor-switched capacitors.

Static var system (SVS) A combination of different static and mechanically switched var compensators whose outputs are coordinated.

Superconducting magnetic energy storage (SMES) A superconducting electromagnetic-based energy-storage system using shunt-connected switching converters to rapidly exchange energy with an ac system.

Thyristor-controlled braking resistor (TCBR) A shunt-connected, thyristor-switched resistor, which is controlled to aid stabilization of a power system or to minimize power acceleration of a generating unit during a disturbance.

Thyristor-controlled phase-shifting transformer (TCPST) A phase-shifting transformer, adjusted by thyristor switches to provide a rapidly variable phase angle.

Thyristor-controlled reactor (TCR) A shunt-connected, thyristor-controlled inductor whose effective reactance is varied in a continuous manner by partial-conduction control of the thyristor valve.

Thyristor-controlled series capacitor (TCSC) A capacitive reactance compensator which consists of a series capacitor bank shunted by a thyristor-controlled reactor in order to provide smoothly variable series capacitive reactance.

Thyristor-controlled series compensation An inductive reactance compensator which consists of a series reactor shunted by a thyristor-controlled reactor in order to provide a smoothly variable series inductive reactance.

Thyristor-controlled voltage limiter (TCVL) A thyristor-switched metal-oxide varistor (MOV) used to limit the voltage across its terminals during transient conditions.

Thyristor-switched capacitor (TSC) A shunt-connected, thyristor-switched capacitor whose effective reactance is varied in a stepwise manner by full- or zero-conduction operation of the thyristor valve.

Thyristor-switched reactor (TSR) A shunt-connected, thyristor-switched inductor whose effective reactance is varied in a stepwise manner by full- or zero-conduction operation of the thyristor valve.

Thyristor-switched series capacitor (TSSC) A capacitive reactance compensator which consists of a series capacitor bank shunted by a thyristor-switched reactor to provide a stepwise control of series capacitive reactance.

Thyristor-switched series compensation A impedance compensator which is applied in series on an ac transmission system to provide a stepwise control of series reactance.

Thyristor-switched series reactor (TSSR) An inductive reactance compensator which consists of a series reactor shunted by a thyristor-switched reactor in order to provide a stepwise control of series inductive reactance.

Unified power-flow controller (UPFC) A combination of a static synchronous compensator (STATCOM) and a static synchronous series compensator (S^3C) which are coupled via a common dc link, to allow bidirectional flow of real power between the series output terminals of the S^3C and the shunt output terminals of the STATCOM, and are controlled to provide concurrent real and reactive series line compensation without an external electric energy source. The UPFC, by means of angularly unconstrained series voltage injection, is able to control, concurrently or selectively, the transmission line voltage, impedance, and angle or, alternatively, the real and reactive power flow in the line. The UPFC may also provide independently controllable shunt-reactive compensation.

Var compensating system (VCS) A combination of different static and rotating var compensators whose outputs are coordinated.

REFERENCE

[D.1] A. Edris et al., "Proposed Terms and Definitions for Flexible AC Transmission System (FACTS)," *IEEE Transactions on Power Delivery*, Vol. 12, No. 4, October 1997, pp. 1848–1853.

INDEX

CPSIA information can be obtained at www.ICGtesting.com
Printed in the USA
BVOW09*0614121214

378513BV00004B/22/A

9 780471 206439